Contents

THE POLITICS OF
EVOLUTION

Morphology, Medicine, and Reform in
Radical London

*

ADRIAN DESMOND

THE UNIVERSITY OF CHICAGO PRESS
Chicago and London

This book has been brought to publication with the generous
assistance of the Publication Subvention Program of the National
Endowment for the Humanities, an independent federal agency
which supports such fields as history, philosophy, literature, and
languages.

The University of Chicago Press, Chicago 60637
The University of Chicago Press, Ltd., London
© 1989 by The University of Chicago
All rights reserved. Published 1989
Paperback edition 1992
Printed in the United States of America

98 97 96 95 94 93 92 5 4 3 2

Library of Congress Cataloging-in-Publication Data
Desmond, Adrian J., 1947–
    The politics of evolution: Morphology, medicine, and reform in
radical London / Adrian Desmond.
        p.   cm.—(Science and its conceptual foundations)
    Bibliography: p. 432.
    Includes index.
    1. Anatomy, Comparative—Research—England—London—
History—19th century.   2. Evolution—Research—England—
London—History—19th century.   I. Title.   II. Series.
QL810.D47   1989
575′.09421′09034—dc20   89-5137
                                                                                    CIP
ISBN 0-226-14346-5 (cloth)
ISBN 0-226-14374-0 (paperback)

∞ The paper used in this publication meets the minimum
requirements of the American National Standard for Information
Sciences—Permanence of Paper for Printed Library Materials,
ANSI Z39.48-1984.

THE POLITICS OF EVOLUTION

SCIENCE AND ITS CONCEPTUAL FOUNDATIONS

David L. Hull, Editor

Illustrations

Acknowledgments

I owe debts of gratitude to a great many people, for both practical help in unearthing manuscripts and help of a more committed kind, particularly in reading lengthy typescripts. In this respect my main support came from Nellie Flexner, J. A. Cowie, and my late mother, Barbara Desmond, all of whom ungrudgingly corrected reams of computer printout. William Bynum was a constant source of encouragement, discussing Robert Grant and his place in radical medicine. The late Richard Freeman at University College too was a mine of information on Grant (some of it handed-down departmental hearsay). James Secord spared time to read the whole typescript and as usual responded with a welter of suggestions, while James Moore sorted me out on the finer points of religious Dissent. My discussions with Evelleen Richards about Robert Knox helped me gain a better perspective on the varieties of radical science. Others who kindly read chapters were David Bloor, Steven Shapin, Roy Porter, Michael Neve, and Fiona Erskine. Some sent unpublished or proof material, for which I also thank David Allen, Pietro Corsi, and Frank N. Egerton. David Hull, Peter Bowler, and an anonymous referee provided some crucial advice on improving the manuscript. Many people replied graciously to a succession of infuriating questions: Did the Rev. Adam Sedgwick know of the pauper evolutionists? Did the young Knox hold views like those of his classmate Grant? Was Charles Darwin scared off by the "fierce & licentious" radicals?

Susan Gove of the Bloomsbury Science Library at University College London provided her usual unflagging assistance, giving me an extended loan of Grant's offprints, lectures, and books. My research was based on University College's extensive collection of early nineteenth-century British and Continental works on comparative anatomy, which once constituted Grant's and William Sharpey's personal libraries. For permission to study manuscript material I should like to thank the following librarians and institutions: Janet Percival and Gill Furlong, University College London; Eustace Cornelius and I. F. Lyle, Royal College of Surgeons of England; M. J. Rowlands, British Museum (Natural History); Jeanne Pingree, Huxley Archives, Imperial College; N. H. Robinson, Royal Society; R. Fish, Zoological Society of London; Gina Douglas, Linnean Society of

London; I. M. McCabe, Royal Institution Archives; the British Library; Edinburgh University Library; Wellcome Institute for the History of Medicine; University of London Library, Senate House; Guildhall Library; John Thackray, Geological Society of London; St. Bride Printing Library; K. Janet Wallace, British Museum Archives; Peter Gautrey, Cambridge University Library; Philip Gaskell, Trinity College Library, Cambridge; Blanche Ballard, Arkell Library, University Museum, Oxford; Patricia Methven, King's College London; Olive Martyn, Royal Society for the Prevention of Cruelty to Animals, Horsham, Sussex; and the New-York Historical Society. For xeroxes, microfilm, and information I am grateful to Janet Browne and Anne Secord, Darwin Letters Project, Cambridge University; M. G. Bassett, National Museum of Wales; Jean Tsushima, the Honourable Artillery Company; Stephen Catlett, American Philosophical Society; the National Library of Medicine, Bethesda; Michael E. Hoare, Alexander Turnbull Library, National Library of New Zealand; and the Muséum National d'Histoire Naturelle, Paris.

The following institutions have generously given permission to use extended quotations from unpublished material in their possession: the Geological Society of London Library, Greenough correspondence; the British Library Manuscript Collections; the Arkell Library of the University Museum, Oxford, for the letters of Richard Owen to William Buckland, and also from Buckland's file "Species Change of Lamarck"; the Royal Society Library for letters in the John Herschel correspondence; the Linnean Society of London Library for the William Swainson correspondence; Trinity College Library, Cambridge, for letters from Richard Owen to William Whewell; the British Museum (Natural History) for letters in the Richard Owen correspondence as well as Owen's Notebook 5 and Manuscript Notes; the Royal College of Surgeons of England Library for Owen's manuscripts—specifically "Notes & Annotations," "General Account of Specimens," Hunterian lectures, and letters—as well as J. H. Green's 1824–27 Lectures, W. Lawrence to R. G. Glynn, and W. Buckland to Owen; the University College London Library for the Brougham correspondence, College Correspondence, the Society for the Diffusion of Useful Knowledge correspondence; the Zoological Society of London for the Council and General Meeting minutes; and the Alexander Turnbull Library, National Library of New Zealand, for the Mantell Family Papers.

Writing a book on evolution in the 1830s without mentioning Darwin is, I suppose, provocative (even if the book's declared subject is comparative anatomy and evolution in the public domain). So for those who cry with one voice, "Where does Darwin fit into the political picture?" see the afterword.

1

Evolution and Society: Setting the Scene

Evolution before Darwin

The "Darwinian revolution"—it is an evocative metaphor. So ingrained has the image become that we all take it for granted. Yet is it anything more than a consciousness-raising slogan bandied about by latter-day Darwinians? Does it really describe events surrounding the publication of the *Origin of Species* in 1859, or even the situation before the new Darwinian synthesis of the 1930s? There is no doubt that evolution became the stock-in-trade of biologists after Darwin published, and yet historians find little evidence of a mass switchover to Darwin's particular theory of natural selection in the nineteenth century.[1] If anything, in Britain Darwin's academic appeal was largely to a small group bent on professionalizing science—to men such as the pugnacious T. H. Huxley, outsiders intent on breaking the gentlemanly Oxbridge grip on natural history.[2]

And what of evolution before Darwin published? Was his a lone voice? Or was there already a "naturalistic" tradition in British society, with certain groups accepting a self-progressing natural organic world? On this question the recent shift among historians is quite evident; one even depicts Darwin's move as simply a "palace coup" among the elite, the final act in a long drama, with the real fight to establish a lawful, evolutionary worldview among the "people" taking place a generation earlier.[3]

But the "Darwin Industry" has hardly begun to acknowledge a naturalistic tradition in British biology. Why not? Perhaps because historians have been looking too closely at the gentlemen of science and their Anglican ministers. Among this governing class there was virtually no concession to evolution before 1859 (and not much to Darwin's brand after).[4]

1. Bowler 1988.
2. Desmond 1982.
3. Secord 1989; Desmond 1987 on the artisan evolutionists.
4. Corsi 1988 on the exception that proves the rule: the Rev. Baden Powell, Savilian Professor of Geometry at Oxford. Powell became an extreme latitudinarian and even warmed hesitantly to evolution, but found himself increasingly isolated as a result.

Darwin's book came as a bombshell (and, as this metaphor suggests, it was attacked by the dons and divines as an act of terrorism against the old wealthy elite). Just what was it about the old or aristocratic Anglicanism that made it so hostile to the progressive, natural, competitive, mobile view of life taught in more radical classrooms? Why did the squires of science consider it utterly irresponsible to view the world in this way?

The situation in the 1830s (when Darwin was secretly devising his theory) raises a whole new set of questions. Could it be that the sorts of evolutionary sciences openly imported from France into Britain at this time were not so much unworkable or old-fashioned (as the gentlemen—and later historians—maintained), but that they had disturbing social and political associations? Was it this that really made evolution unacceptable in the radical thirties? Remember that in 1830, at the time of the July Revolution in Paris, the Anglican elite in Britain was staving off concerted attacks by the radicals attempting to secularize and democratize their own society. Creationist politics were bound to be fierce in an age when French rationalist theories threatened the old subservience on which the ruling class's security depended. Was not France itself a cautionary tale? Had not the Parisian demagogues included Jean-Baptiste Lamarck's execrable evolutionary theory in their arsenal? The British gentry remained constitutionally suspicious of the republican rabble across the Channel, portraying them as the "national enemy."[5] And if France's periodic convulsions were fueled by poisonous, naturalistic, evolutionary philosophies, then the conservatives were determined to keep them off English soil.

This brings us to the "revolution" aspect of the Darwinian metaphor. It is an interesting label to pin on Darwin's particular theory. "Revolution" was no empty figure of speech in Darwin's younger days; it was a real political threat, with constant rounds of violence and repression in England between the time of the French Revolution of 1789 and the European uprisings of 1848. Darwin himself deplored the turbulence of the 1830s and shuddered at the mere mention of revolution. In his notebooks he actually talked of the natural, lawful processes of change in nature and society obviating the need for any sort of violent interruption (see the afterword). Again, a growing number of historians tend to interpret Darwin's beliefs about legislated change and progress through competition not as revolutionary, but as stabilizing. Darwin's might not have been a "conservative revolution," but it did "ratify" the competitive, individualist

5. Gash 1978: 146.

Malthusian ideology of the arriviste merchants then acquiring a share of power.[6]

As a piece of "ratification," Darwin's book came rather late. In 1859, over twenty years after he conceived his theory, society was tranquil, the decades of mass unrest were past, and it was relatively safe to publish, even if he did have to be pushed into it. So what was happening in the meantime, while Darwin's theory was lying dormant in his private notebooks? What sciences were publicly serving a similar function: ratifying the change from the eighteenth-century world of nepotism, privilege, and aristocratic patronage to the more openly competitive, upwardly mobile Victorian society?

This is the cue for *The Politics of Evolution*. This volume surveys the vast social tracts ignored by the Darwin scholars studying the Oxbridge sporting gents. It makes contact with the radical social factions and identifies the audiences for the new dissident sciences. In short, it looks at the social context of public evolution and other naturalistic theories in the decades before Darwin published. But was there really any public support for evolution in the 1830s? Judging by the standard histories one might imagine not.[7] And yet if there was not, why the massive campaign to discredit it by the scientific gentry? Why did worried comparative anatomists and geologists tailor their major works to refute it? The problem, of course, has always been actually to locate it in context. The reason evolutionists have been hard to find is that historians have consistently looked in the wrong place. The curates' classrooms in Cambridge are hardly the best place to start; we need to insinuate ourselves deep into the radical underworld—to explore a totally different set of classrooms, in the secular anatomy schools and radical Nonconformist colleges. This is where we do our hunting. Hence this book is not about polite or "responsible" science—the sort promoted at Oxford or Cambridge—but about angry, dissident views. It is about science to change society.

Nor is it mainly about Darwin's Malthusian brand of evolution, but about a rival, flagrantly radical, anti-Malthusian sort. Darwin had applied the Rev. Thomas Malthus's analysis of society to nature. Malthus saw overpopulation lead to ruthless competition for resources. He believed that population growth would always outstrip food supply, making charity

6. Moore 1986a: 58–59.

7. Even the older histories that did mention the "Minor Evolutionists" (Eiseley 1961) did not trouble themselves with questions of how, why, or where avant-garde biological views were adopted, being more concerned to ferret out Darwin's "forerunners." They paid no heed to the social context and thus were ill equipped to understand how evolution met the intellectual needs of specific groups.

and welfare a waste of time. In 1834 the Whigs (who had Darwin's support) translated Malthus's program into action; they scrapped the old poor laws, ending outdoor relief for the destitute and forcing them either to compete in the labor market or face the workhouse. In 1838 Darwin read Malthus and applied this weak-to-the-wall image to nature, seeing species progress through savage struggle and the elimination of the unfit.[8]

At the other extreme, radical artisans abhorred Malthus's doctrines and the callous anti-working-class legislation passed in their name. By contrast, they envisaged society progressing through cooperation, education, emancipation, technological advance, and democratic participation. Their views of nature were equally distinct. Not for them the powerful and privileged surviving by exploiting and culling the weak, but an inexorable progress for all through harmony and cooperative striving. Cannibalized fragments of Lamarck's evolutionary biology—which provided a model of relentless ascent power-driven "from below"—turned up in the pauper press.[9] Lamarck's notion that an animal could, through its own exertions, transform itself into a higher being and pass on its gains—all without the aid of a deity—appealed to the insurrectionary working classes. His ideas were propagated in their illegal penny prints, where they mixed with demands for democracy and attacks on the clergy. Clearly Lamarckism had some disreputable associations. It was being exploited by extremists promoting the dissolution of Church and aristocracy, and calling for a new economic system. These atheists and socialists supported a brand of evolution quite unlike Darwin's. Moreover, theirs was evolution in a real "revolutionary" context; it grew out of a rival tradition in the 1820s and 1830s and was far more radical than anything Darwin envisaged.

Between the Malthusian Whigs and the socialist demagogues lies terra incognita. It is a territory that should be opened up. In this unexplored terrain all sorts of dissident knowledge flourished: not only varieties of evolution, but a swirling vortex of alternative economic, social, and biological sciences that threatened to wash away the pillars of the establishment edifice. Unlike the gentlemen's polished, expensive treatises, these sciences were spread through radical medical newspapers and inflammatory penny prints. They were not exercises in leisurely intellectual pursuit. Outside medicine, they were angry, working sciences, sometimes half articulated, often half taken for granted. Even inside radical medicine some theoretical sciences became highly speculative. As a result my technique for exploring them is unlike the Darwin scholars' textual analysis. By their very nature, these outcast forms of evolution require a social

8. Ospovat 1981: chap. 3; Bowler 1984a:96–99; Hodge and Kohn 1985:192–93; Young 1985: chap. 2; Moore 1986b.

9. Desmond 1987; Royle 1974:123–25.

understanding. Nothing less can give us an insight into their meaning or hint at the reasons they were so attractive to aggrieved groups. We need to ask new sorts of questions. How could evolution have furthered a group's ends in a rapidly urbanizing, industrializing society? Once we start thinking of the downtrodden seeking greater recognition and challenging the authorities, a fascinating picture begins to emerge. The history of evolution in the past has been a pretty bloodless affair; we need to get some of the grit, humor, and suffering back into it. We need to restore the fine texture of social history and reestablish the proper context of early nineteenth-century scientific naturalism.[10]

A glimpse at the Edinburgh University medical graduates who accepted the natural birth of new species certainly suggests that a political connection must be made. Consider the personalities. That scourge of Oxford and orthodoxy, the rakish Robert Knox, never doubted the *consanguinité* of animal life and had only praise for the anticlerical French. The commercial tree cultivator Patrick Matthew, on hearing of the July Revolution, closed his *Naval Timber* mid-appendix (with its hodgepodge of calls for free trade, popular "self-government," and animal transmutation) to cheer on the republicans. The retiring Robert Grant made slashing attacks on corruption in medicine and society, and brought his evolutionary biology to London, where it was acclaimed by the ultraradicals. (He was actually the first in Britain to talk of animals having "evolved," using this word in 1826 to signify the transmutation of one species into another.)[11] The botanist Hewett Watson, who switched to evolution early in the 1830s, savaged the Tory old guard so viciously that he alienated even moderate reformers.[12]

The common denominator, of course, is that these were all political radicals and scientific materialists,[13] all committed to sweeping social reform. They were also atheists or deists. (Deists accepted God on rational

10. A task already begun by Cooter 1984, E. Richards 1989a, and the various authors in Moore 1989b.

11. Grant (1826b:297, 300) spoke of the spontaneous generation of worms and infusoria and of Lamarck's belief that "all other animals, by the operation of external circumstances, are evolved from these in a double series, and in a gradual manner." Previously the first such use of the word was attributed to the anti-Lamarckian geologist Charles Lyell in 1832 (Bowler 1975:100). First usages are, however, unimportant, for *transmutation, metamorphosis,* and (later) *generation* were much more common terms; I use them interchangeably throughout the text.

12. Wells 1973; Dempster 1983:98–99; E. Richards 1989a; Desmond 1984b, 1989; F. N. Egerton 1979:91–92; and Egerton's forthcoming biography of Watson.

13. Strictly speaking, *materialism* offered a worldview in which there were no spirits (or vital powers independent of matter) and in which the mind was inseparable from the brain; consciousness was simply neural matter in action. Nineteenth-century materialists also as-

grounds but rejected biblical revelation. Hence they tended to favor more lawful, deterministic explanations of nature.) And in a decade of radical demands for the separation (or "disestablishment") of the Church from the state, their venom was reserved for the Anglican placemen inside science and out. In this age of millennial expectations, all saw "a new state" of society "near at hand," in which the ancien régime injustices were to be swept away and a new democratic order instituted.[14]

We ignore these political aspects at our peril. Contemporary protagonists were quite aware of evolution's dark connotations. In pulpits and learned societies, artisans were warned to keep their place in society, and their bestial sciences were denounced for destroying the safeguards of the moral order. Some, like the reactionary country gentlemen, were appalled at the thought of the middle and lower orders laying their hands on any sort of science, let alone Lamarck's abomination. Science had to remain a prerogative of rank to preserve the chain of subordination, which was the "key-stone" of government. Giving ordinary people a taste for it smacked of democracy, "for as distinctions in society arising from wealth & rank form the character of a monarchy, so is the doing away with such distinctions, or substituting talent or science in their place, the character of a democracy," and this would "plunge the country into irretrievable ruin and despair."[15] This was as much a warning to the professional middle classes hoping to establish a meritocracy as it was to the lower orders hoping for a revolution. But it is only when we consider the kinds of self-serving evolutionary sciences favored by many demagogues that the basis of such gentlemanly fears becomes starkly apparent.

---

sumed that force was an inherent quality of matter. They argued that biological phenomena could be explained by the laws of physics and chemistry, and ultimately by the properties of the atoms themselves (this is *reductionism*).

In practice, however, the label was applied much more widely. Deists who expressed reductionist views were also called materialists, in spite of their belief in God. Some teachers (see chap. 4) therefore distinguished themselves as "physiological materialists": while they proposed a material explanation of biological, mental, and even moral phenomena, they claimed that it left their faith in God and the soul untouched. I follow this wider contemporary usage; it captures the feeling of the age and allows me to keep together groups (atheists, deists, radical Christians) that were linked in their struggle against "priestcraft." (Scientific or physiological materialism in these sects was always part of a political ideology; it cannot be dealt with as a philosophical abstraction.) In the text, therefore, *materialism* can signify either an atheistic strategy or a mechanistic explanation of the mind and body.

14. Matthew, reproduced in Dempster 1983:99. While I concentrate in this book on the middle classes, social historians have more generally focused on the working-class atheists and their illegal newspapers (Wiener 1969; Hollis 1970; Royle 1974).

15. Country Gentleman 1826:9, 15, 52–53. Bradfield 1968 on the country gentlemen as a political group.

So we shall be exploring the problem from an unashamedly political perspective. And, unlike the previous histories which focused on the country rectories and comfortable drawing rooms, our spotlight will be on the dirty dissecting theatres. We will examine the reasons why the radicals exploited the doctrines of nature's self-development and how these ideas served their democratic ends. (Why, in Matthew's words, the nobility, with its stultifying customs and "unnatural" privileges, had to yield to nature's law of competition and transformation or risk her revenge.) Ultimately, the radical's new society did not materialize, and the reforms of the 1830s were never enough to satisfy the diehards. As a result the 1840s were years of disillusionment for those teachers who had invested heavily in the cause. There was no payoff. Their letters and articles at this time tell of bitterness and recrimination. Knox and Watson were barred from academic chairs. Knox lost his teaching license and became an itinerant hack. An embittered Grant suffered a financial collapse. Matthew became a Chartist[16] before retiring in frustration to his farm. With the movements for democratic reform defeated, the sciences used to legitimate them receded too. In other words, radicalism's failure sealed the fate of the fiercer reductionist and evolutionary theories riding on its back. This kind of radical evolutionary science did not carry the day.

Of course, this is not the whole story. But the protest movement must be tackled if we are to understand the fluctuating fortunes of transmutation in Britain, at least in the two decades before 1844. In that year the situation changed markedly with the publication of the anonymous *Vestiges of the Natural History of Creation*, a best-selling popularization of a number of dissident sciences by Robert Chambers (a middle-brow publisher in Edinburgh). The book was more than a potboiler; it was a cleverly crafted work that took five years to finish. Chambers modified the view of evolution current in the medical schools and cunningly gave it a providential veneer (see chap. 4). He sold evolution—"development" he called it—as a case of Creation by lawful means. The introduction of new species and the ascent of life were controlled by a natural law preordained by God. In an age rejecting aristocratic intervention and whim and looking for constitutional means of change, Chambers deliberately remade God in the image of a benign Legislator. He dressed up the issue to appeal to the middle classes—those who looked to the law, not noble patronage, for their advancement (and the sort who were now buying educational magazines and novels hot off the press). It was a successful strategy; the

16. Chartism was a mass movement originating in the late 1830s which loosely united large numbers of middle- and working-class radicals. The Charter proposals included universal suffrage by ballot, annual elections, and the removal of property qualifications for Members of Parliament (M.P.s), who were to be paid a salary.

book became the talk of the town, and in seven months *Vestiges* passed through four editions. As James Secord says, Chambers had finally domesticated the science of development and brought it "off the streets and into the home."[17]

The Scientific Context: The New Philosophical Anatomy

To reconnoiter this little-explored evolutionary territory, we must first understand the changing nature of biological theory at this time. And the best places to see the changes taking place are in London's and Edinburgh's cosmopolitan medical schools. Lamarckism actually attracted only limited support, and then mainly on the radical fringes. Far more important for a wide range of reformers was a comprehensive package imported from France in the 1820s known as "philosophical anatomy." This had a much larger medical following. And because it had a convoluted and often contested relationship to evolution, it is central to our story.

Philosophical or higher anatomy was based on the concept of "unity of composition" developed by the professor of zoology at the Muséum d'Histoire Naturelle in Paris, Etienne Geoffroy Saint-Hilaire. By 1830 Geoffroy and his disciples had come to accept a unitary composition for all animals. Not only were all vertebrates built to the same blueprint; in its most extreme formulation, the theory allowed insects, mollusks, and man to be reduced to common organ components. Animal life could therefore be strung into a continuous, related series—rather than broken into discrete "divisions," as Geoffroy's critic at the Muséum, Georges Cuvier, demanded; and it was this that enhanced the prospect of evolution. The series could be used to show the "history" of each organ rising in complexity from snail to man, and that history could be given an evolutionary twist—it could be turned into a real ancestral bloodline. Inside the medical schools, discussions and disagreements broke out over the relationship between higher anatomy and evolution. Some philosophical anatomists (including Geoffroy himself) were transmutationists, and even Chambers's popular work was rooted in the science. So understanding the new anatomy and its supporters is the first step to assessing the place of evolution in the medical schools.

This thumbnail sketch will be filled out as we proceed to look at Geoffroy's and Lamarck's followers in London and Edinburgh, but first one popular misconception has to be swept aside. Toby Appel, reappraising the French situation in *The Cuvier-Geoffroy Debate* (1987), demolishes the myth that Geoffroy was defeated during his famous confrontation with

17. Secord 1989.

Cuvier at the Académie des Sciences in 1830. In reality he carried a bloc of republican sympathizers with him, and a younger generation of comparative anatomists went on to hammer out a compromise between his and Cuvier's extreme views. My conclusion here is broadly similar. The evidence shows that Geoffroy was far more influential in Britain in the 1830s than was previously realized. But again, it is no good searching among the clergy and gentlemen naturalists for his admirers; they detested Geoffroy's coldly deterministic views of animal form and evolution. Look in the medical schools, however, and a wholly different picture emerges. Geoffroy was immensely popular there. Innumerable courses started up, based on his science, after the 1820s, and many were taught by comparative anatomists who had studied in Paris and knew him personally.

To understand Geoffroy's success here, the first half of the book is taken up with characterizing the medical groups that imported the philosophical anatomy. This is absolutely essential if we are to get a grasp on the rowdy audiences for republican science. It must be said straightaway that I am not offering a deep internal study of Geoffroy's anatomy (which can be found elsewhere);[18] rather I am intrigued by the science's social appeal. I have tried to find out what made it attractive to specific medical groups in this period of professional upheaval.

Upheaval there certainly was. The 1830s saw a huge increase in the number of pressure groups demanding that the teachers and practitioners outside the traditional power structure had more say in the running of medicine. It is no coincidence that these marginal men were the staunchest supporters of Geoffroy's science. That there was such a plethora of lobbies, democratic groups, and dissident factions meant that philosophical anatomy spread far and quickly. Its effects were immediately apparent. I would go so far as to claim that the introduction of Geoffroy's controversial ideas and the conservative backlash caused the dramatic flowering of comparative anatomy in London in the 1830s. New teachers, new chairs, and new courses testified to a rapid growth of interest in the subject, which peaked (as did medical radicalism itself) in the mid-1830s (see Appendix A for a list). The result of this struggle to support, refute, or modify Geoffroyism was that comparative anatomy went through one of its most productive phases. Courses at this time were much more varied in style and content than we ever realized. By mid-decade some two dozen teachers and writers were exploring a variety of approaches, ranging from the fiercely materialistic to the doggedly idealistic. To grasp the social changes sustaining this scientific activity, we need to investigate the dem-

18. Appel 1987.

ocratic press, the medical unions, and the secular London University; we
need to tease out the interests of the Dissenters and private anatomy
teachers. In other words, we have to understand how Geoffroyism fitted
in with the campaigns of civil disobedience and how Lamarckism bene-
fited the demagogues.

The Social and Medical Context

To appreciate something as complex as the political appeal of anatomy and
evolution means a close acquaintance with the social context. What was
the situation in England in the 1830s, and how did its problems manifest
in medical society? No decade in the nineteenth century was so racked by
political uncertainty. It opened with the Whigs finally coming to power at
a time of agricultural riots, manufacturing unrest, and pauper press sedition. Lord John Russell in 1832 introduced the Reform Bill disenfranchis-
ing the corrupt boroughs and redistributing seats to London and the industrial towns, while extending the vote to more householders. Hard-line
Tories believed that if the bill became law before they could smash the
working-class unions, then no "earthly power can save this country from a
social Revolution."[19] Conservatives apocalyptically predicted "the *final extinction of the Tory or Church-and-King party*," prophesying that the radicals would sweep to power on the trading and Dissenting vote. The bill
was finally passed (after months of mass marches, riots, and firebombings
in which Nottingham Castle and Bristol city center were sacked). And if
not, as the Duke of Wellington believed, a victory for shopkeepers and
atheists, it did record a significant shift from Anglicanism and agriculture
to manufacture and Dissent.

The activities of the radical M.P.s returned to the House of Commons
in 1832–35 did nothing to allay Tory fears. Over seventy now entered
Parliament, packing the aisles and shattering its decorum with their barracking demands for democracy and disestablishment. By 1835 the radicals had all but captured London, increasing their strength in the Commons at the Whigs' expense. But the doom-laden prophecies—of
abolition of the Lords, appropriation of Church property, working-class
suffrage—were premature. These were the last years of sweeping radical
gains and civic reforms. With the Tories regrouping under their new
leader, Sir Robert Peel, the Whigs maintained only a slender majority in
the later 1830s, and they refused any more concessions to the radicals.

19. Bradfield 1968:736–37; Thomis and Holt 1977: chap. 4; Fullarton 1831:330–31; Halévy 1950:56. MacKenzie (1981:73–74) has pointed out the advantages of studying scientific
production during periods of civil strife (such as this one), when class factors might be less
obscured by the background "noise" of crisscrossing commitments.

The failure of working men to gain the vote forced them to the Chartist conventions, and the decade ended with the Chartist uprising in the Welsh town of Newport and socialists distributing half-a-million tracts at their Birmingham congress.[20]

But a medical "revolution" was talked of too. The gentlemen who opposed the Reform Bill could also be seen in the pews of "the most corrupt of all the Tory trading corporations"—the Royal College of Physicians (RCP) in Pall Mall and Royal College of Surgeons (RCS) in Lincoln's Inn Fields. Here the archbishops were regular visitors, Peel attended anniversary lectures, and Wellington napped through the medical orations; "bishops, judges, and officers of the crown" took tea in the colleges, the social equals of their governing councillors of court physicians and sergeant-surgeons.[21] These comings and goings were commented on caustically by the radical press, intent on exposing the rotten-borough intriguings of the medical elite. It condemned the colleges as hotbeds of nepotism and profiteering—their "Church-and-King system of government" sacrificing talent and suffrage to place and privilege; their mandarin empires protected by "hired assassins."[22] As the Whigs rechartered the old Tory-dominated town corporations in 1835, turning them into elected town councils, the calls for a new democratic medical parliament reached a peak.

The tensions in the medical profession were an upshot of the disturbed state of society generally. Within days of the Reform Bill's reading in March 1831, thirteen hundred general practitioners (GPs), teachers, and journalists held a stormy meeting in London to protest the RCS's "corporate knavery"[23]—its undemocratic practices, nepotism, and discrimination against London's private medical teachers (who competed with the elite teaching in the hospitals). While the radical William Cobbett's *Political Register* attacked "Old Corruption," as he called the existing Anglican aristocratic system of privileges, its medical mirror *The Lancet* denounced the RCP's commitment to "the bigoted, Tory-engendering, law-established *Church*" exhibited by the college's Oxbridge fellowship restrictions.[24] The college councils were "irresponsible" oligarchies that held the city's lucrative hospital posts in their patronage. Reformers were committed to democratic restructuring, to give the ordinary members a

20. Royle 1974:62; Newbould 1980 on the Whig-radical alliance of the mid-1830s.
21. "Medical Reform," *MCR* 1833, 18:583; "The Hunterian Oration," *LMSJ* 1836–37, 10:774–75; "The Abandoned Prosecution," *L* 1831–32, 1:220.
22. "Infamy of Surgical Traitors," *L* 1830–31, 1:568; "Hospital Dinners and Tory Toasts," *L* 1831–32, 2:220.
23. *L* 1830–31, 1:598, 821, 846–65.
24. *L* 1831–32, 2:219; Clarke 1874a: 7.

say in the running of the Royal Colleges. The purged and reformed colleges were to be built on democracy, equality, and merit rather than wealth, rank, and religion. The medical demagogue Thomas Wakley—part of the 1835 radical intake into the House—made his *Lancet* the crusading organ of the cause. He argued in 1830 that England "teems with incendiaries" because of generations of "aristocratic conceit and blindness." The same conceit in the Royal Colleges had to be met with "a revolution in medical government."[25]

Because the debate was fiercest in London, the "Modern Babylon"[26] forms an essential backdrop to the story. London in 1830 was a bustling business metropolis with two million inhabitants and a suspicious secular streak. New medical schools were springing up to meet the demand for increased medical education, and the student population was rising. It was a city of awful delights and terrible temptations. Theaters, pubs, gaming halls, prostitutes: the moral pitfalls for the medical student driven to the nightlife by work and worry were endless (and Mr. Punch considered him a pretty debauched sort in the first place). His bloodied schooling and desecration of the dead did not help either. Dr. Arnold was not the only one to have heard of students degenerating into "materialist atheists of the greatest personal profligacy."[27] London had already stolen Edinburgh's mantle as the medical center of the Empire. By the mid-1820s a tartan army of "Scotch" graduates was marching south (to some alarm), armed with the new French doctrines and a new set of grievances, to staff the secular London University (f. 1826) and start up a second front against the medical corporations.

The Scots' deistic sciences had already been damned by Cuvier for contributing to the anticlerical feeling around the time of the July Revolution. Now they were to be damned all the more urgently by London's medical Tories. The conservative reaction to Geoffroy's and Lamarck's sciences can be gauged by a close study of one besieged institution, the Royal College of Surgeons. The surgeons are important in the story because, among the corporation men, they were generally the ones who cultivated comparative anatomy. The body anatomized and dissected was part of their professional domain, one they guarded jealously, not only

25. "Revolution in Medical Politics," *L* 1830–31, 1:310. Hamburger 1963: chap. 4 on the function of this kind of rhetoric in frightening the corporate power-holders into making concessions.

26. A common term, e.g., in *MCR* 1831, 14:427; *LMSJ* 1833, 2:792. London was even proclaimed "the real capital of the world," rather than Paris (Raumer 1836, 1:7). On the population: J. F. C. Harrison 1979; 23; *LMSJ* 1834, 5:154.

27. McMenemey 1966:145, also 138–39. The disreputable trade in corpses and their dissection in the private schools are discussed in Richardson 1987: chaps. 3 and 4.

against the prescribing physician[28] but against the private and university anatomists. The top surgeons were already at war with these outside teachers and attempting to preserve their teaching monopoly in the hospitals by driving the independent schools out of business. Now the spread of Geoffroy's sciences through these humbler schools added a new dimension to the conflict.

Science at the College of Surgeons played a far more traditional role; it was at once conservative, religious, and supportive of the existing power structure. The RCS Council included a galaxy of courtly, knighted comparative anatomists, all of whom saw perfect animal design chant a hymn to Divine Providence. Following the older baronets in the 1820s came the younger disciples of the poet and philosopher Samuel Taylor Coleridge— the recondite Joseph Henry Green and his industrious protégé Richard Owen. To meet the democratic threat, they promoted an idealist biology based on German *Naturphilosophie*. Theirs was a science of Platonic "archetypes," ideal forms existing only in the Divine mind (chap. 6). They also adopted a traditional philosphy of "descensive" powers: a downward delegation of divine authority through nature and society which illegitimated the democrats' Lamarckian science of "self-developing energies."[29] In an age when radicals were threatening the corporation's very existence, this rival idealist biology was designed to meet the leveling threat. It shifted the emphasis from base nature back to the Godhead, reinforcing the temporal control of the traditional leaders.

Medical Sources

This medical picture has been built up using a little-tapped resource, the proliferating medical journals. I have made extensive use of politically opposing reviews to show the very different perspectives brought to bear on comparative anatomy and the problem of man's relationship to the animals. The medical press at this time was unique for its political aspect.

A welter of new journals appeared in the later 1820s and 1830s, many catering to the burgeoning private schools and the general practitioners. These papers reflected the sharpening divisions of medical labor, themselves related to the wider changes in an industrializing society. The 1820s saw the GPs appear as a coherent group. They were largely the product of the cut-price private schools and London University and were destined

28. There was growing professional rivalry between surgeons and physicians by the 1820s, with many surgeons prescribing (a physician's prerogative) and attempting to extend their intellectual net over the whole body, physiological and anatomical: W. Lawrence 1834:2–9; *LMSJ* 1834, 4:343.
29. R. Owen 1841c:202.

to minister to the tradespeople and middle classes. According to conservatives, the GPs' emergence "completely deranged the natural order of things," throwing a gentleman's profession into turmoil.[30] The Royal Colleges, by contrast, were run by wealthy hospital consultants and used to preserve their privileges. The consultants' refusal to give the GPs' any say in college affairs led to a spate of medical unionizing, to push for change and to prevent the surgeons from placing "their feet in triumph upon the necks of the general practitioners."[31] The new steam-powered presses poured forth a plethora of journals to publicize the practitioners' grievances. The start-up and failure rates were enormous: twenty-seven new titles appeared in the 1820s, thirty-seven in the 1830s—the greatest number in any decade before the 1890s. Most were "losing speculations," as the publisher John Churchill told a student keen to start his own medical paper.[32] Success, as Wakley said, depended on "the *means* of the class" which the periodical served.[33] The elite surgeons had their hospital reports, expensive and safely empirical, while the undermass of GPs was large enough to support a number of weeklies, including Wakley's ribald, radical *Lancet*. Demands for democracy and an end to privilege, then, were as endemic in medicine as in other sectors of British society, where a flourish of prints sprang up to serve the reformers. Even the new term *journalism* was a Tory sneer word taken from the French, which shows how strongly the new press was associated with reform. In medicine, as elsewhere, this literature was met by a Church-and-king counterblast, so that the first cry on greeting a new medical weekly was, "Is he high Tory, Conservative, Whig, Whig-radical, Tory-radical, or out and outer? What has he to say on our *Corn*-laws—what on the *Charter?*"[34]

30. "Present State and Prospects of the Medical Profession," *MG* 1833–34, 13:212. On the decay of the preindustrial "estate system" and de facto segregation of the profession into GPs and consultants by the 1820s: Waddington 1984:15–18; Peterson 1978:6, 17–18; Holloway 1964:307–11; Newman 1957:1–4.

31. "Mr. Green's Views on Surgical Reform," *L* 1830–31, 2:570.

32. The student was Thomas Laycock: Cope 1965:172; "Trash of the Medical Press," *L* 1828–29, 1:659. The figures are calculated from Lefanuc (1937:744–49) and exclude colonial papers. On the effect of the new print technology on the natural history trade: Sheets-Pyenson 1981; Allen 1978: 96.

33. "British and Foreign Medical Review," *L* 1835–36, 1:643.

34. "A New Weekly Contemporary," *MCR* 1839, 30:622; Halévy 1950:18. This is not to suggest an absolute congruence between medical and parliamentary politics; even Wakley admitted that Tory voters could be medical radicals: *L* 1836–37, 2:695. And yet professional grievances often turned into political ones. It was after all in the private teachers' business interests to unite with the radicals to demand rank-and-file suffrage in the RCS, which would give them a foot in the corporation door. It was also revealing that medical reform was championed in the Commons by the radical M.P.s Joseph Hume, Benjamin Hawes, Henry Warburton, and Wakley, while corporation interests were guarded by the Tories Sir Frederick

Given this resource, I singled out the five leading London medical journals with long runs. Scanning the ten-year period 1830–40 (or that part during which the journal existed), I analyzed the periodical's policy, readership, and favored science. These journals can be characterized as follows:

(1) In *The Lancet* (f. 1823), Thomas Wakley's bruising style was based on William Cobbett's, and in the first years the nighttime editorial meetings were attended by Wakley, Cobbett, the surgeon and comparative anatomist William Lawrence, and a libel lawyer. The latter was indispensable; the paper ran into interminable legal trouble over piracy and libel, and was described by its reporters as being carried on at "the point of a bayonet."[35] Only the year before its founding (in 1822) Lawrence had lost copyright control of his comparative anatomy lectures in court because of their alleged blasphemous content, and he wrote many of the *Lancet's* leaders against the college oligarchs. Like Cobbett's *Political Register*, the *Lancet* employed ridicule, with Wakley developing an effective line in caustic mimicry. It savaged the consultants, corporation leaders, Tory and Whig aristocrats, Church, and Anglican universities. It attracted massive support among radical GPs, private teachers, students, and London democrats, and its circulation had topped four thousand before 1830. It promoted the radical unions, anti-Poor Law action, suffrage, secularism, and a mechanistic comparative anatomy that flouted the concepts of Providence and design in nature (chap. 3).

(2) The *London Medical and Surgical Journal* (1828–37) started as a monthly, but turned weekly in 1832. It undercut the *Lancet's* price to capture the GP readership and was still selling for sixpence in 1836. In the early 1830s the *Journal* swung sharply into the radical camp. It catered to London's private teachers and was geared to Dissenting needs, dismissing the Oxbridge-educated physician as a "drawing-room ornament."[36] It promoted an anti-Anglican but still Christianized philosophical anatomy (chap. 4).

(3) The *Medico-Chirurgical Review* (f. 1820) and (4) *British and Foreign Medical Review* (f. 1836) were expensive quarterlies (six shillings a number) that spoke for moderate reform, educational standards, and un-

---

Pollock (Peel's attorney general), Sir James Scarlett, and Sir Philip Egerton, who also acted as patrons to corporation appointees such as Richard Owen.

35. Wakley's "lieutenant" J. F. Clarke 1874a:13, 19, 22–24. Cobbett-Wakley friendship: Sprigge 1899:70–71; Brook 1945:34, 59–60. On Wakley's defense of his style: "Ridicule," *L* 1828–29, 1:241; and praise for Cobbett: "Mr. Cobbett on the Late Trial," ibid., 625–28.

36. "The London University," *LMSJ* 1833, 3:535–36; "What Is to Be Done with the Medical Corporations?" *LMSJ* 1834–35, 6:409–11.

impeded professional access to corporate power. The *MCR* hated vulgarity, decrying the "heroes of the radical press" who "abuse, slander, and vilify" their enemies,[37] but it too became more reformist in the early thirties and entered into a shaky coalition with Wakley against the "medical magnates." The *BFMR* had a strong Unitarian input; it campaigned to extend the powers of the London University, and it promoted a naturalistic science (chap. 5).

Finally the wealthy hospital teachers' organ, (5) the *London Medical Gazette*, was founded in 1827 to oppose Wakley's *Lancet* and unite those who had the profession's "respectability at heart."[38] Castigated by the radicals as "the *chef d'oeuvre* of the medical placemen and corruptionists,"[39] it spoke for rank, responsibility, and conservatism, supported the Royal Colleges, and sported a safe scientific empiricism. The elite surgeons and anatomists who loathed Geoffroy's and Lamarck's sciences felt honor bound to support it. The paper, however, never rivaled the *Lancet* in circulation.

How each journal's science policy was related to its class of readership becomes clearer as we examine the types of science favored by the rival groups. One immediate point, though, cannot be missed; the first four—all reform organs—promoted the new philosophical anatomy. This once again points to the conclusion that Continental higher anatomy was much more prevalent in Britain than has been supposed.

The Importance of Studying Science in the Medical Schools

Evidently the size of Geoffroy's support has been underestimated for so long because the Darwin scholars have sidestepped the medical community. They have counted only those disciples of Geoffroy's who managed to climb the fence into the gentlemanly field of natural history. High-profile lecturers and successful popularizers such as Richard Owen, Peter Mark Roget, and W. B. Carpenter did indeed make a mark here because

37. "Liberty of the Press," *MCR* 1830, 12:438–40; "Medical Reform," *MCR* 1831, 14:573–74. In 1830 the three main journals, *L*, *MG*, and *MCR*, comprised the "dreaded aristarch medical triumvirate" (P. B. Granville 1874, 2:270).

38. "Address," *MG* 1828, 1:3. The *Gazette* was founded by a number of hospital surgeons, including John Abernethy and Benjamin Brodie. Its editor, Roderick Macleod, was well patronized for his conservative defense: he was appointed physician to St. George's Hospital in 1833 and fellow of the RCP in 1836: *MG* 1835–36, 18:534–35. G. J. Bell 1870:299 on Charles Bell's support.

39. "Medical Reform," *LMSJ* 1833, 2:790–93. Born into "the hot-bed of antiquated prejudice" it might have been, but the *MG* by 1833 was showing a more compromising attitude, although there was never any fear of it becoming a "medical *sans culotte*": "The Weekly Medical Press," *LMSJ* 1834–35, 6:440–43; also 1835, 7:89.

of their books. By the mid-1970s the challenge mounted by these men to traditional explanations of animal structure and design was already being noted.[40] More recently, Fritz Rehbock has redirected our sights again, showing the importance of Robert Knox (another private school anatomist) in training a string of philosophical naturalists.[41] Now we need to cross back into medicine to trace the source of this higher anatomy. The wealth of material awaiting study here is apparent from L. S. Jacyna's stimulating papers. Jacyna has added considerably to our cast of characters and, more important, has begun looking at the new anatomists in terms of their professional goals.[42]

It is imperative to track the science back because comparative anatomy was cultivated almost exclusively in the medical schools. Studies of "unity of plan" in natural history can at most only scratch the tip of the medical iceberg. Obituarists, especially after the "emancipation" of biology in the 1860s and 1870s, looked back on these older philosophical anatomists as forming "a link as it were between biological science and medicine."[43] But the teachers in the 1830s did not see themselves as a link. They were an integral part of the academic medical world. And because medical schools were the institutionalized centers of comparative anatomy research, their science reflected "the interests of the medical profession."[44] We have to comprehend these "interests"—the class structure, civil disabilities, hospital monopolies, and so on—in a sectarian and politicized profession in order to appreciate the deployment of the new sciences. In other words, they help us understand why philosophical anatomy took off in the 1830s.

Such an approach allows us to move beyond older historical works in scope and detail. In Dov Ospovat's *Development of Darwin's Theory*, "unity of plan" still comes under the rubric of changing attitudes to design in pre-Darwinian natural history. But notice that Ospovat's four philosophical anatomists—Owen, Carpenter, Roget, and Martin Barry—were all medically trained. (Ospovat categorized them as "non-teleologists" because they did not believe structure was explained by function so much as by morphological laws.) On looking closely at these men, not only do we find few social, religious, or personal ties, but we actually detect a good deal of antagonism. It seems to me that we risk obscuring some deep social divides by lumping them together. Besides, they actually took different approaches to philosophical anatomy. For Ospovat—trying to get away from the stereotyped "Creation vs Evolution" antithesis—this was

40. Ospovat 1978:33–39, 1981:8–23. Also Farber 1976:107–12.
41. Rehbock 1983; Mills 1984.
42. Particularly Jacyna 1984a, 1984b.
43. "Richard Dugard Grainger," *L* 1865, 1:190–91.
44. R. Smith 1977: 218. The same was true of France: Haines 1978: 20; Jacyna 1987.

not important. But for us it is, because it allows us to relocate the teachers in their proper medical contexts, each with its discrete patronage network, professional concerns, and "philosophical" science. Ospovat's "non-teleological" category, in other words, is more complex than was supposed. It can be broken down into socially and scientifically distinct subgroups. And on breaking it down, we find that some "non-teleologists" among the radical democrats supported "evolution," while others who were more conservative repudiated it.

To illustrate this point, I have laid out the book so that it moves across the political spectrum, roughly from "left" to "right" in modern parlance. I start with the anticorporation radicals who championed Grant and his philosophical anatomy, with its atomistic base and unity of plan stretching from monad to man (chaps. 2 and 3). It is in this republican sphere that we tend to find evolution and related naturalistic sciences turning up. I then investigate the sort of science favored by the radical Dissenters and disadvantaged private teachers at war with the surgeons (chap. 4). These teachers too scorned the Cambridge curate mentality with its emphasis on "design" and Creative Intervention in nature. They likewise inserted Geoffroy's science into the medical curriculum partly because it was so overridingly naturalistic—and a deterministic, lawful nature could put the consultants' capricious deity into a legal harness. From these we move to the Whig moderates and the way they emasculated the radical imports for their own professional ends (chap. 5). And we finish with the wealthy surgeons in their besieged college promoting Owen's ideal, anti-Lamarckian anatomy, in which power emanated "from above" and sanctioned a traditional chain of authority.

Keeping the debate within these professional confines has its advantages. It means that all the men we discuss were trained in medicine and employed in teaching. By limiting ourselves to rival medical cliques we get a clearer view of the class, trading, and religious interests at play in the selection of scientific theories. And ultimately, by showing the widespread feeling against the medical oligarchs, whose science was designed to strengthen the bonds of subservience, we begin to make sense of the support for Geoffroy's shackle-breaking doctrines.

On the reverse side of the coin, the lack of anything like this seething republican base in gentlemanly geology or natural history explains why Geoffroy's science was so much rarer here. No revolutionary ferment left the naturalists' world quaking. Quite the reverse; the curates and country gentlemen dreaded Parisian profligacy and had a deep hatred of medical materialists.[45] Periodically throughout the book I move out of the medical

45. Desmond 1985a:164–76.

bastions, following the radicals into the learned societies, watching the sparks fly as these social worlds collided (see especially chaps. 3 and 7). The unreformed Royal Society had a large, cantankerous medical lobby, and the fracas caused when the radicals attempted to democratize the courtly Zoological Society in 1835 made headlines in the *Times*.[46] Medical politics sent a shock wave through the world of polite science, and it was largely the medical republicans who alerted the scientific gentry to the threat to their privileges and position.

Investigating these medical factions enables us to answer a number of leading questions about the fracturing of science following the Napoleonic Wars. Jacyna has already asked how radical anatomy and the surgeons' conservative science were "distinguished in their presentation, style, and content" and how the protagonists' political objectives "manifested in their views of nature."[47] He has done much to answer the latter himself. He has shown that, for the radical physiologists of the Regency period, matter was imbued with active powers and the mind was a function of organization; this explains their belief in a morality deduced from the laws of nature rather than from the canons of Christianity. Such a self-empowered physiology also sustained their faith in a democratic self-determining society, free of all spiritual or aristocratic leadership imposed from "above." Law, morality, and the authority for change emanated from the people below.[48] We can immediately see how the republicans' atomistic physiology prepared the ground for a radical Lamarckism, with its self-developing powers. In fact the artisan atheists, who pirated these physiological works, made the link between their anti-Church insurgence, active-matter theory, and democratic Lamarckism absolutely explicit. This social framework also fits the comparative anatomy community of the period and I adopt it here, contrasting the radicals' belief in nature's unity of plan, serial progression, and self-developing powers with the antidemocrats' opposing science and philosophy.

It is clear from the social alignments that the rival anatomies were sustained by more than medical utility. True, Geoffroy's morphology was advertised as a professional "tool" for the GP. It was promoted as a comprehensive theory that could explain the complex human organs. Geoffroy's disciples declared that man's body could be understood by watching the homologous organs simplify gradually during their descent through the

46. MacLeod 1983:62–63; Desmond 1985a:223–50.

47. Jacyna 1983a:104.

48. Jacyna 1983b: 312–14, 320–27. For studies of the relationship between radical politics and reductionist science outside the British class context, see Mendelsohn 1974, Gregory 1977, and Weindling 1981—all dealing with the more frequently worked issue of German physiology.

animal series.[49] So the Parisian creeds had a medical value, doubtful as this appeared to many conservatives. But there was obviously more to it, with the medical community divided on Geoffroy's worth along social lines.

I am fully aware of the dangers of attempting to map the new science neatly onto social geography, and I shall not paint a rigidly deterministic picture. Comparative anatomy in the 1830s was as subtle and varied as society itself, and it is the cultural warp and weave that has to be caught. To do this means exploring everything from Nonconformist claims, private school grievances, and Benthamite bureaucracies to Unitarian restoration creeds and Methodist attitudes to the animal mind. All shaped medical life in the 1830s; none can be ignored if we wish to cover the "network of causes"[50] sustaining the newer biological sciences. Nor can we afford to miss those emotive issues that tended to draw cross-party support. The alliance of Tory evangelicals and radical Dissenters against vivisection did a great deal to raise comparative anatomy's status in the 1830s. Anticruelty societies, havens for humane Tories and urbane reformers, welcomed a comparative anatomy based on the animal series for providing the means to understand the higher creatures without recourse to the surgeon's scalpel (chap. 3).

A history of biology "from below" is long overdue. If we are to cease being "dazzled by the great,"[51] then we need to pry into those social worlds where the mass of people lived. The scientific gentry and Oxbridge clergy are now very familiar, but they typified only one "class" position on the historical stage. We have detailed accounts of the way they made establishment science a recipe for social stability and Anglican supremacy. We know their reaction to Lamarck's and Geoffroy's "dark school."[52] The expensive, ecclesiastically blessed Bridgewater Treatises, which dwelt on God's goodness deduced from nature, have been studied extensively. (We hear less of the fact that these "Bilgewater" books were pilloried in the radical press.)[53] Yet the radical protagonists of the Anglican dons seem to exist as shadows cast by actors standing offstage. There simply has not been much investigation of their mechanistic sciences which proved so terrifying, or of the extent to which they percolated through to the radical undermass. In the past it has even proved difficult

49. E.g., J. Fletcher 1835–37, 1:78.
50. Barnes 1982: xi.
51. D. Knight 1987: 8.
52. J. W. Clark and Hughes 1890, 2:86; Morrell and Thackray 1981; S. F. Cannon 1978; Garland 1980.
53. Blake 1870–71:334; Rehbock 1983:56; E. Richards 1989a; Desmond 1987:87–88, 90. On Bridgewater science: Gillispie 1959:209–16.

for historians of some disciplines to get a handle on these post-Regency radicals.[54]

A good way to restore the balance is to look at lowlife in the medical schools. What the majority of students were taught in their comparative anatomy courses is hardly known at all. By ignoring the anatomical doctrines circulating among democrats, science history has fallen out of step with social history, where a flourishing tradition of radical studies exists.[55] The time is surely ripe to probe the radicals and their Lamarckian and Geoffroyan imports, sciences whose motif of self-advancement made them immensely attractive to the democrats. If we are not to see science as a monolithic creation of the conservative elite, then we must get this dissident dimension back into the picture. We need to appreciate why the Cambridge clerics projected a total social collapse following the rise of Lamarck's zoology. The medical schools provide our way into the anatomical underworld, where Lamarck's and Geoffroy's doctrines mingled with anti-Church-and-state propaganda. This weighting toward the radicals also explains why the starring role in the story goes to Robert Grant, an intriguing Lamarckian who can be precisely located in a Wakleyan context. Grant's rise in the angry thirties and fall in the hungry forties was symptomatic of radical fortunes generally. As much as anything, this book is a history of these fortunes and the way they told in Grant's career.

Sociology of Knowledge

Although this is primarily a contextual study, using the methods of social history, I also exploit the sociology of knowledge (that is, the study of the social factors affecting the production, evaluation, and use of knowledge—in this case science). Such a mix of contextual and sociological approaches is necessary today more than ever. Few historians see their task any more as reconstructing a rational lineage of ideas through time, picking up gems of foreshadowed "truth" here, ignoring "deviant" approaches there—in short, tracing a path of progressive enlightenment. Nor do they have much truck with the old "internalists" who wrenched science from its social context and wrote ghostly histories of disembodied ideas. With

54. C. A. Russell (1983: chap. 8) has tried to locate them in the London Chemical Society, but without touching on the relationship between radicalism, secularism, and reductionist science.

55. This is not to deny the great strides made by scholars studying phrenology's impact in the 1830s; for more than a decade they have provided sophisticated analyses of the shopocracy's use of this particular self-help science. See especially Shapin 1975, 1979b, and Cooter 1979, 1984.

sociologists and cognitive psychologists teaching that "reality" construction is an active, socially constrained process, they have started examining the network of interests that sustain each community's view of nature. This seachange in historical approach has profound implications. It raises fundamental questions about the status of science as transcendental knowledge whose "discovery" is unproblematic.

Just how problematic it really is becomes apparent as we look at a specific example later in the text: the rival interpretations of the celebrated Stonesfield fossil jaws (chap. 7). These tiny lower jaws, just over an inch long (see fig. 7.8), were found in the Stonesfield slates in Oxford early last century. But there were angry disagreements over their nature. The gentlemen of geology accepted that they belonged to opossumlike marsupials. As such, they were evidence of the earliest known mammals, already living in the "Age of Reptiles." Others disputed this. Where Owen had seen typical marsupial features, Grant saw characteristic reptilian ones. This is the problem: How could two proficient comparative anatomists see the jaws so differently? How could each describe such distinct sets of features?

It would be naïve to cheer one and jeer the other, as if there were inherent rights and wrongs of the case. This would accept that one account was more "objective" because it coincided with our own today. But historians have long ceased reading history backward and assuming that the present explains the past. It does not apply here anyway, because neither diagnosis held up. According to Owen the animals were marsupials; Grant declared them to be reptiles. Thirty years later, however, they were categorized as early, generalized, "sub-marsupial" mammals of an entirely new order.

I have concentrated on the rival social, political, and religious interests of Owen's Anglican and Grant's radical factions. Then I have looked at how these led to diverging presuppositions about nature. And not merely presuppositions. These men represented bitterly antagonistic groups which actually saw the social and natural worlds quite differently. In other words, I am concerned with the way ideological factors influence not merely theories, but even the perception of nature. It has been said that the recent convulsions in the history of science have turned its practitioner "into something of an anthropologist, an explorer of alien cultures."[56] Because British culture during the Industrial Revolution is best treated as "alien," I have devoted a large amount of space to explaining the sectarian contexts in which these contrasting views of nature were held. As a result, by the time readers reach chapter 7 they should find this

56. Hollis and Lukes 1982:1–13.

clash over the fossil jaws quite explicable in contemporary terms. The protagonists' supposedly objective descriptions of nature were in fact socially constrained interpretations. Both Grant and Owen were good comparative anatomists; the reason they came to opposing interpretations of the jaws was that their "good sciences" reflected the contrasting norms, expectations, and perceptions of their respective groups.

This of course raises all sorts of awkward questions about the truth content of science. If other societies are "different worlds," if their sciences make sense "from within," what then of objective knowledge? Does it make all knowledge culture-relative? Can we only talk of local truths?[57] I make no bones about taking a relativist approach here. I am not interested in the eternal verities, only the reasons why rival groups saw them so differently: why one sect's science was another's quackery. As a practical upshot I have looked at a larger number of social groups than is usual in histories of science. The Oxbridge clergyman is not studied exclusively because he was the guardian of "proper," responsible science; the artisan atheist is not ignored because he was writing in illegal penny newspapers. Each is assessed on his own terms; the context is used to elucidate the causes and the reasons why each held a particular view. This leads us back to the idea of Britain in the 1830s as a social patchwork. Once this social diversity and struggle are acknowledged by science historians, we can begin to understand the conflicting opinions over nature that were rife in the period.

I have also taken a largely "instrumental" approach to science, that is, I have looked at the context and uses of competing theories as a means of discerning their local meaning.[58] This can only be done by knowing a subject's social position and group interests precisely. The problem of course lies in detecting the links between someone's implicit social views and explicit polished science. This is one of the challenges still facing sociologists—to expose the "connections between the scientist's social situation and his intellectual output."[59] I have tried to meet this challenge by exploring the medical "class," religious, and occupational structures mediating between the content of comparative anatomy and its context of use. My overall conclusion is that the rival biological doctrines in the thirties were integrated into long-term commercial and political strategies, either to gain or to hold on to privileges. Hence the title of this volume—*Politics of Evolution*—is singularly appropriate: progressive evolutionary theories and related naturalistic sciences, according to this approach, served

57. Barnes and Bloor 1982. Figlio 1976:19 even talks of science as "a naturalized carrier of its context."

58. Shapin 1982:197; Barnes 1977.

59. Norton 1983:305.

to legitimate the radicals' democratic convictions. They were adopted by outsider groups set on breaking the old religious authority and transferring its power to the secular state.[60] As these political strategies were designed to achieve a fundamental redistribution of power, the new sciences were obviously hotly contested. Geoffroy and Lamarck became symbols of resistance; they were the tricolor banners waved by the medical democrats massing outside the corporation porticos.

So my goal is to explain how Geoffroy's and Lamarck's doctrines fitted the reformers' needs. Comprehending the radical milieu is absolutely necessary. We must be sensitive to the new journals and institutions, and the social movements of which they were visible expressions. Foremost among the new institutions was the secular London University (the "radical university" in Gower Street, derided by conservatives as a lecture-bazaar and joint-stock travesty of a gentlemen's seminary).[61] Here the country's first permanent chair of comparative anatomy fell to Grant. Since it was Grant who effectively introduced philosophical anatomy to London, I begin with the science's transmission from Paris to Britain, and Grant's fight to get it established at the university.

60. Moore 1986a:67; Desmond 1987.
61. Coleridge 1972:51; "The Radical University," *MG* 1836–37, 19:463–67.

2

Importing the New Morphology

In 1825 Henry Brougham told the radical M.P. for Westminster, Sir Francis Burdett, that the founding of London University would be "an event of infinite moment." It would, he predicted, "do more to crush bigotry and intolerence than all the Bills either of us will ever see carried, at least until a Reform happens."[1] The indefatigable Brougham, educationalist, lawyer, Whig M.P., man of encyclopedic knowledge (some called it "encyclopedic ignorance"), was adamant that it would provide the "finishing blow to the High Church Bigots."

But how could the new university "crush bigotry"? By offering a secular education and attracting Nonconformist students it certainly cocked a snook at the Tory-Anglicans (hence it was sternly opposed by Sir Robert Peel, then home secretary). But merely founding a non-Christian university would not finish off the bigots. Perhaps we should look deeper, to the sciences being taught, to appreciate the role of knowledge in the "crushing" process.

If we do, we will find that the sciences imported by some of the medical professors were more radical than even Brougham anticipated. This is particularly true of philosophical anatomy, the focus of the present chapter. The chapter charts a circuitous course, for the science originated in Paris and returned with the Scots to Edinburgh, before accompanying them south to staff the new university. But I begin in London by looking at the educational aims of the university's founders, followers of the utilitarian philosopher Jeremy Bentham. Bentham himself—only six years short of his death when the university was inaugurated in 1826—had little personally to do with its establishment. Yet his influence was paramount. And in macabre testimony to this, his badly mummified and clothed body in later decades was brought to University Council meetings, while his better-preserved body of manuscripts passed to the university in 1832: not only the corpus but the corpse came as well, J. H. Burns once joked.[2] We need to understand how the younger Benthamites[3] conceived the

1. Singer and Holloway 1960:13; for the response see Burdett to Brougham, 12 August 1825 (UCL HB 20,031). "High Church Bigots" and "ignorance," in Stewart 1986:35, 198.
2. Burns 1962:2.
3. The Benthamites (or utilitarians) were a heterogeneous group. At their center was John

London University's role in society and what, as corporation reformers, they looked for in a science. Then the scene shifts to Paris, where philosophical anatomy was the rage among republicans. Following the science to Edinburgh—where it arrived in the baggage of the returning Scottish radicals—we begin to see its attraction for these "men who breathed a doubting theism":[4] men at war with the kirk, the corporations, and the country gentlemen. Migrating south, these acerbic Calvinists took their posts in the London University, bringing that brand of savage wit and secular learning so characteristic of intellectual life north of the border. Looking at the use and meaning of the new comparative anatomy in republican Paris and extramural Edinburgh puts us in a good position to see how far it fitted the Benthamite bill in London.

Benthamite Educational Aims

Morris Berman talks of the utilitarians "bending science to entrepreneurial and professional purposes." In this case it was not so much bending as importing the correctly shaped science in the first place. The Benthamites repudiated the aristocratic ideal of polite knowledge as an embellishment to a gentleman's education and sought to create a serious body of knowledge useful to the reformers in government and the professions. Science was to be developed as a "professional tool." In Berman's words, it was to be the basis of "an organized and efficiently administered society," the route to the "upgrading of the medical profession," and the key to "a new type of legal expertise."[5] Armed with this new ideology, the Benthamites founded a number of what Coleridge contemptuously called "lecture bazaars." These educational enterprises were themselves an outward sign of the changing class control of metropolitan science in an urbanizing, industrializing age, and they resulted in a dramatic reorientation in the public presentation of science.

Coleridge's heirs in the old universities deplored these developments.

Stuart Mill's team on the rationalist *Westminster Review*, although Whig statesmen on one side and the more radical democrats on the other supported aspects of Bentham's program. Bentham is remembered for his "greatest happiness for the greatest number" principle, but he wrote extensively on political economy, law, and the constitution. The Benthamites argued for efficient government based on proper expertise (hence the profusion of Whig Select Committees of the 1830s), and their technical education schemes were devised to provide these experts. Believing that the professions must likewise be run by technocrats, Bentham drew support from the medical radicals, eager to oust the old patriachs from their Royal Colleges.

4. Lonsdale 1870:402.
5. Berman 1974–75:123.

The *Athenaeum*—in the late 1820s an organ of Coleridgean romanticism which kept a hostile eye on the Benthamites' books—tied together the "bankrupt" university and Brougham's Society for the Diffusion of Useful Knowledge (also founded in 1826); they were run by the same clique on a "close borough system."[6] It might have added the Mechanics' Institutes and Royal Institution: all were branches of the Benthamite teaching business. They had overlapping managements and flew the same ideological flag. On the other hand, they targeted quite distinct social groups. Where the Mechanics' Institutes sought to provide acceptable science for the querulous working classes, the London University was to turn its middle-class students into a professional elite, a new middle management. Its medical school was to provide an academic education for the improving GPs, the future medical electorate who could carry through Benthamite social objectives in health, welfare, and management reform.

To "upgrade" the medical profession (which meant for Benthamites outlawing nepotism, initiating ballots, and getting trained specialists into office), new anatomical tools had to be honed. Such tools would undermine the authority of the "medical aristocracy"—especially the court surgeons in their corporation "pest-houses"[7]—and enable this new elite respecting professional standards to take over. It was in this context that Geoffroy's "philosophical anatomy" worked as part of the radicals' political strategy. It pointed to the "higher laws" of life, and these laws of form were to be advertised by the radicals as more sophisticated medical aids than anything in the consultant's bag. Those professionally trained to understand and use such laws would be elected through open competition to the most prestigious posts, replacing the old surgeons who had obtained their offices through family patronage and the old-boy network. This was how medical management was to be upgraded; once installed in their executive posts, the new "experts" were to initiate their utilitarian social and medical reforms under the direction of a government health ministry (another Benthamite innovation).

The active Benthamite intrusion into the business of London science is becoming well documented, as are the utilitarians' civic sciences. Simon Schaffer has shown how the nebular hypothesis was promoted by these men as a cosmic correlate of the Whig ideal of social progress. Others have emphasized the growth of statistics in Benthamite hands: how the science was exploited by Dissenters campaigning for the civil registration of births, marriages, and deaths, and how a London University graduate

6. "The London University and the Society for Diffusing Useful Knowledge," *Athenaeum* 1833, 121; S. F. Cannon 1978:49.
7. "Medical Reform," *L* 1830–31, 1:564.

such as William Farr in the General Register Office (f. 1836) used the "ledgers of death" (mortality statistics) to argue the need for government-sponsored civic health schemes.[8] The Benthamite program was also implemented through the creation of specific government offices: from the Statistical Department of the Board of Trade (f. 1832) to the Geological Survey. Statistics might seem an obvious administrative tool, but how far a practical science such as geology could be harnessed for party purposes has been elegantly demonstrated by James Secord. Under the direction of the impecunious former plantation owner Henry De la Beche, even the Geological Survey's program of countrywide mapping and strata identification was brought to the service of municipal reform. Like other Benthamites, De la Beche looked to France for his model of state-supported science. He championed the London University, started the working-men's lectures in the School of Mines, and served on the sanitary commissions of the 1840s. Government geology in De la Beche's hands was expected to help in the political reconstruction. Thus he presented his geological maps, depicting drainage and soil types, as essential aids to local administration. They permitted correlations of disease, soil, and ancient sediments and showed the drainage potential of sites. As a result, they could be used during cholera outbreaks to pinpoint the negligent borough authorities, those who had failed to provide proper drainage or pure water.[9]

To force through the welfare reforms, extend suffrage, and expunge local inefficiency, new electoral procedures had to replace the old practices of family appointment, and that meant ending the Tory-Anglican domination of the municipal and medical corporations. In fact the changes went much further, and the issue of electoral reform touched almost every aspect of civic life in the 1830s. The learned societies were not immune. Indeed the radicals specifically targeted the more aristocratic of London's scientific bodies, attempting to extend the franchise to allow training and talent to compete with title and status. The Zoological Society (f. 1826) had a management structure top-heavy with titled officials. The trustees of the British Museum were still bishops and noblemen. That was as it should be, Tories insisted: noblemen had a public profile, and they could deal directly with ministers.[10] They also guaranteed the protection of rank. The zoologist William Yarrell told the 1836 Select Committee investigating the British Museum that such "public" gentlemen brought "influ-

8. Cullen 1975:43; Eyler 1979:198; Schaffer 1989. The increasing state intervention after 1832 is discussed by Lubenow 1971; Finlayson 1969:65ff.; Halévy 1950:98–129; Morrell 1971a:188–92.

9. Secord, 1986b:224–34, 247.

10. Desmond 1985a:225–27.

ence and power." Tories saw no advantage in letting career scientists even partially replace the landed governors on the British Museum board; the archbishop of Canterbury thought the whole idea positively mischievous. In the same way, the king's physician William Macmichael justified restricting the Council of the College of Physicians to gentlemen who had received an Oxford or Cambridge education on the grounds that this ensured their familiarity with the highest classes. He told the Select Committee on Medical Education in 1834 that the appointment of such gentlemen elevated the whole profession.[11] In his view a deferential membership was to bask in the reflected glory of its aristocratic patrons.

In the 1830s, however, this paternalistic system began to crack under the strain of continuing radical attacks. While Tories looked to their patrons' independence and "disinterestedness" as a guarantee of impartial administration, radicals treated these very qualities as a hindrance to purging rotten-borough practices and getting specialists elected. As we shall see, events came to a head at the courtly Zoological Society with the 1835 hustings, fought on the issues of suffrage, accountability, and the need to keep academic zoologists on the council. Much of the initiative for the Zoological Society reforms, like those at the Statistical Society (f. 1834), came from the small radical faction. Often the same radicals were active in different societies. The East India Company colonel William Henry Sykes, for example—a retrencher and critic of society corruption—served on both the Statistical and Zoological Society councils. Robert Grant, Sykes's collaborator at the Zoological Society, also shared a platform with fellow medical radical William Farr in the militant British Medical Association (f. 1836), and so on. This crossover in personnel allowed a web of personal relations to be spun between activists in different areas. Not even the Royal Society was immune to these new influences. Here too, electoral reformers relying on the trading, professional, and colonial vote made headway, with "traditional loyalties to Crown and Church" being replaced by new allegiances "based upon service to knowledge and utility to the State."[12]

So the "revolution in government" extended far beyond the Whig ministries, statistical offices, and health commissions. It was linked to the educational programs, to the London University, to scientific bodies such as the Geological Survey and Royal Institution, and to new electoral demands in the learned societies. While many of these institutions are now

11. *Report SCME Pt. 1*, 35. Yarrell's evidence: *Report SCBM*, 167–71; Gunther 1980:79–82. Yarrell was the Zoological Society's leading comparative anatomist in the 1820s before Grant and Owen introduced their newer approaches (Vigors 1830:208).
12. MacLeod 1983:57.

coming under historical scrutiny, the sciences they promoted need to be better understood to assess their role in the reform strategies.

For the London University radicals, the new French approaches to comparative anatomy symbolized progress and expertise. Geoffroy's "philosophical anatomy" pointed to natural law rather than craft lore; with its arrival, anatomy was to become a truly theoretical science at the heart of a modern, democratic medical administration. Elsewhere in the Benthamite educational empire—in the Mechanics' Institutes and the Society for the Diffusion of Useful Knowledge (SDUK)—the situation was very different. Here more traditional sciences were to be put to alternative work, for example, in disguising class friction and reducing civil disturbance. Indeed, because of their Mechanics' Institutes and attempts to "naturalize" capitalist relations and wage restraints, the Broughamites'[13] aims have largely been interpreted in terms of labor control at a time of union formation and social unrest. The institutes have been pictured as middle-class moralizing bastions in urban working-class environments. Steven Shapin and Barry Barnes see their Broughamite managers mating science to a property-conscious morality, teaching the laborers acquiescence before the divinely ordained natural and moral order.[14] The sixpenny tracts put out by the SDUK, promoting an acceptable science and natural theology, were designed to penetrate working-class areas and to proselytize in ways unimaginable to the bearded Methodist missionaries.

13. Broughamites were the reformist Whigs and their allies who shared Henry Brougham's views on secular education. Many had Scottish credentials: Brougham himself was an Edinburgh-educated lawyer who had helped found the Whig *Edinburgh Review*. They also had Benthamite leanings, although unlike Bentham (a freethinker), Brougham wrote extensively on natural theology. This, like his working-class educational schemes—the Mechanics' Institutes and Society for the Diffusion of Useful Knowledge—was designed to turn the thinking artisans and middle classes away from the gutter-press infidels and prevent a repetition of the godless violence seen during the Regency.

14. Shapin and Barnes 1977:32, 34, 41. Cf. C. A. Russell 1983:160–73; Vincent 1981:138–65. Also Grobel 1932, 3:815; Berman 1978:112. Cooter (1979, 1984) suggests that phrenology too, by presenting images of organic interdependence and validating an orderly growth and progress through melioration, diverted minds away from revolution.

Cobbett denounced the SDUK tracts as educational sops devised to depoliticize the working classes, "diverting their attention from the cause of their poverty and misery" (Harris 1969:75). And the Mechanics' Institutes' ban on politics was said to be driving "all the most sober, industrious & reflecting of the handicraftsmen . . . to Socialism": T. Coates to Brougham, 27 September 1839 (UCL HB 95). The pauper presses responded with a rash of unstamped penny papers propagating "really useful knowledge"—in the most extreme cases an amalgam of Lamarckian materialism, d'Holbachian atheism, and socialist environmentalism. This independent labor-oriented science, it was feared, would disrupt discipline, destroy subservience, and inflame wage relations, making it all the more urgent for the middle classes to regain the educational reins: Desmond 1987; Royle 1974; Johnson 1977:87–93.

They were to cover "the country with knowledge," Brougham once said, and not only the home country but also the colonies.[15] The institutes became part of a program for the containment, control, and education of the poor because more overt control mechanisms were incompletely developed at the time; indeed, it is no coincidence that the next few years saw the creation of the police, poor laws, workhouses, madhouses, and new punishment procedures.[16]

The London University, however, was established for different purposes and promoted different types of science. Moreover, the radicals in the university and statistical offices repudiated many of the Broughamites' anti-working-class policies. Farr and Grant were vehemently opposed to the Whigs' New Poor Law of 1834, which ended financial relief for all but the most indigent poor, those desperate enough to endure life in the workhouse. Wakley was opposed to it as well and he always remained suspicious of the "Whig Lordlings",[17] even though he was defended in court by Brougham himself. (Brougham would be kicked upstairs as lord chancellor when the Whigs took office in 1830.) The medical radicals wanted something more than moral equality between the classes (not the least because an increase in poverty depressed the GP's pay), and many supported the Chartists and denounced the workhouses and Police Bills. So too their sciences were different. While natural theology, which sought Divine beneficence in nature and gave a conservative moral stature to science, appealed to the older Whigs and Mechanics' Institute managers worried by artisan atheism, it appalled the more freethinking radicals. In the final analysis, rival sets of sciences were being pursued in different departments of the educational empire. At the university after an initial Whig-radical power struggle, the ascendancy of anti-Whig radicals ensured the success of the more reductionist, secular sciences. The university harbored medical radicals who imported Lamarck's zoology and taught the new French sciences—radicals who despised the Whigs' Poor Laws, yet who shared the Benthamites' corporation-reforming ideals. Natural theology might have been integral to the campaigns designed to stamp out working-class atheism, but here it had no such appeal. The antidemocrats in the medical corporations were themselves Paleyite[18]

15. Brougham 1827:524.

16. Philips 1983:65–66.

17. L 1830–31, 1:763; Sprigge 1899:117, 139 on Brougham's defense of Wakley in cases brought by the hospital surgeons. On the relationship between working-class wages and the GPs' pay: "Emigration of Professional Men," L 1829–30, 1:474–75.

18. *Paleyite* refers to the doyen of natural theologians, the archdeacon of Carlisle William Paley. Although Paley had died in 1805, his *Natural Theology* (1802) remained immensely

natural theologians. Something stronger was needed to undermine their authority. It was in this context that the radical morphologies imported into the university made political sense; they invalidated the surgeons' pious science and with it their reactionary hold on the profession. So at the end of the day, the radicals' Lamarckian and Geoffroyan biology served quite distinct middle-class professional ends.

We will return to this anticorporation aspect of radical science time and again. Corporation reform was high on the Whig-radical political agenda. The Tory-Anglican oligarchies had to be ousted from their municipal strongholds, just as the nepotistic sergeant-surgeons had to be dislodged from their medical seats. At the civic level, the 1835 Municipal Corporations Act was designed to replace the undemocratic corporations by elected town councils. Derek Fraser has characterized local politics following its passing as a "power struggle between rival élites within the bourgeoisie"—between the Tory-Anglican dynasties that had long held civic power and Dissenting reformers ambitious for the trappings of official rank.[19] The radicals saw no difference between the medical and municipal corporations: both were party instruments operated for patronage and profit, run by self-perpetuating oligarchies and staffed through family connection.[20] If the Municipal Act was pure "poison to Toryism," then the radicals wanted nothing less virulent to wipe out the medical Tories.

The medical elite in 1820 comprised the knightly physicians and surgeons attending a few wealthy or noble patients. The physician, in particular, was still judged more by his breeding and "moral" education than his medical expertise. His initial Oxford or Cambridge studies were in classics, these being more important than medical knowledge to place him on a cultural par with his noble patrons. But by the 1820s reformers were putting a very different slant on this traditional education, picturing it as a sacrifice of scientific expertise on the aristocratic altar—deference to the detriment of "professional" standards. Radical censure on this point was unmitigated: as a "savage exchanges a pig for a few beads," so "natural philosophy, chemistry, and physiology have been bartered at the Colleges

---

popular and had passed through numerous editions. Paley had sought evidence of purposeful design in nature. In his view the adaptations that fitted an animal to its niche were means to an end: enabling the animal to survive, reproduce, and enjoy life. Such design in his view implied an omnipotent, caring Designer. Thus he was using evidence from Nature to prove the existence and attributes of God. "Paleyism" or "Paleyite" in the text refers only to this design aspect.

19. Fraser 1976:115–16, 1979:13, 1982:5.

20. Tirades were directed almost weekly in L and LMSJ against the RCS elect. For an airing of the collected grievances see, e.g., L 1830–31, 1:846–65. Compare Fraser 1976:116 for parallel denunciations of the civic oligarchs.

for the pageantry of Latin learning."[21] The pure surgeon was no better, having gone through an apprenticeship rather than an academic education in science. It was not only ultraradicals who detested the corporation "toad eaters"; calls for the reform of these "conservatories of bigotry and ignorance" and the institution of a standardized scientific education dominated all the medical journals in the 1830s.[22]

There has never been an attempt to understand the new London University curriculum in this social context—hardly surprising, given that the establishment of the newer Parisian anatomies in the university was itself practically unknown. So we need now to discuss the French origin of the scientific imports and the political shuffling to get them established at the university, before showing how they were used to undermine the moral claims of the corporation elders.

The London University

One of the radical and most conspicuous blunders in the London University consisted in trusting the arrangement regarding the Medical school to those . . . ignorant of the science of medicine. . . . The governing body is made up almost entirely of lawyers and merchants, nor would it be easy to select a class of men less qualified by the nature of their pursuits and occupations to regulate the business of a medical school.

The antagonistic *Medical Gazette*[23]

The conservative *Gazette* was partly right on one point: the composition of the University Council. Nor was its antagonism misplaced: as a supporter of the Royal Colleges it knew that a school founded by corporation reformers boded no good. The University Council was arranged around Brougham and embraced many prominent campaigners for religious and civil liberties. Its members had a common interest in overhauling the electoral system and giving the Dissenting professionals and industrialists a greater say in Parliament and the boroughs. The Benthamites and radicals were a powerful caucus and included the lob-

---

21. "Medical Education," *L* 1838–39, 1:908. On literature, breeding, and the physician: *Report SCME Pt. 1*, 4–13, 32–43; Holloway 1964:301–2; Newman 1957:5; Peterson 1978:4, 8–9, 153.
22. Grant 1841:6; "The Assaulters," *L* 1830–31, 1:867.
23. "London University—Mr. Bell," *MG* 1830, 7:305.

byists for medical reform. Almost half the members of Council held reforming briefs in Parliament. Many were fellows of the Royal Society. The
legal profession was well represented, as were the financial houses, trade,
and the colonial service,[24] enabling the council to appeal directly to the
capital's Nonconformist merchant and professional population for support.
This composition also explains the council's wider goal, announced in its
first *Statement* in 1827—that is, the educational upgrading and reform of
the legal, medical, and administrative services—a goal that was reflected
in its recruitment policy and in the medical school's scientific orientation.

From the first the university had the support of the Dissenters. In the
mid-1820s they still suffered social and professional disabilities as a result
of Anglican control of the rites surrounding birth, marriage, and death.
They were banned from matriculating at Oxford or graduating at Cambridge, handicapped at the Inns of Court, excluded from the Fellowship
of the College of Physicians, and barred from public office. But the religious exclusiveness of existing higher education was only part of the
problem. Oxford and Cambridge universities were expensive and
overcrowded, the quality of medical education was inferior to that in
Edinburgh, and they turned out few medical graduates, even if these
often took the key hospital posts and most prestigious offices in the College of Physicians.[25] Worse, the old universities still cherished the
eighteenth-century ideal of the leisured, classically educated physician.
Radicals saw them as ill-adapted "to make men fitted for the real world,
or for the real business of life."[26] So the London University's founders
never intended to emulate the Anglican universities. It was not simply

24. The council's twenty-four members were overwhelmingly Whig to radical, with a
common interest in religious and electoral reform. They included eight Whig M.P.s and
noblemen, and two radical M.P.s. Nine had Scottish backgrounds or educations, five were
trained in the law, and four represented finance houses. There was a strong East India Company presence (including James Mill, who worked in India House), and the council included
two former colonial administrators. Nine were fellows of the Royal Society, and many were
active in the Royal Institution, Mechanics' Institutes, and scientific societies. Only three had
received (Scottish) medical educations, including George Birkbeck and the radical M.P. Joseph Hume (who had also been an East India Company assistant surgeon). Two members,
Hume and Henry Warburton (whose background was in the timber trade), spoke for medical
reform in the House of Commons.

25. In 1801–50 Oxbridge produced only 273 medical graduates, during which time 8,000
doctors were trained in Scotland. Of the graduates in London in 1850, 65 percent were
Scots– and 20 percent London University–educated, yet 55 percent of the hospital teachers
in London had Oxbridge educations (Robb-Smith 1966:49–52); *Statement by the Council* 1827:7–8. On the superiority of Edinburgh medical education, *Report SCME Pt. 1*, 93,
104–5.

26. "The London University," *LMSJ* 1833, 3:535–36.

that education was to be cheap, nonsectarian, and "enlightened" in London.[27] Rather it was to be of a different kind: London University was projected as a teaching factory turning out trained medical and legal personnel for bureaucratic state service.

To avoid sectarianism, the school was kept avowedly secular, to the disgust of Tories. Some actually questioned the legality of calling a non-Christian, nonchartered body a university, while others deplored the council's refusal to offer courses on the Evidences of Christianity. But this secularism appealed to the school's radical backers and left Wakley crowing that the scheme "afforded not a single compliance with the demands of the 'Church and State' bigots of the day." With the council containing so many promoters of science and medical reform, it was widely believed that the secular sciences would be central to the curriculum and, as Wakley continued in his bellicose way, that the new chairs would "be unpolluted by those pestilential vapours which had ever surrounded a certain class of professors in the 'ancient' Universities."[28] There was already growing criticism of the Latinity and lack of science at Oxford and Cambridge. The liberal Charles Lyell (himself a barrister-turned-geologist) pointed out in 1827 that the French and Italian universities now incorporated chemistry, physiology, and zoology into their medical degrees. The radicals for their part dismissed Oxford's classics orientation as useless "to the agriculturalist, the manufacturer, the merchant." They urged that physics, chemistry, animal physiology, and the "sciences of every day application in the business of life should form the basis of instruction."[29]

The council therefore conceived the kinds of professional courses in law and medicine that were either unavailable or inexpertly provided at Oxford and Cambridge. Its target clientele, too, was quite distinct: not the sons of the Anglican eminent (Oxbridge was still the training ground for judges, physicians, and clergymen), but those destined for the legal and medical middle rungs, the solicitors and GPs "in whose hands the whole ordinary practice of England is placed." Of the six thousand members of the College of Surgeons and eight thousand solicitors, it pointed out, not one in a thousand was a university graduate. The object was to give the new legal, medical, and administrative middle management an "enlightened," secular, and scientific education. The council was in effect aiming to create a new constituency of Benthamite "experts" in science and the professions. Thus recruitment was to take place among the Dis-

27. *Statement by the Council* 1827:8.
28. "London University," *L* 1830–31, 2:689–90; Burns 1962:5; Bellot 1929:25, 29; New 1961:365–67.
29. "Medical Education," *L* 1838–39, 1:908; C. Lyell 1827:229.

senters of "easy yet moderate" means, among professional families and civil servants, and through the colonial service (the Bombay colonial and medical staffs were strongly supportive).[30]

The reforming Whigs were offering a professional education to the sons of the Nonconformist merchant population. Hence the metaphors used to describe the school: it was an "emporium for the supply of intellectual goods" in the "Queen of cities—the Empress of the commercial world."[31] There is no doubt that the council's professional goals had political ramifications (not least the democratic reform of the municipal and medical corporations) which clearly met the Dissenting merchants' needs. Without this democratization, the repeal in 1828 of the Test and Corporation Acts (which had barred Dissenters from public office) would have remained hollow: permitting Dissenters to run for civic posts, while denying them the chance to vote out the Anglican encumbents. At the same time, by giving the GPs and solicitors a secular scientific education, the council hoped to create a new reforming electorate in the professions.

The type of knowledge bartered in the intellectual "emporium" was no less a reflection of this Benthamite ideal. Students destined for government service were to be grounded in law, jurisprudence, and political economy and taught the "true" wage-labor relations to combat the demands of the working-class agitators—in short, armed with irrefutable economic laws legitimating the Benthamite state.[32] In medicine, too, a new-style academic education based on the wider laws of life was to further the Benthamites' ends in undermining corporation control and establishing a democratic medical government. The reformers demanded an administrative overhaul of the Royal Colleges and insisted on a wholly new approach to medical teaching. They attacked the "superannuated system" of old-school surgery, with its expensive apprenticeships, "showy" demonstrations, and "rant concerning the folly of study as opposed to

30. *Statement by the Council* 1827:8, 9, 10, 15, 29, 37. The council's colonial interest manifested in other ways besides the founding of a chair of "Hindoostanee." The Zoology Museum was to act as a repository of colonial specimens, medical graduates were to meet East India Company requirements, and pupils were to be prepared for imperial service. Grant, whose own brothers were military men (two were captains in the Madras army), taught a number of Hindus; his favorite pupil and companion was Soorjo Coomar Chuckerbutty (his 1846 gold medalist), himself to become professor in Calcutta's Government Medical College. Sharpey 1874:ix; *Distribution of the Prizes*, College Collection A 3.2, UCL; on the Army Medical Board's sanction of Grant's courses, Grant to C. C. Atkinson, 18 February 1836 (UCL CC 3609); and "Biographical Sketch of Robert Edmond Grant," *L* 1850, 2:695 (hereafter "Biographical Sketch").

31. "University of London," *LMSJ* 1832, 1:210–11.

32. *Statement by the Council* 1827:8, 35–36; Burns 1962:8–9. Halévy 1972:489 on the Benthamite understanding of the laws of nature and society.

practice."[33] Since the Gower Street professors were to be "unfettered by the habits of [these] previously existing schools," the university attracted outside support from the medical reformers rather than the corporation conservatives. Indeed the school's more radical councillors—especially that "teasing, biting flea" on the Tory rump, Joseph Hume, and the medical free-trader Henry Warburton—were in close contact with Wakley. They attended his public meetings to expose the "irresponsible" Council of the College of Surgeons, and as M.P.s they also aired the GPs' grievances in the House. Warburton's Select Committee on Medical Education in 1834 was even said to have been the outcome of Wakley's "successful agitation."[34]

Catering to the merchants, the school was financed like other city institutions: the business sector was approached for support. The capital of £150,000 was raised by selling shares (and squandered, groaned the *Gazette*, on the "showy façade and theatres" in Gower Street; see figs. 2.1 and 2.2).[35] Brougham tried to restrict shares to those with an interest in the institution, devising a scheme that permitted each shareholder to enroll one student per share owned. This was to keep out the jobbers (the share traders hoping to make a profit). Others were more concerned to keep out the fierce democrats. The improving nobility wanted to maintain the school under tight Whig control, many sharing the marquis of Lansdowne's fears that as it prospered, efforts would be made by the "various classes of methodists, radicals, dissenters & jobbers to wrest it to their own particular purposes."[36] But prosperity proved elusive; the whole scheme fell into immediate financial difficulty, and although the money was raised, it never paid a good dividend. All this proved grist to the *Medical Gazette's* mill. But then the idea of a joint-stock medical school "established for converting science into a matter of traffic" was anyway obnoxious to conservatives defending rival interests. The *Gazette* believed it to be the "grossest perversion of the desires of the more soberminded members of the community."[37] The paper editorialized incessantly against this merchandizing of medical science, appalled by the idea of a teaching factory which threatened the privileges of the corporation and hospital elite: "it is sickening to think what medical teaching would thus become. Twenty years ago it was the province of the best of a profes-

33. "Recent Improvements of Medical Education," *Quart. J. Educ.* 1832, 4:4, 5, 10, 15.

34. Sprigge 1899:278, 222. On Hume at Wakley's mass meetings: *L* 1826, 9:804; 1830–31, 1:846; cf. Newman 1957:149; and Cowen 1969:36; "flea" quoted from Wiener 1969:55.

35. "Ways and Means at the London University," *MG* 1832–33, 11:806; also 1833–34, 13:918; Bellot 1929:33.

36. Lansdowne to Brougham, 17 August 1825 (UCL HB 38,908).

37. "The Conversation in Gower Street," *MG* 1833, 13:49, also 918.

Figure 2.1. A lampoon of Henry Brougham, himself a leading lawyer, peddling shares in the proposed university around Lincoln's Inn. By Robert Cruikshank, 1825. (Courtesy The Library, University College London)

Figure 2.2. The "showy façade" of the London University in Gower Street. It is pictured here in 1833, at the time of the foundation of the North London Hospital across the road. (Courtesy The Library, University College London)

sion of gentlemen—now it is descending every year into lower hands. It is a trade, and worse than a trade—stock-jobbers buy and sell it."[38] These sentiments reflected the Tory anger at the attempts to exploit new business markets and shift the control of medicine from the Oxbridge or hospital-educated gentlemen to a new class of urban professionals. The *Gazette's* jibes were indicative: it pictured shopmen entering the medical manufactories, exchanging apron for gown, and emerging as "*bourgeoises gentilshommes* . . . mimicking the manners of the great." At times these class fears were barely concealed: Grant's attack on the "imperfect" Oxford and Cambridge schools was countered by conservative claims that physicians, like Anglican clergymen, must be "educated in the same manner, and in the same classes as the highest rank of society" in order to preserve moral standards and noble patronage.[39] What really galled the critics, though, was the radicals' praise for "that *low* place in Gower-Street."[40] Wakley infuriated the *Gazette* by his support for the "politico-

38. "Another Joint-Stock School of Medicine," *MG* 1838, 22:474.

39. "Dr. Grant and the College of Physicians," *MG* 1833–34, 13:120; cf. a physician's rejoinder, ibid. 165–66.

40. Clarke 1874a:319.

medical establishment." The *Gazette* also basted the radical professor of midwifery David Davis for inviting Wakley to the school's Saturday conversaziones (where he was lionized by the pupils). It warned the professors to "repudiate his patronage" or "lash" their journalistic "watch-dog" into line to preserve their self-respect.[41] But radical professors such as Grant remained Wakley's staunchest supporters, while Wakley himself drew great ideological strength from Grant's "brilliant" science.

Given the university's medical aims, Wakley's support is not surprising. Doctrinaires were quick to seize on available pro-French, proscience sticks with which to beat the court surgeons, and those supplied by the new professors were perfect. In recruiting, the council had largely avoided the local anatomists and brought men in from the outside, mostly from Edinburgh. The governors wanted teachers with an academic medical education, preferably Edinburgh-Paris based to ensure a broad Continental understanding. Not for nothing was the enterprise denounced by conservatives as a piece of "Scotch" jobbery. Over half of the first Education Committee were Scottish born or bred, and a third of the professors had Scottish credentials. An academic "New Edinburgh, an Educational New Jerusalem" was being created at the top of the Tottenham Court Road.[42] As Burns observed, the Benthamite influence which "irradiated the new University . . . did so through a prism shaped largely by these Scottish forces." So we must examine the sort of refraction produced by this "Scotch" prism and the way it bent the light of knowledge into ideologically useful shapes.

The university's curricular innovations are poorly known. Pauline Mazumdar has argued that its policy was to push medical education beyond the "purely practical" limits of the old-school anatomists. She also suggests that the council's recruitment of teachers inclined toward Continental methods caused a "withering away" of the older approaches to anatomy based on functional design and natural theology.[43] We do know that, within a decade of the university's founding, philosophical anatomy had become widespread among the medical reformers, and Jacyna has suggested that these marginal men were harnessing its powerful laws in an attempt to raise their professional standing.[44] Both of these insights are developed as we proceed. What we have to show first is the meaning of the new anatomy and zoology in republican Paris and its attraction for the visiting Scottish radicals. Only by understanding philosophical anatomy's

41. "London University," *MG* 1834–35, 16:53; also 1830–31, 7:372–73; 1831, 8:218; 1831–32, 9:21–23; and Bellot 1929:159.
42. New 1961:375; Burns 1962:7; Bellot 1929:47.
43. Mazumdar 1983:231, 233; Bellot 1929:144.
44. Jacyna 1984a:60, 62–63.

role in France around the time of the 1830 Revolution can we really offer an explanation of its value to London's anticorporation radicals.

The French Morphologies and Their Scottish Importers

The great majority of manuals and systems of anatomy or physiology of the present day are either avowedly or in reality taken from the French, and our students would seem to imagine, like our modern play-goers, that nothing whether large or small, opera or interlude, system or manual, can possibly be worth a fig unless it is imported from the other side of the water. God knows we share nothing in common with that anti-Gallican, exclusive, and bigotted party whose gorges rise at the bare mention of French or Frenchman, and who tickle their John Bullish prejudices by holding all which is not their own in profound contempt. At the same time we are sufficiently patriotic to desire that in knowledge, as in war, we should not succumb before the sons of Gaul, but rather endeavour in generous rivalry to lead the van.

—*Medico-Chirurgical Review*, in 1830,
commenting on the trade deficit in knowledge.[45]

If we wish to view the study of animated nature in a form truly worthy of occupying a philosophic mind, we must direct our attention to the French school.

—Robert Grant, a yearly traveler to Paris,
urging devotion to Parisian science in 1830.[46]

Large numbers of British medical students visited Paris in the 1820s. Originally they were mostly Scottish graduates completing their training. Paris offered better hospital and postmortem facilities, with its availability of cheap, legal cadavers. In the city's hospitals thirty thousand patients a year were treated; four-fifths of those who died were dissected (a situation unheard-of in London).[47] Here these foreign students were introduced to the preeminently French sciences of pathology and comparative anatomy; indeed, attendance at zoology and comparative anatomy lectures was almost de rigueur at the Muséum d'Histoire Naturelle, itself at a pinnacle of international prestige.

45. "Lectures on Anatomy," *MCR* 1830, 12:95–96.
46. Grant 1830a:343.
47. "Dr. Armstrong's Reform Principles," *L* 1830–31, 2:403.

When the foppish radical Robert Knox arrived in 1821, there were over thirty British students enrolled in the Paris Faculty of Medicine. Numbers increased throughout the decade, swollen partly by events in London. The Royal College of Surgeons (RCS) in 1822–24 altered its by-laws to discriminate against London's private medical teachers, who were undercutting the hospital teachers' fees. By refusing to accept their course certificates (which a candidate presented at the examination for his license), the RCS intended to divert students to the hospitals where its own councillors taught. But this policy ended up driving many students to Paris, where the number of émigrés had risen to two hundred by 1828.[48] In 1822 the Dublin-educated James Bennett established an "English" school in Paris, renting a dissecting theater in the Hôpital de la Pitié. The College of Surgeons, alarmed by this further threat to its medical jurisdiction, refused to acknowledge Bennett's school or accept his certificates. It justified this decision on the grounds that to encourage Bennett was to invite the "decay" of English teaching and methods, such that "when a time of war came" the English would no longer be able to educate their own sons.[49] When the school ran into difficulties in 1825 (facilities were withdrawn during the Ultra-royalist reaction when anti-British feelings were running high), the council dissuaded Foreign Secretary George Canning from taking up Bennett's cause with the French government. The school collapsed, and Bennett's case became a radical cause célèbre. Wakley continually used it as another noose to hang the RCS Council, while Warburton's Parliamentary Select Committee referred to it as an act of gross injustice.

When the first wave of Scottish students arrived in Paris after Waterloo (Grant was among the earliest to cross the Channel, in 1815), they found Jean-Baptiste Lamarck, at seventy-one, tetchy, pessimistic, and losing his sight (see fig. 2.3). As professor of "insects and worms" at the Muséum, he was still lecturing on the "invertebrates" (his word), largely to medical students. (He was only to relinquish his post in 1820 after going completely blind.) He was also still publishing, and 1815 saw the first of seven volumes of his taxonomic tour de force *Histoire naturelle des animaux sans vertèbres* (1815–22).

Lamarck had been influenced by the *idéologues*[50]—a group of ration-

48. Maulitz 1981:491; Rae 1964:23; Limoges 1980:221–25.

49. G. J. Guthrie's testimony, *Report SCME, Pt. 1*, 80, also 36–37; *L* 1830–31, 1:402–7, 854; 1833–34, 1:363.

50. The *idéologues* included the historian and deputy to the National Assembly, Constantin de Volney, the mathematician Marquis de Condorcet, and the physician Pierre-Jean Cabanis. With others, they met in a salon in Auteuil, where they also welcomed Baron d'Holbach. As rationalists they placed morality and justice on a naturalistic (rather than

Figure 2.3. Jean-Baptiste Lamarck in 1821. By A. Tardieu after Boilly. (Courtesy Wellcome Institute Library, London)

alist historians, scientists, politicians, and educational reformers writing in the turbulent years before and after the 1789 Revolution. The *idéologues* and their fellow travelers (the most extreme of whom was the atheist Baron d'Holbach) accepted that matter contained the potential for sensation, which was realized in animal life, and on this assumption they went on to interpret morality and behavior in terms of natural law. Believing that ideas were the refined product of sensory associations and ultimately derivable from external stimuli, they stressed the importance of the environment for shaping mind as well as body. They maintained that man could be perfected through control of the social environment, and they campaigned for political, medical, and educational reforms to remove the religious obstacles in the way of this progress. The rationalists' claim that unaided matter had the power to produce everything from suns to starfish had potent political implications in the ancien régime. D'Holbach for instance denied all supernatural existence and, in *System of Nature*, called for thinking men to "make one pious, simultaneous, mighty effort, and *overthrow the altars of Moloch and his priests*," a rallying cry so inspiring to British working-class insurgents in the Regency period.[51] Indeed, conservatives in Britain and France continued to insist for many years that the French Revolution had been largely the product of this kind of poisonous philosophy. Coleridge complained in 1817 that the democratic "Ruffians" were still using a science of self-empowered atoms to justify their struggle for a society of self-governing individuals.

This kind of materialism remained pervasive well into the nineteenth century, especially in the Paris Faculty of Medicine. It also shaped the collateral sciences of life, largely because aspiring comparative anatomists routinely took medical degrees at the time. At the turn of the century influential *idéologues* such as the transformist (and former revolutionary activist) Pierre-Jean Cabanis were still arguing for the animality of man and the qualitative similarity of human and animal minds. In about 1800, too, Lamarck had begun extending this kind of approach to the entire organic realm. (Although Lamarck did have his differences with the older

Christian) footing. Their works were still being pirated by the pauper presses in Britain in the 1820s. This is true of Volney's *The Law of Nature* (1793), in which natural law was promoted as egalitarian and universal. Even more inspiring in radical Britain was d'Holbach's critique of Christianity in his uncompromising *System of Nature*. Here order and organization were emergent properties, a function of matter's self-organizing atoms; no scientific metaphor better served the agitators' struggle for a secular democracy.

51. Desmond 1987:95–96; Griggs 1956–71, 4:758–62 on Coleridge. R. J. Richards 1982, 1979:93–94; and Jordanova 1984:77–78 on Lamarck's *idéologue* debt. R. W. Burkhardt 1977 on Lamarck's science generally.

idéologues: for him life was a property of organization rather than of matter itself, even if this organization had ultimately resulted from "mechanical causes, regulated by laws".)[52] In his speculative *Philosophie zoologique* (1809), and again in the *Histoire naturelle,* Lamarck tackled what he called the "march of nature." His earlier studies in meteorology and geology had convinced him that the earth's surface was ceaselessly changing. To explain how this affected animal life he resorted to a notion, current among the *idéologues,* that new needs in a changing world called for new habits. These then forced the relevant organs to be used to a greater or lesser degree, which caused them to increase or decrease, and these structural modifications were inherited. For Lamarck (unlike his British admirers in the 1820s), these global changes were not progressive. Yet his own taxonomic studies had shown that living animals could be arranged into a finely graded, progressive series. This series could have arisen, he suggested, if the microscopic infusorians (the simplest known animals) were spontaneously generated, and the subtle fluids coursing through them had then carved out new channels and caused new organic arrangements. Without any external distorting factors, the action of these fluids would result in a perfect series from monad to man. But environmental exigencies forced the series to be continually deflected, as animals adapted their habits to fit specific niches. So the "march of nature" was not ideally uniform. Indeed, there was not even a single series; by 1815 Lamarck had come to accept the spontaneous generation of gut parasites as well, and in the *Histoire naturelle* he depicted two major invertebrate streams rising in parallel fashion—one originating in the parasitic worms, the other in the infusorians.

It must have struck the visiting students that the most voluble of Lamarck's supporters were the republican materialists: men such as Jean Baptiste Bory de Saint Vincent, editor of the influential *Dictionnaire classique d'histoire naturelle* (1822–31), and the republican physician Françoise Raspail. Bory, by the mid-1820s, was extending Lamarck's system. He argued that the animal series was the product of the mechanical laws of aggregation and transmutation, that thought was a necessary outcome of this increasing organization, and that man's relationship to the ape was closer than even Lamarck realized. Raspail, once jailed as a subversive and destined to take to the barricades in 1830, denounced the nepotism and privilege of the conservative scientific cliques and included Lamarck in his pantheon of visionaries who stood outside this world of intrigue and corruption.[53] To appreciate this republican sympathy we must remember

52. Hodge 1971:348–49; R. W. Burkhardt 1977:140; Conry 1980.
53. Corsi 1978:227–29; Bory 1827; Appel 1987:171, 176; Outram 1984:110, 1980:35–36.

that the *idéologues'* enterprise had never been purely theoretical; as Jacyna says, for two generations—through the revolutionary period—it had been espoused by groups antagonistic to the Church and monarchy. Materialist physicians in the 1810s and 1820s were still using physical theories of mind to justify their claim on the moral terrain occupied by the "ignorant and presumptuous theologians."[54] They were also trenching on the clergy's domain in other ways—trying to reduce the Catholic church's control of the hospitals and nursing orders. At the same time, the materialists' insistence that the sciences of life were essential to the correct formulation of social policy enabled them to angle for posts as government advisers on questions of health, the environment, and education.

There was, of course, a theological revulsion against this monistic materialism. Royalists developed the rival notion of an inert matter that was dependent on external agents for its action. For these conservatives life was no self-creation of matter, but depended on a higher authority, ultimately a supreme Will. This idealist physiology was tied to a diametric social ideal, in which authority was delegated downward through the Church and king. The royalist philosopher Louis de Bonald—accepting the Platonic definition of man as an intelligence served by organs—abhorred d'Holbach's "insane" *System* and deplored Lamarck's transformism, which led to the brutalizing of man.[55] Bonald's arguments against the subversive notion that individuals, like atoms, were sovereign authorities became prominent in Ultra-royalist circles after 1815 (much as Coleridge's similar arguments were favored by the antidemocrats in Britain). His spiritual reaffirmation, too, was welcomed by the Catholic clergy reasserting its old authority. With the revival of clerical power, bishops announced once more that it was God who empowered brute matter, just as it was He who sanctioned the authority of "masters over their servants, magistrates over the city, and governments over the people."[56]

When the British radicals arrived, then, Lamarck's deistic transformism was under attack from the antidemocrats, and the medical debate over the sources of authority and morality was becoming heated as the royalists gained in strength. For the conservatives, giving man back his soul, subjugating matter once more to a guiding intelligence, meant that "duty" could again be dictated by clerical authority. The visitors could not have ignored these attacks on "republican" science. Nor could they have mistaken their antidemocratic meaning, with Bonald announcing that atoms could no more produce intelligence than a "mutinous populace"

54. Jacyna 1987:114–15. I have leaned heavily on Jacyna's interpretation of French medical politics in this period.
55. Ibid., 123, 130–32; R. W. Burkhardt 1977:188.
56. Jacyna 1987:133.

could usurp the king's power, and with Abbé Fraysinnous accusing "false *savants*" of carrying "democracy into nature, as false politicians had carried it into society," to dethrone God and the king.[57] Indeed, similar political analogies were made in Britain as the radicals brought the offending sciences home with them.

Lamarckism by no means dominated republican science, however. In France, as in Britain, it attracted only a tiny doctrinaire minority—the noisy artisan atheists, socialists, and medical democrats. But by the 1820s it was being recommended by the zoology professor at the Muséum, Etienne Geoffroy Saint-Hilaire. Indeed, Geoffroy's own attempt to trace a "unity of plan" throughout the animal kingdom was seen by some as a prerequisite for a theory of transmutation. Geoffroy's own anatomical enterprise, in contrast to Lamarck's, appealed to a wide spectrum of medical reformers on both sides of the Channel. His deistic science was far and away the most important medical landmark of the republican cause in the 1820s. The recent study by Toby Appel has made it much easier to interpret the political appeal of the rival comparative anatomies encountered by the British in this period: Geoffroy's speculative, progressive science of "unity of plan" on the one side, and Georges Cuvier's more conservative, factual, and safe science on the other. I will briefly pick out the leading points of Appel's revisionism, because the estrangement and eventual public clash between Cuvier and Geoffroy was a talking point among the Scottish students no less than among their Parisian contemporaries.

Georges Cuvier was scientifically and politically powerful, the permanent secretary of the Académie des Sciences and professor of comparative anatomy at the Muséum (see fig. 2.4). By the 1820s he was bedecked with medals: a councillor of state, and influential in formulating public education policy as a member of the Committee of the Interior. He feared a return of the revolutionary turmoil and advocated a strong centralized state, responsible science, and popular education as a way of maintaining law and subordination. Essentially he was a pragmatist-turned-conservative, whose acquiescence to the Ultra-royalist ministries of the 1820s alienated many republicans.

How Cuvier's approach to science supported his stand in this politically sensitive period we shall see shortly. As regards substance, his comparative anatomy was quite unlike his predecessors'. First and foremost, Cuvier advocated a totally functional explanation of animal structure. This meant that, while he distinguished four *embranchements* or divisions of animal life—vertebrates, mollusks (snails, cuttlefish), articulates (insects, crustaceans), and radiates (starfish)—he envisaged no abstract "plans" as

57. Ibid., 127, 131, 133.

Figure 2.4. The doyen of French comparative anatomy Georges Cuvier. This engraving was based on a painting by H. W. Pickersgill (1831) which was owned by the leader of the British Tory party, Sir Robert Peel. By G. T. Doo, 1840. (Collection of the author)

such. The structural similarities within each division were due solely to similar functional needs. Take the comparable leg skeletons of lizards, mice, and men. These were not a result of some preestablished regulative law producing the same organization in each; rather, they were similar structural responses to the same requirements of walking. In Cuvier's extreme formulation, the common pattern indicated no more than that the mammals and reptiles played a similar "role." Indeed, because function was the sole arbiter of structure, the shape and number of the elements in any organ could be varied by God, or new elements could be added to suit individual needs. Also, because every organ was integrated and functioning perfectly, there were no rudimentary or useless organs, introduced solely to complete some vertebrate "plan."

Cuvier characterized the *embranchements* by their nervous systems. The animals in each division had a discrete arrangement of nerves. And because all the other organs were subordinate, the distinctness of these nervous systems meant that the *embranchements* of animal life were themselves separated by huge gaps. No intermediate forms existed, nothing transitional between, say, vertebrates and mollusks or the radially symmetrical starfish and the insects (articulates). This in itself made Cuvier's nature profoundly different from Lamarck's, with its continuous chain and fine gradations. But then his methodology was quite unlike Lamarck's and Geoffroy's speculative approaches. Cuvier believed that the comparative anatomist's job was to look closely at an animal's immediate functional needs. It was not to conjure up sweeping laws in order to give some spurious legitimation to the animal chain or "unity of plan." Because of these methodological strictures, Cuvier understood anatomical "law" differently from the Enlightenment deists. For him the notion that some overarching "law" could cause animals to conform to a plan was an absurdity. God had no such restraint imposed on Him; He acted contingently in nature to adapt and harmonize as conditions required. The only legitimate law of comparative anatomy was that describing the correlation or interrelation of organs resulting from His action.

Cuvier spearheaded the trend away from the older Enlightenment systems of nature toward a new specialization and professionalism, which concerned itself with facts, description, and low-level laws of correlation. This move had its conservative dimension, shown by the fact that Cuvier's empiricism became exaggerated when the political temperature rose in the 1820s and rival deistic theories of form and progress began to be purloined as republican ammunition.[58] For their part, popularists led by Geoffroy and Raspail criticized this new specialization and myopic concentration on facts as an attempt by the conservative elite to circumscribe and monopolize scientific power.

Geoffroy was temperamentally different from Cuvier. Where Cuvier was austere and authoritative, Geoffroy was romantic, intuitive, and like Lamarck given to flights of theorizing in areas outside his own (see fig. 2.5). Also a teacher at the Muséum, Geoffroy developed a science completely at odds with Cuvier's. In the first instance he ignored function and denied that a structural element was in any serious sense determined by

58. Appel 1987: chap. 3, also pp. 108, 193, and *passim*.

Figure 2.5. Geoffroy caricatured as a sad orangutan, with Cuvier looking on. (From Hetzel 1842: opp. page 201)

its role. Instead he looked for the resemblances between similar organs in different animals—resemblances underlying and often obscured by local adaptive modifications. For example, in the leg bones of lizards, mice, and men, Geoffroy assumed a constancy in the number and the connection of the parts of the skeleton, irrespective of usage. The femur was always connected to the tibia, the tibia to the tarsal (ankle) bones, and so on. The connections were invariant and could be related precisely to an organizational blueprint, a vertebrate "plan." Because of this constancy of connections, the femurs in humans, mice, and lizards were homologous; they were the same part of the "plan" manifesting itself in different animals. True, homologous bones were sometimes modified, fused, or rudimentary, but they could usually be identified, often by looking at their embryonic development.

Recognizing this constant connection of parts, anatomists could, following bone by bone, determine all the homologies between, say, mammals and fishes. The surprising results of just such a search were revealed by Geoffroy himself. In the *Philosophie anatomique* (1818–22), where he first argued for the "unity of composition" of all vertebrates, he announced his success in identifying the homologies of the opercular plates in the gill cover of fishes. Cuvier had considered these unique, a singular adaptation to the fishes' mode of respiration in water. But Geoffroy announced that they were none other than the homologies of the inner ear ossicles of mammals. This was a totally unexpected result, and once again it bore out Geoffroy's belief that homological relationships transcended functional modifications. Many saw it as a stunning piece of anatomical detective work, even if not everyone accepted his precise homological determinations.

Geoffroy was a deist; like Lamarck and the Enlightenment rationalists, he viewed nature as a working out of preexisting laws. In this case he argued that a set of morphological laws was responsible for the ground plans of animal types. These plans had nothing to do with function, but everything to do with certain fundamental, predetermined laws of organization, and it was the anatomist's job to investigate these higher laws. In his view, nit-picking description had to be superseded by bolder, more imaginative attempts to unveil the causal laws responsible for the similarities among animals.

Before 1820 Geoffroy considered only a common vertebrate plan, but his work inspired a welter of studies on other animal groups. After 1816 a number of zoologists, including Lamarck's successor at the Muséum, Pierre André Latreille, sought homologies between insect and crustacean limbs, mouthparts, and exoskeletons. Then in 1820 Geoffroy himself extended the vertebrate plan to include insects, effectively joining two of

Cuvier's "discrete" divisions (this is what caused the furor). Ultimately he was to adopt the idea of a single "plan" for all animals. Latreille followed suit; from 1820 to 1823, already sympathetic to Lamarck's animal series, he investigated continuities between crustaceans, mollusks, and fishes. The excitement generated by Geoffroy's researches cannot be underestimated. Appel goes as far as suggesting that almost all the leading zoologists around 1820 experimented with this new "philosophical anatomy"— even if Cuvier's threat to withhold his patronage eventually forced some (including Latreille) to abandon their efforts to connect the embranchements.[59] Others broke irrevocably with Cuvier. As early as 1816 Henri de Blainville, professor of zoology at the faculty of sciences, had adopted Lamarck's animal series, with its spontaneously generated base (although without accepting his transformism). Appel contends that the idea of a connected animal series was better supported than is generally believed, and a study of the Scottish students tends to confirm this. Grant sympathized with Blainville's attack on Cuvierian discontinuity in nature and applauded his theory of the animal series as "luminous and philosophic."[60] Since Blainville was yet another to trace homologies between vertebrates and articulates, the visitors were clearly exposed to a wealth of research on philosophical anatomy and non-Cuvierian approaches to morphology.

Geoffroy's leading disciple, though, was Etienne Serres, senior physician at the Hôpital de la Pitié, where Bennett taught. Serres in 1824–26 published a book on comparative brain development in which he provided the embryological dimension to philosophical anatomy. He showed that, while it was difficult to sort out the homologies between the brains of some adult vertebrates (those in which the lobes had ballooned out of shape), they could be discovered in the young embryos. In fact the brains of all vertebrates at an early age were very much alike. He explained this similarity in terms of "recapitulation theory": the organs of the higher classes start in a simple condition, and during development they pass through the stages where those of the lower classes stop and become permanent (the human embryo, for example, passes through a "gilled" stage). This theory was ultimately based on the premise of a single scale along which all animals developed, some stopping at a certain point (they showed an "arrest of development"), with the higher ones carrying on. It also depended entirely on the idea of a unity of composition; it was only because all the animals had homologous organs that an embryo could recapitulate those of lowlier organisms as it climbed the scale during devel-

    59. Ibid., chap. 4, also pp. 110–19, 140.
    60. Grant 1833–34, 1:96; Lessertisseur and Jouffroy 1979; Appel 1980:304–5, 1987:66, 94–95, 119.

opment. As a consequence, the doctrine of "recapitulation" quickly became one of a cluster bolstering Geoffroy's central premise of a unity of structure.

The doctrine also allowed Geoffroy and Serres to explain "monsters" or malformed infants. The abnormal organs of, say, a human baby were literally retarded in their development: they retained certain features of the lower animals, which they should have grown beyond. Serres's work prompted Geoffroy to experiment in a commercial hatchery with the incubation of chicken eggs, altering their environment to see if he could produce "monsters" at will. From this Geoffroy went on to speculate that the past transformations of life on the planet had been affected not as Lamarck supposed, but in a similar way—through the changing environmental conditions of each age acting on the embryo to alter its development.[61] The fact that embryos, "monsters," and the series of animals from monad to man were now subsumed under Geoffroy's unity of plan only served to heighten Cuvier's fears—fears that the philosophical anatomists were subordinating man to an autonomous lawful nature.

The anticlerical implications of Geoffroy's anatomy made his relationship with Cuvier difficult. He also used republican rhetoric, predicting that his new anatomy would sweep away the zoological ancien régime.[62] The Scottish radicals watched the ensuing debate with partisan interest. Knox, himself an orator of d'Holbachian power, was already prophesying difficulties for Geoffroy at the time when "Louis the Fat and Gross festered and rotted in the Thuleries; [and] the priests were gradually acquiring their lost influence."[63] At the outset of Charles X's reign (1824–30), relations between Cuvier and Geoffroy finally collapsed. Cuvier's attacks on Geoffroy's "progressive" anatomy became increasingly politically edged, hardening the Edinburgh radicals against him. In 1825—a year in which anticlerical feeling flared up among the opponents of Ultra-rule—Cuvier denounced the romantic morphologists for promoting the sovereignty of material laws. As he later said, Geoffroy's homologies "would reduce Nature to a sort of slavery."[64] So the debate had clear religious and political overtones, with Cuvier rooting his opposition in the belief that nature was incomprehensible without God. He tactically linked unity of composition, the animal series, transformism, and recapitulation as dangerous theories founded on the false premise of nature's autonomy. He made them logically inextricable, warning naturalists that to accept unity of composition was to pave the way for the transformation of life. He was

61. Appel 1987:121–31; Serres 1824–26; E. S. Russell 1916:79–83; Gould 1977:47–52.
62. Outram 1984:111; Appel 1987:98.
63. Knox 1852:20.
64. Geoffroy 1830:160; Appel 1987:7, 108, 130, 137–40, 151, 154; Cuvier 1825a:261–68.

clearly worried that these deistic doctrines would be used to fan the anti-clerical flames. Geoffroy did nothing to allay that fear. He responded by ridiculing Cuvier's science and appealing over the heads of the savants to a wider reform audience. He also began arguing for the serial transmutation of fossil forms, in papers that were abstracted in the Edinburgh journals.[65] He praised Lamarck's transformist laws and advised students to consult his *Philosophie zoologique*. Lamarck himself died in 1829, and Cuvier (obviously with one eye on Geoffroy) delivered a blistering *éloge*, condemning his speculative methods and transformist fantasies. But the Geoffroy-Cuvier fracas only served to give Lamarck's work a new topicality, and the publishers reissued *Philosophie zoologique* in 1830.

The visiting students found themselves on the firing line after 1820 as French society polarized and a succession of royalist ministries took on the professors. In 1821 Abbé Fraysinnous was made maître of Paris University and began to crack down on the dissidents. The following year the authorities actually closed the faculty of medicine after an anticlerical demonstration. There was a clear-out of its republican professors (arranged by a committee that included Cuvier, who, alongside his offices in the state, was also the university chancellor), and they were replaced in part by royalists. Getting Bourbon sympathizers into the key teaching posts and dispersing the riotous students enabled the authorities to substitute a biological curriculum more in keeping with their clerical ideology.[66] But this medical shakedown and the growth of ecclesiastical and royalist power engendered a republican backlash, and Geoffroy tailored his anatomical rhetoric, his attacks on Cuvier, and his support for Lamarck to this radical audience.

Geoffroy's advocacy of unity of structure remained a needling point until February 1830, when feelings finally erupted during a public confrontation at the Académie des Sciences. The acrimonious exchanges between Cuvier and Geoffroy here lasted for two months. They were sparked by Geoffroy's defense of a paper supporting a relationship between fishes and cephalopod mollusks (squid and cuttlefish), then believed to be the highest invertebrates. The paper went so far as to suggest that a fish bent double over its back, with head touching tail, had a similar organizational plan to the cuttlefish. More specifically, the squid's cartilages were said to be homologous to parts of the vertebrate spine and girdles. Although the

---

65. "Of the Continuity of the Animal Kingdom by Means of Generation, from the First Ages of the World to the Present Time," *ENPJ* 1829, 7:152–55; see also 1830, 8:152–54; and "Memoires du Muséum d'Histoire Naturelle," *Zoo. J.* 1825, 2:424–27. Geoffroy (1825:151) discussed Lamarck's laws of nature. On the *éloge:* Cuvier 1835; Outram 1978; and Appel 1987:168.

66. Jacyna 1987:134–37.

debate opened on this topic, with Cuvier denying that even a doubled-up vertebrate could be made mollusklike, it quickly spread to the validity of Geoffroy's whole program. The two men defended familiar positions. Geoffroy was at pains to point out the utility of the theory of unity of composition. In particular, he recalled one of his triumphs: having assumed that the mammalian hyoid (a small bone near the root of the tongue) comprised a standard number of elements, he had been led to search for certain pieces missing from the hyoid in humans. He located them in the process connecting the bone to the skull. This process, in other words, though on the skull, was not actually part of it at all; it was an element in the hyoid apparatus. His point was that from a starting assumption—unity of composition—he had uncovered the most unexpected relationships and derivation of bones. But Cuvier retorted that, if anything, the multitude of differently shaped hyoids in vertebrates actually undermined the idea of a standard composition for the bones. They were rather the product of contingent, functional needs; they should not be related to one another, but to different individual requirements.

The debate finally closed in April 1830 amid charges and countercharges: that new horizons were needed in science, not old facts; that Geoffroy's autonomous and enslaved nature squeezed out the Creator; that Cuvier's ancient philosophy subjugated structure to function; that behind the theory of homologies lay the specter of transformism, and so on. The exchanges had continued for two months—anxious months for those in the debating chamber, simultaneously watching the bigger events unfold in Parisian society. For these were also the months preceding the July Revolution, and the wider political tensions in this period help explain the growing scientific extremism on both sides. The revolution itself was caused when Charles X, having appointed a reactionary and incompetent ministry in August 1829, refused to accept its defeat at the polls in June 1830. On 25 July he dissolved the new Chamber of Deputies, restricted press freedom, and limited suffrage to the wealthiest quarter of the population in an effort to get his ministers reelected with a majority. The republican opposition, which had been organizing steadily since 1829, called for resistance, and by 29 July, after three days of street fighting, Paris was in the hands of the workers. Charles fled to England; the army and bureaucracy were purged of Ultra diehards, and the more moderate duke of Orléans was placed on the throne as King Louis-Philippe. The Académie debate took place even as the storm clouds were gathering. Appel suspects that it was the upsurge in republican activity in 1829–30 which finally forced Cuvier into confronting Geoffroy publicly (something he had been loath to do before). As the crisis deepened, he hoped once and for all to "destroy the basis of theories that he regarded as a threat to the well-being of society."[67] Geoffroy's supporters, like Lamarck's, now included the younger materialistic medical men, romantic writers, and republican activists such as Bory and Raspail, all of whom praised his attempt to give science a progressive vision. As Appel says, he was providing republicans with a self-developing alternative to an autocratic religious universe in which change and reform were blocked.

It was partly because this higher anatomy had become the scientific cutting-edge of anticlerical politics that it was so attractive to the visiting radicals, eager to undermine the theological foundations of the unreformed corporations and ancient universities at home. The science was even more appropriate in radical Britain. Cuvierian explanations of structure had become the core of the British "design arguments," particularly at the Anglican seats of learning. Here natural theologians had changed the tenor of Cuvier's science. He was primarily concerned with the correlation of organic functions and the way they fitted an animal to its niche. British theologians were aware that perfect functioning was the raison d'être of an animal's perfect adaptation. But they went further to consider each and every adaptation as the clearest sign of intelligent design, which they attributed to the actions of a caring Designer. They were proof of God's immediate attention to every aspect of nature—His personal tailoring of each animal to its niche. This theological superstructure made Cuvier's science an attractive target. With the Established Church in Britain already under radical Dissenting attack, an imported Geoffroyism could be pressed into immediate political service.

Like Knox, Grant visited Paris regularly during his summer vacations in the 1820s and got to know Geoffroy well (see fig. 2.6). Even before 1820 Grant had crossed the Alps seven times on foot, visiting the old German and Italian university towns, but it was to Paris that he continually returned. Of course, even though he backed Geoffroy, he was not blind to Cuvier's advances or ungrateful for his patronage. In fact, Cuvier had guaranteed Grant "unlimited access" to the Muséum, and Grant in 1822 completed an abridgement and translation of Cuvier's *Règne animal* (never published), while in 1830 he chronicled Cuvier's life and works in the *Foreign Review*. He attended Cuvier's soirées and introduced young English anatomists such as Richard Owen to Cuvier's coterie. Yet as radicals, Grant and Knox showed little ultimate fidelity to Cuvierian principles, nor were they overly impressed with Cuvier's sycophants such as Pierre Flourens (shortly to become professor of the anatomy of man at the

---

67. Appel 1987:144, also 9, 156–57, 171, 176, 193. The Académie debate is analyzed in Geoffroy 1830; Appel 1987:144–49; and Rieppel 1984:17–32.

Figure 2.6. Robert E. Grant (c. 1837–40), a fierce Francophile and probably Geoffroy's most consistent British supporter. (Courtesy The Library, University College London)

Muséum). Owen, in Paris with Grant in the summer of 1831, recorded in his pocket book:

Attended a lecture of Flourens—with Dr Grant—who said that if he had given one like it in 20 minutes his pupils would have put on their hats & walked off. On the structure & function of the Resp. Organs in Mammifi, Oiseaux, Reptiles—the most superficial & well known facts.[68]

On the contrary the Scottish radicals became Geoffroy's leading British disciples. Edinburgh's extramural schools were turned into Geoffroyan citadels, propagating the morphological doctrines of the Paris anti-Cuvierians. Knox bought out John Barclay's Surgeons' Square school in 1825–26 (where Grant had taught invertebrate zoology in 1824) and made it his Geoffroyan platform, immodestly informing "all reflecting men acquainted with the fact that a new philosophy had appeared."[69] Grant launched anonymous Lamarckian papers from Edinburgh University's zoology museum until 1827, when he began instilling homological principles into a succession of London University students. He imported the new philosophical anatomy lock, stock, and barrel, supporting even Geoffroy's more controversial claims. He promoted Geoffroy's single plan, his specific homologies, his successful hyoid studies, his "philosophic nomenclature" for the homologous bones (which Grant transliterated and introduced to the English), blended in Blainville's theory of the animal series, and adopted Serres's embryology. Grant visited Geoffroy almost yearly, sent him information from the London museums, and announced in class that Geoffroy had surpassed all his contemporaries as a result of "his profound, philosophical, and original views" of the homologies and development of the higher animals. Geoffroy was not unappreciative; visiting England in 1836, he praised Grant in turn as "le premier entre tous les savans."[70]

Knox saw philosophical anatomy approach Newtonian physics in status. He defended Geoffroy in Paris, considering him "a man of genius and original powers of thought, beyond the logical mind of the celebrated [Cuvier]." Knox read fiercely antifunctionalist papers and attacked Cuvier's much-vaunted "principle of correlation." Cuvier had boasted that, because an animal is functionally integrated and adapted to a specific lifestyle, from any one part—a leg bone, the skull, or an organ—he could reconstruct the whole beast. But the anti-Cuvierians were scathing about

68. R. Owen, Notebook 5 (1831) (BMNH). Cuvier supplied the details for Grant's (1830a) biography: see *BFMR* 1843, 15:24. Grant's translation of *Animal Kingdom* was presumably made redundant by Griffith's edition, which appeared in 1824 (Cowan 1969).

69. Knox 1852:212, 1850:442, 1843:638 *L* 1839–40, 1:5. Grant entered Barclay's class on returning from the Continent in 1820. His 1821 notes on "Dr. Barclay's Lectures on Comparative Anatomy" are bound in his "Essays on Medical Subjects" (UCL MS Add. 28).

70. Geoffroy to Grant, 10 September 1836, in "Biographical Sketch," *L* 1850, 2:691–92; Grant 1833–34, 1:767, also 572–74, 539, 624, 703, 735, 767, 1835–41: 57, 1839:40; "Development of the Vertebral Column," *MG* 1833–34, 14:425–26. Grant also followed Geoffroy in his determinations of the composition of the sternum, hyoids, and vertebral skull.

this and, like Blainville, Knox ridiculed Cuvier's "imaginative and fantastic" claims.[71] He pointed to the identical dentition of fruit- and flesh-eating bears, and to the difficulty of predicting the camel's stomach from its teeth, to show the impossibility of correlating organs accurately. This iconoclasm immediately thrust the radicals into a maelstrom of debate. Nor was Knox's explanation for orthodox sensibilities guaranteed to ease the tension. So extensive, he later wrote,

had the Cuvierian mania and party become in this country, that to doubt the correctness of any of the views of Cuvier amounted to a personal attack upon thousands of his satellites, who . . . placed him precisely in the same position as the Monkish writers of the middle ages placed Aristotle.[72]

Knox harped on about the structures, especially rudimentary organs, that defied Cuvierian functional analysis. And, following the Paris Geoffroyans, he explained cases of human monstrosity—the appearance of harelips and humeral ridges—as arrested developments, even if he was often unable to locate the particular animals in which these human deformities were permanent features.[73] In his view, function and Creative design as causes of structure represented one of "the most lamentable failures in human reasoning." Only Geoffroy's "unity of composition" provided a satisfactory explanation for the structural similarity of all vertebrates.

Lamarckism in Edinburgh: The Social Geography of the Debate

As old species perish, do new ones rise up? Is there some secret law of animal reproduction by which there is a succession of species in the course of ages, as there is of individuals in the course of years!

—John Playfair[74]

Cuvier's fears were not unfounded. A serial stacking of organisms, all reducible to a common composition, did leave nature susceptible to a Lamarckian explanation. Worse, the concept of a self-determining series had an obvious appeal to democrats. Hence in Britain we find Lamarckism turning up among working-class atheists and socialists, among medical demagogues and radical phrenologists,[75] with rival theories of nature's

71. Knox 1831:481–84, 1850:440–41, 1852:17, 19, 96–97; Blainville 1839–40:217, 222n.
72. Knox in his edition of Blainville 1839–40:222.
73. Knox 1843:500–501, 529, 530, 532.
74. Playfair 1812:382.
75. Phrenology was an unorthodox science of the brain fashionable among dissident factions in the 1820s and 1830s. It particularly attracted the Edinburgh traders seeking a scientific power base from which to challenge the Kirk-dominated university and civic authorities.

self-development proferred by Chartists and free-trade radicals. Among
the artisan atheists (who were already distributing pirated copies of
d'Holbach's *System of Nature*) Lamarckism was being used to legitimate a
priest-free democratic republic. Socialists, such as the "red republican"
William Thompson at London's Co-operative Society, began exploiting
Lamarck's theory of the inheritance of acquired characters in 1826 to justify the education of women. (With improving character traits inherited
from both sexes, women had to be educated for society to progress.) In
other words, Lamarckism was taken up by groups that flatly rejected aristocratic authority. All of these agitators accepted the old *idéologue* notion
of life as an inherent property of matter; all therefore believed that animals had developed and changed through the operation of natural laws.
They were moreover fierce materialists, and it is telling that in Edinburgh
an upsurge in materialistic thought occurred during the period (c. 1826–
33) when the medical radicals—men such as Grant, the future Chartist
Patrick Matthew, and the phrenologist Hewett Watson—first broached
the idea of organic self-emergence. It should come as no surprise that
these radicals attacking "Priestcraft" and espousing notions of self-
determination and popular power welcomed a left-wing Lamarckian science or its equivalent. Knox provided the perfect political metaphor in
his zoology lectures; it was, he said, "a self-created, self-creating world—
ever alive, never decaying, never old."[76]

As a conchologist, it is fair to say, Lamarck had long been appreciated
by those with cabinets to organize. In the early 1820s even that haughty
Tory John George Children, the assistant keeper of natural history in the
British Museum, had kept abreast of the latest numbers of Lamarck's *Histoire naturelle* in order to arrange the museum's shells. An amateur's cabinet could equally be turned from a dilettante's fancy into something
scientific using Lamarck's classification. Hence the *Histoire naturelle* insinuated itself into some strange places: the police magistrate William
Broderip used it to arrange the shell collection in his Lincoln's Inn chambers, while the Linnean Society taxonomist W. S. MacLeay actually rated

---

Phrenologists believed that the mental faculties—corresponding to love, hate, reverence,
and so on—were located in discrete "organs" of the brain. (And could be "read off" from
bumps on the head, which is what made phrenology an accessible people's science.) Since
the faculties were inherited, behavior and morality were determined; this, by seeming to
deny accountability, worried the ruling orders. The hereditarian aspect weakened after the
mid-1820s. Social reformers began to argue that only mental tendencies were inherited and
that, since these were susceptible to educational molding, it was possible for individual characters to be reclaimed.

76. Knox 1855:218; W. Thompson 1826:250, 253, 254; Desmond 1987:89–95; Royle
1974:123–25; Moore 1982:173–78.

Lamarck France's finest zoologist for his invertebrate work. Systematics apart, though, Lamarck was loathed by these conservatives for what they considered his insane transformism. Children, as a regenerate Christian, was not alone in seeing Lamarck's "blasphemous" claims about nature's productivity and God's impotence threaten society at its very core.[77]

Explaining why Edinburgh produced the radical theorists is not so easy. (Knox, Grant, Matthew, and Watson were all products of the university and private medical schools.) Obviously, the Edinburgh students were exposed to Geoffroy's and Lamarck's anatomies while finishing their education in Paris. But more was involved. The professional options of the returning students seem to have been greater in Edinburgh than, say, London; they certainly found outlets for their radical science that might not have existed in the south. Institutional science was also less constrained by rigid methodology in Edinburgh. Unlike the geological community in London, which tacitly prohibited divisive cosmologies and discussions of origins, Edinburgh geology remained philosophical and fractured, and not merely into the Huttonian and Wernerian camps (clashing over the plutonic or aqueous origins of the rock strata). There was a further evangelical cross-fold, which lead to sharp rivalries over the question of geological catastrophes.

Moreover, theories of the earth—often with a strong philosophical content—were not uncommon in Edinburgh. When Grant's mentor, the professor of natural history and head of the museum Robert Jameson, annotated Cuvier's *Discours préliminaire* (the introduction to his multivolume work on fossils), he retitled it *Essay on the Theory of the Earth* (1813). In casting these broader philosophical nets, Scottish geologists obviously intended to rope species into any final explanation, and Jameson concluded his course on natural history in 1826 by discussing the "Origin of Species."[78] By this he probably meant Cuvier's discovery of the successive appearance of new faunas in the rock strata. Cuvier had argued that the abrupt changes from one geological stratum to another had been caused by marine inundations (which also wiped out the local terrestrial life), or by the seas retreating (when new species migrated in from adjacent continents). But Jameson equated the last inundation with the Flood, and after his translation appeared many conservative English geologists interpreted the breaks in the strata as signs of much more violent global

77. Children to Swainson, 11 July 1831 (LS WS); MacCulloch 1823:415; Desmond 1985a:167, 163; Corsi 1988:232ff. The conchological parts of Lamarck's *Histoire naturelle* were well translated, e.g., by Children 1823, Dubois 1824, and Crouch 1827.

78. Ashworth 1935:100; Porter 1980:145–56; Rudwick 1972:132. Cuvier's *Essay* ran to five editions in fourteen years (1813–27): Grant owned the fourth (1822). Others (Playfair 1813–14) were sniping at Cuvier before Knox and Grant came along.

"catastrophes," while arguing that each new stratum marked the stage of a fresh Creative act. In this way geology was seen to testify to God's personal intervention in nature and to His progressive "updating" of life to meet new conditions. Few of the medical rationalists accepted this "Creative" inference; indeed some Scottish evangelical naturalists disputed it. Anyway, this failure to outlaw high theory, the feeling that species must be discussed, and the divided state of geological society meant that medical Lamarckians possibly found more freedom to maneuver in Edinburgh's scientific society.

Grant first broached Lamarckism publicly in 1826. He had graduated from Edinburgh University in 1814 with Knox, and like Knox he was a fierce radical whose deconsecrated Calvinism left him a zoological determinist. He also had the same satirical wit, which often turned on the Scriptures. He was a man of contrasts: gentle and humorous, yet with a hefty political punch; a quiet recluse who could pillory his teachers for their antimechanistic views.[79] Grant was thirty-three in 1826, and after twelve years of seasonal study in Paris, Rome, and the German states he was back in Edinburgh advocating a more positive paleontological role for Lamarckism.

We know that transformism was abominated by his patrons, the polymathic editor of the *Edinburgh Journal of Science* David Brewster and the minister and naturalist John Fleming. This has led to the suggestion that Grant was somehow an "anomaly." But such conservative exponents of high scientific culture as Brewster and Fleming give little indication of the new subcurrents after the turbulent Regency period. These currents produced some of the most materialistic, anticlerical scientists of the 1830s—not only Knox and Grant, but a host of radical anatomists, phrenologists, and medical union organizers. Grant's transformism is better seen as part of the wider radical reappraisal of science at a time of a growing division in Scottish society. By the 1820s Edinburgh had a strong radical subculture among the artisans and shopkeepers, while its commercial classes were harrassing the city's old elite and looking for ways of "translating wealth into political and cultural influence."[80] We know that groups

79. At the Royal Medical Society as early as 1814 Grant had lampooned his anatomy teacher John Gordon: Grant, "An Essay on the Comparative Anatomy of the Brain," ff. 25–26 in "Essays on Medical Subjects" (UCL MS Add. 28). He was a stand-in lecturer for Barclay in 1824, even though Barclay (1827:xii, 140, 168; 1822:30, 36, 126–52, 522, 526) abhorred skepticism, reductionism, and the "idle hypotheses" prompted by paleontology.

80. Shapin 1979a:56. See also Shapin 1975; Cooter 1984; and on the academic politicking Morrell 1971b, 1975, 1972. Working-class radicalism in Regency Scotland is discussed in Thomis and Holt 1977: chap. 3. On Grant as an "anomaly": Eiseley 1961:145; cf. Desmond 1984b:195.

trading under the phrenologist George Combe's banner, for example, were intent on taking science into their own hands, and that the sciences they favored were shaped by a variety of Dissenting, class-based, and anti-Kirk attitudes. So it is unprofitable to dissociate Grant's biology from its political base and paint him an "anomaly," when it is through this base that he can be related to the activists in adjacent scientific areas.

It is not certain that a Calvinist theology was inimical to all parts of Grant's science. Some of his anti-Cuvierian leanings might actually have been encouraged by it. This would also begin to explain how he could have been patronized by evangelicals such as Brewster and Fleming— indeed, why it was so often the lapsed Presbyterians who became the transformists. Brewster hated transmutation, though he agreed that Lamarck had been an able naturalist. Brewster even admitted having "seen and admired this handsome descendant of the Monads" at the time when he was elaborating his "ingenious" theory. Oddly, while abhoring Lamarck's transformism for degrading the "godlike race,"[81] Brewster openly patronized Grant. He supplied him with "zoophytes" (the sponges and polyp-bearing sea fans, Grant's speciality), took seven of his papers between 1827 and 1829 for the *Edinburgh Journal of Science* (most on the free-swimming ova of marine invertebrates), commissioned his article "Zoophytology" for the *Edinburgh Encyclopaedia*, cited him as an authority, and recommended him for the zoology chair at London University. And when Grant was nominated in 1827, Brewster assured the council that it would have no regrets.[82]

All of this suggests a strongly supportive role. Yet Brewster's theology was so orthodox that the latitudinarian Baden Powell could accuse him of biblical literalism. So where was Grant and Brewster's cultural common ground? When John Brooke studied the differences between liberal Anglican and Presbyterian theology in order to understand the Cambridge-Edinburgh disputes over life on other worlds, he was able to highlight revealing differences over the question of organic adaptation. He contrasted Brewster's evangelical image of "Divine Resourcefulness" with the Anglicans' Paleyite God of Precision. William Paley's natural theology concerned itself with the perfect and unvarying fit between organisms and niche; indeed, Cambridge divine William Whewell argued that organisms could not exist under changed conditions, the fit was so perfect. But, said Brewster, what value Whewell's theology if naturalists one day proved

81. Brewster 1845:472, 500–501.
82. Brewster to Brougham, 24 July 1827 (UCL CC 445); Grant 1830c; Brewster 1833–34: 445–46, 1834:147.

that plants and animals could adapt themselves? He offered a counter-view, presenting an intriguing Presbyterian perspective on flexible adaptation. He argued that this limitation to a fixed niche cannot be regarded as evidence of design. Quite the opposite: "the *very want* of this limitation, or the existence of an elastic energy in organic bodies by which they could accommodate themselves to a residence on every planet in the system, might be held to be a proof of divine wisdom and power."[83] Brewster was not underpinning some sort of transformist "accommodation"; his "energy" was presumably not elastic enough to transmute a species (and anyway he was talking of another planet). But the episode gives a tantalizing glimpse into the potential of Presbyterian theology to do more heretical work. So even if Grant's nature was too resourceful for comfort, it seems that he could derive some support from a theology of organic accommodation.

Aspects of evangelical geology, too, were integrated into Grant's Lamarckian science. To illustrate these we have to look at another of his unlikely patrons—the dour John Fleming. Fleming was a Presbyterian minister (his parish lay across the Firth of Forth, in Fifeshire). He was also a gifted naturalist who was to take the chair of natural history at King's College, Aberdeen, in 1834. He shared Grant's obsession with the Firth's marine life and presented his young friend with over thirty species of sponges for study. The two evidently got on, for Fleming even named a new sponge genus *Grantia*.[84] Because Fleming and Grant studied the local sponges and polyps, Lamarck's *Histoire naturelle* (the main guide in the field) figured extensively in their writings, and on questions of Lamarckian taxonomy Fleming and Grant were in perfect accord.

But this sympathy extended further. For an evangelical minister, revelation overshadowed natural theology; it made any attempt to prove God's actions from nature not only redundant but actually pernicious. Arguing this point, Fleming was led into some revealing controversies. Thus in criticizing the Oxford geologist William Buckland's evidence in 1826 for a catastrophic Flood, Fleming stood securely on scriptural grounds. The Mosaic account was inconsistent with the notion of a series of geological cataclysms. The Bible, he said,

83. Brewster 1833–34:435–36. Here Brewster also cites Grant as an authority on infusoria, while attacking Whewell's belief that the microscope had revealed "the *finity* of animal life" (pp. 445–46). For a discussion of Brewster's views see Brooke 1977a:231; Ospovat 1978:33.

84. Fleming 1828a:524; 1822, 1:14; 1829:321; Grant 1827a, 1826d. Fleming (1822, 1:121, 310–12; Corsi 1978:222–24) derived his taxonomic criteria, vertebrate/invertebrate distinction, and dichotomous system of classification from Lamarck.

permits me to believe, that the waters rose upon the earth by degrees, and returned by degrees; that means were employed by the Author of the calamity to preserve pairs of the land animals [allowing the continuity that Rev. Buckland denied]; that the Flood exhibited no violent impetuosity. . . . With this conviction in my mind, I am not prepared to witness in nature any remaining marks of catastrophe, and I feel my respect for the authority of revelation heightened, when I see on the present surface no memorials of the event.[85]

Geologically speaking, the Flood could be dismissed as an invisible overlay—an event that made no impact on the fossil record. From an evangelical perspective, Noah's mission was a matter of faith; underpinning it with a shaky science of catastrophes could only loosen dependence on the Bible. This struck at the very heart of Buckland's teachings at Oxford. After the translation of Cuvier's Theory into English, Rev. Buckland— from 1819 the reader of geology at Oxford—led the way in interpreting the Flood as a scientifically provable event. Unlike Cuvier, he made it the most recent of a series of global catastrophes, which left evidence of its action in the diluvial gravels. In 1823 he published Reliquiae Diluvianae (Relics of the deluge), arguing from English cave deposits containing the bones of extinct hyenas and from similar bones found high up in the Himalayas that the entire earth had been drowned. (This, like earlier catastrophes, had been followed by an act of "Creative Interference" to repopulate the globe.) Nicholas Rupke suggests that, by making the Flood a geological event and renouncing the Calvinists' "quiet deluge," Buckland was able to justify including the suspect science of geology in Oxford's clerical curriculum.[86] In other words, it was partly a professional move; he was securing the future of geology by making it relevant to the Old Testament scholars in a classics-orientated university. But this kind of clerical appropriation infuriated the secular radicals. They were to deny all catastrophic acts of "Interference" and remove this need for an ordained, classically educated elite to interpret them.

Fleming's support for geological gradualism and his attention to the ecological conditions of life were well known to the Edinburgh radicals. The minister read his anticatastrophist paper to the university's Wernerian Society on 25 March 1826, when Grant was on the council (in fact only a month after Grant's own paper on the transmutation of sponges).[87]

85. Fleming 1826:214, 208, 215, 1859:xxxviii, xl; Page 1969:264; Gillispie 1959:123–24.
86. Rupke 1983:21, 24–25, 58; Rudwick 1972:135–38.
87. Wernerian Society Minutes, 1: f. 249 (EUL Dc.2.55); Rehbock 1985:133. Fleming (1828b) argued against a formerly hot earth. But Grant (1833–34, 1:480), drawing on the physiology of the living nautilus, concluded from ammonite distribution that the northern seas had once been tropical.

Strong ideological differences admittedly existed between the two men; they also disagreed on specifics, for example, on the existence of a formerly warm Arctic (which Fleming disputed, but which was crucial to Grant's transformist model). The point, however, is that Grant shared this distaste for primordial turbulence. As a rationalist, in fact, he had even less sympathy for Buckland's biblical gloss on the French stratigraphic work. Moreover he knew that in Cuvier's view, the sudden changes in strata signified nothing in the way of divinely inspired global disasters. Cuvier visualized something far more natural, a regular series of local, albeit swift, marine incursions. Grant, knowing Cuvier personally, was quite *au fait* with this original French interpretation. Moreover, like Geoffroy, he was adamant that these sudden regional changes had not actually affected the continuity of life. At any rate, in discussing the paleoclimatic causes of transmutation a few months after Fleming's talk at the Wernerian, Grant too was scathing about this English catastrophism. A theory of ecologically controlled gradual evolution, he said, must prevail, "for it is in harmony with the natural laws of order and permanency which rule the universe," whereas "science, facts and human reason" were opposed to the idea of geological "cataclysms."[88]

Belief that the earth's temperature had been declining peaked in the mid-1820s, and it provided Grant with the basis of an explanation for life's progressive change. He pictured a slow migration of life away from the poles, with the loss of uniformly warm global conditions and the onset of climatic zoning (caused by the sun assuming an increasing importance over the residual planetary heat). The appearance of seasons and changes in temperature, tides, volcanic activity, and weather systems provided "the regular, general, and continued natural causes of the modifications which life has undergone." These progressive climatic changes caused life to develop successively; in the fossil record it appears as a finely graded chain, showing no "line of demarcation between the different terms of that series," as might have happened if "life has been once or oftener totally renewed on the earth."[89] Likewise Geoffroy came to talk of the "continuous filiation" of life and of its "unidirectional changes" being a result of evolving climatic conditions (now emphasizing an environmental drive rather than his original theory of embryonic transformation). For him too life was left as "so many links in a progressive series."[90] Grant's conception

88. Grant 1827c:300–301.

89. Ibid. His later theory, of decreasing planetary temperatures driving life toward higher "hot-blooded" forms, was built onto this 1827 platform: Desmond 1984c:405–6; Grant to C. Babbage, 30 April 1856 (BL Add. MS 37,196, f. 489). On central heat: P. Lawrence 1977, 1978.

90. Appel 1987:184.

of nature as connected and continuous, then, was very close to Geoffroy's. Grant taught in the 1830s that the "constant and progressive" forces from Cambrian times to the present day had produced a "continuity of the [fossil] series through all geological epochs" and that "the gradual transitions which connect the species of one formation with those of the next . . . indicate that they form the parts of one creation, and not the heterogeneous remnants of successive kingdoms begun and destroyed."[91] This "one creation" image was common coinage among the Edinburgh Geoffroyans, and transcended Grant and Knox's growing differences over progressive transmutation.

Grant's transformism now went far beyond anything Brewster or Fleming could condone. Fleming, knowing Grant, realized that Lamarck had "succeeded in making . . . converts," and in 1829 he openly attacked Lamarck's theory. He said that, as a mechanism for "the formation of Man," it was so impossibly "complex and circuitous" as to be "obviously a dream." He also began to downplay the notion of a uniform fossil ascent, possibly, like the geologist Charles Lyell, after recognizing the brutalizing threat posed by a progressive transmutation.[92] Fleming pointed out that the Scottish Old Red Sandstones, rather than housing the simplest plants and animals, actually contained fish. He also insisted, like so many other antitransformists about the time of the Geoffroy-Cuvier debate, that the gap between the mollusks and the vertebrates was unbridgeable.

Fleming himself was deeply worried by the new medical materialism, and the free-for-all discussions in the student societies show the sort of thing that must have scared him. The worst offender was the Plinian Society. This had been founded by Jameson in 1823, although it was largely autonomous by 1826. Grant was active here and its secretary until 1826. The minute books record that reductionist philosophies of mind were hotly debated by its student members in 1826–27.[93] A number of Plinian officials held flagrantly reductionist views, although William A. F. Browne, one of the presidents, was the most outspoken. Browne was a disciple of George Combe and actually toasted by Combe at the Phrenological Society for his success in proselytizing the students.[94] Browne's activities after graduating in 1826 gave an indication of his allegiances. He lectured on physiology and zoology to the clerks and shopkeepers at the

91. Grant 1839:60; 1833–34, 1:351; Desmond 1984c:410, n. 54. The "one creation" image was invoked by Knox 1850:443, 1852:109; Blainville 1839–40:189; Appel 1987:184.

92. Fleming 1829:320–21; cf. his earlier view in 1822, 2:96–97, 102–4; Di Gregorio 1982:227, 233ff. Bartholomew 1973 on Lyell's similar shift.

93. Plinian Society Minutes MS, 1: ff. 37, 57 (EUL). On the society's independence: *Evidence, Oral and Documentary* 1837, 1:145–46.

94. *Phrenol. J*. 1829, 5:141–42.

new Edinburgh Association, formed in 1832 by the town's tradesmen. In 1834 he was appointed medical superintendent at the Montrose Lunatic Asylum, where his enlightened views on the treatment of the insane attracted support from the radical community.[95] He advocated a material basis for madness, considering it a fault of the brain, not of the mind. And he was a committed secularist, with a fine line in sustained religious satire, studying inmates from his asylum to prove that the men and women canonized by the Church for their hyperactive organ of veneration would today be diagnosed as insane.

The intersection of the mental reductionism of younger phrenologists with the developmental zoology of senior students gave the Plinian meetings a lively aspect in 1826–27. During discussions, Charles Bell's pious *Anatomy and Physiology of Expression* was pulled apart, the materiality of the mind was asserted, visions were assigned physiological causes, and brutes accorded the range of human mental states.[96] So strong were some of these propositions that attempts were made at censorship, the record of Browne's propositions on mental materialism being struck out. Possibly this was on Jameson's orders, for he was antagonistic to the radical sciences and had already used his influence to bar phrenologists from his museum.[97] But the formal striking from the minutes of the offending propositions was more ritualistic than effective. The climate within the society can be gauged from its melancholic effect on another Plinian president, the evangelical John Coldstream. Before matriculating at Edinburgh, Coldstream had been a writer for the Leith Juvenile Bible Society, and Darwin found him still prim and religious at Edinburgh in 1826. At Plinian meetings Coldstream was closely associated with the materialists; he had proposed Browne for membership, and allowed Grant to study his zoophytes and publish a description of his rare *Octopus ventricosus*. But the materialism and irreligion of the radical activists deepened the spiritual crisis he was already suffering. After months of prayer he had a break-

95. On the Edinburgh Association: *Phrenol. J.* 1834, 8:571–72; Shapin 1983:154–55. On Montrose Asylum: *Phrenol. J.* 1834, 8:662–63; Combe 1850:228–29; also W. A. F. Browne 1837, 1836; and Cooter 1976:11–12.

96. Plinian Society Minutes, 1: ff. 11–12, 34, 51, 57 (EUL); Gruber and Barrett 1974:39.

97. Jameson denied students interested in craniology access to the museum's skull collection. William F. Ainsworth, another Plinian president, complained to the commissioners investigating the university that in 1827 he too had been refused permission to see some skulls; in his own *Edinburgh Journal of Natural and Geographical Science* he attacked Jameson's mismanagement. Interestingly, by 1830 another of Ainsworth's grievances concerned the inaccessibility of the invertebrates presented by Grant. Jameson's autocratic control is discussed in: "On the Present State of Science in Great Britain. No. 1. Edinburgh College Museum," *Edin. J. Nat. Geog. Sci.* 1830, 1:275; "Phrenology and Professor Jameson," *Phrenol. J.* 1824, 1:55–58; Chitnis 1970:90–93; *Evidence, Oral and Documentary* 1837, 1:632.

down in Paris in 1827, which a visiting Glaswegian doctor put down to his "doubts arising from certain Materialist views, which are, alas! too common among medical students."[98]

So the radical materialism was a potent ideology. It encouraged naturalistic attitudes and mental reductionism, allowing men and brutes to be treated as qualitatively similar, and it left its adherents a short step from accepting a shared ancestry for man and the animals. Grant in 1826 took this step; in fact he went further, speculating on a transmutatory relationship for all organisms—men, animals, and plants. He realized Lamarckism's relevance to Jameson's claim that the age of rocks and the scale of life corresponded—a relevance that was already worrying English geologists.[99] He argued in fact that the successive strata housed a progressive, naturally evolved series of fossil animals. As he explained, these fossil "forms have been evolved from a primitive model" as a result of "external circumstances." And the series was pushed upward as a result of the "pressure" from below due to the spontaneous generation of life and "aggregation process of animal elements."[100] On a crucial point, as Philip Sloan has demonstrated, Grant actually went further than Lamarck. He accepted, like his friend the coral expert August Schweigger of Königsberg, and the embryologist (and transformist) Friedrich Tiedemann of Heidelberg, a convergence and common origin for the plants and animals. Evidence for this common origin of the algae (plants) and zoophytes (animals) came from his own researches on the sponges and marine polyps. He studied their free-swimming ova, which he considered analogous to the infusoria on the one hand and the "gelatinous globules" composing all plants and animals on the other.[101] Since, as Schweigger observed, the algae could also resolve themselves into free-swimming infusoria, the basic organized units of life, the "monads," must ultimately be interchangeable between plants and animals.

Grant believed that these monads were spontaneously generated at the

98. Balfour 1865:7, 39; J. P. Coldstream 1877:10–11; Barlow 1958:48; Grant 1827d. Coldstream advised Darwin in 1831 to see Grant in order to pick up hints before setting off on the *Beagle* voyage (F. Burkhardt and Smith 1985, 1:152).

99. Jameson had studied under Abraham Gottlob Werner in Freiberg. Werner taught that the successive rock formations had been precipitated gradually from a changing oceanic solution. This meant that the strata were laid down sequentially, and since each had its characteristic fossils, these too could be put into historical sequence. This temporal correlation of rocks and fossils left the fossil series amenable to a Lamarckian environmental explanation. On English worries over Werner, Lamarck, and the fossil record: Greenough 1819:281–82; Porter 1980:169; and on the Wernerian tradition: Laudan 1987, esp. 100–101.

100. Grant 1826b: 296, 298. Hodge 1972:140–46 discusses the upward "pressure" on life due to the spontaneous emergence of monads at the base.

101. Sloan 1985:78, 83–84; Grant 1828:110, 1829:18, 20–21.

zero point, and this had profound implications. Deriving life from chemicals by no other means than physical law was reductionism writ large. It raised the specter of design without a designer, causing clergymen to thrust it with transmutation into the same "sinkhole of human folly."[102] Hence the German rationalists, who substituted an inorganic generation of life at each epoch for "a *Deus ex machina*," were indicted for atheism when their works were extracted in the English press.[103] To understand how a young teacher could sustain this position we must look again at the medical community, for it was here, in the debate on the origin of gut parasites, that the subject was most thoroughly aired.

Grant's writings show that his views on spontaneous generation were partly shaped by these medical discussions. He believed that the simplest infusorians, "originating from atomic nuclei, almost verging on the mineral kingdom, and sole inhabitants of the heated waters of our primeval globe," paved the way for the "higher tribes." This they did in two ways. First, they provided a continuous, self-originating source for the diverging animal and plant series. And, second, because these infusorians also existed as "independent germs" in "the living fluids of organized beings," they produced, through "their development, the various tissues of the body," its epithelial linings, and all its internal secretions.[104] It is in this second aspect that we see his universal monadism connecting directly with the question of gut parasites, which were believed to emerge from these living infusions in the intestines. This subject was hotly debated in medical circles, both in Germany and Britain; even the cautious physiologist Johannes Müller admitted that spontaneous generation received its strongest support from the presence of these intestinal parasites (though he did not accept it). Many physicians were equally skeptical, but evidence for the de novo origin of intestinal worms was frequently presented. And often it appeared in the reform press, largely because it was here that a philosophy of "living atoms" was implicitly accepted as part of the secular strategy. So again, Grant could derive support from the wider radical community, with its Enlightenment faith that, as one atheist put it, "a power or energy in *nature*" compels chemicals to combine into organic bodies.[105]

102. J. W. Clark and Hughes 1890, 2:83; Sedgwick 1834b:305. For producing design without a designer, spontaneous generation was dismissed by Drummond 1841 and Keith 1831.

103. Weissenborn 1838:370.

104. Grant 1844:355, 358.

105. Desmond 1987:98. On the medical evidence for the spontaneous emergence of gut parasites, see, for example, "Rhind on Worms," *L* 1828–29, 2:693; "Dr. Scouler on Worms," *MCR* 1830, 13:221–23; "Mr. Bushnan on Animals Found in the Blood," *Medical Quarterly*

It was not strange that Grant should have chosen this moment to speak out, given Geoffroy's praise for Lamarck in France. But Grant's brand of progressive, environmentally driven transmutation was contentious, even within the Scottish Geoffroyan community.

Some failed to see how it could actually be squared with the tenets of philosophical anatomy. Take the merchant's son John Fletcher, a Londoner who taught comparative anatomy and physiology in Edinburgh's Argyle Square from 1828 to 1836 (see fig. 2.7). Fletcher's own Continental-style lectures were rated by one radical journal as Europe's best in terms of "research, erudition, science, and comprehensiveness."[106] They were published as *Rudiments of Physiology* (1835–37), which was the first book to be completed on higher anatomy in Britain. Fletcher was one of the most astute teachers of the science in the city, but he was always more temperate than Knox or Grant. In outlook he was closer to Brougham's group, and like them was far more cautious on the question of transmutation. He knew that progressive "generation" was believed to be the corollary of unity of composition. Yet for him the notion of beings "built each upon the one below it" had nothing but "vague and rambling presumptions in its favour." If, he reasoned, "men and toads are descended from the same original parents," then the "one common nucleus" from which "all the various tribes" developed must have had a real primordial existence. The problem with this is that we can only talk of the "prototype" of individual organs, not whole organisms. He freely admitted that, for instance, the livers in all animals "are fundamentally the same"; they are all homologous, only becoming more complex as we ascend the scale. But this unity of organs does not imply the descent of organisms because in each animal the relationship of the organs to one another is unique. Every organ has its own "lineage," as it were, but these "lineages" progress in the animal series at different rates. In the same way, the human embryo (which in its development was supposed to recapitulate the animal series) is never absolutely identical to any lower animal because its organs too develop differentially. Thus when its liver resembles that of one animal, its heart might have reached the stage of another. In other words, "the fetus collectively is never formed upon any model but its own."[107] So each

---

Review 1834, 1:364–66. See also Müller 1837, 1:16; and for an overview Farley 1972, esp. 110–11. Secord 1988 investigates the gentleman radical Andrew Crosse's success in 1836 in creating test-tube life in the laboratory of his Jacobean mansion.

106. "Dr. Fletcher's Rudiments of Physiology," *LMSJ* 1835–36, 8:756–58. Fletcher had his protégés, for example, J. S. Bushnan (1837:x), though rarely of his caliber, as Carpenter (1841a:103) notes.

107. J. Fletcher 1835–37, 1:12–16, 78. Many adopted a similar position: "This [recapitulation] must not be taken . . . quite literally; for the resemblance or analogy is only seen

Figure 2.7. John Fletcher, one of the most incisive philosophical anatomists in Edinburgh. By W. O. Geller, 1838, after W. S. Watson. (Courtesy Wellcome Institute Library, London)

organ was a part of a progressive series traceable throughout the animal kingdom, and as such proof of a higher unity. But organisms, being assemblages of organs each with its independent "history," were immutable. Even if, as Fletcher argued elsewhere in his lectures, the various organs—heart, lungs, liver—were themselves ultimately derivable from the same "model" or prototype, no animal could transform itself into one higher in the series. Its organs would have to develop synchronously for Lamarck's theory to be true. And this they did not do.

Fletcher was on many counts a moderate: a man, unlike Knox, who was courteous toward natural theologians (however many others were lashed by his tongue.) But even on the anti-Christian wing of the Geoffroyan community sharp differences were to emerge over Lamarck's belief in the transformation of adult characters. Knox, despite his savage radicalism and satires on design, came to promote developmental doctrines quite distinct from Grant's. We actually know very little about Knox's early views, except that he was a partisan Geoffroyan and accepted the "consanguinité" of animals, believing that all creatures were "descended from primitive forms of life."[108] At that time he was probably more of a lineal progressionist than in his later years. At least in 1843 he was still talking (guardedly) of monstrosities as "arrested developments," which betrays his one-time belief in a recapitulationist embryology, even if he later claimed that he had been "forced to teach it for want of a higher generalization."[109] But, as Evelleen Richards shows, Knox's thought eventually changed. He came to repudiate recapitulation, arrested developments, and a progressive ancestry for existing species. His was still a "self-created" zoological world, only now nature's "deformating powers" operated only on the "generic embryo," which he interpreted idiosyncratically as a reservoir of new species. These species, when they emerged, were immutable; nor did they emerge in a necessarily progressive way.[110] So Knox, despite his fiercely naturalistic views, also came to reject Lamarck's "wild conjectures" on the progressive transformation of adult species, one from another, although on very different grounds from Fletcher.

In another way Knox was to move out of step with his radical contemporaries—his denial of environmental explanations. Grant was commit-

---

between individual organs, not entire beings": "Saint Hilaire's Treatise on Teratology," *BFMR* 1839, 8:5.

108. Knox 1852:109; E. Richards 1988.

109. Knox 1855:25.

110. E. Richards 1988; Rehbock 1983:50–51; Knox 1843:555, 1855:625–27. Note that Grant (1826b:300) too talked of species evolving from an "original generic form," although he embedded this in an explicitly Lamarckian context. Knox comments cryptically on Lamarck in his notes to Blainville 1839:189n.

ted to the *idéologue's* view of nature. A similar, inexorably progressing, environmentally controlled nature was defended by the working-class extremists. They, in fact, paraded their Lamarckism much more prominently, glorying in its subversive implications. Uncompromising atheists such as the Bristol printworker William Chilton and ex-socialists Charles Southwell and George Jacob Holyoake were jailed in the early 1840s for the blasphemies in their gutter print *The Oracle of Reason* (1841–43), which used Lamarck's self-evolving nature to legitimate the establishment of a secular republic. The Jacobin notions of social conditioning and human perfectibility so prevalent in the Enlightenment were still immensely influential among these radical subcultures, reaching an apotheosis in utopian socialist thought. The socialists taught an undiluted environmentalism in their schools, and their leaders espoused Lamarck's theories.[111] The middle- and working-class radicals exploited the same *idéologue* sources. Wakley's bookshelves, for instance, contained copies of works known to be favored by the radical artisans, including pirated French revolutionary tracts. Wakley's coterie also held strict environmentalist views: they shaped Farr's account of the cause of slum diseases as much as his teacher Grant's explanation of new species.

Richards, however, portrays Knox as growing apart from this tradition. Though his Jacobin father had supported the French Revolution, and he himself promoted the new trades unions and used Paineite rhetoric in his attacks on "Priestcraft"—indeed though he, like the *Oracle* atheists, talked of man as "absolutely nothing" in an impersonal and deterministic universe—still by the 1840s we find him crying off notions of environmental control and perfectibility. He attacked the Benthamites' efforts to manipulate the social environment through sanitation and welfare programs.[112] By then he had migrated totally into the free-trade camp. We have seen how much his science began to diverge, and this loss of his faith in environmentalism left him politically distanced from the Lamarckians after the 1830s. He kept up his swingeing attacks on the old corporation elite, but on issues like state provision and welfare he was to oppose Grant and the Wakleyans.

111. Desmond 1987:85–104; Royle 1974:123–25; Budd 1977:26–34. Cullen 1975:36, 136 on Farr's environmentalism. Volney's *The Ruins: Or a Survey of the Revolutions of Empires* was listed in the inventory taken after Wakley's house was firebombed (Sprigge 1899:65). Pirate editions of this work also circulated among the working classes. The radical artisans' exploitation of d'Holbach, Volney, and Lamarck gave their movement, as Gwyn Williams (1974:8) says, "a 'Continental' bite," and shows it to have been part of the European-wide revolutionary trend.

112. E. Richards 1988. Desmond 1987:87–88 for Chilton's similar declaration that "Life is nothing."

This, however, was in the future, and Fletcher's, Knox's, and Grant's diverging approaches to the development of life should not blind us to their shared faith in higher anatomy in the 1820s. All were passionate advocates of the French sciences fashioned by Geoffroy, Serres, and Blainville and had only withering words for the critics of the new morphology. None doubted that body and brain were bound by strict laws—laws that, some believed, were simply a reflection of the inherent properties of matter. This returns us to the question of exactly who these anatomical sciences appealed to in Edinburgh.

With industrialization and commerce picking up after Waterloo, new audiences were emerging for these secular sciences. Steven Shapin has shown that the shopkeepers and small businessmen started retailing their own sciences at the Edinburgh Association in the 1820s without awaiting the benevolent dispensation of the Whig diffusers of official knowledge. These upstart traders were seeking a political voice and a way into the Tory-held corporations; they were threatening "the old security of the Edinburgh 'Select' based on inherited position."[113] The sort of sciences the traders favored can be understood if we keep sight of this struggle for municipal power. The way one type of transformist biology worked for them is illustrated by the demands of the laissez-faire radical Patrick Matthew. As a commercial tree grower and fruit farmer near Perth, Matthew complained in 1831 that the merchants were overburdened by taxes and tariffs, which went to support an idle aristocracy. His argument for abolishing tariffs and freeing trade is revealing. In nature, he maintained, competition and the selection of the best-adapted individuals transformed life to fit a changing environment. But this natural process was blocked by the system of "hereditary privilege" in British society, which was consequently deteriorating. Privileges and sinecures (which also permitted a supine priesthood to "disseminate" its "darkness") were "unnatural customs" sapping the nation's strength. As such they required "natural" correctives, and Matthew warned that unless the nobility bowed to nature's laws of competition and transformation, she would "avenge" herself. (He undoubtedly meant by reducing man to decrepitude. As a future moral-force Chartist, who resigned at the London Chartist Convention of 1839 when confronted by the violent demands of Feargus O'Connor and the physical-force agitators, Matthew would not have seen nature sanctioning a bloody revolution.) In short, society must accept unfettered commerce

113. Saunders 1950:87; Shapin 1979a:56, 1983:153–59. On the "outsider" status of the audiences for these "rank breaking" sciences, see also Cooter 1984, Shapin 1975, 1979b, and Cantor 1975.

and capitalist self-government for its own sake. The town corporations had to be cleansed of their entrenched Tory elites and turned into commercial bastions. And a democratic Parliament must be established, where the manufacturers and merchants, those whose intellects had been honed by "commercial adventure," might rule more beneficially.[114] Only then would society become progressive and "self-regulating" again, after the fashion of nature. Only then would mankind continue on its path to perfection.

So the shopkeepers and merchants were trading in politically useful commodities, including transmutation, phrenology, and the reductionist physiologies. These anti-Establishment sciences were incorporated into the democratic strategies of the disaffected groups outside the civic power structure, those resenting the privileges of the university, corporations, and Kirk. Radical graduates from the medical schools helped spread the new doctrines. Browne taught the Edinburgh Association's traders a reductionist physiology. Matthew invoked transmutation, talking of the "plastic quality" of life with its power to change for the better. He praised the "beauty and unity of design" of this "continual balancing of life to circumstance," showing how a self-modifying, improving, purging nature suited the tradesmen looking for new sorts of legitimating knowledge.[115] Also moving in this direction was Hewett Watson, who studied medicine in Edinburgh from 1828 to 1831 and was (like Grant before him) a president of the student-run Royal Medical Society. A former solicitor's apprentice, Watson despised his high-Tory father (a justice of the peace and anti-Jacobin). In the 1830s Watson joined Browne at the center of Combe's coterie, writing the kind of caustic leaders for the *Phrenological Journal* that caused even Combe to flinch. As a student Watson had already become a democrat and freethinker, and by the early 1830s he too was seeking transformist solutions to the problems of plant distribution.[116]

Grant's "democratic" evolution—in which life, empowered from the base, swept upward under its own impulse—was a more radically environmentalist science than phrenology. But a number of reformers, accepting what Shapin calls a "'Lamarckian' social determinism,"[117] sharing the same faith in social progress and the law-bound nature of body and mind, managed to blend the two doctrines. In some senses they could be visualized together: while Grant's zoological escalator moved inexorably to-

114. Dempster 1983:98–102 reproduces Matthew's relevant appendix; Wells 1973:229–30, 236–39, 241, 253 provides the best discussion; Wright 1970:71–72.
115. Dempster 1983:106–7.
116. F. N. Egerton 1979:90; Cooter 1984:95, 105, 185; Rehbock 1983:126–27, 176; J. Browne 1983:65–68; and Egerton's forthcoming biography of Watson.
117. Shapin 1979a:60; Cooter 1984:131–32.

ward its human apex, Combe's phrenologized man was aspiring to still greater intellectual heights. Socially, too, the medical transmutationists and phrenologists were often close and running mutually supportive civic campaigns: Browne, for example, fought for asylum reform, Grant for an antimonopolistic medical democracy. Hence they were praised equally by Wakley, as much for their defiance of the rich as for their moral therapy and materialistic science.

Geoffroy's Rise and Fall in Extramural Edinburgh

What did separate the phrenologists and Geoffroyans was the latter's success in medical management. While Combeites exported their science to the hinterland, capturing the institutes and proselytizing shopmen, Geoffroyans, being anatomy teachers, made their greatest gains in Surgeons' Square. The anti-Christian radicals now ran Edinburgh's most popular extramural school (John Barclay's). Knox's class here in 1826–28 was almost twice the size of his nearest rival's. By 1830 five hundred pupils were enrolled, making it the largest anatomy class in the country.[118] Nor was his a lone Geoffroyan voice. In Argyle Square Fletcher lectured on the organ "prototypes" and their progressive expression from polyp to man. He too praised Geoffroy for giving the "fundamental identity" of composition among organs its "scientific character," and while he agreed with Grant that Britain languished a century behind the Continentals, he saw London and Edinburgh racing to catch up.[119] Then in 1831, the quietly proficient William Sharpey set up next door to Knox's school. Sharpey had just returned from Germany, having traveled extensively on the Continent for a decade; he spoke German and French fluently, had heard the transcendentalist Lorenz Oken in Berlin, and had studied under Tiedemann. He knew Knox (dedicating his doctoral thesis to him in 1823) and embraced the new anatomy. But temperamentally Sharpey stood in sharp contrast to the fiery radical next door, and he cautiously restricted his discussions in class to the vertebrate type.[120] Given the caliber of these lecturers, conversions were common: the "unmeaning grin of buffoonery" was wiped off opponents' faces to be replaced by the "smile of conviction" (Fletcher's stiletto of a pen was as sharp as Knox's).[121] With the "despised

118. Struthers 1867:92.
119. J. Fletcher 1835–37, 1:36–38, referring to Grant 1833–34, 1:96.
120. Sharpey 1840–41:489–93; D. W. Taylor 1971:129, 131, 150; Cathcart 1882. Sharpey lectured at 9 Surgeons' Square in 1831–36 with Allen Thomson, also recently returned from Germany (on whom, see Jacyna 1983c:82); Knox was at Old Sugeons' Hall (No. 8) until 1840, when he went to the Argyle Square School.
121. J. Fletcher 1835–37, 1:78.

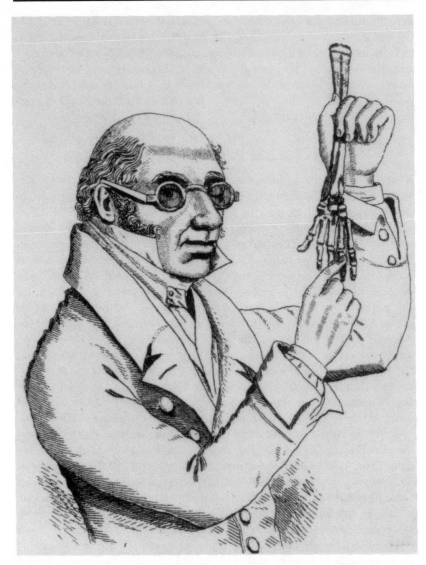

Figure 2.8. Before 1830 Robert Knox was the most popular extramural anatomy lecturer in Edinburgh. With one eye and a pox-marked face, he also proclaimed himself the ugliest. He overcompensated in his gaudy dress, teaching in a puce coat and dripping with jewelry. (Courtesy Wellcome Institute Library, London)

philosophy of St Hilaire" reaching such huge audiences, it is not surprising that students graduating from Edinburgh at the end of the 1820s, like the suave Whig reformer John Addington Symonds, revered Geoffroy's "transcendental Anatomy" as one of the achievements of the age.[122]

Yet however exhilarating Knox's lectures, however responsive his class, Henry Lonsdale's intimation in his blustery *Life* of a "Knoxite" masonry among the students was misleading.[123] Few shared the master's cynical and often malicious radicalism, and most were considerably more conservative transcendentalists. That anatomical "double star" in the ascent, the reserved John Goodsir and his lodging companion, the Manxman Edward Forbes, typified Knox's newer conservative students. Both were enamored of German *Naturphilosophie,* with its Platonic ideal forms; both followed the idealist zoologist Carl Gustav Carus[124] and supported Coleridge (himself using German romantic philosophy for antiradical social ends). They were religious—Goodsir's Edinburgh chair in 1846 testified to his Presbyterian piety; Forbes was a High Anglican unhappy with Lamarck and his radical followers (he was later scathing on the subject of Grant's shadowy reputation).[125] Goodsir muted Knox's Geoffroyism, blending it with teleological elements to forge a "higher physiology"; from this it was only a short step to idolizing the London Coleridgean Richard Owen. And Forbes, from 1842 professor of botany at the quintessentially Coleridgean King's College in London, became a colleague of Owen's allies in the school. Forbes's attempt to Platonize Knoxism was a typical Tory domestication ploy—turning genera into "divine ideas" and the world into an incarnation of Thought. It was an attempt to counter the radicals' materialism and recapture the transcendental standard for conservative Coleridgean ends.[126]

Knox's star itself had waned dramatically by the mid-1830s. No doubt his involvement in the Burke and Hare scandal had chronic career implications. (With cadavers constantly in demand for dissection classes, two grave robbers, William Burke and William Hare, had murdered some six-

122. Symonds 1871:80; "despised": Struthers 1867:82.

123. Lonsdale 1870:283, 278.

124. Carus, the king of Saxony's physician, followed Lorenz Oken in bringing Schelling's *Naturphilosohie*—with its ideal forms and ground plans—into comparative anatomy, although he was less wildly speculative than Oken. More is said on German *Naturphilosophie,* Platonic idealism, and the younger English surgeons in chap. 6.

125. Forbes to Huxley, 16 November 1852 (IC THH).

126. Secord 1985c:192; Mills 1984:375–85; Rehbock 1983:91–98; J. Browne 1983:144ff.; Jacyna 1983c:90ff.; W. Turner 1868, 1:32, 33, 121, 185. Desmond 1982:chap. 1 on Forbes and Owen. Another of Knox's religious assistants was Thomas Wharton Jones, who moved to London in 1837 to lecture at Charing Cross Hospital.

teen people in 1827–28 to supply the trade. Unwittingly Knox had bought
the bodies.) Certainly *Blackwood's* Tory wit turned vicious at his expense,
and his university rivals whom he had so mercilessly stigmatized in class
made a show of desertion. For the same reason he was evidently obliged
to resign his curatorship of the Edinburgh College of Surgeons. But it was
Knox's egotistical attitude and savage radicalism that really generated the
virulent opposition. His persistent ridiculing of the "canting provosts" and
"whining town-councillors," combined with his anti-Christian wit, made
him the "subject of intense hatred" among the clergy and council leaders.[127] Still his school suffered from more than a settling of the corporate
debt. Personnel changes in the private schools, the failing supply of cadavers, even John Fletcher's sudden death in 1836 (from a lung infection),
all took their toll. (Knox's best demonstrator, the bucolic John Reid, left to
replace Fletcher in Argyle Square.) Yet a fierce radical prepared to damn
Calvinism and Kirk and to defame rival applicants for posts instead of presenting testimonials was ill-equipped to secure municipal patronage or
official sanction for his sort of science. Edinburgh's town council, finding
itself continually slandered, naturally took a jaundiced view of his application for a university chair. Given, too, the jingoism of the age—the rail-
ing against profligate Paris ("the Mother of Whoredoms," in Coleridge's
words) and that "dessicating" materialism which had spawned the
Revolution[128]—one can understand why patriotic hackles rose at Knox's
preference for the "less doltish" French and Germans.

His classes dwindled drastically after the mid-1830s, but by then wider
factors had already acted to shift the center of medical education to London. The northern Athens no longer shone in its Enlightenment splendor.
The university was in decline; its curriculum had failed to keep pace with
the times, and the quality of its teaching staff was dropping as a result of
town council intriguing.[129] The rise of the university and private schools
in London was already accelerating the "Scotch" exodus to the capital.
Now Edinburgh, the imperial producer of what Roger Cooter calls "crisis
knowledge," was to ship its Lamarckian and Geoffroyan commodities to
London, to be sold in the Gower Street "emporium." Henceforth London

    127. "Dr. Knox," *MT* 1844, 10:246; E. Richards 1988; Lonsdale 1870:190–93, 195–96,
408. Knox had been building the collection in the museum of the Edinburgh College of
Surgeons since 1824 (Rae 1964:31, 36–37). His position there was not incongruous for a
radical. The Scottish medical corporations had a different history from the London RCS;
having been more concerned with teaching, they were less troubled by calls for reform (R. S.
Roberts 1966:78–79).
    128. Knights 1978:50; Lonsdale 1870:72ff., 262; Rehbock 1983:43–45.
    129. "Medical Faculty of the University of Edinburgh," *MCR* 1833–34, 20:315–19; Mu-
die 1825:210, 220; Morrell 1972:41; Cooter 1984:85.

was to act as the focus of Geoffroyan defiance—the movement fueled as disaffected Scots became further radicalized by their brushes with the London monopolists.

Establishing the New Morphology in London

The student is no longer content with a knowledge of the forms and processes of the bones, the origin and insertion of muscles, the ramifications of arteries, veins, and lymphatics, and the structure of glands. He looks for the philosophy of anatomy.

—*Quarterly Journal of Education*, the Broughamite house organ, commending the new London University curriculum[130]

Grant arrived from Edinburgh to take the London chair of zoology in 1827. In many ways he fit the bill: the council was intent on recruiting younger men, academically trained, with Continental leanings and Benthamite flair. A solicitor's son and Edinburgh educated, Grant was recognized in the Parisian empire of comparative anatomy and shortly would be acclaimed by Geoffroy as a "master" of higher anatomy. He was also a gifted linguist, taking an interest in even the minor European languages. Brougham looked to "zeal and experience in teaching as everything";[131] Grant had both, as Barclay's stand-in and a prolific publisher of papers on sponges and other invertebrates in 1826–27, some of which were in the process of being translated into French (an important pointer to one's European standing).

Local anatomists and clerical naturalists had been passed over in favor of polished Francophile learning. Grant himself recognized the value of Continental credentials, and he went on to support Edward Turner's candidacy for the chemistry chair on the grounds that his Göttingen education and proficiency in German would enable "him to keep progress with the rapid march of the science."[132]

130. "Recent Improvements," *Quart. J. Educ.* 1832, 4:12.
131. New 1961:376; Mazumdar 1983:235ff. Grant's appointment: Grant to Council, 27 May 1827 (UCL CC Applications); J. Thomson to Horner, 9 July 1827 (UCL CC 445), for reports on Grant's morals and manners. See also "Biographical Sketch," *L* 1850, 2:689, 691–92; and Desmond 1984c:400. Grant's earliest papers on sponges were translated in *Annales des Sciences Naturelles* 1827, 11:150–210. This French recognition was important in a decade when, as one critic of British medicine complained, not a single book had been found worthy of translation into French (McMenemey 1966:138).
132. Grant to Horner, 21 September 1827 (UCL CC Applications: Chemistry). The elderly Joshua Brookes, Royal Institution professor John Harwood, and missionary-naturalist Rev. Lansdown Guilding were passed over for the zoology chair.

In London Grant and Turner were now joined by James Bennett, recruited to radical applause after the shutdown of his Paris school.[133] Bennett came in as the anatomy demonstrator, flushed with fresh ideas and French innovations. It was he who amassed the anatomical preparations and acted as adviser to the warden Leonard Horner: he advocated designing the dissecting theaters along French lines and using paid *prosecteurs* as in Paris (students, elected by "concours or public trial," who would help prepare the classes). And he saw his own office embrace that of the French *répétiteur*, whose duty was to retrace the most difficult parts of the professor's course. But his French expertise embraced far more contentious areas. From the first he urged a curricular emphasis on "L'Anatomie Générale," which "teaches us to distinguish the different tissues or textures which enter into the composition of the several organs and treats their generic characters and functions," and he was adamant that this science of structure could only "be elucidated by constant reference to the organs and functions of the lower animals."[134] Bennett, having failed to set up an English school in France, was establishing a French school in England. It was quickly apparent that his capabilities extended far beyond those of a demonstrator—even beyond those of his superior, the pedestrian professor of anatomy Granville Pattison. So in 1830, after considerable politicking, Bennett was given his own chair, while a furious Pattison was sacked. This is indication enough that establishing the new morphology was never a straightforward affair. It took an active radical effort, which alienated many moderates and resulted in a series of resignations. But the higher anatomy was established. In his inaugural address as professor, Bennett proclaimed (like Grant and Knox) that a "new science" had arisen illuminating the "darkness that hitherto has clouded all our knowledge of the phenomena of life."[135]

In line with its insistence on prestigious European science before local London practice, the council had originally offered comparative anatomy to the distinguished developmental morphologist Johann Friedrich Meckel. Meckel had studied at Göttingen, visited Paris, and settled in the Prussian city of Halle. He specialized in comparative embryology and had independently developed a version of the "recapitulation theory" (it subsequently became known as the Meckel-Serres law). Securing his services, it was agreed, would have guaranteed the university a "European reputation."[136] But his exorbitant demands (including a £1,000 annuity for

133. "The London University," *L* 1827–28, 2:818; Bellot 1929:148–49, 160–61.
134. Bennett to Horner, 25 March 1828 (UCL CC 611), May 1828 (CC 614), 3 November 1828 (CC 617).
135. J. R. Bennett 1830:9.
136. "Meckel and the London University," *MG* 1828, 2:494; also *MG* 1828, 1:539. The

his museum) led to a collapse of the negotiations, and the chair was re-allocated to Grant in 1827, who now added it to zoology.

Grant began building his museum from scratch (he had brought a hundred or so invertebrate specimens from Edinburgh). Although he had to pay for the students' dissection material himself, the council covered the cost of the permanent museum exhibits, and it provided him with funds to attend the local auctions.[137] Hopes ran high in these early days, and the museum was even projected as an empire showcase. Edinburgh University then trained about a third of all the medical officers of the armed forces and East India Company, and Jameson received a wealth of colonial exhibits for his museum. (Knox and Grant had dissected the marsupials and monotremes shipped here by the governor of New South Wales, Sir Thomas Brisbane.) Many of Grant's London students were also destined for the services, and his lectures were recognized by the army medical board (which required its surgeons to attend a five-month course on natural history). So he confidently expected his museum likewise to "be inundated with donations from all quarters at home, and from our scientific countrymen in the most distant colonies."[138] But the early optimism quickly gave way to cutbacks and recrimination. By 1833 the university had run up a debt of £3,700 and was operating with an annual deficit of £1,000. Grant's letters to the council became increasingly carping. The students protested about the lack of books, and Grant complained that the museum's contents, constantly reduced by Gower Street's hungry rats, were more likely to "repel" than attract auditors.[139]

It was not only the rats that were getting hungry. Some professors too felt the pinch. Because the investors expected a return for their outlay,

---

council had sent Pattison to negotiate, but Meckel believed him incompetent to assess his museum's worth. The council also negotiated for S. T. Sömmering's museum in Frankfurt, but he was unwilling to let it go for less than £4,000: "Soemmering and His Museum," MG 1828, 2:496. Meckel's standing was then high; he had published his seven-volume System of Comparative Anatomy in 1821, although he was to die in 1833 leaving the work incomplete. On Meckel's science see Lenoir 1982:56–61 and Gould 1977:45–47.

137. Grant to Horner, 20 December 1828 (UCL CC P149). He bid against William Buckland, William Clift, and Gideon Mantell at the auctions of Joshua Brookes's anatomy museum in 1828–30: Mantell to Clift, 22 July 1828 (RCS SC); Grant to Horner, 22 July 1828 (UCL CC P148; also CC 759–60, P139–41). Donations to the museum were received from J. E. Gray (British Museum), N. A. Vigors (Zoological Society), and Lady Raffles, who, through Vigors, presented skins brought back by Sir Stamford Raffles from Sumatra.

138. Grant to Horner, 20 June, 5 July 1828 (UCL CC 759–60). There was even talk of hiring a keeper to cope with the influx, and Jameson's curator, William Macgillivray, told Grant that he was willing to take the job. But retrenchment soon stopped such talk. Chitnis 1973:174ff. discusses Edinburgh and the military.

139. Grant to Horner, 28 May, 29 May, 20 October 1830 (UCL CC P132, P136, P138).

the proprietors had modified the Edinburgh fees system. The professors were provided with "guarantee money" (£300 a year) for a tiding-over period (1828–31), after which they were to become self-reliant on fees. More than self-reliant; the teachers had to generate enough income to provide a return on the shareholders' capital (something the Edinburgh professors were spared), and any earnings over £100 had to be shared with the proprietors. These arrangements left the professors of noncompulsory subjects vulnerable. It might not have been so bad had the university's intake matched early projections. About 2,000 pupils a year were forecast, putting London on a par with Edinburgh. At Edinburgh there were 900 students in medicine alone, and Jameson was assured of a big class—200 sat his course in 1826. But at London quotas fell far short. The total intake in 1829–30 was only 630, and it would fall before it rose.[140] However, after a poor start, Grant's classes picked up, holding at an average of 30 students a year for the rest of the 1830s (a figure that included a number of teachers and London literati, who also audited his course).[141] And within a few years he had put together a teaching collection, taking his students to the Zoological Society or borrowing fossils from the Geological Society when it proved inadequate.

Bennett and Grant were well matched. They were passionate Francophiles and open opponents of the corporations, and they were of one mind on the direction that medical science should take. Bennett articulated the aims of the Gower Street higher anatomists. They were to move beyond the merely practical "Anatomy of Man" taught by the corporation surgeons and embrace "the wide field of animated beings, observe the common laws which govern their existence, and make organization generally the subject of . . . research." Bennett's own course was an exciting illustration—a blend of Geoffroy and Meckel, those twin "Newtons," the new lawgivers of morphology. This was to be London's first school of "General Anatomy," like its Continental counterparts devoted primarily to a study of the laws or organization common to all life. Such a study had "been lamentably neglected in England" for "want of that legislative protection

140. Jameson's class size: Morrell 1972:49; Ashworth 1935:100. On Grant's first sessions: Grant to Horner, 18 November 1828, 16 May 1829, 29 January 1830 (UCL CC P128, P142, P147). He had to drop his fees in these years to £2. London University medical student numbers climbed slowly in the early 1830s, from 248 in 1831, to 353 in 1833 (Lindley 1834–35:87).

141. Among Grant's early students were William Farr (1832), George Newport (1832), Thomas Laycock (1833), W. B. Carpenter (1834), Edwin Lankester (1836), John Marshall (1842), and W. H. Flower (1847). His auditors ranged from Henry Hallam (1831) and the Swedenborgian C. A. Tulk (1828–32), to medical men such as P. M. Roget (1832), J. F. Royle (1832), T. Southwood Smith (1836), and the Beagle's erstwhile surgeon-naturalist Robert McCormick (1836).

which, in other countries, has supported and fostered its cultivation as an object of paramount utility to society."[142] Grant too praised the bureaucratic structure of French science and lamented that, by depriving zoology of "public support," England was failing to exploit its potential as a colonial power. He wanted zoology brought within a national medical administration headed by the secretary of state (to keep it out of the hands of the corporations) and taught by an efficient salaried staff.[143] One of the "great designs" of the university was to act where the local surgeons had failed. It was projected that, through the "united efforts" of Bennett and Grant, the students would acquire a wider scientific appreciation of man and his development in relation to nature. From Bennett's perspective, comparative anatomy and the new medicine were inextricable. Grant's science was not peripheral, as the surgeons supposed, but central to any investigation of the "common laws" of the formation and functioning of life.[144] And of its *mal*functioning; since "deviation" from these laws produced physiological or structural defects (arrests of development), the general practitioner had to be initiated into these comparative techniques to give him a more philosophical comprehension of medical disorders.

So the doctrines were sold in a way that would attract wider medical attention. Teachers were soon using these comparative tools to unravel the more arcane aspects of human anatomy. The skull architecture, hearing apparatus, and nervous system could all be understood by watching the analogous organs develop through the animal series.[145] Moreover, practitioners could now explain structures that defied the surgeon's func-

142. J. R. Bennett 1830:8, 9. Lenoir (1982:60) identifies Meckel as an antagonist of Schelling's *Naturphilosophie* and one of the new "vital materialists" of German embryology. That is, he explained the development of the embryonic germ in terms of an "unfolding" set of chemical relations; this constituted the special "developmental force" that guided embryogenesis. While I look at the influence of Schelling's *idealist* philosophy at the RCS in chap. 6, it is clear that German nature-philosophers (such as Oken and Carus) and "vital materialists" (such as Meckel and the transmutationist Tiedemann) had a much broader influence in Britain on the philosophical anatomists and their allies. (Many—Knox, Sharpey, Grant, Turner, Marshall Hall, Martin Barry—had attended courses in Germany.) Although I have not gone into the debt here, it is a subject that deserves further study. In particular, given the impact of Geoffroy, we now need to know how far Meckel's and Tiedemann's studies of the vertebrate "morphotype" or plan, and their embryological criteria for defining it, also influenced these younger Scottish-educated anatomists in the 1820s.

143. Grant 1833–34, 1:97; 1833b:10; 1841:25, 37–39.

144. J. R. Bennett 1830:23.

145. As early as 1828, Jones Quain (1828:3–6, 1831:9–12) made great play of the value of the series in elucidating the functioning of human ears. See also: "On Philosophical Anatomy," *MCR* 1837, 27:84. On malformed and rudimentary organs: "Saint Hilaire on Teratology," *BFMR* 1839, 8:1–36; Sharpey 1840–41:489–93; Knox 1843:499–502, 529–32, 554–56; Jacyna 1984b:36–37.

tional approach. Vestigial organs such as the appendix became the rudiments of structures fully developed in other animals in the series. Malformations such as harelip were simply fetal "retardations": according to the recapitulation theory, the baby's lips had "frozen" at a level representing an earlier mammalian stage. Indeed it was as an aid to human embryology that the new doctrines generated the greatest interest. Tyros crooned over the "stupendously beautiful fact" that the ascent of the entire animal kingdom was duplicated in the first nine months of human life.[146] This recapitulation doctrine not only gave embryology its peculiarly comparative flavor in the 1830s, but it led to a spate of new studies on the animal nervous system aimed at elucidating the brain in man.

The animal series in effect presented the practitioner with a ready-made "ancestry" for the human organs, and it was partly because radicals such as Grant gave this ancestry a literal Lamarckian twist that these doctrines were so contentious. Grant's own course was modeled on that of the French Geoffroyans. The four theories Cuvier had seen usurping God's power formed the leitmotiv: unity of composition, recapitulation, transformism (called by Grant and Geoffroy "metamorphosis"), and the animal series. But it was the relationship between his Lamarckism and extreme interpretation of unity of structure that provided critics with their chief target. Grant taught that all animals shared a common plan. It can, he said, be detected "through all the great divisions of the animal kingdom, from the highest to the lowest."[147] And his lectures were devoted to tracing its progressive expression from monad to man. He also sided with Geoffroy on specifics. Most notably, he defended Geoffroy's paradigm case, the homology of the fish's opercular bones and mammal's ear ossicles, and in so doing made this "the chief battle-ground of homological controversy" in Britain, as it had been in France.[148] At the same time he shunned Cuvierian functionalism, insisting, for example, that Cuvier's rival explanation of the opercular plates as adaptive features could be overcome "by exceedingly minute and careful examination" of their fetal development.[149]

Just as Grant ruled out function as a determinant of deep structure, so he criticized a Cuvierian taxonomy based on discrete *embranchements*.

146. Anderson 1835–36:74; Flourens 1834–35:273–75. On the nervous system: Parker 1830–31, Anderson 1837, Solly 1836. All were indebted to Serres's *Anatomie comparée du cerveau* (1824–26). Critics of this approach included Joseph Swan (1835), although he was given short shrift by the reform journals: "Mr. Swan on the Nervous System," *BFMR* 1836, 2:192–96.

147. Grant 1833–34, 1:121, 89; 2:1.

148. R. Owen 1846c:231.

149. Grant 1833–34, 1:573–74, 703, 770.

He regretted that Cuvier had always "remained fettered by his earliest views of classification." One of its weak points was that each of the divisions had been named according to a different criterion: the vertebrate for its spinal column, the mollusk for its fleshy foot, the articulate for its exoskeletal joints, and the radiate for its radial symmetry. The Geoffroyan program by contrast demanded a single criterion in line with its postulate of homological continuity; in Grant's words, comparative anatomy was far enough "advanced, as to afford the means of distributing the animal kingdom on some more uniform and philosophic principles."[150] And this meant that it was ripe for a new terminology. Grant therefore redefined the primary groups according to the architecture of their nervous systems. Vertebrates became "Spini-Cerebrata," mollusks "Cyclo-Gangliata" for the ring of oral ganglia, articulates "Diplo-Neura" for the double ventral nerve chord, and radiates "Cyclo-Neura" for the nervous filaments circling the mouth.[151] It was a tentative step toward altering the classification, even if it did little to tackle the problem of actually unifying Cuvier's four divisions into a continuous classificatory series.

In the 1830s Grant's new terminology generated local interest, turning up in some of the radical medical schools. But more important was its heuristic function. Because organ systems seemed to disappear completely on descending the animal scale, it was problematic in what sense man, in all his complexity, could be said to have any higher structural relations with a granular zoophyte. Geoffroyism now stimulated the search for the diagnostic nervous structures ever lower in the scale. If the definition of the "Cycloneura" held good, for instance, then the onus lay on Geoffroy's disciples to detect the nerve collar around the mouth of simpler radiates, such as the jellyfish. Grant's eventual detection of a filamentous nerve ring in the tiny gelatinous *Beroë*, a transparent comb jelly he had caught in the Thames, was therefore taken as a vindication of the program. His observation was initially welcomed by the Scots-trained zoologists,[152] who expected better microscopic techniques to turn up evidence of the nerve ring in these lowly forms, and the search was switched to the still more primitive medusae.

Grant went beyond Serres's two-fold parallelism, between the animal

---

150. Grant 1830a: 368, 1836:107.

151. Grant 1833–34, 1:155–59; 1836:107–8; 1835–41:chap. 4. Those adopting Grant's classification included Jon Pereira (1835–36:246–48) of the Aldersgate Street anatomy school and the Paris-trained Samuel Solly (1836:4ff.). It was snubbed by the moderate reviews: "Outlines of Comparative Anatomy," *MCR* 1835, 23:381.

152. Grant 1835b:9–12, 1835–41:183. On its acceptance: W. B. Carpenter 1839:9; Bushnan 1837:69–70; J. Coldstream 1836:38, 40–41; R. Owen 1836:48; and Owen, Hunterian Lecture 4, f. 39 (BMNH MN 1828–41). See also Winsor 1976:34, 39ff.

series and the stages of individual fetal development. He added a third dimension, the continuous progression of fossil life, which was an extension of this scale back through geological time. Because these were expressions of the same series in different modes, he was able to swing freely between comparative anatomy and fossil history, telling his students:

When we speak of animals low in the scale, it is equivalent to our speaking of animal forms that have existed in the primitive conditions of this planet; for everything shows, that this kingdom itself has had a development from the most simple forms, and that in the first condition very likely nothing existed but myriads of animalcules swimming in the heated ocean that encompassed this cooling planet.[153]

Geologically, Grant's maxim that "Nature begins by simple forms" and proceeds gradually was extremely contentious in England. The oldest-

---

The subsequent history of *Beroë* is interesting from our standpoint of the developing rival coteries. Soon the younger Coleridgeans began to question Grant's sighting. Forbes (1839:149) turned the comb jelly upside down to leave the nerve collar at the wrong end. Owen (1843:106–7; and 2d ed., 1855:174–75) changed his mind about Grant's correctness. And the junior Guy's anatomist John Anderson (1837:5, 1839–47:601–2, 626) returned *Beroë*'s nerves to a "diffuse" (that is, invisible and nonfilamentous) condition.

Owen (1836:47, 1843:15) went on to break the Radiata into two groups. Those with nerve rings (starfishes, rotifers, etc.) were placed in the Nematoneura; the rest (sponges, polyps, jellies) were divided off into the Acrita, characterized by their lack of nerves or by their diffuse condition. He did however cautiously acknowledge a gradation, conceding that the "nervous globules" of polyps "begin to manifest the filamentary arrangement" on passing into higher groups.

But Grant's protégé W. B. Carpenter (1839:3–6; 1840–41, 27:858–59; Grant 1835–41: 182) remained convinced that it was simply the "transparency of the nervous filaments" in the acrites which rendered them difficult to see. And under the right conditions, Carpenter claimed to be able to see *Beroë*'s nerves quite distinctly with the naked eye. But then he was adamant that all radiates must have nerve filaments; to him Owen's "granulated" nerves were an absurdity. ("As well might we [talk of] a 'diffused' circulating system"!) Coordinated action would be impossible, he believed, in a polyp whose nerves were shattered into "isolated globules."

Owen and his pupils (Jones 1841) subsequently renamed all of Grant's divisions, calling Grant's Diploneura "Homogangliata," and his Cycloneura "Heterogangliata." The fit was not quite precise because Grant (1836:108–9; followed by Carpenter 1839:53–54) had shifted the rotifers and gut parasites from the radiates to the articulates, leaving him free to construct a continuous series within the Radiata from the single-celled Infusoria→Porifera (sponges)→Polypifera (polyps)→Acalepha (jellyfish)→Echinoderma (starfish). Such a lineage would have appeared more natural to Grant for meeting his serialist canons.

153. Grant 1833–34, 1:276. Like all recapitulationists, Grant did continue to elaborate Serres's two-fold parallelism. For example, at the Royal Institution in 1837, he described the appearance of the liver in the animal series—follicular in infusoria, glandular and opening directly into the digestive sac in the "lower animals," but connected by a duct in higher ones.

known fossils—crinoids, crustaceans, and mollusks—notoriously were not the simplest. He had tried to get round this empirical anomaly by arguing that the most ancient rocks had long ago metamorphosed through heat and pressure into "crystallised limestone" with the loss of their fossil infusoria.[154] Or perhaps not total loss: the very nature of some Cambrian silicate rocks could, he guessed, be evidence of the activity of these first silica-forming protozoa.

The extent to which Grant equated taxonomic rank with antiquity can be gauged from his work on Scottish sponges. He published twenty papers on invertebrates between 1825 and 1827, describing half a dozen new sponges (and coining the name Porifera for the group). He described the sponge's canal structure, the small entrance and raised exit pores, and current flow. By making this pore structure a defining characteristic of sponges, he was able to absorb into the porifera an anomalous organism, the flat freshwater *Spongilla*, which became the basis for his transformist speculations. *Spongilla*'s structure was then unknown; there was not even a consensus as to whether it was a plant or an animal. Grant sectioned specimens, observed the unprotected and unraised pores, and measured the current. By treating it as a "primitive" sponge, with "imperfect" pores, he created a base on which to construct a graduated series. Then, by making appropriate paleoenvironmental assumptions, he was able to draw conclusions concerning its mutation into more advanced marine forms. From its simplicity, he wrote, "we are forced to consider it as more ancient than the marine sponges, and most probably their original parent." And he explained this ancestry by assuming that its marine "descendants have greatly improved their organization" as a result of the "changes that have taken place in the composition of the ocean, while the spongilla, living constantly in the same unaltered medium, has retained its primitive simplicity."[155]

The forking image implicit in the split between the unchanged "parent" living contemporaneously with its advanced offspring was never exploited. Even though Fleming, looking at the problem as an exercise in

---

He likened this to the stages of liver development in the human embryo, where it is first "a mere thickening of the intestine," which becomes successively "follicular, glandular, lobular," and finally develops a duct: "Dr. Grant on the Glandular System," *MG* 1836–37, 19:749–50. He also pointed out the novelty of this observation to Michael Faraday, 13 January 1837 (RI Faraday Folio II, f. 135).

154. Grant 1833–34, 1:433, 276, 195; 1839:5–6; Desmond 1984c:403–6; Grant, Palaeozoology Lectures (BL Add. MS 31,197, ff. 26, 94, 144).

155. Grant 1826c:283–84. Fifteen of Grant's papers were read first to the Wernerian Society of Edinburgh, beginning in February 1825, and he sat on its council in 1825 and 1826 (Minutes, 1: f. 248, EUL).

pure taxonomy, pictured Grant's findings in another area in terms of a common *Spongia* "stem" splitting into silicious and calcareous "branches,"[156] the metaphor was not followed up. Fletcher too observed that according to transformist theory an increase in complexity could only have occurred among some forms at each level, leaving the rest unchanged to the present day. But in 1826 Grant's focus remained on the serial aspect rather than on the point of divergence or persistence of unchanged "parents," possibly because it was more amenable to Lamarckian explanations.

Grant added to this sponge series throughout 1826. He described a new Scottish form *Cliona*, a fleshy spongelike animal found inside oyster shells, as a "connecting link" between two formerly distinct classes: it had sponge characteristics, but it also possessed delicate polyps like *Alcyonium* ("dead men's fingers").[157] This enabled him to detail "a regular and beautiful gradation," from the primitive *Spongilla* via the marine sponges and polyp-bearing *Cliona* to the true colonial polyps. Because of Grant's environmental assumptions he could also portray this as a historical sequence. As *Spongilla*'s unprotected pores suggested its emergence at a time when the primordial lakes were still "unpeopled," so *Cliona*, armed with complex defensive spicula, had evolved later when predatory grazing animals "swarmed in the heated ocean."

But this sponge ancestry took up only a tiny portion of one Cuvierian *embranchement*. To prove the unitary structure of all four *embranchements* Grant had to bridge the larger structural gaps and detail far more extensive sequences of intermediate forms. So in the London lectures he extended the poriferan pedigree from *Alcyonium* through the crinoids to the worms and lowest articulates, producing a series rising by "imperceptible gradations" through the supposedly discrete divisions.[158] This radical "march of development" stoutly defied conservative tradition. In France a zoology that portrayed all life as reducible to a single plan and the product of evolutionary laws stood condemned as socially irresponsible. But Grant continued to follow the path laid down by the republican deists. He also supported Geoffroy in his Académie clash with Cuvier over the supposed homological relations between mollusks and fishes. Cephalopods

156. Fleming 1829:321–32; J. Fletcher 1835–37, 1:13. Grant's position in 1826 is less surprising, knowing that Lamarck envisaged no common descent (Hodge 1971:345), only a series of parallel lineages (so that today's reptiles are the ancestors of tomorrow's mammals, which will be unrelated to the living mammals). On the other hand, both Charles Lyell (1830–33, 2:10) and William Kirby (1835, 1:xxviii) were shortly to attribute a genealogical tree to Lamarck himself, showing that it could be done.

157. Grant 1826a:79–81, 1826c:283; Wernerian Society Minutes, f. 261 (EUL).

158. Desmond 1984b:205–7.

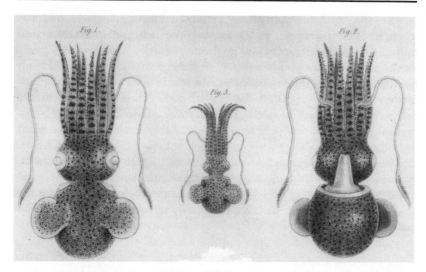

Figure 2.9. Grant dissected cephalopods in Edinburgh and London and took Geoffroy's side on the question of their relationship to fishes. While it was easy to obtain a small local squid such as the *Sepiola vulgaris* (center), at the Zoological Society Grant was also able to study tropical species such as the *S. stenodactyla* from Mauritius, which he was first to describe (figs. 1 and 2 in the illustration). (From Grant 1835a, pl. 11)

(the cuttlefish group of mollusks) were another of Grant's specialities (see fig. 2.9). He had dissected Coldstream's rare octopus at Edinburgh and in London studied the Zoological Society's tropical squids. In his lectures he traced the "successive degradation" of the invertebrate shell from the snail through the squid to its "remnant" state in the scale armor of sturgeonlike fishes. He echoed Geoffroy's views, adamant that diagnostic vertebrate traits were already present in the mollusks. For example, he interpreted the neck cartilages of the squid *Sepiola* as the "first rudiments of the cranial vertebrae, [and] of the rest of the vertebral column, although not yet divided into distinct vertebrae." Cephalopod fins with their supporting laminae were identified as incipient pectoral fins, and he believed that the oesophogeal ganglia of the highest mollusks "already approximated in form, position, and texture, to the brain of the fishes." Indeed, in his view "all the great systems of nerves of the vertebrata were already developed in cephalopods."[159] As in France, so in London these claims quickly moved to the center of the Geoffroyan debate, largely accepted by reformers and challenged by the anti-Lamarckians, especially in the corporations.

159. Grant 1833–34, 1:520, 512–13, 505, 508, 537–38; 1835–41:56; 1833c, 1833a, 1835a:79, 83.

After the Académie clash, Grant's statement had a loud partisan ring to it. He had aligned himself with the anti-Cuvierians in Paris and was committed to structural continuity, fetal recapitulation, and species "metamorphosis." But though his lectures provided a strong program which attracted praise from the reform press, his severe serialism alienated many moderates. Some conceded the cephalopod's "close affinity" with the fishes, but even they baulked at placing the armored articulates below mollusks in a strict unilinear series. Instead they mooted multiple paths from the invertebrates to the vertebrates, with mollusks united to fishes, crustaceans to turtles, and worms to snakes.[160] Many, including Owen, inclined to similar multiple passage models, believing nature simply too intractable to be straitjacketed into a simple series, worried also that the "march of development" lent itself too easily to Lamarck's iron-law explanations of ascent.

Internal Opposition: The Whig Moderates

Not only outside the school was there unease. The new-breed morphologists intent on usurping the craft-surgeon's role also caused consternation among the older Whig teachers. During the first three or four years, the university was itself racked with dissension, and its scientific direction became a heated issue. These financially disastrous years saw a prolific turnover of teachers, so much so that the *Gazette* in 1834 could conjure up a picture of educational mayhem. Most "respectable men," it observed, had resigned "in despair or in disgust"—"the Warden and the Professors of Medicine, of Surgery, of Anatomy, of Medical Jurisprudence, of Clinical Medicine, of Natural Philosophy and Astronomy, of Mathematics, of Greek, of German, of English Literature, &c. &c., having all changed within an inconceivably short period."[161] Clearly the pressures on the school were enormous. But any explanation of these medical losses must take into account the conflict over a Benthamite ideology which would relegate the older surgeons' design arguments and substitute the theory of homologies for a lucrative craft practice.

Take the case of the professor of anatomy and surgery Charles Bell, the medical school's most famous early recruit. With the university "going fast to the dogs," Bell resigned "unwillingly, and mournfully" in November 1830, midway through his course on "design" (a course that highlighted

---

160. "Outlines of Comparative Anatomy," *MCR* 1835, 23:382; R. Owen 1843:15–16.
161. "Memorial of the Medical Teachers," *MG* 1833–34, 14:242.

the Divine craftsmanship of human anatomy).[162] That he was using a framework of natural theology itself shows how distant he stood from the younger morphologists. Bell in truth had represented a distinct moderate faction within the school. He was one of the older Edinburgh Whigs: a friend of Leonard Horner, a collaborator in Brougham's schemes to promote Paley's theological works, a Scot who courted and commented on high society. Even as a young surgeon in London he had feasted with the literary lions: breakfasting with the president of the Royal Society, Sir Joseph Banks, and calling on Astley Cooper, soon to be the most famous surgeon of the day. Above all Bell admired the successful careerists, praising Brougham in particular for getting "popular and rich by his profession," as he himself was now doing among his wealthy patients. Actually it was through Brougham's intercession that Bell, despite opposition from the radicals on the council, acquired his chair in the first place. Bell never truckled to the rowdy democrats. His was an unflustered, gentlemanly Whiggism; a world of pious anatomy, salmon fishing, and Old Masters. Not for him Grant's and Bennett's hurried production-line mentality and fierce Francophilia. Bell loathed the radical declamations. He hated Lamarck's godless transformism and Geoffroy's stark reduction of animals to a set of common elements.

Bell's old-school mentality, liking of natural theology, disliking of the French, and despair of morphology all served to distinguish him from the new men. He even saw himself as an "exception, the only old lecturer" recruited from the local schools, as he said on resigning.[163] He stood defiantly outside Bennett and Grant's "philosophical" orbit. He was at odds with them in urging the "superiority of English physiology to French, which is so improperly popular," and at the outset he had opposed the council's attempt to buy a Continental anatomy museum. More crucially, at the time of his recruitment he was also the professor of anatomy at the College of Surgeons. True, he was uneasy at the college. But however "strange" the "old gents" found his presence there, it was no stranger than the medical school's radical body now finding itself with a Paleyite head. After all, here was a man who had praised the "good cheer and old London habits" of the local surgeons just as they were bringing down Bennett's French school. This sympathy made his position tricky in Gower Street. On resigning he told the council that the London surgeons had "carried their Science to the highest state of perfection" and that he was proud "to

162. Bell to Lord Auckland, 12 November 1830 (UCL CC P46); "London University—Mr. Bell," *MG* 1830–31, 7:305–11; G. J. Bell 1870:199, 316, also 20–21, 53, 92, 295; Kelly 1957:157, n. 2.

163. Bell to Auckland, 12 November 1830 (UCL CC P46); Mazumdar 1983:238; G. J. Bell 1870:263; Geison 1978:22–23; Gordon-Taylor and Walls 1958:251.

follow their steps";[164] he chided it for ignoring this proven system and taking a Continental course. The corporation gentlemen he emulated all interpreted anatomy in terms of natural theology: for them function was the arbiter of structure and adaptation the sign of Creative Design. Bell had believed that the university would also follow this royal road. He had even laid a set of plans before the council "for instituting a school of Design; all [of] which were received in good part, and neglected."[165] But it was precisely the older surgeons' system of medical education, run on nepotism and financed by expensive apprenticeships, that the Benthamite lawyers and merchants backing the medical school (Bell's "ignorant" "gentlemen of another profession") were committed to replacing. As such most had supported a secular school propagating a "higher" science. They had deliberately looked to the new morphologists, with their pro-French and naturalistic anatomy. Such a system Bell could never countenance. He remained ideologically opposed to the Continental coalition of Bennett and Grant. In his parting shot he described Bennett as an able demonstrator, but

avowedly French in all his medical opinions, modes of teaching, and technical terms;—to him has been assigned a department under the term 'Anatomie Generale'. Thus gentlemen not educated in the principles or practice of those schools of London, which have gained the approbation of the world, and which I have felt it my duty with every effort to maintain, have been put together to teach in modes and systems at variance with our own most approved opinions . . . [while a curriculum] is authorized where all that was excellent in the London schools is lost sight of, and modes of teaching and views of Practice, at variance with our general tenets, are maintained.[166]

This curricular reform inevitably worked against Bell. The self-styled "captain of anatomists" was galled at his lack of superior status in the school; he suffered a ruinous overlap of courses, and his "design" efforts were undermined by the morphologists. Not all were sad at losing the celebrated surgeon. Wakley predictably saw his academic excellence marred by a "sickening affectation and mannerism."[167] Nevertheless Bell's resignation damaged the university's standing, and at the worst possible moment, with the school eating up its capital and desperately trying to boost enrollment.

All of this, however, was soon overshadowed by the furious row in 1830–31 over Granville Pattison's competence. Politicking now became

164. Bell to Auckland, 12 November 1830 (UCL CC P46); G. J. Bell 1870:284, 288–89.
165. Bell to Auckland, 12 November 1830 (UCL CC P46).
166. Ibid.
167. "London University," L 1830–31, 2:693; G. J. Bell 1870:251.

rife as old scores were settled, clouding what has been portrayed as a clear-cut issue—the anatomy professor's proficiency. Historians have tended to take the contemporary accusations of Pattison's incompetence at face value. But this is a dangerous thing to do, given the rival anatomical programs being pursued in the school's early years. The case never was straightforward. That there were problems with his teaching is obvious from the fact that he was castigated by Bell and Bennett alike. But a closer study again shows that the more serious accusations were part of the drive to establish the new morphology.

As a former lecturer at the Andersonian Institute in Glasgow and sympathizer of the Mechanics' Institute movement, Pattison had seemed an obvious choice for the Broughamites. True, they had to recruit him from Baltimore, where he had gone after being cited as co-respondent in a sordid divorce case. Nor was this the only skeleton rattling in his closet: he had already been indicted for body snatching, accused of malpractice, and cursed with syphilis. He was pugnacious to a degree, even leaving a pair of dueling pistols on his desk as a reminder of how he obtained satisfaction. But irascibility was not the main issue, nor moral turpitude. Pattison came to London, Turner complained, with no European reputation, and this was a large part of the problem. In Gower Street his lectures were pedestrian and practical, whereas Bennett and Grant were firing the imagination of students with the latest developmental morphology. Pattison saw Bennett as a threat, with good reason, for Bennett was clearly taking classes on the side. Pattison now became involved in a drawn-out demarcation dispute to protect his teaching "interests."[168] His fears were barely concealed: he accused his demonstrator of trying to "ruin" his reputation and "undersell" his lectures, while Bennett in turn jibbed at his "menial duties." Bennett despised Pattison, needling him about his factual "blunders" and deploring his indifference to General Anatomy. This was the crux. In Bennett's original projection, General Anatomy was to have been central to the curriculum, a showcase subject. It was also one, said Bennett sarcastically, aware that the students were complaining about Pattison, which gives "the Professor an opportunity of displaying his learning and acquirements, [and] enables him to command and win the popularity of a class."[169] Since Pattison was branded an anatomical philistine by his pupils, this was designed to cut to the quick.

Pattison now came under intense pressure. Student militancy was al-

168. G. S. Pattison 1830:4–8; F. L. M. Pattison 1987:31ff., 64ff., 75ff., 122, 150–56, 159, 179. For the radical mauling of Pattison see L 1830–31, 2:693–95, 721–27, 747, 753–57. The details of Andrew Ure's divorce were dredged up to smear Pattison's character in L 1829–30, 2:741, 847–48.

169. Bennett to Council, 27 August 1829 (UCL CC 1228).

ready increasing at the time of the July Revolution. The radicals encouraged it, jubilant at events in Paris. For example, Thomas King, a higher anatomist who had formerly been a house surgeon at the Hôtel Dieu in Paris (a notoriously republican institution in which almost all the medical staff had once boycotted a visit by the king), infuriated conservatives by arranging for the London students to hold a public meeting—"a mutual expression of surgico-military propensities—an interchange of physico-political sympathies—with their *confrères* in Paris!" And, what is more, to hold it in the university's largest classroom.[170] With feelings running high, the students now turned on the anti-French Pattison, and in August 1830 he gruffly bowed to demands that Bennett be made an "adjunct Professor," with responsibility for the "Anatomy of the development of the Animal Structures, the Anatomy of Tissues or what is termed by the French 'L'Anatomie Générale'."[171] Even then he galled the militants by trying to keep Bennett a "supplemental" professor. The students—led by Grant's gold medalist Nathaniel Eisdell—recognized Bennett as incomparably Pattison's superior. Pattison's class became ungovernable; near riots ensued, and one student demagogue, Alex Thomson, was expelled (and promptly went off to Paris). Eisdell led a delegation of seventeen student medal winners to demand Pattison's dismissal. Others slated the "parcel of potters and haberdashers" constituting the council and outraged Tories by mixing "political feeling with insubordination" and distributing tricolor leaflets in the anatomy theater.[172] (The French tricolor was an evocative symbol in 1830–31, and flown in London's radical quarters during the July Revolution and on the fall of Wellington's ministry.) But the politicking at this point became very confused. Because the disliked warden came out against Pattison, a number of professors and councillors moved to defend him, which totally split the radical camp. In the end, though, what really horrified many university watchers was the idea of a professor holding his chair at the discretion of his class (a feeling that must have intensified when Thomson produced a massive, self-justifying pamphlet claiming that a professor was "only the hired servant of the pupil").[173] But Grant, Turner, and Anthony Todd Thomson (professor of materia medica, father of the banned Alex), who were asked to investigate

170. "Proposed Medico-Political Meeting," *MG* 1830, 6:800. King was supported by Wakley. Jacyna 1987:135 discusses the snubbing of Charles X by the Hôtel Dieu's staff.

171. Bennett to Horner, 2 August 1830 (UCL CC 3340); A. Thomson 1830:441.

172. "London University," *MG* 1830–31, 7:117–18; G. S. Pattison 1831:18; "London University," *L* 1830–31, 2:744–50; F. L. M. Pattison 1987:163ff. On the tricolor elsewhere in London: Halévy 1950:5; Wiener 1983:169. Eisdell was Grant's first gold medal winner (in 1830): *Distribution of the Prizes*, UCL College Collection, A 3.2; Bellot 1929:197, 199, 204.

173. F. L. M. Pattison 1987:179.

the student complaints, saw little to fear in such accountability. All were sympathetic to the students (a fact not lost on the critics) and saw "ruin impending over the University, if Mr. Pattison remained."[174]

Closer examination of Eisdell's allegations shows how carefully one has to interpret this dispute. True, the students could cite Pattison's Latin slips and odd questionable facts. But they always ended on his "*total ignorance* and *disgusting indifference* to new anatomical views and researches."[175] Eisdell was devoted to the new developmental anatomy and dismayed at Pattison's contemptuous attitude. He made it plain that Grant's lectures had only served to point up Pattison's "ignorance." He told the *Lancet* that he had "obtained sufficient insight" into the new anatomy in Grant's class to convince him that Pattison,

by almost wholly omitting to treat of that department of the science called "general anatomy;" in neglecting to indicate the pathological changes to which the various tissues are subject, and in failing to reveal to us the researches of Tiedemann, Meckel, Serres, Geoffrey [*sic*] St. Hilaire, and others, into the laws of organization,—did a wrong to the cause of science, which could only be obviated by his removal from the chair of anatomy. This conclusion was forced upon me more particularly by one circumstance amongst others, viz. the "*deplorable ignorance*" Mr. Pattison manifested of the stages through which the brain passes in the progress of its development, when he gave his class to understand that every part was developed simultaneously. Dr. Grant was present when this statement was made, and has confirmed the truth of the allegation.[176]

Eisdell was hardly an impartial witness: he went on to become Grant's collaborator, helping him with lectures and sitting alongside him on medical union committees. Further, he echoed Grant's feeling on the degraded state of anatomy in the corporations. Pattison was quite aware of the thrust of this attack. He told the council: "I am complained of . . . because I do not teach 'French anatomy.'" He proudly admitted ignoring its "idle, extravagant, unintelligible theories" and made no bones about his practical approach. "I teach anatomy for the purpose of educating useful medical practitioners. I teach anatomy as it is taught by the most dis-

174. "London University," *L* 1830–31, 2:749, 746. The divisiveness can be gauged from the fact that George Birkbeck, who hated Horner and praised Pattison, denounced Eisdell: Eisdell to Horner, 15 July 1830 (UCL CC 1784); Kelly 1957:155–59. The radical David Davis even demanded his expulsion. On the other hand, Grant backed Eisdell completely. The students stayed solid, and Eisdell collected a petition signed by ninety-five pupils protesting Thomson's expulsion: Eisdell to Horner, 14 October 1830 (UCL CC 1930); A. Thomson 1830:437–39, 443–47; *L* 1829–30, 2:623–24.

175. A. Thomson 1830:437, 443–47; Bellot 1929:197–98.

176. "London University," *L* 1830–31, 2:763. Eisdell as Grant's assistant: Grant to T. Coates, 25 July 1834 (UCL CC 3253); as fellow BMA councillor: *L* 1839–40, 1:182.

tinguished of my countrymen; and, however '*low*' Mr. Eisdell '*may consider anatomical science in this country,*'" for three decades the "splendid discoveries and improvements in the healing art" had all been made by these gentlemen.[177] It was the practical corporation approach that Pattison was championing, not Grant's "idle" higher science. Pattison, playing on conservative feeling (he was championed by the *Gazette*), accused Grant, Turner, and Thomson of "caballing with the students," siding with "riotous pupils" in a "wicked conspiracy." He held Grant to be an incompetent judge—a teacher who could waste his time discussing Geoffroy's laws and the theory of the vertebral skull, recalling in his own defense:

One day when I was in the habit of visiting Dr. Grant . . . he directed my attention to a large work which was lying on his table and made the following observations:—"Pattison, if you could only produce a work like that, you would render your name immortal. The GREAT MAN who has published it devoted his whole life to its preparation" (I think he mentioned forty years), "and I should be content to die if I could only leave such a legacy behind me." Anxious to examine the nature of the book which had excited so warmly Dr. Grant's admiration, I opened it, and to my amazement I found that the single subject treated by the author was the anatomy of the beetle!!! In Dr. Grant's opinion the man who spends his whole life on the anatomy of the beetle renders himself immortal, whilst in mine he is convicted of a wilful waste of his existence, which was surely bestowed on him for other and more important purposes. Dr. Grant may therefore honestly believe that because my lectures on anatomy were all made to bear on the great, and to the medical practitioner, the all-important doctrines of practice, that I am an incompetent teacher of anatomy. That I had devoted my time and attention to the idle and unprofitable speculations of some of the German anatomists, and for example, spent nearly the whole session in the attempt to prove an absurdity, viz. that all the bones of the skull are vertebrae, I should then have merited and received the mead of his approbation.[178]

This again shows how central the new anatomy was to the indictment of Pattison, and that much of what he was "ignorant" of was really what he was opposed to. Turner in this instance took up the cudgels on Grant's behalf. The two were now close friends, and with Turner hailing Grant as the future "*Cuvier of this country*"[179] (referring to his stature in science, not the direction his anatomy would take!), he was not above throwing the smears back in Pattison's face. Turner deplored Pattison's "feeble attempts" at ridicule and saw him condemn himself by admitting his indifference to a "principle [the vertebral theory of the skull] recognized by all

177. Conolly et al. 1830:20–21.
178. "University of London," *L* 1831–32, 1:86, also 82, 84–85.
179. Turner to James Mill, 20 April 1831, in *L* 1835–36, 2:844.

modern authorities."[180] In the same vein, one student was appalled that Pattison had not actually heard of Hercule Straus-Durckheim's *Anatomy of the Melolontha Vulgaris*, the deprecated "beetle" book, which the pupil claimed was "the best monograph" ever written on the articulates. The students rallied, testifying to Grant's "amiable character," and finding him the unlikeliest "conspirator" in Pattison's delusion of a "cabal."

The students were successful in their fight for Pattison's impeachment. He was sacked with a golden handshake and returned to the United States. The professors of mathematics, Greek, and Oriental literature immediately resigned in protest. The *Lancet* demagogues, reveling in the "victory," used the episode to justify their demand for the public competition for chairs.[181] While the university never went this far, it did institute a series of constitutional changes. The warden's post, prodigiously overpaid at £1,200 a year and an affront to the poorer professors, was abolished. The lecturers too gained a new voice. They had been denied any collective power by the original council; even so, the medical professors had early combined into a faculty (with Grant its secretary from May 1831). Following Pattison's dismissal, faculties were recognized and a senate established to give the professors a proper say in the future government of the university.

But the more important upshot was that with Bell and Pattison went the exponents of the "practical" approach and adherents of natural theology, strengthening the hands of the higher anatomists. From now on professors were to be picked for their morphological credentials. Bell's successor in 1831, the competent Jones Quain, had already introduced his *Elements of Descriptive and Practical Anatomy* (1828) with a paean of praise for the new French approach.[182] And his own replacement in 1836, William Sharpey, came complete with Scottish testimonials to his knowledge of "Transcendental Anatomy" and the French and German writings.[183] Such outward-looking Europhiles as Grant, Bennett, Quain, and Sharpey suited the university's internationalist pretensions, while their

180. "University of London," *L* 1831–32, 1:188–89; "Defence of Professor Grant," ibid., 199–200.

181. "London University," *L* 1830–31, 2:689–95, 749; *MCR* 1831–32, 16:v; *Statement by the Council* 1827:18. The *concours* was, however, instituted for the house surgeon's post at the North London Hospital attached to London University (Clarke 1874a:314–15).

182. Both J. R. Bennett (1830:9–21) and Quain (1828:1–27) advocated using the animal series, Geoffroy's laws of unity and teratology, and Serres's laws of embryogenesis (eccentric and double development) to elucidate human anatomy.

183. D. Craigie to Sharpey, 22 July 1836 (WI ESS/b.1/5). Sharpey (1840–41:489–93) praised the new morphology in class; D. W. Taylor 1971:134–35; Mazumdar 1983:242–44; Jacyna 1984a:55, 1984b:31–32.

naturalistic anatomy fitted the Benthamite educationalists' designs on science and society. The radical press was quick to approve, claiming that the professors "as a body stand unequalled in the medical schools of the metropolis."[184]

The university medical curriculum had come to reflect the secular ideology of its more radical backers. Utilitarian journals praised the "higher character" medical education was attaining in Gower Street through this emphasis on organization and development.[185] With reformers in Parliament after 1832 planning to legislate against the corporations, this Benthamite intrusion into the business of education can be seen as an essential prelude, a drive for theoretical standardization to serve new administrative ends. The result in Gower Street was a syllabus asserting the intellectual authority of the new professional middle classes in an industrial age: as steam-engine theory was discussed in physics and analytic methods in mathematics, so a law-bound anatomy replaced the surgeons' craft and a "higher" zoology staked its claim to the gentleman's conchological province.

184. "The London University," *LMSJ* 1830, 5:85. Bennett would die in 1831, and Grant delivered a "feeling" eulogy: *LMSJ* 1832, 1:210–11.
185. "Recent Improvements," *Quart. J. Educ.* 1832, 4:1. Cardwell 1972:46 on the syllabus in general.

3

Reforming the Management of Medicine and Science: The Radical Perspective

[Because the corporations are] part of the organization of the State, and were made for public, not private advantage, they must be preserved fit for their functions, or removed as morbid excrescences. The most effective and fundamental changes in a torpid constitution, are effected, not by spontaneous action, but by foreign agents, through the "primae viae," which produces a healthful re-action to remove the distemper; or, by the weakness of the *vis medicatrix*, induce the more violent, but more curative paroxysm of a radical Reform.

—Grant's medical metaphor for political change[1]

Who supported the new comparative anatomy and why? These questions dominate this and the following two chapters, which focus on the context of medical practice in London. Here I look at the ultraradicals in medicine—at their union activities, attacks on natural theology, and leveling campaigns aimed at the unreformed corporations. We have to understand the friction in the profession and the dissidents' grievances if we are to explain why only certain groups were receptive to Geoffroy's approach.

It was not only the medical corporations (the Royal Colleges of Physicians and Surgeons in London) that were targeted by the "destructives" in the 1830s. By the time the town halls were democratized in 1835, ancien régime paternalism was under strain in the whole legal, medical, and scientific establishment. The aristocratic and Church trustees of the British Museum reacted indignantly that year to the suggestion that science specialists should be put on the board, while the squirearchy at the Zoological Society was appalled at demands for its financial accountability. The radicals sought to substitute paid administrators and professional specialists for these "public" gentlemen—noblemen and Church dignitaries—running London's institutions of science. The sternest critics of the

1. Grant 1841:7.

old elites remained the Wakleyans. Having been locked into a long struggle with the surgical oligarchs, the medical radicals had had ample opportunity to organize strategy, found campaigning journals, and identify borough mongering. They often formed noisy cliques inside the learned societies, where they demanded professional standards, paid boards, and rank-and-file mandating. The very fact that the new lawful anatomy appealed to these democrats made it extremely suspect to the respectable gentlemen.

It was Grant's own calls for medical reform that first brought him to the attention of the metropolitan radicals. Opening the 1833 session at the London University medical school, he delivered a passionate speech, a mixture of Paineite rhetoric and Benthamite bureaucratic demands, pitting the inalienable "rights of man," the new medical man, against the "custom and power" of the corrupt corporations. He deplored the "exclusive and obnoxious power gradually usurped by the College of Physicians" and its restriction of the fellowship to Oxbridge (and therefore Anglican) graduates. All this was "contrary to reason, justice, expediency, and public good" and made a purge imperative if the RCP was to meet the needs of the Nonconformist practitioners. The "existing evils" extended to the Council of the College of Surgeons, composed almost entirely of eminent hospital consultants.[2] As we shall see, the consultants were using their legislative powers on the RCS Council to discriminate against their trading rivals, the teachers in the private medical schools, while barring the general practitioners from any say in RCS affairs. To end this kind of abuse, the radicals called for the medical colleges to be made responsive to a larger electoral community, to include the private teachers, GPs, and military practitioners. The radicals also sought to divest the medical corporations of their licensing powers and give these to a new state licensing board. This was not strictly a cry in the wilderness. There were already signs of state intervention in adjacent areas. The Whig government (to the chagrin of the surgeons) had created a new independent inspectorate in 1832 to administer the Anatomy Act (principally to apportion cadavers for teaching), while the Poor Law Commission showed the extent to which the Benthamites were prepared to reshape existing welfare arrangements. The radicals welcomed the setting up of Warburton's Select Committee on medical education in 1834, and it was widely expected to recommend sweeping democratic changes in line with Whig civic policy. Radical activists urged Warburton to extend the concept of centralization to medical administration generally. Grant, like so many doctrinaires,

2. Grant 1833b: 4–5, 17, 19.

wanted it brought under the home secretary, with the government supporting and financing medicine directly, as on the Continent.[3]

But conservatives defending the corporations' licensing and legislating
powers denounced the French system of state control. Frightening images were evoked—of John Bull's freedoms being crushed underfoot, of a
"revolutionized" England suffering the repeated turmoils of her French
neighbor. An alarmed *Gazette* noted that in Paris the

> whole *Ecole de Médecine* may be regarded as an engine of the state, the most
> minute portions of the machinery of which are committed to the surveillance of
> police officers. Hence the close connexion of politics with the repeated alterations
> in the medical establishments of that country: hence the frequently disturbed con-
> dition of the Parisian students, carrying the violent and factious conduct of politi-
> cal partisans into the theatres of the schools; and hence, we may add, the frequent
> presence of *gens d'armerie* in the saloons, for more than a guard of *honour* to the
> professors. How far would the admirers of our Gallic neighbours go? . . . would
> they prefer seeing our lectures attended by military—as we have seen Cuvier,
> during a whole course, address his audience having a sentinel, with fixed bayonet,
> mounted beside his chair?[4]

Some radicals took personal stands against the Royal Colleges. Grant,
for example, refused to kowtow to the College of Physicians by taking out
a London license, and in so doing he effectively severed his financial life-
line. He portrayed the college as a "kind of aristocratic high Church Establishment" engaged in the "public fraud" of extracting money from
those already licensed. (He was a fellow and license-holder of the Edinburgh RCP.) He declined to "disgrace" his existing fellowship by submitting to the London College's "arbitrary, illegal and ignominious" demand
for retesting.[5] Others such as the Baptist private teacher John Epps practiced illegally without a license.[6] Grant would not. As a result he was without the means to augment his scanty teaching fees and was to suffer financial hardship in later years. Grant's principled stand endeared him to the

3. Grant 1841:25–26, 88; Durey 1975:208–15.

4. "Medical Reform—Education," *MG* 1832–33, 11:89–92.

5. Grant 1841:10, 54–61, 97–98; see also *BFMR* 1843, 15:24; and Grant's bitter comments in the *London Medical Directory* 1845:64–65. The grievances of Scottish graduates in
London were in fact much broader than this. In 1833 attempts to placate them resulted in
the Apothecaries-Act Amendment Bill, which allowed a practicing Scottish physician to supply medicines: *L* 1832–33, 2:409–12.

6. Despite practicing, Epps still considered that he was injuring himself financially by
refusing to take out a license: "London College of Medicine," *L* 1830–31, 2:215; also *L* 1830–
31, 1:180. Practicing with a Scottish degree but without an RCP license carried risks—practitioners could not, for example, sue for damages: "Value of Scotch Degrees in England," *L*
1833–34, 14:391–93.

ultraradicals. Even moderates in 1833 thought that his university address had "truly represented" the RCP "as injurious, unjust, and oppressive."[7] Wakley of course was ecstatic. He was in wholehearted agreement on the RCP's "fool's play" in restricting its fellowship, and he saw in Grant's "brilliant" censure of the supine sergeant-surgeons another proof of the professor's "moral courage."[8]

Local events in 1831–33 show just how typical Grant's attitude was of "one faculty" agitation (for a profession run democratically and no longer divided by "estate," that is, into physician, surgeon, and GP)—and, as a consequence, why it was the radicals who were reading and praising his scientific works about this time. For example, there were massively attended meetings at the Crown and Anchor in 1831 to institute a rival, democratic London College of Medicine (LCM) (Wakley's brainchild). The delegates heard a succession of speakers recite a list of complaints against the existing colleges: concerning their self-elected councils, the "servile" status of the licentiates and their lack of voice, the RCS's discriminatory regulations and the council's treatment of the library and museum as private property, the RCP's Oxbridge restrictions, and the nepotism in the hospitals.[9] Although the antagonistic *Gazette* dismissed the speakers as "the humbler and more obscure" members of the profession, many were active private school teachers and GP lobbyists. Even the *Gazette* was soon wagging angry fingers at university lecturers such as the professor of midwifery David Davis for joining the LCM's steering committee.[10] The meetings were called at this time partly because the radicals thought that the new Whig ministry might be receptive to their demands. But events had also been accelerated by the forcible ejection, just before the first LCM rally, of prominent radicals from the College of Surgeons' theater. (A meeting to protest the exclusion of naval surgeons from the king's levées had been broken up by the police, with Wakley and the "rioters" ending up in Bow Street police station.) Wakley, Thomas King, the harddrinking teacher George Dermott, and George Walker (known as "Graveyard Walker" for his churchyard reforms) were all threatened with prosecution,[11] and at the second LCM meeting they turned up waving their

7. "The Introductories. 1. Dr. Grant," *MCR* 1833–34, 20:152–53.

8. *L* 1833–34, 1:73; also 1830–31, 1:180.

9. At the two meetings, speeches were made by Wakley (who began his ninety-minute address "amidst waving of hats and the loudest cheers"), Thomas King, John Epps, George Walker, and George Dermott among others: "Public Meeting," *L* 1830–31, 1:821–23, 846–65; 1830–31, 2:212–22.

10. "London University," *MG* 1831, 8:218; also 1830–31, 7:792–93.

11. "Second Naval Surgeons' Meeting," *L* 1830–31, 1:785–97; also 1830–31, 2:273–77, 305–11. Cf. the conservative view of the riots: "Outrage at the College of Surgeons," *MG* 1830–31, 7:760–67, also 787–91.

summonses. Protest at the court cases and the ensuing chorus of confrontationist demands were carefully orchestrated by Wakley, in his effort to whip up support for the new college. This was to be a wholly democratic institution. Its officers and senate were to be elected by annual ballot.[12] The cost of the diploma was set very low, greatly undercutting the charges of the existing corporations. Those already licensed to practice, by whatever body, were to be admitted as fellows (so Scottish physicians would be received without reexamination), and all fellows, whether surgeons, physicians, or apothecaries, were to carry the title of doctor (a leveling aspect even moderates abhored). Premises were obtained in the Strand, and the issuing of diplomas was begun. In 1833 the senate was lobbying the under secretary of state,[13] although few members could have been sanguine about the chances of obtaining a charter.

By 1833 even former moderates were turning radical, and London's medical societies found themselves reverberating to "one faculty" calls. For example, at the prosperous Westminster Medical Society over four hundred members attended in November 1833 for the first in a series of major debates on reform, which were to run until March 1834. Less exclusive than some societies, the Westminster had nevertheless coldshouldered Wakley's reporters in its early days, although by 1833 personnel changes and the influx of GPs were having an effect.[14] Here too the litany of charges against the corporations was chanted by a succession of speakers, culminating in motions calling for the merger of the three estates into one democratic faculty. These motions were passed by massive majorities, despite stonewalling tactics by the diehards. Appalled conservatives raised the specter of "entryism"—of a fanatical clique seizing control and using the respectable body as a cover for its radical program. The *Gazette* pointed out that the principal "performers," the "same little *corps dramatique*," were also playing at other medical venues, spreading their radical contagion to the London Medical Society and Medico-Botanical Society. It accused demagogues of pushing the societies toward anarchy by adopting the tactics of trade unionists and Newspaper Tax re-

12. "Scheme of Government for the London College of Medicine," *L* 1830–31, 1:865–66; also 1830–31, 2:177–83; and 1831–32, 1:456–57. The original committee included Joshua Brookes, David Davis, George Dermott, John Epps, Thomas King, Thomas Wakley, and George Walker.

13. "London College of Medicine," *L* 1830–31, 2:502; and 1832–33, 2:475; "The London College of Medicine," *Brit. Med. J.* 1926, 1:8; Sprigge 1899: chaps. 2–4; Brook 1945:87–91.

14. Clarke 1874a: 146–47, 239; *L* 1828–29, 1:121–22, 468–76; 1831–32, 1:336. Speakers here included Epps, King, Walker, and James Johnson: "Westminster Medical Society," *L* 1833–34, 1:363–67, 464–69. On President George Gregory's attempts to quash the reform vote in December 1833: *L* 1833–34, 1:605–7. The later debates attracted an audience of less than a hundred: "Politico-Medical Societies," *MG* 1833–34, 13:483.

pealers (the newspaper Stamp Duty, designed to stifle the pauper press, had inflamed artisan and bourgeois activists alike and led to a proliferation of illegal prints and prosecutions). "Borrowing a torch from their brother destructives," the republicans were setting it "to the foundation of all our medical establishments." The *Gazette*, upturning Wakley's rhetoric, lampooned the "medical millennium" and considered the demand for "one uniform Utopian democracy of medicine" so "wildgoose and wanton" that it rivaled the ludicrous attempts to revolutionize science in Marat's France. It called for conservatives to regain control, depoliticize these once-moderate societies, and return them to their proper medical business.[15]

This purging was imperative. The repercussions for the corporations and their favored hospital schools should Warburton's committee recommend such a democratic scheme would have been enormous. (And Warburton was known to be a supporter of Wakley's new college.) The corporations would lose their licensing revenues (about £7,500 annually in the RCS's case),[16] but more critically, hospital posts would cease to be rotten-borough gifts and would have to be competed for in open elections. Hospital revenues too would be less secure. Each candidate appearing before the LCM was to be placed in front of a cadaver and subjected to a lengthy public examination (this sort of practical test was unheard of in the RCS, where the question-answer sessions were simple and ritualized). The rigor of the test was designed to obviate the need for the examiners to see any course certificates. This itself would have had important repercussions. As things stood, the RCS councillors were protecting their own hospital courses by refusing to accept certificates from "unrecognized" schools or from private-school summer courses. By ignoring certificates altogether, the radicals could put the hospital and private schools on a par, forcing them to compete fairly and encouraging students to choose courses on their merit alone.

The doctrinaires' intent was to transfer control from the college oligarchs ministering to the gentry to the new practitioners serving the shopkeepers. The GPs would become the constituency for the new medical commons, where the Nonconformist and "Scotch" elite, relieved of their disabilities, could run for office. This too was attractive to moderates: out of the ballot and *concours* (public competition for posts) would come a new professional order based on talent. Unlike the RCS courtiers, who placed their protégés in posts, the radicals insisted that better-

15. On unions and anarchy: *MG* 1833–34, 13:327–29; on Marat, pp. 373–75; "Utopian democracy," pp. 402–7; entryism, performers, the torch, and depoliticization, pp. 482–86; medical millennium, pp. 955–57.

16. "The Money Rifled by the Council," *L* 1832–33, 2:696–98.

qualified men could be created through a correct scientific education and fair competition. This stitching of science and *concours* to the reform banner reflected the French sympathies of ultras such as Wakley and King, radicals who were "ever appealing to French arrangements as the *beau ideal* of everything that is excellent about medical affairs."[17] Wakley's emphasis on professional molding was reinforced by the cultural determinism of Dissenting teachers such as Dermott and Epps. All were leading "actors" in what the *Gazette* saw as the subversion of the societies. All believed that the LCM was founded on such principles that "it will *make* great men."[18]

Thus the political ground was prepared for the London debut of Grant's French-based science. His Lamarckian and Geoffroyan biology meshed well with the radical understanding of nature and society. It was defiantly secular and revealed laws of form where the surgeons had seen only functional relationships. His Lamarckism was environmentally controlled and ultimately rooted in the same Enlightenment soil as the radicals' own Jacobin philosophy (reflected in their faith that the social ills could be ameliorated through environmental manipulation—through better sanitation, welfare, housing, and health care). So there was a certain congruence between this kind of radical science and social theory. Indeed, Lamarckism provided a "natural" legitimation for democratic self-development, for power stemming from the base and mandating "upwards," rather than for the aristocratic ideal of a "downward" delegating authority. In 1833, medical Tories were already shivering at talk of the new "laws of medical science," based on "truth" and "reason," being "*universal* and *republican*," and at demands that medical society should therefore emulate the French revolutionary model, becoming "one and indivisible."[19]

There was no fear of Grant's science going unnoticed. His democratic speeches in Gower Street ensured that it was brought to the attention of the agitators. His 1833 medical address was wildly applauded, and Wakley (see fig. 3.1) praised the professor's "extraordinary mental powers." Grant reciprocated in class and commended the *Lancet's* "indefatigable" editor as a "castigator of evil-doers."[20] Conservatives were furious at this mutual backslapping. The *Gazette* abhorred Grant's toadying attitude toward Wakley, and even more his recommending a scurrilous print like the *Lancet* to his pupils. In a slamming indictment, the *Gazette* asked what part of "his patron's exertions" Grant applauded most: Wakley's "insolent mock-

17. "The One Faculty," *MG* 1833–34, 13:406.
18. "London College of Medicine," *L* 1830–31, 2:218.
19. "Pranks of Certain Radical Orators," *MG* 1833–34, 13:645.
20. *L* 1833–34, 1:279, 73.

Figure 3.1. Thomas Wakley, editor of the *Lancet* and a political pugilist. His and Grant's mutual backslapping infuriated the conservative medical press. By W. H. Egleton, after K. Meadows. (Courtesy Wellcome Institute Library, London)

ery" of eminent surgeons, his attempts to suppress the rival Anglican King's College in the Strand, or his "blasphemous derision of the sacred truths of Christianity."[21] Grant lashed back in kind at those "captious hirelings" who would impugn Wakley's character, and he ended on a rousing defense of the *Lancet*'s record in fighting corruption.[22] Thus by 1833 Grant had fallen afoul of the conservatives in the corporations and hospitals. That feelings were running high against him was evident from the *Gazette*'s warning. It never doubted his competence in comparative anatomy, but it persisted in warning him that he could only injure himself by siding with the violators of public decency. No teacher could maintain any "pretensions to respectability" while abetting "a publication which sets all morality, courtesy, and even decency, at defiance,—which outrages every feeling acknowledged among gentlemen, and violates every principle held sacred in society."[23]

Wakley wasted no time; a week after publishing Grant's address he began printing his comparative anatomy course, running the entire sixty lectures as they were delivered at the university in 1833–34. He declared at the end that he was astounded at the "depth and extent" of Grant's learning. The state of the subject before Grant began teaching in London had made England a laughing stock; now the shoe was on the other foot, and "translations of his lectures into the French and German languages are already demanded in the foreign schools of medicine."[24] Wakley drew great ideological strength from Grant's naturalistic views: he considered them unique, enduring, and "brilliant." He was also blowing his own trumpet in 1836 when he claimed that Grant's Geoffroyan philosophy had attained "universal diffusion in the profession" through its promotion in the *Lancet*. Obviously we have to be cautious of Wakley's hype, although all reviewers agreed that Grant's *Lancet* course was the first "comprehensive and accessible" exposition of philosophical anatomy in English.[25] His

21. "Professor Grant and Mr. Wakley," *MG* 1833–34, 13:292–93; "Dr. Grant and the College of Physicians," ibid., 120, and 165–66.

22. *L* 1833–34, 1:644–45.

23. *MG* 1834, 13:677. On his comparative anatomy, ibid., 22, 677; and 1834–35, 15:809, where the *Gazette* admitted on reviewing Grant's *Outlines* that he was "perhaps the most competent person in England to write a manual on the subject."

24. *L* 1834–35, 1:689. *Outlines* was translated into German in 1842 as *Umrisse der Vergleichenden Anatomie* (Leipzig: Otto Wigand).

25. *L* 1835–36, 1:586. Others had or were to deliver Geoffroyan courses, particularly Knox and Fletcher in Edinburgh, Henry Riley in Bristol, and R. D. Grainger and J. R. Bennett in London. But none had published his course yet, and most included the subject under the rubric of human anatomy, whereas Grant's was the first "complete and entire course of comparative anatomy": "Outlines of Comparative Anatomy," Renshaw's *LMSJ* 1835, 7:206.

technical mastery and lawful approach were widely praised in the reform press, and Wakley was congratulated for having published the course and introduced its new "high and philosophic principles" to the profession.[26]

One can understand its attraction for Wakley. Grant's lectures were Parisian, naturalistic, and eschewed all natural theology—fitting well with the radicals' pro-French, secularist onslaught on the cruder teleologies of the "'Church and State' bigots." Indeed, being acknowledged by Geoffroy as the British authority on philosophical zoology only increased Grant's republican reputation. Radical channels were immediately opened to receive his anti-Cuvierian anatomy. This kind of progressive science of morphological reform was to be used time and again by the radicals to confront the Oxbridge-educated medical elite, with its static Creationism and Paleyite natural theology. It was to be integrated into a wider radical program, which aimed at weakening Anglican power and Tory corporation authority. At the same time a Geoffroyan unity and Lamarckian ascent, powered from below, could be presented as nature's sanction for an emergent democratic authority, enabling the clamorous *canaille* to justify its claim as the new medical electorate—a claim expected to be legalized by the parliamentary Benthamites after Warburton reported back to the House in 1834.

Paley and Old Corruption

This nepotism or private influence operates in every appointment, from the Crown down to the beadleship at the doors of our halls and churches.

—An observer, despairing of eradicating nepotism in the hospitals when corruption was so prevalent in society.[27]

Such a discussion of political legitimation runs the risk of becoming facile if not given more concrete form. To understand the potency of this lawful morphology we need to see it actually deployed. So consider just one of its applications: its use to undermine Paleyite natural theology. Armed with the new science, radicals could deny that an animal reveals evidence of Divine design or, ultimately, that nature and society were perfectly functioning hierarchies, fashioned and sustained with Divine care. English natural theologians had long held that the adaptation of an animal or plant to its niche was a sign of intelligent design. They also taught that the purpose each organ was to serve, its role in life, was the sole explanation

26. *MCR* 1837, 27:85; *LMSJ* 1836–37, 10:250.
27. *MCR* 1833–34, 20:186. Rubinstein 1983 for the wider context.

of its structure. Paleyites saw this perfect adaptation as part of the proof that God had created the best of all possible worlds. In the 1820s and 1830s they used Cuvier's immense authority to rubberstamp this view. He, after all, had argued that an animal's construction allowed it to function perfectly in its given habitat (and that this functioning was sufficient to explain its structure). It was this connection between Cuvierian functionalism and British design approaches that made a rival Geoffroyism so attractive to medical radicals and the attacks on Cuvier so politically edged. Rather than function determining structure, the reformers now followed Geoffroy in arguing that all structures are, first and foremost, predicated on a common plan. The long bones in the mole's arm or bat's wing were not designed from scratch for digging or flying. Rather they were homologous bones conforming to the vertebrate blueprint, and modified only secondarily for a particular function. They are explained by Geoffroy's "unity of composition" and by its associated set of morphological laws.

The working-class atheists—particularly those contributing to the inflammatory *Oracle of Reason*—lampooned this "design" argument simply because it supported that poisonous "monster" priestcraft, and with it the iniquities of the undemocratic state. The agitator Charles Southwell derided all notion of "creation" and "design" in his epistles from prison. That eyes were made to see was as absurd as "to say that stones were *made* to break heads, legs were *made* to wear stockings, or sheep *made* to have their throats cut."[28] Holyoake, his fellow editor, actually wrote *Paley Refuted in His Own Words* in jail after being presented with a copy of Paley's *Natural Theology* by the prison chaplain, a nephew of the famous surgeon Sir Astley Cooper.

Unlike the working-class militants, the medical radicals had professional goals, and their approach was less brusque and more focused. True, as rationalists and (often) atheists, Wakley's men had cause to deride "design" as a prop to priestly power. But another reason they took a jaundiced view of Paleyism was that it was the corporation surgeons, the "self-elect," who were its leading medical exponents. Gentlemen such as Sir Astley Cooper, Sir Anthony Carlisle, Sir Charles Bell, and John Abernethy all interpreted comparative anatomy from a functional aspect, seeking proofs of divine care in the adaptation of organs. Wakley persistently ridiculed the baronets for indulging in such Paleyite "levity." Bell "never touches a phalanx and its flexor tendon, without exclaiming, with uplifted eye, and most reverentially-contracted mouth, 'Gintilmin, behold the winderful

28. Southwell 1842; Desmond 1987:88, 108, n. 123. Perfect adaptation is discussed by Ospovat 1978:33–34, 1981:33–37.

eevidence of *desin!* "[29] (Nobody was safe against the *Lancet* mimics, least of all Bell.) Radicals had attempted to butcher the sacred cow as early as 1826 while sending up the old hardliner Anthony Carlisle's Hunterian Oration at the College of Surgeons. Wakley's sarcasm knew no bounds:

whilst tearing asunder its bivalves [Carlisle was dissecting an oyster], lacerating its ligaments, and inflating its rectum, [he] piously observed, that the benevolence of an omnipotent power is exhibited in all the works of nature. Without questioning the propriety of this remark, we may, we think, be permitted to say, that it was at least ill timed, and we are inclined to believe, that had the oyster spoken, it would have given a flat denial of the Orator's proposition.[30]

These leading medical Paleyites controlled the reins of corporate power in the city. As hospital surgeons they sold apprenticeships (for £500 to £1,000 apiece), often taking on relatives, thus treating their posts as a form of "invested property" or inherited wealth.[31] Consider one of the radicals' prime targets, Sir Astley Cooper's network in the hospitals. Astley Cooper (see fig. 3.2) had amassed great wealth and fame. At the close of the Regency his name was said to have been as popular as Wellington's, and his income reached £24,000 in some years, rivaling the fees received by the great lawyers.[32] In the 1820s he had no less than seven godsons, nephews, and other apprentices holding posts at St. Thomas's and Guy's hospitals, all of whom were to become long-standing members of the RCS Council. The "family system" was underwritten by the hospital bylaws, which stipulated that surgical posts could only be filled by those who had been indentured to an existing officer. Moreover, these rich apprentices, by purchasing hospital posts, were effectively buying future RCS Council seats, and, as councillors, they obviously spoke in the hospital interest at a time when the private schools and university were beginning to pose a threat. For the reformers this was the sort of nepotism that made the profession so morally inferior to that of law.[33] Wakley, his ideology dominated

29. *L* 1832–33, 1:154–55. John Brooke (1979) has suggested that the design arguments also served a mediating function between the conservative religious factions in science.

30. "Hunterian Oration," *L* 1826, 9:689–95. Even Bell believed that Carlisle had "miscalculated the subject, and the time, and the audience" (G. J. Bell 1870:293).

31. Dermott 1835–36:94.

32. Clarke 1874a:34–35, 89, 116, and 296–99 on the purchase of posts at St. Thomas's— on which see also Singer and Holloway 1960:5–6; Brook 1945:17–18, 22–23, 43–44; Sprigge 1899: chap. 9, also pp. 109–10; Burn 1965:205–6n. On the prestigious position of the surgical elite even in the eighteenth century: Porter 1985:15–16.

Cooper's coterie comprised at St. Thomas's his apprentice B. Travers, godson J. H. Green, and nephews F. Tyrrell and C. A. Key; and at Guy's apprentices J. Morgan and T. Calloway, and nephew Bransby Cooper. All were to be appointed council members, and Cooper, Travers, and Green were to become RCS presidents.

33. *L* 1830–31, 1:564; on the comparison with law: 1828–29, 1:722–25.

Figure 3.2. Sir Astley Cooper, controller of a vast hospital patronage network. He was harangued by the radicals for his nepotism and natural theology. By J. Cochran, 1831, after T. Lawrence. (Collection of the author)

by a hatred of the rich, referred time and again to Cooper's "medicogenealogical" tree. At meetings he parodied Cooper's brusque manner to thunderous applause. And he used this illustration of corruption to turn the tables on the self-professed "patriots" in Lincoln's Inn Fields by urging that they themselves were indulging in "treacherous" acts: that for living upon "the fruits of corruption" they were "the bitterest enemies of mankind," the real "spies, traitors, villains." In an age dominated in both science and medicine by Tory talk of patriotism, it made sound sense to portray these surgical "traitors" as the real fifth columnists among the people.[34]

So the surgeons' design arguments, testifying to a personally attentive Creator, were anathema to the radicals. Knox savaged Cooper's "Guy's Hospital" theology and his attempts to provide a functional explanation for rudimentary organs such as male nipples. He also harangued Cooper and Bell for seeking some purpose for deformities such as harelip and skull crests; such functional explanations of arrested structures were "sometimes very pompous and imposing, as in the Bridgewater Treatises, but still downright nonsense." For the anti-Christian radicals, Paley's natural theology was a dead letter, "a vile patchwork, almost peculiar to British physiology, a jumble of expedients and contrivances to meet difficulties."[35] The ability to provide a rival, lawful explanation of organs, deformed or otherwise, made Geoffroyism politically useful to the secularists. Wakley was a longtime critic of Sir Astley's close-borough proceedings at Guy's; now, by disseminating the new morphology, he was able to discredit this "Guy's theology," further undermining Cooper's moral claims to medical leadership.

A reductionist science also provided something more positive—a basis for morality, founded on a secular understanding of matter's inherent properties. The medical democrats understood nature quite differently from the old Anglican order, which saw design and the Divine fiat ensure a stable social order by teaching moral subservience. The aging entomologist and country rector Rev. William Kirby was stung by the "utter irrationality" of Lamarck's autonomous nature. Kirby proposed instead a system of "*inter-agents* between God and the visible material world," a cherubic chain of spiritual vice-regents. These powers initiated every event in nature, and in society realized God's Will through His Church, which became the source of all civil authority. In L. S. Jacyna's words, the social and natural worlds were linked through a spiritual "command structure": their hierarchic ordering was a necessary consequence of the Divine Government.[36] With the natural and ecclesiastical status quo ordained by God's Word, Anglican ministers were ipso facto the legitimate authority on His Work. The idea of Anglican society and stratified nature

34. *L* 1830–31, 1:565, 856; *MG* 1830–31, 7:793. Wakley's attacks on Cooper's "family party" began in *L* 1824, 3:240–43. On patriotism in zoology and geology: Desmond 1985a:153–78 and Secord 1982.

35. Knox 1843:501–2, 529–31; 1831:486–87. Cf. Cooper 1840, 1:13–14, 160–61.

36. Jacyna 1983b:325–26; Kirby 1835, 1:xxiv, xxxiii–iv, xxxviii, c.

upheld "by the word of his power" was common in conservative philosophy. Radical Dissenting tracts openly attacked this view. John Epps complained that the Church presided "like a harpy" in Bow Street police station over those who could not pay the tithe money, seizing fields only to leave them fallow, yet "she claims to be the BLESSING OF THE LAND—the *channel* through which the Almighty pours his goodness around us—the source of all our 'national prosperity.'"[37] What proved the irresistible target in Kirby's case was his plum-in-mouth sermonizing and unabashed scriptural literalism. Even moderates lampooned his manikin nature manipulated by invisible demigods. Such was the "silly and superstitious" nonsense to which men "of Mr. Kirby's class" are reduced "when they attempt to unravel *final* causes!! To the credit of the medical profession, its cultivators have long abandoned such fruitless speculations." By "materializing the furniture of the Holy of Holies," mining the Bible for theological rocks to throw at Lamarck, Kirby actually sent distraught critics delving into the new Continental sciences. Medical moderates might have found Lamarck's spontaneous creation "absurd and atheistical," but no more than Kirby's Pentateuchal zoology. To some, a cherubic hand guiding human actions was worse than "blind chance and fatalism," a denial of free will and future rewards:

No, no.—The Almighty laid down general laws at the beginning, for man, for animals, for the elements, and even the Universe around them. The laws, being founded in infinite wisdom, require neither revision nor supervision. They are eternal and immutable. The animal creation he has bound by instinctive laws, from which they cannot deviate. To man he has given REASON, and, consequently, liberty to transgress some of his laws, both moral and physical, with the certainty of punishment for the transgression—thus making him a responsible BEING.[38]

This judicial view of natural law was common among medical moderates in the 1830s. But the secularists went much further. They not only repudiated talk of Divine powers vivifying an "inert" matter, but like the Regency radicals William Lawrence and Southwood Smith they portrayed life as an emergent property dependent only on organization. Many were committed reductionists. Grant saw in the "principle[s] of chemical and mechanical science" the complete explanation of an animal's "complicated functions." The complexity of the animal machine and plethora of "counteracting agents" might make it difficult in practice to unravel the physico-chemical operations, but this is no reason "to conclude, as many do in the

37. J. Epps 1834:5.

38. "Kirby on Instinct," *MCR* 1835, 23:400–413; 1836, 24:79–93, 358–65, quoting pp. 401, 413, 87. The reviewer was censured for his irreverence: "Kirby's Bridgewater Treatise," *MG* 1835–36, 17:318–19.

present day, that the laws by which complex substances are governed, are something altogether different from the laws which regulate inorganic nature." Grant saw development and "metamorphosis" "as much a part of the great system of nature as the movements of the celestial bodies." The whole constituted "one grand and harmonious system of the material world."[39] For him the march of fossil life was a function of Lamarckian and Geoffroyan laws, themselves higher-order cases of basic mechanico-chemical relations. And like earlier Paineite deists and d'Holbachian atheists (whose works were still being pirated by the pauper presses), he promoted an ethical naturalism. Only the "correct perception" of the "harmony" of these laws, he believed, would enable us to lay the true foundation of "morality and virtue."[40] The radicals, no less than the Enlightenment rationalists, saw this natural morality, resting on the truths of physics and biology, as superior to a Christian one.

So the new naturalistic anatomy had a wide cultural application. Corporation apologists took the professional brunt, but it also struck at the root of the Bridgewater tradition. In 1829 the eccentric earl of Bridgewater, in partial atonement for an impious life, had left £8,000 in his will for the publication of a book or books on the wisdom of God deduced from nature, in a fund to be administered by the archbishop of Canterbury, bishop of London, and the president of the Royal Society. Because the authors (four clergy, four medical men) were to use science to prove the power, wisdom, and goodness of God and to underscore a set of conservative Christian values, these Bridgewater Treatises received a rough ride from the radicals. Even the parceling out of titles and payments, largely to the Oxbridge divines and their allies, smacked of "jobbery," while the lavishness and price of the volumes led one reviewer to conclude that they were destined for the irreligious upper classes. Bell's book on the providential structure of *The Hand* elicited a yawn from the Wakleyans; Kirby's "extravagent and imprudent" treatise on animal habits (the most backward-looking of the Bridgewater books) was derided as an attempt to cast "the whole material world" after "the contents of the Jewish Tabernacle!!"[41] Others rated even harder words. Since the Divine hand was seen as much in the success of British industry as in the adaptation of animals, these books often raised issues of more immediate concern to the radicals. Geological patriots, for example the Oxford geologist William Buckland and the conqueror of the new Silurian territories Roderick Murchison, well known to land and mine owners, saw the hand of providence

39. Grant 1833–34, 1:127, 198, also 275; 1829:5.

40. Grant 1844:353; Desmond 1987:95.

41. "Kirby on Instinct," *MCR* 1835, 23:400, 401; "Bell on the Hand," *L* 1833–34, 1:165–69; Gillispie 1959:210.

in the proximity of Britain's coal and ore deposits. But the divine benefits were showered only on the gentry and entrepreneurs. Miners and their children working long hours underground counted themselves less blessed, and the agitators treated this appeal to providence as nothing short of moral blackmail, a justification for toil and sweat. Nor did this divine sanction stop radicals proposing that the squires be fiscally penalized. Since the days of Tom Paine, demagogues had called for the landowners to be taxed to indemnify the landless poor for their original dispossession, and Wakley continued the practice.[42]

Ultimately, then, a naturalistic anatomy in Wakley's hands posed a threat to ecclesiastical power. In the radicals' confrontationist interpretation of history, medical science has been continually impeded by the "chilling patronage of the church."[43] One case in particular gave Wakley a direct insight into the threat posed by a French-derived materialistic science and the authorities' reaction—that of his former lead writer and friend William Lawrence.

Lawrence (see fig. 3.3) had been expensively apprenticed to John Abernethy at Barts, and by the late 1810s he was a promising young surgeon at the Bridewell and Bethlem hospitals. But he was brash and sarcastic. He was also well versed in Continental thought, and in his 1816 lectures to the College of Surgeons he dismissed all vital principles and mystical life-forces as poetic personifications worthy of the benighted savage. (Vitalists—and Lawrence was targeting the great eighteenth-century surgeon-anatomist John Hunter—postulated the existence of an active power or principle, distinct from matter, which controlled the life processes, resisted bodily decay, and directed embryonic development.) Perhaps this was imprudent—slating Hunter's belief in vitalism—given his venue, for Lawrence was addressing the college just as Abernethy himself was praising the orthodoxy of Hunter's doctrines here. (It was even more reckless when one considers that Hunter's own books, manuscripts, and preparations formed the basis of the college library and museum. As a result, the surgeons were fiercely protective of Hunter's reputation, to the extent that the yearly Hunterian Orations were often little more than self-serving panegyrics of his work.) Lawrence used the animal scale to argue that the manifestations of life depended on structure, not mysterious vital agents; moreover, he believed that the ordinary laws of physics and chemistry were quite adequate to explain this life-giving organization. But Abernethy saw life and mind as something "superadded" to mat-

42. Wiener 1983:105; Buckland 1837, 1:524–47; Gillispie 1959:200–201; Secord 1986a:34; Rupke 1983: chap. 18; Desmond 1987:88–91; and Porter 1973 on the condescending attitude of the elite geological theoreticians to miners and the mining industry.

43. L 1830–31, 1:470–72.

ter, and he was convinced that such a view was essential to keep humanity on a "virtuous" course.[44] He upbraided his former pupil for his skeptical views, insinuating that if mind were simply some endogenous expression of neural matter, the soul would share in man's mortality and the masses would have no reason not to rise up to obtain redress. This was a common fear, and one the Cato Street conspirators—revolutionaries who hoped to start a general uprising among the oppressed masses in 1820 by assassinating the Cabinet—were about to realize with a vengeance.

One can understand how the issue should have resolved itself into one of social responsibility. This was a period of economic depression, unemployment, and high prices, with the sort of mass violence that was causing the authorities to crack down, both physically (as at Peterloo, where in 1819 a large Manchester crowd at a reform meeting was massacred by cavalry) and legally (with antisedition and antiblasphemy laws). Lawrence rejoiced in what the French and American democrats had achieved, but now he feared that Europe was falling again into despotism. He was appalled by the British government's clampdown on the freedom of expression. He opposed this attempted regimentation of society and standardization of behavior by referring to the differences of individual biology. With nervous Tories branding the French as Britannia's enemies "in science, as well as in politics,"[45] Lawrence's politics and anatomy stood condemned at both ends of Lincoln's Inn: at the Court of Chancery and the College of Surgeons. Lawrence noted in *Lectures on Comparative Anatomy* (1819) that the argument had shifted to one of "motives," with him standing accused of supporting French physiology "for the purpose of demoralizing mankind"—of loosening moral restraints in a period of popular unrest. But faith in vital principles cannot make us "good and virtuous," he retorted, or "impose a restraint upon vice stronger than Bow Street or the Old Bailey [London's Criminal Court] can apply."[46] Nor should souls be sought in the "blood and filth of the dissecting-room." He reasserted that life and intelligence were inseparable functions of organization and that nature's series—in which mental attributes can be seen changing by degrees—was the best evidence that man's mind was the product of his superior organization.

The "evil consequences" of these doctrines were quite evident to *Quarterly* reviewers: materialism in metaphysics, faction in politics, and infidelity in religion.

44. D'Oyly 1819:3; W. Lawrence 1816:164–77; Figlio 1976; Wells 1971. On Hunter's vitalistic views: Cross 1981:36–41; and Jacyna 1983a on the surgeons' vested interest in praising Hunter.

45. W. Lawrence 1844:4.

46. Ibid., 2–3, 8, 77.

Figure 3.3. William Lawrence in 1839, by now more circumspect on radical questions in science, even though his "blasphemous" lectures continued to be pirated by the pauper press. By C. Turner. (Courtesy Wellcome Institute Library, London)

Their tendency to impair the welfare of society, to break down the best and holiest sanctions of moral obligation, and to give a free rein to the worst passions of the human heart, is fully admitted even by those who embrace them. Voltaire, it is well known, checked his company from repeating blasphemous impieties before the servants, "lest," said he, "they should cut all our throats."

Lawrence's republicanism and impiety were cited as consequences of his French rationalism. To *Quarterly* Tories, thinking medullary substance was an absurdity—the endogenous vitality of man and cabbage a blasphemy. Were they not, there would still be insuperable problems to teaching such doctrines. Not to be sidetracked by demands for the inviolability of Truth, the *Review* inquired whether such "truths" for Lawrence could "outweigh the feeling of what he owes to the welfare of his fellow-creatures."[47]

The *Quarterly* urged the college to review the terms of Lawrence's contract, but Lawrence himself resigned and withdrew his *Lectures* from sale, writing to the political satirist William Hone—himself triumphantly acquitted of blasphemous libel in 1817—explaining his expedience and commending Hone's "greater courage" in these matters.[48] This ensured the book's notoriety, and two pirate editions appeared in 1822, one printed by the militant shoemaker-turned-bookseller William Benbow (who financed his political activity through his pornographic print trade). Lawrence tried to stop the working-class reprints, taking out an injunction in 1822, only to have Lord Eldon rule that an author had no property rights on a blasphemous book. This increased its attractiveness to the atheists. Two new pirate versions appeared in 1823—one issued by the doyen of artisan activists Richard Carlile, who organized the reprinting while serving a sentence in Dorchester jail for sedition. Thus Lawrence's *Lectures* appeared amid a profusion of antimonarchy, antigagging, and increasingly subversive and blasphemous prints and were channeled through established pauper-press outlets. Nor was there any letup in radical exploitation; in 1836 the sixth pirate edition was preceded by an advertising flysheet "sneering at the clergy," the hospital governors, and Lord Eldon.[49]

June Goodfield speculates that Lawrence suppressed the *Lectures* to protect his practice. However Lawrence's letters suggest rather that it was his position at the hospitals of Bridewell and Bethlem that was at

47. D'Oyly 1819:3, 5–6, 8, 9, 33. George D'Oyly was to become one of the founding fathers of the Anglican King's College in the Strand.

48. Temkin 1977:357. Wiener 1983 and E. P. Thompson 1980:791–803 on Hone, Carlile, and Benbow. The latter's pornographic sideline is discussed in McCalman 1984. Lawrence's editions are listed in Goodfield-Toulmin 1969:307–8.

49. "Sixth Edition of a Denounced Notorious Medical Work," *MG* 1835–36, 17:783.

stake. He was suspended from his surgeon's post on the publication of the pirate editions and obliged to write a retraction. His mea culpa was not wholly ritualistic. He had been deeply disturbed by the Chancery charges and the threat to his hospital career. He told the Bridewell governors of his "regret" at having published these "highly improper" passages and of his resolution "not only never to reprint them, but also never to publish any thing more on similar subjects."[50]

Bloodied he might have been, but Lawrence was not bowed enough to refuse Cobbett and Wakley's offer in 1823 to write for the *Lancet*. He contributed pungent leaders and went on to chair the mass meetings at the Freemasons' Tavern in 1826, called to draw attention to the corporate abuses of the surgeons. Wakley worked closely with him in these years and knew his case well. Lawrence did subsequently moderate his public views. But Wakley, as a journalist outside the hospital career structure, remained committed to a campaign of confrontation. He continued his denunciation of the church establishment in the *Lancet*. No sooner, he wrote, had medical science begun investigating "the mysteries of nature, than the facts which it brought to light were construed into so many contradictions of revealed religion" because they harmed "the interests and stability of the church." He demanded medicine's "complete emancipation . . . from the fetters" of the Church-run universities.[51] He wanted to slash the state funding of divinity schools and release the money to "endow Professorships in cities where students are numerous" in order to "bring science within the reach of the sons of the humbler classes." Like so many doctrinaire radicals, he saw the "regenerative" power of materialist science combat a corrupted clergy profiting through church-rate and tithes. Priests were cautious "not to risk the temporal advantages" which establishment had brought them, hence their "crusade" against science and infidelity. "Tithes and benefactions, church-lands and mortgages on the living and the dead, were of too divine an origin and of too earthly a value to be put in jeopardy by the diffusion of intelligence and improvements in education."[52]

---

50. Lawrence to Sir R. G. Glynn, 16 April 1832 [1822 or 1823] (RCS MS Add. 194). Lawrence obviously misdated the letter because he talks of suppressing his book three years previously, and he mentions a new "piratical act of a bookseller in the Strand, named Smith" (the J. and C. Smith edition eventually appeared in 1823). The letter could have been written in April 1822, three weeks after he had sought the original injunction against Smith. This would explain why he refers to the "charge of irreligion again hinted at in the Court of Chancery." My explanation of Lawrence's suppression supports that of Brook (1945:35); cf. Goodfield-Toulmin 1969:319; Temkin 1977:357.

51. *L* 1830–31, 1:470–72.

52. *L* 1837–38, 2:260–61.

Grant's and Wakley's anti-Christian wit and attacks on Anglican privilege united them at a deeper level within the radical movement. Wakley satirized the Old Testament stories and Christ's miracle of the loaves and fishes, "which some of the profane 'band of modern sceptics' have had the audacity and folly to deny."[53] He was antisabbatical, opposing the Lord's Day Observance Bill in Parliament on the grounds that working men had only Sunday free to spend their wages. Time and again the *Gazette* denounced his "scurrilous jibes" at Scripture and accused him of the vilest blasphemy.[54] Grant adopted the same skeptical tone in class, teasing his students with "satirical references to Providence."[55] He slammed the "monastic ignorance" of the Anglican universities and was warned against treading Wakley's path to sacrilege and social abandon. Grant's evolutionary science, acknowledging no gods but material laws, was compatible with this wider anticlerical movement. While his Lamarckism reflected a more extreme form of political dissent, his reductionist zoology was nonetheless typical of the anti-Paleyite science of the medical democrats.

After 1833 Wakley lost no opportunity to promote Grant: publicizing his activities, supporting his candidacy for chairs, smiting rivals (particularly the RCS trainee Richard Owen), and posting notices of his lectures. Both the *Lancet* and radical *London Medical and Surgical Journal* were now hailing Grant as "one of the most highly-gifted physiologists in Europe" and well titled the "English CUVIER."[56] This became a radical slogan during the university hustings in 1836, when Wakley's men tried to get Grant into the better-paying physiology chair. Wakley knew the canvassing tricks, exaggerating Grant's physiological attainments ("not surpassed by those of any professor in Europe") and crying "treachery, envy, and fraud" when the post went to William Sharpey.[57] The *Lancet* thus became the promotional organ for Grant's science and extramural activities in the 1830s. The paper might have begun sobering up by 1833, but it still reveled in editorial abuse. Whatever the merits of its lectures and hospital reports, it remained, in the words of one of its own reporters, among "the very lowest of the political prints of the day."[58] This left no doubts in Tory minds about Grant's appeal.

53. "Hunterian Oration," *L* 1826, 9:693; "Jonah's Residence in the Whale's Belly," *L* 1824, 1:305. Sunday closure: *Hansard* 1835, 28:505; Sprigge 1899:304; Brook 1945:120.

54. "More Consistency; or, Dr. Grant on the Use of 'Low Epithets,'" *MG* 1834, 13:676.

55. Godlee 1921:102. Grant to Horner, 5 November 1830 (UCL CC P30); *L* 1838–39, 1:909.

56. *L* 1835–36, 2:676; Wakley's "puffing": *L* 1835–36, 2:566, 610–11, 646–48, 675–78, 789–91, 844; 1836–37, 1:21.

57. *L* 1835–36, 2:789–91.

58. Clarke 1874a:68.

The hospital surgeons took a jaundiced view of Wakley's "particular patronage," and they did not care much for Grant's "heavy course."[59] Revealingly, conservative reviewers tended to avoid all philosophic commentary on his lectures, unlike the appreciative reformers. The *Gazette* sported a safe empirical and practical approach, the sort fostered by the conservative zoologists who equally feared profligate philosophy. It pictured the French-inspired teachers as tainted by theory, implying an unpatriotic extravagance. Thus the *Gazette* ran editorials not merely against the Frenchness of the university, but against academic anatomy per se. The university's Continental teaching arrangements and curricula were anathema to the London surgeons. They pointed to Germany, where anatomy had degenerated into philosophical zoology. "There, while the theory of medicine and all its collateral sciences are carried on to an extent that is never thought of here, yet a more efficient body of practitioners cannot be found in any intellectual country in Europe." Patriots pinpointed the precise aspect of theoretical anatomy that they held objectionable. In the *Gazette*'s words, the "collateral" sciences should be taught only if they can be rendered "useful in practice."

In anatomy, for instance, the illustrations should be drawn from the sick bed, and from the operating theatre and dead-house; and to make way for these, let all the doctrines of *homologues,* and *heterologues* . . . be banished with the imaginary analogies of *ptereal* and *herisseal* bones, and the theories of *morphological* and *histological* development; for in what mortal sick-room would these find their sphere of usefulness.[60]

The paper was promoting the hospital surgeons and their practical on-site approach. The Paleyite surgeons, whose position and earnings depended on an expensive apprenticeship system, stood to lose financially from an extension of the university's academic approach, politically from the surgical franchise, and morally from an anti-Oxbridge morphology. Since all three were embodied in the Wakley-Grant alliance, they had sound reasons for opposing not only the professorial system, but even the esoteric aspects of Grant's "impractical" anatomy.

Materialist-transformist lectures emanating from the godless college might be expected to have had a political appeal, and the fact that they

59. "Outlines of Comparative Anatomy," *MG* 1834–35, 15:808–9; "Medical Professorships," *MG* 1835–36, 18:782.

60. "Necessity of Lectures Being Practical," *MG* 1838, 22:778, 776. The demand for practicality cut both ways, and radicals in turn slated the Oxbridge schools for leading "English education out of this practical direction"—for using "ornamental decorations" such as the dead languages to prop up "antiquated systems of education" divorced from the real business of life: *L* 1838–39, 1:908. On zoological empiricism: Desmond 1985a:167 *passim*.

were published and praised by radicals confirmed the conservatives' worst fears. It all goes to explain Grant's basting in the *Gazette* for luring students down the radical path. Just as Wakley as part of his political campaign lauded Grant's "brilliant course," so the elite surgeons were defending their civic interests by impugning it. It was precisely because these political lines were so clearly drawn that we can relate the scientific issues to these larger social movements.

The Radical Medical Unions

We have failed to assist each other. Tradesmen, and members of other professions, help themselves, for the general good of all. Why do not we?

—An organizer of the British Medical Association urging joint action[61]

The London College of Medicine was not the only democratic forum open to radical anatomists. GP activists and demagogues from the university, private schools, and statistical offices also worked closely in the proliferating medical unions. These became melting pots of Continental science and democratic politics, and a perfect milieu for the new mechanistic morphologies. To a large extent the initial success and eventual decline of an extreme Geoffroyan science can be correlated not only with the rise of the London University and private schools in the 1830s, but with the first flourish and final collapse of the radical medical unions.

After 1830 new GPs' associations came and went with the frequency of their working-class counterparts. "Union is power" was not a slogan restricted to the burgeoning trades organizations following the repeal of the Combination Acts in 1824. It was an evocative cry in radical medicine.[62] (Tellingly, Wakley's first major speech in the House was an impassioned plea for the repatriation of the Tolpuddle martyrs—the Dorchester laborers transported in 1834 for union oath-taking while resisting a wage cut.) The early associations ran the gamut from respectably reformist to aggressively radical, with the latter resorting more to union muscle to force through corporation reforms. Since the militant associations also housed the British Geoffroyans and Lamarckians, it is these I concentrate on here.

The early moderate associations sought only to raise the GPs' status.

61. "Public Meeting," *L* 1836–37, 1:599.
62. "County Medical Association," *L* 1832–33, 2:574. Wakley's Tolpuddle speech: *Hansard* 1835, 28:1235–71; Sprigge 1899: chap. 29.

They were concerned with professional goals, the practitioner's social image, and the "dignity of the art."[63] The Metropolitan Association of General Practitioners (f. 1830), for example, was formed to promote "the prosperity and respectability" of GPs by capitalizing on Lord Chief Justice Tenterden's ruling that practitioners could charge customers for their labor.[64] This was lauded by radicals, who went further and urged that GPs be paid for attending inquests, signing health certificates, and advising at dispensaries. But they were worried by the association's lack of political bite and urged it to become a real "COMBINATION" and "organize COOPERATIVE branches throughout the country."[65] Talk of cooperatives was of course anathema to gentlemen practitioners, sounding as it did of socialist collectives. But it was typical of Wakley's language in the *Lancet* and at the militant National Union of the Working Classes, which he chaired in 1831. The Metropolitan Association's peripatetic equivalent, the Provincial Medical and Surgical Association (PMSA) (f. 1832), was more successful. Being a professional gentleman's reformist club and concerned with "honour and respectability," it was supported by the *Gazette*,[66] which also promoted its traveling twin, the British Association for the Advancement of Science (BAAS) (f. 1831). Like its richer relation, the PMSA invited local dignitaries from the cities it visited, for instance, providing a platform for the geologist Rev. William Buckland when it visited Oxford in 1835. Radicals made no bones about detesting this deferential approach. As the thirties progressed, they began accusing the PMSA of betrayal, slating it as a "disgraceful ABORTION"[67] and picturing the provincials as the unwitting dupes of the corporations.

What did bring the provincial moderates and metropolitan radicals together, and cause a massive wave of medical combinations in every county, was the passing of the Poor Law Amendment Act in 1834. The act was designed to end outdoor relief and force the "genuinely" sick poor into the workhouses. These were made so abominable through physical discomfort, family breakup, and prison regime that none but the chronic

63. "Association for the Promotion of Science," *MG* 1832–33, 11:187.

64. "General Practitioners," *L* 1829–30, 2:738; 1830–31, 1:52–54; *MCR* 1830, 13:270–71; also 1830, 12:486–87, 518–20. On radical demands for higher pay for the GPs: *LMSJ* 1833, 2:342; 1835–36, 8:56, 88; 1836, 9:633–34.

65. "Society of General Practitioners," *L* 1829–30, 2:505–6. Its more radical members were recruited by the LCM in 1831. Wakley's role in the NUWC: Lovett 1920, 1:72, 75.

66. "Association for the Promotion of Science," *MG* 1832–33, 11:186–89. The PMSA (the origin of the modern British Medical Association) is discussed in Little 1932.

67. "Provincial Medical Association," *L* 1836–37, 2:697, also 593–96, 722–25; and 1835–36, 2:608–10. On Buckland at the 1835 meeting: *L* 1834–35, 2:551–59; Morrell and Thackray 1981:288.

would endure them. By keeping more people at work, it was reasoned that competition would increase and wages decrease, in line with the low workhouse relief. Cobbett railed against this "Malthusian bill designed to force the poor to emigrate, to work for lower wages, to live on a coarser sort of food."[68] Resistance to the "Starvation Law" among laborers was violent, and riots broke out in the Home Counties when the commissioners arrived to organize the workhouses. Wakley remonstrated that "the strength of the labourer is *his* property," yet it was the only form of property "denied the protection of the law," in the form of wage-adjusted relief and medical provision. And he prophesied that the act would produce "a convulsion of society from one extremity of the kingdom to another," for no land or "property can be secure in the rural districts, when it is surrounded by an enraged and starving population."[69] But the act also affected the medical community, since putting medical contracts for workhouse consultancy out to tender forced down bids. Medical reformers of all shades deplored the workhouses for degrading pauper and medical attendant alike. Wakley presented GPs' petitions to the Commons, pointing out the "ruinous and cruel consequences to the poor who are farmed out to the lowest bidder."[70] The act led to a burst of union activity among parish GPs, with anti-Poor Law associations springing up throughout the country, followed by attempts at amalgamation into a General Association to lobby Parliament.

The major radical union of the period was the British Medical Association (BMA), launched in Southwark in 1836 by Grant's friend from Edinburgh days, the ex-army surgeon and Dulwich practitioner George Webster. From the start it had a strong caucus of South London radical GPs, many workhouse or prison attendants, and those closely concerned with the poor.[71] It was originally designed to protect GPs from "the assaults of Poor Law Commissioners," but it soon came to espouse more militant aims: the establishment of democratic medical government, destruction of all "degrading distinctions," the "perfect union of all ranks," and abro-

68. Edsall 1971:1, 7, 14, 18, 21, 27–31. E. P. Thompson 1980:294–95, 334–35, 379–80, 904.

69. "Lord Brougham on Poor Laws and Dispensaries," *L* 1833–34, 2:667; *MCR* 1835–36, 24:590–93.

70. *Hansard* 1836, 34:667. Anti–Poor Law associations sprang up immediately in towns all over Britain: *L* 1834–35, 2:363–64, 388, 395, 454–55; 1835–36, 2:393–94.

71. The BMA swallowed up the older Southwark General Medical Practitioners' Society, which had been formed in 1832 to prevent "interference" to GPs from the physicians and surgeons: *L* 1832–33, 1:27–28. Some BMA stalwarts such as Edward Doubleday were workhouse medical attendants; others, for example R. L. Hooper and John Lavies, were prison surgeons.

gation of corporation privilege.[72] As a result it quickly adopted a confrontationist stance toward the councils of the Royal Colleges.

Being a "class"-based union (in the medical sense), the BMA barred corporation men and consultants from its ranks, and in doing so it alienated the press and university moderates (who caused raucous laughter at the first meeting in Exeter Hall by warning delegates against "plunging at once into a sort of radical reform").[73] The tide had turned against moderation: one speaker said that he hated the very name "physician"; others (to more laughter) urged the "annihilation" of the higher ranks. The *Gazette* not surprisingly denounced it as "a mere political union, distinguished by all the vices of ultra radicalism," and composed of a knot of disaffected and disagreeable GPs.[74] But the *London Medical and Surgical Journal* applauded it and agreed that its council should be restricted to GPs. However, the union's publicist par excellence was Wakley (who also sat on the council). The *Lancet* quickly assumed the role of propagandist organ for BMA policy. Wakley contrasted the association's bold demands for one faculty and equality of title with the "truculence and servility" of the PMSA "prattlers."[75] In turn, Grant toasted Wakley at BMA dinners for resisting intimidation and insult, and the *Lancet* for its decade-long crusade, defending principles now enshrined in the association's manifesto.

The union soon attracted university and private school teachers. Scots-trained secularists and Nonconformists joined the GPs on the council and often acted as their spokesmen. High-profile recruits included Grant, his close friend (and Webster's partner in practice) Marshall Hall, the Webb Street teacher Richard Grainger, statistician William Farr, London University surgeon Robert Liston, and that cosmopolitan Italian nationalist Augustus Granville. Among the members, boasted Webster, were "men, second to none, as anatomists, physiologists, pathologists, and surgeons, and who have acquired European reputations."[76] He denied charges that the union

72. "Public Meeting," *L* 1836–37, 1:595; also 173. Webster was one of Grant's "oldest friends," and as secretary to the fund-raising committee for Grant's "testimonial" in 1853 spoke of his "moral and intellectual worth": *L* 1853, 1:141.

73. Anthony Todd Thomson speaking: *L* 1836–37, 1:601–2, 603–5.

74. "The *British* Medical Association," *MG* 1836–37, 19:660–63; *LMSJ* 1836–37, 10:737–40.

75. "British Medical Association," *L* 1837–38, 1:89–90. Grant flattered Wakley at the 1838 and 1839 dinners: *L* 1838–39, 1:84; 1839–40, 1:98.

76. "British Medical Association," *L* 1837–38, 1:103. Farr, Grant, Hall, Liston, and Wakley were elected onto the council in 1837, and Granville and Grainger shortly after. Hall, Grant, and Granville became Webster's vice-presidents in 1839: *L* 1837–38, 1:103, 174; 1839–40, 1:18; 1840–41, 1:158. Grant acted as chairman at the 1838 dinner: *L* 1838–39,

did nothing for science; that nothing was talked of but abuses. He would put forward a bold challenge to those who made the charge; he would engage to produce out of the Association six members who had done as much for the science and practice of medicine as any six gentlemen belonging to any other society in London. It was ridiculous to assert that science and a desire for reform could not exist in the same person. Their worthy Chairman [Grant] was a living instance to the contrary. (*Cheers*.) Who had more thoroughly devoted himself to science than Professor Grant? Who had been more forward in the cause of medical reform than Professor Grant? By reforming the profession they would promote the cause of science in an eminent degree; they would remove the trammels which the corporations had placed upon it.[77]

The rhetoric reached a pitch on the subject of corporation science. The Augean stable had to be cleansed before science could be properly pursued: "The monopolists regard their legal privileges only as hereditary stepping-stones to personal fortune, and the means of family advancement—(*applause*)—prostituting science to private purposes."[78] Rolling privilege, heredity, wealth, and the perversion of truth into one rhetorical bundle had its polemical advantage, but the demagogues were wielding a double-edged sword. The danger lies in seeing the Tories fighting to preserve their commercial advantages, with the radicals actuated by higher ideals. In fact the radicals' rival science was itself so intertwined with the political disputes that protagonists could portray it just as easily as prostituted in the pursuit of power. The Wakleyans' reforms would have given the private teachers and GPs the economic edge in the profession, and few Tories doubted that the radical anatomies, which undermined the moral claims of the medical elite, were subordinated any the less to political exigencies. The social and scientific goals of BMA members, their close personal ties, and the cross-linking of issues easily enabled protagonists to conflate the BMA's political and scientific stands. This is the most interesting aspect. Of course, all of this presumes that BMA science was quite distinct from that taught in the corporations. It was. The BMA rad-

1:80–82 for his reform speech. He drafted his pupils into the union; one, Nathaniel Eisdell, was sitting alongside him on the council in 1839. Grant was elected an honorary councillor in 1840.

77. "The Second Anniversary Meeting," *L* 1838–39, 1:83. Grant was deeply embroiled in BMA politics by this time. He was part of a BMA deputation lobbying Warburton at the Commons in 1838: *L* 1837–38, 2:336. And with Granville, Hall, Webster, Farr, and others he called on Lord John Russell at the Home Office to urge a reform of the Royal Colleges in line with the 1835 municipal reforms: *L* 1837–38, 2:410–13; P. B. Granville 1874, 2:274–75.

78. "Report of the First Anniversary Meeting of the British Medical Association," *L* 1837–38, 1:64.

icals were championing three new approaches, all related and all eliciting an antagonistic response from the conservatives. These were as follows:

1. There was a general insistence on French morphology. Grant and Grainger led the university and private school campaigns to institute a new-style academic anatomy based on unity of plan. As we have seen, they met opposition from the hospital surgeons, whose power, patronage, and profits depended on the existing apprenticeship system. The science was denounced at the College of Surgeons as impractical, mechanistic, and irresponsible in shutting out "design."

2. Related to this was a united defense of Marshall Hall's reflex arc. It was related because Grainger (like Grant) used Geoffroy's theory of homologies to establish the identity of the vertebrate and invertebrate nervous systems, and then went on to show that researches on the insect's "cerebro-spinal axis" and reflex system supported Hall's evidence for a spinal reflex arc in man.[79] Thus the new comparative anatomy became a cornerstone of Hall's neurophysiology (see fig. 3.4). Hall's neural arc was mechanistic in its operation. He advocated a structurally distinct reflex system, which required no consciousness for its control. His provitalist critics found this abhorrent, and the debate became intimately connected with controversies over the ultimate explanation of life.[80] With no Divine powers or principles "activating" neural matter, no consciousness guiding neural function, the soul became suspect, and worse, self-regulation became a prospect. BMA radicals were the major source of social support for Hall's doctrine. Grant championed it in class and at the Royal Society, Wakley in the press, while Grainger sought direct anatomical evidence for the arc in the gray matter of the spinal cord. These men had strong personal ties: Webster, Grant, and Hall had been at Edinburgh together and remained "true and affectionate" friends;[81] Grant sometimes traveled to France with the Halls, and Webster would meet them for bowls on the lawn. They were also close colleagues, with Grant teaching at Hall's Sydenham College, and Hall himself seconded to Grainger's Webb Street school at the time of the reflex controversy (1837–39). There was, then,

79. Grainger 1837:105–9; Jacyna 1984a:75–78.

80. Leys 1980:8, 34, 51, n. 13; Manuel 1980:146–53; C. Hall 1861: chaps. 4 and 5. R. Smith 1973:84–86 discusses the interest in the reflex arc on the antiestablishment fringes, and its relevance as a "regulative principle of the continuity of nature" to the theory of evolution.

81. C. Hall 1861:224, also vii, 21–22, 443; Grainger 1837: chap. 3; Grant's and Wakley's support: *L* 1846, 1:391–93, 418–20; 1850, 1:88. Another of Hall's BMA supporters was Grainger's Benthamite brother-in-law George Pilcher (1840–41, 1:666). On Sydenham College: Grant to C. C. Atkinson, 22 September 1837 (UCL CC 4166). BMA activists exhibited a classic high grid/group profile (Oldroyd 1986).

strong mutual support within the union, and a shared group defense against external "threats" (usually emanating from the *Gazette*).

3. Finally, a radical environmentalism manifested in different ways among members of this group, most obviously in Grant's Lamarckian biology, Farr's social policy, and Wakley's cultural determinism. Grant imported an environmentally based zoology, teaching that climatic changes caused by the earth cooling had triggered the serial generation of higher, more hot-blooded forms. His pupil William Farr applied this environmentalism to human social conditions. Farr was a Shropshire farm laborer's son; he had received a Dissenting education and had visited Paris on a legacy at the time of the 1830 Revolution. While at London University in 1830–32 he was inducted into Wakley's coterie. Like all the radicals, he argued that misery and disease were the consequence (rather than the cause) of urban squalor, and he condemned the government's failure to check these environmental defects. He used his work in the statistical office to argue for greater state intervention to ameliorate working-class conditions.[82] This "social Lamarckian" outlook also influenced his later anthropometric studies, designed to prove that racial degeneration was an inevitable result of poor living conditions.

BMA science and politics were thus closely aligned, which is not surprising. For a long time mechanistic physiology, Lamarckian environmentalism, and organizational autonomy had been associated with radicalism. That "superannuated . . . old dame the Quarterly"[83] had not been the only Tory review to see flaming democracy heralded by the arrival of a reductionist physiology. But what finally riveted the contingent political and scientific aspects of BMA policy together (at least in the eyes of detractors) was the disputes engaged in by BMA members with the conservatives. These disputes kept the camps polarized, ensuring, for example, that the *Lancet* and *Gazette* supported the rival socio-scientific packages in their entirety. The souring of personal relations in this period so alienated the *Gazette* Tories that they were prejudiced against all BMA science even before examining it. For example, when Hall and Grant in the *Lancet* denounced a former university pupil for plagiarizing Grant's work on the insect nerve column, the *Gazette* immediately leapt into the breach.

---

82. Cullen 1975:36, 38; Eyler 1979:1–4, 6, 23–24, 27, 124, 155–58, 198–200. Interestingly, the census work of the Department of Public Records (which was carried out to provide Benthamites with the data necessary if they were to manipulate society correctly) itself provided a model for botanical "statists" such as Hewett Watson in his studies of plant demography. Watson expected his botanical statistics to highlight the environmental causes of plant distribution which would in turn allow a more accurate Lamarckian explanation (J. Browne 1983:66, 77).

83. A. B. Granville 1830:2.

Figure 3.4. The ambitious, Edinburgh-educated Marshall Hall joined the chorus of cries against the RCP's Oxbridge exclusivity. In the 1830s he became a fellow traveler with the Scottish radicals, who championed his reflex arc. His political broadsides from a BMA platform did much to clear away obstacles to his own career advancement. By J. Holl, after J. Z. Bell. (Courtesy Wellcome Institute Library, London)

The *Gazette's* own knee-jerk reflex was set off solely by personal dislike—the editor admitting that he had long deplored the conduct of the gentlemen in question.[84] Only subsequently did the *Gazette* shift to an outright attack on the originality of Hall's doctrine (which Grant's work on insect nerves supported), extending the controversy into the realm of neural science. The conservatives supported a traditional interpretation, accepting a conscious reflex more compatible with the vitalist philosophy of Hunter's disciples in the hospitals. These political coalitions, with the conservatives protecting their hospital interests and denouncing the democrats' radical anatomy, highlight the degree to which personalities, politics, and science had become inseparable—how indeed professional and commercial advantage was seen by all as the prize for the successful physiological-political strategy.

The BMA's annual orations reflected the anti-Poor Law, anti-corporation, anti-Malthusian interest of its members: Granville in 1838 teased apart the testimonies received by Warburton's committee; Farr in 1839 statistically disputed the Malthusian principles on which the New Poor Law was based. The suave Hall, his reflex discoveries slighted and his Scottish credentials a bar to advancement, pleaded for greater rewards for talent and an equal deal for Scottish graduates. With so many of the BMA's leaders trained in Edinburgh, London, and Paris, its bitterest criticisms were reserved for the RCP's Oxbridge fellowship restriction, particularly as Oxford and Cambridge were "notoriously [the] most inefficient medical schools in existence."[85] Despite rumors that the RCP was to extend its fellowship eligibility, reformers inside the college failed to get the bylaws amended in 1836. As a result the BMA placed the college's discrimination high on its hit list. Clearly some, notably the ambitious Hall, were attempting to dig out channels for their own advancement; but others, like Grant and Webster, were doctrinaires and took their opposition to the colleges to tragicomic lengths. Webster snorted indignantly that the fellows of the RCP "are taught to plume themselves upon the supposed superiority in having been educated under the *moral restraints!* and *pious discipline!* of Oxford and Cambridge."[86] Grant's address in 1841 was a bruising mixture of satire, solicitation, and invective aimed at those "hot-beds for the development of all the higher vices," the corporations. He too saw the RCP's exclusivity designed to "crush medical dissenters,

84. *MG* 1838, 22:72, also 40–47, 93–96, 128, 160, 248–49, 252–54; Manuel 1980:152–53.

85. Grant speaking: *L* 1838–39, 1:81. Addresses: "Dr. Granville on Medical Reform," *MCR* 1839, 30:282–84; Farr 1839–40; Hall 1840–41; Grainger 1842–43d.

86. *L* 1836–37, 1:595; *LMSJ* 1835, 7:152–53; 1836, 9:153–54.

or Scotch Graduates . . . and to forward the interests of the English Church."[87] Hall's flourish was characteristic:

Can anything be imagined more preposterous, more iniquitous, more *immoral*, than this mingling of sacred things with profane, of religious with medical distinctions and privileges? Of religion it is a mockery; it is hypocrisy; it is the offering of bribes and temptations to act with insincerity; it is intolerance; it is, in a word, the same fire which consumed the bodies of our fellow-men in Smithfield![88]

The impassioned speeches were monotonous at times for their unrelenting demands. At others they were pure comedy. A delegate would have the audience in stitches, marching up and down, twitting that Waterloo veteran George Guthrie, the new president of the College of Surgeons:

One goose, alone, of the name of Guthrie, who (they say) loves siller or pelf more than principle, waddles about the citadel, and gabbles out *no surrender*, and, fancying himself a duodecimo edition of the illustrious Wellington, declares that no reform is needed; that the corporations are mirrors of administrative excellence; that the oligarchy of which he is head is the least rapacious of its species. (Laughter.) Advance to the attainment of your rights; cry justice, reason, experience. Halt! says Corporal Guthrie; stand at ease: it would be democratic, it would be dangerous; it would lessen my fees, limit my power.[89]

The pauper press had no monopoly on ridicule.

The later 1830s and early 1840s were not only a period of widespread unrest (173 petitions calling for medical reform were presented to Parliament during the spring 1840 session alone). They were also one of radical cooperation, marked by the series of sixteen conventions hosted by the BMA at Exeter Hall in Spring 1841 and attended by representatives of eleven reform associations (meetings to coordinate strategy and advise the M.P.s Henry Warburton and Benjamin Hawes, both of whom were preparing "leveling" medical bills). But the doctrinaires' remedies met with a mixed reception from the reformers generally. When Grant called for suffrage, annual ballots, and bylaw approval by the membership he was backed by the free marketeers, who agreed on the need to open up posts to competition. Yet when, like other Francophiles (Farr, Wakley, King, and Grainger), he suggested placing medical affairs under the home secretary, he alienated medical Tories and freetraders alike. The prospect of another Benthamite bureaucracy appalled many. The *Medical Times* applauded the talk of inalienable rights, but deplored such "visionary" solu-

87. Grant 1841:6, 49, 56–57, 73; *L* 1840–41, 1:144–48, 556; 1841–42, 1:163.
88. "Poverty and Religious Bigotry of the College of Physicians," *L* 1840–41, 1:557.
89. Jordan Lynch, in *L* 1840–41, 1:142.

tions: monopolistic practices had to go, but making medicine an appendage of "the State in the same way as the Army and Navy" was arrant nonsense.[90]

The radicals were isolated on the question of state control. But on wider electoral matters the free traders and bureaucratic radicals worked together both inside medicine and out. The medical demagogues did not restrict their attacks to the Royal Colleges. They were also active inside the learned societies. Here there was no GP rank and file to provide automatic support. Free-trade alignments were therefore all the more important. At the courtly Zoological Society, for example, medical reformers joined with Whig retrenchers and other critics of the gentrified council. Unlike the older corporations, the newer societies had ballots (at least for lower-tier posts), yet in the Zoological Society aristocratic domination, accountability, and scientific direction remained causes of contention—as did the Tories' priorities at the British Museum and Zoological Gardens, where wealthy promenading needs were put before science and utility. Knowing Grant's political allegiances and Lamarckian proclivities, we are now in a position to evaluate his clashes with the gentlemen of science: the patriarchs of the Zoological Society and British Museum in 1835–36, the Royal Society placemen from the later 1830s, and the Geological Society conservatives in 1838–39. These gentlemen and noblemen were concerned that the right sort of science should go on public show. Hence those institutions that specialized in scientific exhibition—the Zoological Gardens and British Museum—make good subjects for study. The following two episodes show how the oligarchs of old science confronted the mandarins of the new, how they contained the radical threat with minimal compromise, and how the new Lamarckian sciences were identified as legitimators of the republican order.

Reforming the Zoological Society, 1835

Had Professor GRANT confined his invaluable labours in the Society solely to the scientific departments of the institution, it is possible, just barely possible, that his presence might have been endured by the jobbers; but the honourable zeal of this really great man for the general interests of the Fellows, having led him, with other conscientious gentlemen, to take a part in the pecuniary management of the establishment, he has brought down upon himself the hatred

90. *MT* 1841, 5:79. Grant 1841:89, 90; Grainger 1842–43d:231. The 173 petitions: *L* 1840–41, 1:116. The 16 conferences ran from 1 March to 6 April 1841. Grant was greatly in evidence and spoke for the host union here: *L* 1840–41, 2:56–58.

and malignity of the entire band of mercenary speculators in the
funds of the Society.

—Wakley fulminating at Grant's removal from the
Zoological Society Council in 1835[91]

The "inner cabinet" of the Geological Society has been the subject of historical focus for many years, and rightly so, for it was extremely influential. From its ranks, for example, were drawn the spokesmen for that
self-styled "Parliament of Science," the British Association for the Advancement of Science. Jack Morrell and Arnold Thackray have characterized this star chamber as a close-knit group of Oxbridge dons and wealthy
London allies, gentlemen specialists who channeled funds to their friends
to ensure an acceptable Anglican science. The lesser known Zoological
Society (f. 1826), although its personnel and policies were very different,
provides an insight into the sort of democratic challenges that caused
these established elites to tighten their grip on science. Events at the
Zoological Society reveal the tensions and frustrations as medical and
merchant reformers invaded the zoological preserve, pushing, like their
Commons heroes, into an aristocratic arena. Roy MacLeod pictures the
Royal Society as moving from absolute to constitutional monarchy during
these years. For the Zoological Society too it was a case of increasing the
members' power and making the managers' more accountable. The metropolitan societies experienced dramatic changes in their social composition in the 1830s. New professional and trading interests were represented. Benthamite lobbies began to be heard. Nonconformist gains
threatened old Tory loyalties with the newer gods of utility, competition,
and capitalism.[92] The Zoological Society shows how passions could flare
up as these urban reformers now tried to reformulate the society's policy.

Sir Humphry Davy conceived the society about 1824 as a cultural
showcase for Britain's "Colonial Possessions" and a visible affirmation of
London's global preeminence.[93] He envisioned a vocational club for both
the sporting nobleman and zoological specialist. Top among his priorities
was the importation of exotic game, to tempt the palate of his aristocratic
patrons with foreign delicacies. In exchange for their dues, Davy told the
home secretary Sir Robert Peel, country gentlemen would get their pick

91. L 1834–35, 2:263, 199.

92. MacLeod 1983:56–57. Morrell and Thackray 1981; Morrell 1976:139. On changing
social compositions: Desmond 1985a:234, 248; Berman 1978: chap. 4; and Allen 1986:41–42.

93. Davy to Peel, 13 and 21 December 1824 (BL Add. MS 40,371, ff. 96, 201); Miller
1983:36–37; Bastin 1970:370–71, 1973. Raffles at the Linnean Society: LS Zoological Club
MSS A: 28 April 1825; Raffles 1830:590. The older histories of the Zoological Society (Mitchell 1929; Scherren 1905) have aged badly.

of the society's fish and fowl to stock their private parks. This combination of game conservation/consumption and rational recreation appealed to the improving nobility, whose patronage got the infant society off the ground: Lords Lansdowne (elected president in 1827 after Sir Stamford Raffles's death), Stanley (president, 1830), and the enthusiastic Auckland played leading roles in negotiating government grants of land, acquiring animals, including His Majesty's Tower menagerie, and gaining the society a charter in 1829. Davy's cofounder, Raffles, home from the East with a cabinet of tropical beasts, was more interested in the consolidation of imperial gains. His recruitment of naturalists at the Linnean Society in 1825 meant that many of the working officers were narrow systematists, jingoistically urging the claims of the new "British zoology" against the French savants.[94] These working zoologists took office in the society's new museum (Lord Berkeley's town house was rented in Bruton Street). Here the imports were to be collated, and dissectors and catalogers were to provide a permanent record of the breeding and domestication program. So there was a diversity of interests among the gentlemen themselves, and those of the Linnean Society systematists were unlikely to be served by a nobleman's game park. There was a danger that the taxonomists' needs would come in a poor second to the gentry's fancy.

Davy's original plan, an ornamental game park tailored to the squirearchy, was never put into full effect. From the first his attempt to provide "practical and immediate utility to the country gentleman"[95] was compromised by the reformers' professional demands. There were attempts in April 1829 to implement Davy's program by buying a farm in Surrey, and breeding experiments were begun to improve the table quality of fish and fowl. The inspector general of taxes, Joseph Sabine, a horticulturalist and sympathizer with Davy's gentrified goals, became the farm's director. But he met with stout opposition. The decreasing numbers of gentry and clergy on the society's rolls and the influx of merchants, M.P.s, East India officers, and medical professionals led to a vociferous reform lobby that "maliciously" taunted Sabine. The new men objected that the farm was too costly, too far, and at odds with the "scientific purposes" of the society, and they urged that resources be shifted to the society's museum instead.[96] They harried Sabine so mercilessly that within a year he was

94. E. T. Bennett 1831:201; Desmond 1985a:174. The Windsor and Tower menageries (ZS MC 1: f. 462; 2: ff. 332–33). The imperial background to the development of the metropolitan menageries is explored in Ritvo 1987: chap. 5.

95. Raffles 1830:592. Gentlemen farmers are discussed by Secord 1985a.

96. ZS MM 1: f.75. The "malicious" attacks, Sabine's resignation, and the farm's collapse: MC 1: ff. 313, 407; 2: ff. 445, 453; 3: f. 217. Farm's origins: MC 1: ff. 151, 155; *Reports of the*

forced to resign, and by 1834 they had succeeded in shutting down the farm entirely.

The council's defense of the farm was halfhearted because its own priorities had almost completely reversed by 1830. Official efforts were by now concentrated on developing the landscaped Zoological Gardens in Regent's Park as a promenading space (see fig. 3.5 and 3.6), where the wealthy might poke their parasols at the king's beasts. Its success can be judged by the hyperbole of contemporary travelogues: "the most delightful lounge in the metropolis," one called it. Gate takings soared, making the society rich. The council embarked on a massive program of capital investment and expansion (spending over £100,000 between 1830 and 1836). For a while the gardens became synonymous with elegance and curiosity. The "carriages of fashionable London" would line its entrance. "At least as many," a foreigner remarked in 1835, "as drove up and down Longchamps" or paraded round Hyde Park.[97] But giving the gardens priority as a "raree show" for this "select and fashionable company" drew angry taunts from the reformers. The expenditure too raised a storm. Whig retrenchers were furious that the council in 1835 could spend £2,400 on importing four young giraffes (two-thirds of which went into refitting the steamship *Manchester* to bring them to London). The council's exhibitionist and stockbreeding emphases thus suffered mounting criticism through the early 1830s. Lack of accountability, and of any constitutional means to block these purchases, finally brought matters to a head at the turbulent 1835 council elections.

In 1830 the active zoologists formed their own management organization inside the museum. Named the Committee of Science and Correspondence, it reflected the strong East India Company interests of the museum's backers and their desire to coordinate the émigré network worldwide. The Bruton Street building was to become a model India House ruling over its own faunal empire. The committee's original brief was to elicit information and specimens from the colonies, establish military and diplomatic contacts, and advise naval officers embarking on surveying voyages. So its priorities too were at odds with those of the landed backers. Thus when it was asked to draw up guidelines for the importing of animals for "utility or exhibition," it reported that all animals and not

---

Auditors of the Accounts of the Zoological Society for the Year 1829, and of the Council, Read at the Anniversary Meeting, May 3, 1830 (hereafter cited as *Reports*).

97. Raumer 1836, 2:111; "lounge," C. Knight 1841–44, 5:257. Carus 1846:62–64. Giraffes/expenditure: Desmond 1985a:230, 245–46, n. 151.

Figure 3.5. The Zoological Gardens in 1831, when only members and their guests were admitted: "the most delightful lounge in the metropolis." By James Hakewill. (Courtesy Zoological Society of London)

just the exotic or edible were "desirable as objects of science."[98] The list it drew up, ranging from snails to platypuses, was obviously intended to whet the new specialist's appetite more than the old gentleman's palate. It is clear from the minutes that the zoological specialists wanted funds diverted to the museum. They desired a more scientific society, more "philosophical," with the radicals further demanding that this systematization of nature and collation of imperial gains should be used to benefit health and education. Zoology was no longer to provide the gentry with conversational gambits, or subjects to discuss and dine off. For the new codifiers of nature in Bruton Street, science was an evocative symbol—a sign of modernization, retrenchment, seriousness.

This "professional" lobby—medical zoologists, reforming M.P.s, merchants, and military empire-builders—now demanded a larger museum to replace the crowded Bruton Street premises. (With forty thousand specimens by 1829 its doors were already bursting.) The council was ada-

98. *Reports* 1831. On the colonial connection: Fish and Montagu 1976; Desmond 1985a:229–30.

Figure 3.6. Contrast the early tranquil scene with the situation later when the public was admitted to the gardens. (From *Illustrated London News*, 1866, 48:509; Courtesy Illustrated London News Picture Library)

mant that the new building should be sited within the gardens—to make it cheaper, but also to stop any geographical split that might benefit the dissident fellows. Reformers wanted it distanced from the "raree show" and closer to the libraries and clubs of central London. They forced a referendum and defeated the council. They then pushed it in 1836 into spending £2,000 refurbishing John Hunter's house in Leicester Square, giving them a museum with 460 feet of cabinet space right in the heart of the West End.[99]

The Zoological Society, now one of the richest sources of animal cadavers in the country, was naturally attractive to the medical teachers. Unlike the Geological Society (with its Oxbridge-trained "star chamber"), in Bruton Street the active academics and early lecturers were largely local reformers, many of them antimonopolists at war with the medical baronets. Elderly Joshua Brookes, his Blenheim Street school broken up as a result of the RCS's refusal to recognize private school certificates, was an early lecturer here. Marshall Hall first announced his reflex arc at the society

99. *Reports* 1830, 1833, 1836; ZS MC 2: f. 176; 3: ff. 397–409; 4: ff. 396–97.

(in 1832). Grant, as the university zoologist, was one of the society's most active managers in the early 1830s, sitting on and chairing a number of committees.[100] Time was costly to a teacher giving two hundred lectures a year, yet Grant delivered long gratis courses to the fellows. That he was willing to devote considerable time to the society's affairs while struggling financially at the university testifies to his commitment. Of course the menagerie and museum were very useful to him. He took his classes to Regent's Park, dissected squids in the museum, and in 1833 submitted eleven papers for publication in the society's journals.

His program, however, diverged profoundly from the conservatives'. The young College of Surgeons assistant conservator Richard Owen was also intent on making the society more scientific, but he had in mind a different kind of science. Moreover, while the radicals were concerned to make the managers more accountable, Owen opposed the kind of drastic democratization that would make them subject to electoral vagaries. In short, he saw the society moving in a different scientific and political direction. Grant and Owen were already competitors in the medical marketplace. They were also the only medical men to manage the Committee of Science and the only comparative anatomists to publish extensively on the society's specimens, making a direct comparison rewarding. Owen himself was a prominent administrator. Like Grant, he sat on the Publication and Museum committees, and he joined the council a year before Grant, in 1832.[101] Owen published even more extensively—fifty or so papers in 1831–35—on the zoo's dead mammals. So for these medical men the Zoological Society was a source of institutional power and prestige. Here if anywhere, as reformers tried to democratize the membership in opposition to the gentrified backers, we should be able to see how Owen's Tory politics and Grant's radical rhetoric translated into action.

Grant's Lamarckian and Geoffroyan anatomy were well known in Bruton Street. For four months, beginning in January 1833, he delivered the society's first major lecture course, on the unitary structure of animals, to a packed museum. It had a mixed reception. The reformers were presumably supportive, but the clergy echoed the *Gazette*'s fears. Grant was "amiable" and "conscientious" enough. But the Anglican entomologists who reveled in beetle minutiae saw little to recommend in his "philosophic" tone or transliteration of Geoffroy's terminology. He could be "eloquent & animated." But generally he was too pedantic, "too much given to *coin* hard words," the Rev. Frederick Hope wrote to Darwin,

100. ZS MC 3: ff. 107, 354, 436–37. ZS Minutes of the Committee of Publication (January 1833–37). "The Zoological Society," *L* 1826–27, 12:132–34, 421–23 for Brookes.

101. ZS MC 2: f. 383; 3: ff. 89, 107, 354.

then in South America.[102] The course was evidently as naturalistic as its university equivalent, for he spoke on the origin and duration of species and the changes wrought by environment and domestication. He talked of the "unity of plan" extending throughout the animal kingdom, the "gradual development" of animal forms, imperceptible gradations in the chain, and a confirming recapitulatory embryology.[103] Again, in his "Fossil Zoology" lectures to the society the following January, he probably discussed the continuous serial ascent of fossil life, although whether he mentioned his belief in a cooling earth "motor" or the Lamarckian "metamorphoses" of life is not known. Nonetheless the fellows' exposure to his thought was lengthy, and its scientific, political, and religious impact would have rendered him triply obnoxious to the Tory managers.

Even the subjects he broached would have caused Owen concern. In Bruton Street Grant tackled a favorite theme: the homologies of cephalopods and cartilaginous fishes. By positing a rudimentary vertebral column and limb skeleton in squidlike mollusks, he was endorsing Geoffroy's transition between Cuvierian *embranchements* and underpinning a transformist continuum. Owen by contrast weighed into the Geoffroy-Cuvier debate on the opposite side. In 1832 he used the RCS's newly acquired pearly nautilus to refute "the theory of the simple and unbroken series" and Geoffroy's doctrine of kingdom-wide unity, contradicting those who "have endeavoured to produce a semblance of conformity between the *Cephalopoda* and the *Vertebrata*."[104] Owen's opposing scientific stance made it no accident that his first public antitransmutatory statement, a lengthy refutation of Lamarck, Geoffroy, and the Parisian materialist Bory St. Vincent on the question of the ape's transformation into man, was ready for the society's *Transactions* in May 1835, the month Grant was balloted off the Zoological Society Council.[105]

The 1835 council elections finally brought a number of simmering issues to boil. Opposition to the patrician councillors had been growing for some time. That testy conchologist William Swainson denounced them as despotic "presiding judges."[106] The reformers' main grievances were the management's financial priorities, lack of accountability, and downgrading

102. F. Burkhardt and Smith 1985, 1:363. Even some in the medical profession thought Grant's lectures too gravely "philosophical." Few knew the subject better, one reviewer griped, but "no one could have made the subject less interesting, less attractive": "Outlines of Comparative Anatomy," Renshaw's *LMSJ* 1835, 7:208.

103. Grant 1833d. On the full house, K. Lyell 1881, 1:397. Also *Reports* 1834; *Zoological Magazine* 1833, no. 2, 61. Fossil course: ZS MC 3: f. 290.

104. R. Owen 1832b:1.

105. ZS MC 4: f. 158.

106. Swainson 1834:305; Swainson to Babbage, 10 January 1832 (BL Add. MS 37,186, f. 210.

of science. The attempt by the vice-presidents in April 1835 to oust the most outspoken critics, Grant and the M.P. Robert Gordon (a powerful secretary to the Board of Control governing the Indian colonies), forced such a surge of popular feeling that Grant's case became a cause célèbre. Wakley deplored the efforts of the "odious junto" to unseat the men.[107] He accused Sabine, a vice-president, of engineering the removal of opponents, and his Tory "clique" of administrative abuse. Critics had a case, for funds destined for the museum had been siphoned off for the gardens. As a tax inspector Sabine was, it is true, a natural Wakley enemy. Also he was already suspect, having been forced to resign from the Horticultural Society after an inquiry there, chaired by Gordon, had found him guilty of financial irregularity.[108] But the crux, of course, was that Sabine spoke for the Tory backers, who were more interested in a promenading park than research. Grant and Gordon had deplored Sabine's extravagances, urging greater exploitation of the society's scientific potential (that is, of their stronghold, the museum). Grant, the taxonomist and liberal M.P. Nicholas Vigors, and the retrenching East India Company colonel William Sykes were removed from the Publications Committee in April 1835. But the "junto's" attempts to remove Grant and Gordon from the council met stiff opposition. At the turbulent April meeting twenty members objected and substituted the names of three vice-presidents—two because of slack attendance (including the Tory magistrate William Broderip).[109] The third was Sabine. Reformers justified including him by arguing that the length of service should also count, and he had been on the council since 1826.

There was uproar as the officials aborted the meeting on a technicality. This gave the "junto" time to publish a self-justifying *Statement* in which they exempted the higher officials from democratic control and announced that the vice-presidential names would not be substituted. The officers thus remained able to direct the voting without themselves being subject to election. Accountability of course was an emotive issue, and this attempt to stave off management reform infuriated many. Grant was voted out in a ballot so rowdy that it caused comment in the *Times*. He was inclined to withdraw quietly, but Wakley stepped in:

As for Dr. GRANT, his character, his discoveries, and his great fame, render him the main pillar in the institution. Of course that gentleman will consult his own dignity by not attempting by any personal exertion of his own to thwart the mach-

107. *L* 1834–35, 2:389: Sprigge 1899:304.
108. H. R. Fletcher 1969:67–68, 120–24.
109. Lord Stanley et al. 1835: Appendixes A and C; ZS MM 2: f. 15; MC 4: f. 150.

inations, or expose the intrigues, of the despicable faction who have acted so ruinously for the interests and progress of the Society.[110]

In an over-the-top editorial Wakley lashed out at the "clique" for attempting to infuse the "satanic spirit" of the mob that had burnt Priestley's house "into the minds of the Fellows." At a stroke, he had conjured up an image of Church-and-king riots, Priestley's martydom, and Old Corruption, and he went on to tar Owen with the same brush for voting with the "junto."

In June 1835 a group of fellows and opposition councillors convened a Special General Meeting and again suggested that in the interests of harmony the executive should in future be guided in its choice "by a combined principle of length of appointment and non-attendance of its members."[111] The motion exempted the president, treasurer, and secretary, although it did leave the vice-presidents subject to the ballot. The conservatives now found this an acceptable compromise and it passed into law, despite filibustering tactics by some diehards. Thus a compromise was hammered out after Grant's removal. He was formally thanked for his services. Turning anything to advantage, Wakley interpreted this unconvincingly as a belated rout for the "junto," which had "been exposed and defeated."[112] In fact, after 1835 it was the reformers who were retreating. Grant, disgusted by the council's "offensive" behavior, began his withdrawal from the society. Marshall Hall resigned his council seat later that year. Sykes, too, had threatened that if the "junto" won he would resign, although in the event he stayed to fight on.

Rank-and-file criticism persisted through the summer of 1836. The new council immediately created a furor by its treatment of livestock as party gifts. It presented Lord Stanley with a pair of the zoo's ostriches for his own park, causing an outraged Sykes to try blocking the gift, arguing that the council had no right to pay off its friends with society property. To check this kind of abuse, he moved time and again for the right of members to nominate a number of councillors. The fellows too tried to obstruct the council's major purchases, insisting that all large disbursements must be subject to general ratification by the membership. But these fiercer demands were consistently ignored, leading frustrated fellows at the

110. *L* 1834–35, 2:199; *Times*, 29 May 1835; *Literary Gazette*, 2 May 1835, p. 280; 30 May 1835, p. 344. ZS MC 4: ff. 141, 145, 150; MM 2: f. 10. Owen's support for the "junto": *L* 1836–37, 1:766.

111. ZS MC 4: ff. 165–68.

112. *L* 1834–35, 2:390. "Biographical Sketch," *L* 1850, 2:694; on Hall's resignation, ZS MC 4: f. 254. Sykes's fears: *L* 1834–35, 2:200.

packed Annual General Meeting in April 1836 to table a motion denouncing

> the irresponsible powers assumed by the Council, and the unconciliatory and unbusinesslike proceedings which . . . have shaken the confidence of the society, and fully justify the nomination of the Candidates for the Council & officers, recommended by the Fellows for the year ensuing.[113]

Visitors commented on this continual barracking and the electric atmosphere at meetings. Charles Darwin, then fresh from the *Beagle* voyage, was appalled at the "snarling" when he visited the society late in 1836.[114] He refrained from playing any major part in its proceedings, opting instead for the more gentlemanly Geological Society. Yet 1836 effectively marked the passing of radical agitation, and further demands through the year fell on deaf ears. With the reform ranks depleted, the hand of Tory amateurs such as Broderip and anatomists such as Owen was strengthened, making the society at an executive level more representative of its antiradical backers. Owen and Broderip gave short shrift to the remaining "malcontents," bemoaning the "jaundiced eye" with which the radicals "look upon everything belonging to the Society" and warning in the *Quarterly Review* that their demands must cease.[115]

By 1836 a stable bureaucracy in Bruton Street had been imposed and the trade-off between landed, Broughamite, and radical interests was at an end. It was perhaps inevitable that the fight for electoral reform should have been so violent, with the society home to such antagonistic factions. The country gentlemen and the querulous radicals were not a good mix, here any more than in Parliament. Partly they were kept separate. It is significant that the society's holdings were so diversified, with the farm, museum, and zoo geographically as well as politically split, providing each faction with a discrete power base. Many radicals might have withdrawn in mid-decade, but their gains were tangible: the fellows had imposed further constitutional restraints on the executive, democratic mandating had reached the vice-presidential level, and the scientific fellows had a large, centrally located museum. The society of 1836 was not Davy's pocket borough of 1826, and even if ostriches were still acceptable party gifts, this kind of ancien régime corruption was disappearing.

113. ZS MM 2: f. 74, ZS. Gifts: MC 4: f. 253.
114. F. Burkhardt and Smith 1985, 1:514. Darwin presented his *Beagle* animals and birds to the Zoological Society (Sulloway 1982). Rudwick 1982 discusses Darwin at the Geological Society.
115. Broderip 1836:331; R. S. Owen 1894, 1:96.

The Exhibition of Science: Radicals and Museum Presentation

Management reform of a public body such as the British Museum proved far harder. The Zoological Society existed by subscription and offered its paying members professional advantages—use of the gardens, museum, rooms, journals, and so forth. Even if the officers were above election, the fellows formed a powerful lobby able to influence council action. Not only was there nothing equivalent in Montagu House (home of the British Museum), but there were even political disagreements over its actual purpose. The Church prelates and noble trustees considered that they held its literary, artistic, and zoological treasures in trust for the nation. The radicals, by contrast, looked to Paris to appreciate how a museum should function. Warburton's claim that their lordships had neither the leisure nor the understanding as trustees to promote the museum's scientific interests revealed his different understanding of the museum's role—as a research institution. Here we see how the country gentlemen fought all attempts to replace the noble trustees, introduce a Lamarckian serial display, and turn the museum into a Parisian-style teaching school. Believing that the museum should house national treasures, not necessarily advance knowledge,[116] they insisted that the nobility, as guardians of morals, manners, and Mammon, had a right and duty to include trusteeship in their public calling. The museum was therefore run like a pocket borough: appointments were in the gift of the principal trustees, led by the archbishop of Canterbury, and its librarians and curators were recruited "from the inferior departments of the church and public offices."[117] The reformers' demands for access for non-ticket holders, greater scientific utility, and above all a Benthamite board of directors trained in science had wider political implications. Boardroom specialists responsive to the new professional classes, respecting not wealth and rank but competition and talent, would break the Tory-church control on one more civic institution. Radicals and Dissenters, holding bourgeois briefs at odds with the rights of the hereditary peers, would bring the museum more squarely within the utilitarian orbit.

How demands for this professional autonomy were related to wider political goals can be seen from the machinations within the all-party Select Committee set up in 1835 to examine the running of the museum. This committee was chaired by the radical Lambeth M.P. Benjamin Hawes (a soap-boiler turned magistrate), and the politicized interrogation

116. Gunther 1980:75.
117. Warburton speaking in Parliament: *Hansard* 1836, 31:308–12.

of the expert witnesses and their tactical responses reveal the complex nature of the demands for scientific self-management.

The high Tory Sir Robert Harry Inglis, M.P. for Oxford University, examined for the Church-trustee party. He believed that wealth and rank were essential for trustees soliciting patronage for a public museum. He defended the titled trustees, their efficiency and record, and disputed the advantage of putting men of science on the board. He was committed to halting the erosion of Tory Anglican privileges in face of Dissenting merchant and metropolitan demands. He had voted against Catholic emancipation, the repeal of the Test and Corporation Acts, and parliamentary reform. By the time the committee published its report he was leading the Commons Church party against the "Dissenters'" bill for the civil registration of marriages (which would remove another Church monopoly)—observing that not since "the great Rebellion" had there been such an attempt to secularize the sacraments.[118] And in 1836 he resisted the first Whig bill on Church reform as "fatal" to the clergy's interests.

The radicals derided Inglis as a "sleek, oily, capon-lined man of God."[119] By contrast, his opposite number on the committee, Benjamin Hawes, was spokesman for the medical reformers. He concentrated on the benefits of scientific governors at the museum, in terms of bettering display, taxonomic accuracy, and systematic arrangement, presenting this as sound strategy to raise Britain's prestige abroad. Grant was Hawes's star witness, and as such the chief Tory target. Egged on by Hawes, Grant extolled the Parisian system with its savant-specialists, each in control of his own department. He agreed that the archbishop and noblemen—even had they time from more pressing public duties—were incompetent to head a scientific institution.[120] Fellow witness Nicholas Vigors deplored the lack of a single technically trained "commoner" on the board, and could point to only one trustee, Lord Stanley, who was a proficient naturalist.[121] Grant and Vigors here, as at the Zoological Society, were clearly putting the career zoologist's needs before those of the sporting nobleman. They argued for paid professional trustees and increased Treasury funding. Indeed they wanted natural history to have its own national museum. During Tory cross-questioning, and to allay fears that scientist-administrators might be biased toward their own departments, Grant proposed bringing in outside managers, while conceding that "men of high

---

118. *Hansard* 1836, 32:162; 1836, 34:491.
119. *L* 1840–41, 1:803.
120. *Report SCBM*, 22, 27, 29, 30–31; Gunther 1980: chap. 8.
121. *Report SCBM*, 118.

rank and station" might still conduct the "financial affairs."[122] Vigors wanted the trustees to be appointed by the learned societies. But Lord Stanley's son instanced the Bruton Street debacle as evidence that the zoologists were incapable of ruling themselves. This reinforced Tory claims that noblemen and gentlemen in the public eye were more responsible, being disinterested (in the other sense), while providing what no science specialist could—a social link to the museum's society patrons. The interrogation thus exposed the obvious party loyalties of both committee members and witnesses: Tories were preserving Church-and-Crown authority, with reformers intent on devolving management onto the professional classes. Again, the radicals were attempting to associate specialist hegemony with social progress; they were juxtaposing their ideal of open, competitive, scientific professionalism with the closed Church sinecures and hereditary privileges, seeing science as symptomatic of utility, talent, and national improvement.

What was expected to follow after the traditional managers had been deposed was even more revealing. Grant believed that with a salaried board, systematic rearrangement, and a set of instructors in the museum teaching along Parisian lines, "decently supported by our Government," we might yet bring forth our own "Lamarcks, our Latreilles, our Cuviers, and our Geoffroys."[123] But the country gentlemen would be damned before seeing any Crown institution spontaneously generate two of those named. The evangelical Old Etonian John Children (his own appointment as assistant keeper standing testimony to the value of manners over accomplishment) deplored Parisian profligacy. He protested "against the abominable trash vomited forth by Lamarck and his disciples, who have rashly, and almost blasphemously, imputed a period of comparative imbecility to Omnipotence, when they babbled out their puerile conditions about a progression in nature."[124] If by putting science on the board, and progressive nature on display, the museum was to raise up a generation of Lamarcks and Geoffroys, then radicalism condemned itself. Many of Grant's explicit criticisms fell at Children's feet: the museum's failure to exhibit a single fossil bivalve, coral, or cephalopod (even though it pos-

---

122. Ibid., 92. It was an article of Tory faith that, if the existing "social pyramid" were to be maintained, rank and wealth must never be subordinated to "intellectual qualifications." This was one reason why many opposed working-class education (Country Gentleman 1826:14).

123. Report SCBM, 127, 136.

124. Children to Swainson, 11 July 1831 (LS WS). On Children as a placeman: T. S. Traill to Swainson, 22 April 1822 (LS WS); and Traill 1823 on the clerical patronage in the BM; Desmond 1985a:171–72; Gunther 1978:82–84; A. B. Granville 1830:24.

Figure 3.7. A decade after the Select Committee took evidence about the state of the British Museum, the new "Coral Room" was finally opened. (From *Illustrated London News*, 1847, 10:221; Courtesy Illustrated London News Picture Library)

sessed the best ammonite collection in the country), the lack of labels and the Linnean antiquity of those remaining, and not least the haphazard arrangement. A systematically ordered collection was essential, and for Grant this meant an attempt to display "the whole continuous chain of beings, from the lowest corals up to the highest animal forms that exist,"[125] which was hardly a prospect to appeal to the Cuvierians with their discrete *embranchements* of life.

Grant's allegations provided Hawes with his leverage point. Hawes now questioned a succession of specialists on the best arrangement for display. (Only in the 1830s were zoological systems, managerial policy, and political control so ideologically intertwined as to become the leitmotiv of a parliamentary inquiry.) Grant tried to slip the Cuvierian yoke by suggesting that groups be ordered according to the findings of contemporary specialists. But J. F. South, teacher of comparative anatomy at St. Thomas's Hospital, believed that systematic neatness came from following

125. *Report SCBM*, 21.

Cuvier throughout, while Owen advised against any hasty concessions to taxonomic novelty.[126] (Although he was not averse to appointing specialists to the board.) Specific axes were also being ground. Grant and James Scott Bowerbank, a wealthy London distiller sharing Grant's interest in sponges, agreed on the need for a separate fossil department. In support Grant pointed out that in Paris a "stratological" series was used to great effect in illustrating the history of life.[127] Bowerbank also wanted a network of collectors on the payroll, those who could evaluate, barter, and assemble specimens. The Paris Muséum was far ahead in this respect, financing foreign collecting expeditions. (When Darwin arrived in Patagonia in 1832 he was chagrined to find that Alcide d'Orbigny had already been collecting there for six months.)[128] But the very idea of "trading" in shells and fossils was appalling to the gentlemen trustees; Sir Philip Egerton, at one with Inglis on the need to maintain Anglican privileges and standards, feared that such merchandizing would compromise the museum's dignity. (For this reason, the "uncultured" William Swainson, a customs collector's son and specimen seller, had originally been rejected in favor of Children for the assistant keepership, even though Swainson was an accomplished systematist and Children was not.)[129]

Even to talk of Montagu House in the same terms as the Paris Muséum was "ludicrous" in Grant's eyes:

In the Garden of Plants all the collections are extensive, well preserved, well exhibited, classified, and named by the first authorities in Europe; they have been increasing for more than a century; they are supported by large annual grants from Government, and they are directed and superintended by many of the most eminent zoologists living, who have each their particular departments, with numerous assistants under them. . . . The Zoological department of the British Museum is miserable in funds, miserable in science, miserable in materials, and its collections are for the most part without either classification or nomenclature; so that the Museum more resembles a store-house than a school of zoology.[130]

This so infuriated Inglis and Egerton that they went to great pains to discredit Grant's testimony. They belittled his achievements—Egerton claiming that he had published no "complete work upon zoology"—making any concession to "so slight an authority" unthinkable. Hawes angrily

126. Ibid., 49–55, 64.
127. Ibid., 73, 78–79, 130–33. Bowerbank later (1864–82, 1: dedication) acknowledged Grant's pioneering work on the Porifera; see also Bowerbank to Owen, 3 May 1863 (BMNH RO); and Grant equally enthused over Bowerbank's monographs: Grant to G. G. Stokes, 22 July 1861 (RS Referees' Reports 1863–65, RR.5 26).
128. F. Burkhardt and Smith 1985, 1:280.
129. Gunther 1978:84–85, 94–99; Desmond 1985a:171–72.
130. *Report SCBM*, 133.

reaffirmed Grant's preeminence, pointing to *Outlines of Comparative Anatomy* (the first four parts having been printed in 1835–36), only to see it weighed up against a risible Tory rag: Egerton dismissed the work as comprising "four small numbers not the thickness of a single number of Blackwood's Magazine."[131] The Tories called in Children to reaffirm London's advantages, a dealer for a more condescending assessment of the French collection, and Zoological Society officials to support Cuvier, the titled trustees, and the status quo.

The Tories had the better of it. Egerton and Inglis (a barrister by training) were forceful counsels, despite Hawes's periodic flourish and Grant's cutting accusations. Partly the conservatives' success reflected the strength of the patronage system: some of the naturalists questioned owed their livelihood to the museum or had relatives there; others were integrated into the conservative zoological or medical communities, were patronized by the Tory gentry (Owen was the favorite of Egerton and Inglis), or were from the gentry themselves (Egerton and the earl of Enniskillen both owned cabinets of fossil fishes). Obviously none of these had any truck with Hawes's or Grant's political radicalism or Lamarckian demands. Apart from the rather tetchy Vigors (who supported a "quinarian" or circular system of classification), almost all endorsed a Cuvierian conservatism. And while many conceded that scientific gentlemen might usefully assist the board, most took the line that titles guaranteed patronage and protection. Thus when the reformers at the end put forward a motion that twelve trustees should be elected by the London societies, the aristocrats defeated it. Lord Stanley proposed a (literally) patronizing compromise— that the trustees should themselves take the "opportunity . . . of conferring a mark of distinction upon men of eminence in literature, science, and art."[132] The Tories slipped a debilitating "occasionally" into the clause and carried the resolution. So the committee's recommendations, while they resulted in departmental rearrangements, left the power of appointment in traditional hands and did nothing to upset the administrative status quo.

It comes as no surprise that the more extreme recommendations of the Lamarckians and medical republicans were rejected at the museum and the Zoological Society, which were after all two of the courtliest institutions of science in London. What this episode does show, apart from the fact that these scientific exhibition centers should have been buffeted by

131. Ibid., 182.
132. Ibid., vi-vii. The only changes made in 1837 were a splitting of the Natural History Department into three branches (Mineralogy, Zoology, and Botany), a revision of salaries, and new allocations for purchases (Gunther 1978:99, 1980:84–86).

the new political winds at all, is that their respectable elites were now sensitized to the Lamarckian democratic threat. On the other hand, the greatest activity in reconstruction and scientific rationalization remained at the grass-roots level, with the growth of the radical unions and democratic colleges. These must remain our focal point if we wish to understand the appeal of the imported sciences. In this regard, there is one side to the democratic movement still to be considered—radical Dissent. So I now move on to a group of little-known Nonconformist "lecture bazaars" and continue the theme that it was these radical "manufactories" which provided the most fertile ground for the new materialist morphologies.

4

Nonconformist Anatomy in the Private Schools

Now these low fees are a great eye-sore to the Monopolists, because they see it is the practical commencement of knocking down *the golden bar of exclusion,* and they begin to fear that they do not stand on such a firm footing, upon their purchased ground as heretofore; it has been said by them, that I wish to bring tinkers and chimney-sweepers into the profession. . . . [But] when we recollect that a HUNTER, was a Carpenter, a NEWTON and a COBBETT, Plough-boys, and that the students least possessed with money are generally those most gifted with industry and talent, I am at once justified in lowering as much as is in my feeble power the golden bar, which checks the progress of that knowledge so intimately connected with the salvation of the lives of the people.

—Radical teacher George Dermott discussing his low fees and the
social class of his students[1]

London University was not the only teaching institution founded at this time. A number of new "private" or nonhospital medical schools emerged after the late 1820s. Wakley's "lieutenant" J. F. Clarke, in his gossipy account of medical lowlife in the 1830s, described London's private anatomy teachers as "outsiders" to the system.[2] He was perfectly placed to record their fears and foibles, even if, as George Dermott's tavern amanuensis (Dermott dictated notes in his local gin palace), his record does have a rather tipsy ring. But on his major point Clarke was right: London's private schools stood defiantly outside the hospital-corporation orbit. Surprisingly little is known of this sector of radical medicine. No studies exist of the schools' proprietorial ideology, influence, or clientele, essential as these are to understanding the shifting scientific patterns in medicine. The historical gap appears the more glaring now that these patterns are themselves beginning to show up. Pauline Mazumdar has pointed out

1. Dermott 1833:20–21.
2. Clarke 1874a:37, 296; Mazumdar 1983:231.

that the whole anatomical approach of the schools differed from the clinical instruction available in the hospitals. Here I suggest that the radical, ethical, and chapel ideologies of the disadvantaged and Dissenting proprietors made these schools breeding grounds for the new Continental anatomies established at the university. Later in this chapter I look at Dermott's school in Soho, one of the smaller establishments, staffed by radical ex-Methodist and chapel anatomists. It is ultimately to teachers such as these that we must turn if we are to appreciate how the new reductionist and serial zoologies were exploited; how they were given a special meaning and a new use within the "outsiders'" anti-Anglican strategies.

By 1830 the old eighteenth-century schools were dead or decaying, the survivors being forced out of business by the College of Surgeons and the arrival of the university.[3] Their place was taken by a new group of aggressively politicized schools, founded in private residences or converted blocks near the London hospitals. These medical manufactories with their cheap courses and humbler clientele captured a discrete corner of the medical market. Their proprietors formed a professionally and sometimes religiously distinct group. Many were active Nonconformists, handicapped in society and medicine, and suffering from religious and professional discrimination. Even after the repeal of the Test and Corporation Acts in 1828, which finally allowed Dissenters to hold public office, Nonconformist ministers were still barred from solemnizing marriage and burial services.[4] And the failure of the Universities Bill to pass the Lords in August 1834 meant that Dissenters remained excluded from the ancient universities and therefore from certain offices of the medical establishment in London. Their medical demands thus formed part of a larger Nonconformist agitation for religious equality. For many medical Dissenters the campaign for the removal of religious disabilities implied a further commitment—to the disestablishment of the Church and democratization of the corporations. Like the secularists, the Christian democrats were pounding on the corporation doors, demanding recognition and a say in the government of medicine: in effect, medical "emancipation." This only came in 1843 after twenty years of agitation, and then only in

3. The oldest survivors were the Great Windmill Street school (f. 1766), Joshua Brookes's Blenheim Street school (f. 1787), and Joseph Carpue's Dean Street school (f. 1800). Brookes retired in 1826; Carpue ceased teaching in 1830; and the Great Windmill Street school was finally dismantled in 1830 (Cope 1966; Power 1895; S. C. Thomson 1942, 1943).

4. John Epps was forced to marry in an Anglican church and bitterly resented it (E. Epps 1875:216–17). "London College of Medicine," *L* 1830–31, 2:214; "University Bill," *LMSJ* 1834–35, 6:57–59; Halévy 1950:168–69, 200.

the diluted form of a limited RCS fellowship. By that time many of the schools had collapsed, partly as a result of competition from the reorganized hospitals and university, but more obviously as a consequence of the RCS's manipulation of the law.

The Professional Grievances of the Private Teachers

The College of Surgeons decreed that when the Sun approached the tropic of Cancer, all anatomical knowledge then acquired possessed no kind of value, and that when that luminary was in the other hemisphere, anatomy was excellent.

—A sarcastic comment on the college's refusal to accept
summer course certificates[5]

The private school boom stemmed partly from the massive population growth of the early nineteenth century (which itself meant more sick people) and from the need of the expanding middle classes to place their sons in professional positions. But it also reflected the changing educational requirements of the profession itself. A demand for more teachers and courses was created by the Apothecaries Act of 1815, which stipulated lecture attendance for students intending to take up general practice. The student influx led to complaints in the 1830s that the profession was bursting at the seams. At the same time the cut-price enticements by the private schools caused some critics to call for a tax on matriculating students in order to cut their numbers and keep the working classes out.[6] Expansion characterized all aspects of the medical market. The new steam-powered presses issued a flood of new journals and books. According to one reviewer in 1830,

The number of works upon anatomy at present pouring from the press, argues an increasing taste or necessity for the study, despite . . . the impediments and difficulties that now beset it. We say an increasing call for anatomy, for no principle in political or other economy is better established than this, that the supply will always be regulated by the demand. There may indeed be an occasional glut, but a few bankruptcies or failures soon trim the balance.[7]

5. James Johnson speaking: "Westminster Medical Society," *L* 1833–34, 1:467.
6. "Tax on Entering the Profession," *MCR* 1830, 13:440–41. Tories decried the levelers who would remove all barriers and leave the profession open to "the irruption of the Gothic hordes": *MG* 1832–33, 11:90.
7. "Lectures on Anatomy," *MCR* 1830, 12:95–96.

HOSPITALS AND MEDICAL SCHOOLS OF LONDON 1836 37.

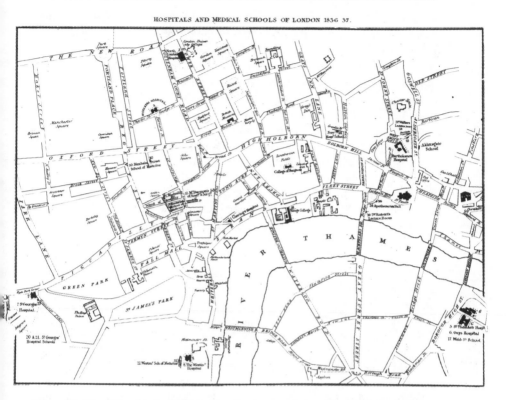

Figure 4.1. Map of London in 1836, showing the close proximity of the hospital and private medical schools. The College of Surgeons is near the center, the London University to its north. Among the private schools discussed, the Aldersgate Street school is east, near St. Bartholomew's Hospital, Webb Street is south of the river, and Dermott's and Brookes's are in Soho, between Oxford Street and Leicester Square. Comparative anatomy was taught in all of these schools. (From *Lancet*, 1836–37, 1: opp. p. 5)

The new schools were a case in point. They proliferated to meet the new needs in the 1820s and 1830s, only to collapse in the 1840s. The private teachers chose their venues carefully, usually siting their schools close to the established hospitals (see fig. 4.1). Indeed, some proprietors had been goaded into founding their own schools in the first place by the governors' refusal to let them lecture in the local hospital. Edward Grainger's offer was declined at Guy's, so in 1819 he set up in a dilapidated Catholic church in Webb Street. Frederick Tyrrell was turned down at St. Thomas's, leading him to found the Aldersgate Street School in 1825. Samuel Lane knew that the St. George's governors wanted no school on their grounds, so he opened rooms in Grosvenor Place in 1830. Seventeen private schools were recognized by the inspector of anatomy in 1832.[8] The proprietors had a variety of family backgrounds, many in trade, manufacture, or the professions. Some had humble origins. Samuel Lane was a tailor's son. Others came from the better-off middle classes: John Epps's father owned a chain of ham and beef shops in London. Still more came from the industrial regions. The Graingers were sons of a Birmingham surgeon and had "unpleasant" Brummie accents.[9] Dermott's father was a Northamptonshire Wesleyan minister. Another of Methodist origins was Marshall Hall. His cotton manufacturing family had joined Kilham's New Connexion in Nottingham, and their "Methodist Jacobinism" guaranteed their immunity from Luddite attacks. Exceptions of course existed. The cultured Dean Street proprietor Joseph Carpue was the son of a wealthy gentleman. But as a Catholic he too suffered civil handicaps, and he was equally snubbed by the College of Surgeons.

The presence of so many independent schools made competition with the hospitals inevitable. The private teachers undersold the pure surgeons, often cutting their charges in half. George Dermott admitted that he deliberately underpriced his lectures to draw students away from the hospitals.[10] Cost cutting was not restricted to courses. The new voice of the private teachers, the *London Medical and Surgical Journal*, was cheaper on a page-for-page basis than a penny magazine.[11] This cut-price medical education directly threatened the livelihoods of the hospital consultants. They now feared that, if not controlled, these down-market schools would vulgarize the profession by opening the floodgates to the lower orders. There was some basis for this fear. The teachers, like their radical heroes in Parliament, showed a greater sensitivity to the plight of the poor. If nothing else, there were economic reasons for this. The radical M.P.s, for example, were elected by the shopkeepers who were dependent on working-class custom. Thus Wakley was derided in 1835 as "the *honourable* (!) member who represents the Jew clothes-venders of Finsbury."[12] So the M.P.s and smallholders remained sensitive to their

8. Durey 1975:200. Feltoe 1884:109; Cope 1966:97, 98, 99, 103; "Aldersgate Street Medical School," *L* 1829–30, 1:745–47.

9. Mazumdar 1983:244; Richardson 1987:311, n. 59; C. Hall 1861:1–4. Peterson 1984 also considers social origins. On Primitive and Kilhamite Methodism: Hepton 1984:67–69, 104; Vidler 1974:42; E. P. Thompson 1980:48ff.

10. Dermott 1835c:366.

11. It was the cheapest medical journal on the market: "Diffusion of Useful Knowledge," *LMSJ* 1834, 5:249–50. In 1836 it was still selling for sixpence a week, or twenty-six shillings a year, compared to thirty-four shillings and eightpence for the *Lancet* and *Gazette: BFMR* 1836, 1:1.

12. *MG* 1834–35, 15:562; P. Richards 1980:62.

customers' complaints. In a similar way the teacher's livelihood depended on his pupils' prospects, which gave him a direct interest in the financial health of the tradespeople and poor. Sympathy was one thing. It was another to admit tinkers into the profession, as the school-tax lobby accused the teachers of wishing to do. Anyway, the price-cutting and proliferation of the schools led to a brisk trade war with the hospitals. Competition for cadavers, always scarce, now became fierce and exacerbated the situation. Before the Anatomy Act of 1832 the grave robbers had exploited this seller's market. (The arrest of seven London gangs in fifteen months in 1830–31 shows how many "resurrection" men must have been operating.) The act was designed to ensure that the bodies of the parish poor were divided among the schools according to their enrollment figures. But rows broke out even after the act. Hospital teachers continued to negotiate separately with the parishes for their dead paupers, forcing a furious Dermott into falsifying his own student register and threatening violence.[13] The net effect of all this was that militancy and union activity increased dramatically in the schools at this time, as the new proprietors grappled with the old privileges of their hospital rivals.

As early as 1822–24 the RCS had tried to control the growth of the schools by means of new discriminatory bylaws. How far the college's misuse of power for party ends affected the private teachers might be gauged by its effect on Joshua Brookes's long-standing Blenheim Street school. Dermott was Brookes's pupil and forty years his junior. He had acquired his taste for comparative anatomy in Brookes's class and witnessed his teacher's treatment by the RCS Council. Dermott was among the most aggressive of the new school owners: a physical-force radical, hater of the clergy, and defiantly materialistic. In contrast to Brookes's antiquarian comparative anatomy and acquiescence before the old order, Dermott fought aggressively for a French package of reforms: *concours,* new approaches to anatomy, and government regulation of medicine. Dermott was not alone; all the independent proprietors had similar experiences with a hostile College of Surgeons and by the 1830s had come together in the societies and unions, supporting the sort of science and democratic ideals that had led to Lawrence's public lashing a decade before.

In the 1820s the elderly Brookes was acknowledged as one of the best practical anatomy teachers in town (see fig. 4.2). Like so many private proprietors, he was a working anatomist and spent the whole of his day in the dissecting room with his students; he would tolerate no rote learning for exams. In forty years Brookes had taught some five thousand students, turning out a string of good zoologists, veterinarians, and comparative an-

13. Dermott 1833:19–20, 22; Durey 1975:211–12, n. 36, 65.

atomists.[14] The main reason for the zoological bent of his students was the completeness of his Comparative Osteological Museum. It was considered "the only one of its kind in London of any value" and a rival to the RCS's own Hunterian Museum (which was in fact closed in the early 1820s and its collections relegated to the college cellars).[15] Over the decades, £30,000 from his profits had been ploughed into this museum. He had obtained carcasses from every source, including his brother, a keeper at the Exeter 'Change menagerie (a wild beast show in the Strand which shut down after the establishment of the respectable Zoological Gardens). Brookes's school was a four-storey brick building, with the top two floors given over entirely to the museum, "and so crammed with skeletons and other zoological specimens that it was hardly possible to move without knocking down something with one's coat-tail."[16] Before the university's coming, press advertisements promised students wanting cheap courses "uncommon opportunities" at Blenheim Street for "prosecuting their researches in comparative Anatomy."[17] But in the late 1820s the school went into a precipitous decline. Although Brookes's museum was hailed in 1825 as an "imperishable monument of his genius," imperishable was one thing it was not.[18] Within five years it had been auctioned off and the school faced closure.

The collapse of the Blenheim Street school had a number of causes. Bad health dogged Brookes, forcing him to relinquish his class to his old friend Carpue in 1826.[19] But the main reason for his financial crisis, as all observers (including Brookes himself) agreed, was the passing of discriminatory bylaws by the RCS. In 1822 the Court of Examiners had ruled that it would no longer accept the summer course certificates of students wanting to sit for the college diploma. An even more draconian ruling followed in 1824, when it was announced that only certificates from the universities and London hospitals—or those countersigned by London surgeons—were valid.[20] Teachers and radical journalists denounced these laws as

14. Including zoologists Thomas Bell and E. T. Bennett, and veterinary surgeon William Youatt. The others are listed in: "Dinner to Mr. Brookes," L 1830–31, 2:441; Cope 1966:95; Feltoe 1884:103.

15. Feltoe 1884:105. But others recollected that, however "many good preparations," Brookes's museum also contained "an inconceivable quantity of rubbish": MCR 1836, 25:412n.

16. Feltoe 1884:105.

17. L 1825, 9:26–27.

18. "The Annual Dinner Given to J. Brookes, Esq. by His Pupils," L 1825–26, 9:452–54; 1826–27, 11:297; King 1834:3; Feltoe 1884:104.

19. "Mr. Joshua Brookes," L 1826–27, 11:83, also 214, 292, 299.

20. "Royal College of Surgeons in London," L 1824, 3:117. Cope 1959:43–44; Durey 1975:201; Maulitz 1981:481, 490–91.

Figure 4.2. Joshua Brookes in his Blenheim Street museum. By H. Cook, 1816, after T. Phillips. (Collection of the author)

coercive and corrupt pieces of legislation, designed to drive the students into the hospitals (where eight out of the ten examiners held posts). The 1822 bylaws were framed to stop the summer classes, which had become a successful feature of the private schools. Hospital anatomy courses were restricted to winter, when cadavers decomposed more slowly. But the competition between private and hospital schools had provided a strong incentive to shift to summer lecturing. Brookes had been able to dissect in summer after developing a niter solution that arrested the deterioration of tissues during hot weather. He was even awarded a fellowship of the Royal Society largely for this work. But the reaction of the medical establishment was far from enthusiastic. His action after all had considerable financial consequence; it meant that, although his classroom smelt like a ham shop, he was able to attract huge numbers of students during the summer months. Classes of 150 or more were common in Blenheim Street before the new bylaws.[21] Astley Cooper might have protested that dissection during warm weather increased the risk of infection, but the teachers knew that the preservation of corporation monopoly rather than student life was the real reason for the rulings. (Brookes had lost only one student in a quarter of a century, and that from a winter accident.) Brookes's revenue dropped dramatically after the legislation. Threatened by Brookes with a court case, the college made an exception: the council surreptitiously agreed to accept his summer certificates provided the dates were left off. Still this was not enough to stop the collapse of his class; the students' faith remained shaken in the validity of his certificates.[22] He retired in 1826, offering his museum to the university as "the most elegant, extensive, and celebrated in Europe," constructed "regardless of trouble and expense."[23] Unable to sell it intact, he auctioned off the skeletons in a series of spectacular sales beginning in July 1828,[24] with representatives from London and Oxford universities, the Zoological Society, College of Surgeons, and private museums bidding against one another. But even though the bidding was fierce for the best specimens, Brookes never recouped anything but a fraction of his costs.

Brookes was no isolated casualty. Carpue's school crashed in the early 1830s, and he too "turned popular politician."[25] The Webb Street school was thrown into similar difficulties. Edward Grainger too had started a summer course to give him the edge over the nearby St. Thomas's. At the time of his premature death in 1824 the school was immensely successful, with 250 to 300 pupils a year. But the bylaws led to a drop in admission just as his brother Richard was taking over the business. The *Lancet*

21. *Report SCME Pt. 2*, 202.

22. "London College of Medicine," *L* 1830–31, 1:855.

23. Brookes to Birkbeck, 11 March 1826 (UCL CC 1826:53).

24. "Brookesian Museum," *L* 1827–28, 2:441; Brookes 1828, 1830. On the bidding: Grant to Horner, 22 July 1828 (UCL CC P148; also CC 759–60, P139–41); Mantell to Clift, 22 July 1828 (RCS SC); Gordon 1894:88; Vigors bought the birds for the Zoological Society (ZS MC 1: f.361).

25. Feltoe 1884:166–68. Carpue organized Joseph Hume's return as the radical M.P. for Middlesex (E. Epps 1875:215); Carpue's testimony, *Report SCME Pt. 2*, 202.

pointed out that the new laws improved the chances of a dull lecturer, such as J. H. Green at St. Thomas's, drawing students away from Grainger's cheaper course.[26] In the event, again, strong representations were made and his certificates were quietly accepted,[27] but without permanent and public recognition the Webb Street school underwent a steady decline in the 1830s.

Sympathizers rather unrealistically expected the retired Brookes to be offered a seat by the RCS in the Court of Examiners (the preserve of senior surgeons). Of course it never happened, giving the excuse for radical recriminations and anger. The private teachers made great play of his fate before Warburton's committee. Carpue and Grainger suggested that there was no man so well qualified for the council yet "so ill used," adding that his treatment was deplored by the Continentals, who rated him highly.[28] Wakley too lost no opportunity in highlighting the "brazen injustice" of the councillors in blackballing Brookes, a teacher "immeasurably their superior in a knowledge of anatomy." Wakley saw financial factors as the cause, arguing that Brookes, Carpue, and Grainger were broken and excluded because their fees were half those charged by the hospital surgeons on the council.[29] But the threat was social as much as financial. Carpue put Brookes's exclusion down to his failure to make allies among the hospital elite and his preference for the company of private teachers and GPs—at a time, remember, when practitioners could be written off with aristocratic diffidence as a "low-born, cell-bred, selfish, servile crew."[30] The council considered Brookes "to be not a gentleman, and very dirty." One observer rubbed it in: "Joshua Brookes was without exception the dirtiest professional person I have ever met with; his good report always preceded him, and his filthy hands begrimed his nose with continual snuff. In his ordinary appearance I really know of no dirty thing with which he could compare—all and every part of him was dirt."[31] This grubbiness was itself a tangible reminder that, like his brother at that "disgusting" menagerie, the Exeter 'Change,[32] he was not wholly reputable, but a caterer to the medical tradesmen. Like Dermott he was accused of wishing to induct chimney sweeps into the profession. Brookes might have

26. "London College of Surgeons," L 1824, 3:266; 1825–26, 9:738–39.

27. The council was effectively buying off the bigger schools while checking the spread of the smaller ones. Grainger's testimony, Report SCME Pt. 2, 191–94; Clarke 1874a:321; Cope 1966:97–98.

28. Carpue, Report SCME Pt. 2, 203, 193. "Farewell Dinner," L 1826–27, 11:298; 1830–31, 2:442.

29. "London College of Medicine," L 1830–31, 1:855, also 2:407.

30. "Provincial Medical and Surgical Association," MCR 1832, 17:574.

31. Feltoe 1884:106.

32. Desmond 1985a: 226.

excelled as an anatomist, but this was not the point. His grimy, guinea-grabbing attention to the lower orders, while it produced some of the best GPs of the day, left him far from the Lincoln's Inn ideal, which was preeminently a social ideal.

More important, the council was unlikely to appoint members who scorned its privileges and practices. All the proprietors condemned the council's self-election as "bad for the members of the College."[33] They demanded "equality of rank" for all practitioners. Richard Grainger was prepared to discuss a system of incentives and rewards, but believed that all posts must be won through *concours* and competition rather than remain in the council's gift. Carpue used Brookes's case to illustrate the social benefit that would accrue from enfranchising all the members and holding a free council election. The resulting board would include private teachers, military surgeons, and GPs. It would thus be more directly responsive to rank-and-file demands. This is what alarmed the hawks on the council, and Carpue's grudging admission that some hospital surgeons might be returned "in consequence of their connexion with their pupils" was hardly designed to assuage their fears.[34]

Brookes's case became a cause célèbre. In radical propaganda it was linked to James Bennett's as an example of the crushing financial defeats awaiting an unorganized medical commonalty faced with corporate aggression. It was cited at the 1826 rally at the Freemasons' Tavern, at the inauguration of the London College of Medicine, and at the "one faculty" debates in the medical societies. At dinners in his honor Brookes was flanked by radicals like David Davis and Carpue, and toasts to the College of Surgeons were defiantly refused.[35] Thus the ailing Brookes edged into the radical camp, pressured by corrupt regulations and wooed by reformers. In 1831 he was warmly welcomed onto the first LCM Committee, where he was accorded the status of senior anatomist. But he lived only two more years, dying in 1833 "the most illustrious of many victims to a bad system."[36] Warburton was told that the amiable old man whose anatomy class had once numbered hundreds had died "without a single shilling." Brookes had been so successfully ruined that a public subscription had to be got up to pay for his funeral.

Some surgeons wanted even stricter controls of the schools. Guthrie suggested that no teacher should be recognized by the college unless he

33. Grainger's testimony, *Report SCME Pt. 2*, 98.

34. Carpue's testimony, *Report SCME Pt. 2*, 203.

35. "Annual Dinner," *L* 1825–26, 9:452–54; also 738–39, 781; 1826–27, 11:292, 295–301; 1830–31, 1:855; 1830–31, 2:407, 441–42.

36. "Joshua Brookes," *L* 1832–33, 2:722. Carpue, *Report SCME Pt. 2*, 202.

owned a museum valued above £500 or £1,000.[37] Dermott replied tartly to Guthrie, who, as president of the college and a consultant at the nearby Westminster Hospital, was both a class and business competitor. Guthrie complained to Warburton that the private teachers charged too little and imparted as much, with a result that standards in the profession had declined dramatically. But Dermott resisted all efforts to exclude the poorer pupils. Guthrie's position might have been secured by wealth, but for the industrious poor "knowledge is power begotten by a lawful birth."[38] Dermott also slammed attempts to establish this fiscal barrier (few teachers could afford £1,000 museums). He insisted that his standards were as high as any hospital surgeon's, indeed that his mode of lecturing while dissecting a cadaver not only obviated the need for an expensive collection of spirit-hardened museum preparations, but was being increasingly followed in the hospitals themselves. He then carried the fight into Guthrie's camp by insisting on more rigorous hospital standards and the opening of posts to talent. He compared Guthrie to Peel for his eagerness to lend an eloquent tongue to a corrupt cause and ended wagging his finger with a typical warning: "You, conservatives, have yet to learn the first rudiments of politics . . . that coercion and despotism are incompatible with the spirit of the age."[39]

So resistance in the schools was fierce. Having undercut the surgeons and been stymied by new laws, the proprietors joined the radical coalition pitted against the RCS Council. The newer teachers defied the college and kept their prices low, to suit a humble clientele aiming to trade among the storekeepers. They taunted the medical virtuosi, suggesting that their days were numbered—pointing out that their lectures, once restricted to the privileged few, were now freely available through pirated versions in the free press. The new cheap weeklies gave the shopkeepers' GP access to the latest information to tackle any medical emergency. Now each general practitioner could cover the ground of the whole host of specialists required by the "higher ranks."[40]

By the 1830s many of the schools were staffed by Benthamites and Wakleyans. The teachers supported the gamut of Benthamite reforms, moving beyond calls for the state control of medicine or a minister of health (as mooted by Bentham). The Webb Street school was the most

37. Guthrie's testimony, Report SCME Pt. 2, 71–72. The comparative anatomy preparations accompanying each of Astley Cooper's lectures were said to have cost £1,000 (Feltoe 1884:75).

38. Dermott 1835c:366.

39. Dermott 1835–36:315–17, 1835c:365–66.

40. Ryan 1836–37:412–13.

obviously Benthamite, and its anatomists publicly examined the master's constitution in more ways than one.[41] In 1832 Grainger's forensic lecturer Thomas Southwood Smith actually dissected his old friend Bentham in the Webb Street theater, flanked by a phalanx of philosophical radicals. Grainger himself was to become an inspector for the Children's Employment Commission in 1841, while Southwood Smith drew up reports on factory children, sanitary improvement, and the health of towns. Elsewhere the "destructives" were equally in control. The Aldersgate Street school, cofounded by those enfants terribles William Lawrence and James Wardrop (whose "Intercepted Letters," brilliant incriminating fakes purportedly from the medical courtiers, increased the *Lancet*'s reputation for scurrility), was one of the most radical in London.[42] With LCM councillors on its staff, George Birkbeck from the local Dispensary on call, and a cluster of union activists, it remained a center of medical agitation and Parisian materialism.

As the schools campaigned against the college oligarchs, they fell increasingly under what Bell called the cold shadow of French science. When the French-educated republican Thomas King reopened Brookes's school in 1833, he did so with a transcendental trumpet. Like the university lecturers, he promoted philosophical anatomy, and taught the best French comparative embryology and serialist zoology. Grainger's Benthamite brother-in-law, the aural surgeon George Pilcher, likewise delivered anatomy lectures in Webb Street that were comparative in approach, based largely on Blainville's work, and rested on a Lockean or sensationalist epistemology. Grainger's own progress in this direction was discernable. His 1829 *Elements of General Anatomy*, while it noted the new embryological laws (serial development, recapitulation, monsters, and so on), failed to come to grips with that unity of structure on which they all depended. Within ten years, however, this unity had become central and explicit: it was the "one grand principle," "the great law of the organic world."[43] It had become an indispensable tool, a justification for the "evo-

41. Webb Street reformers included John Armstrong, David Davis, John Elliotson, and Marshall Hall, and Benthamites Southwood Smith and George Pilcher: "Richard Dugard Grainger," *L* 1865, 1:190–91; "The Late R. D. Grainger," *Medical Times and Gazette* 1865, 1:157–58; Grainger 1842–43d:231; Pilcher's Benthamite speech to the BMA is reported in *L* 1839–40, 1:97; Armstrong was said to have attracted the largest class in London at one time: "Dr. Armstrong's Reform Principles," *L* 1830–31, 2:401. On the dissection of Bentham: T. S. Smith 1832:62–63; Poynter 1962; Lewes 1898:42–43.

42. "Medical Education," *LMSJ* 1836–37, 10:483. Early teachers included Jones Quain and LCM councillors Charles Waller, John Epps, and Thomas King. Later came Hall (1834–35), Birkbeck (1835–36), Grant (1835–38), and Southwood Smith (1837–41).

43. Grainger 1837:106, 1842–43b; cf. 1829: 77–81; King 1834; Pilcher 1840–41.

lutionary" series of animal organs which shed so much light on the origin and structure of their complex human analogues. This trend was strengthened as the schools exchanged personnel with the university. Grant, forced by financial difficulties to eke out a secondary living in the private schools, gave the teachers' developmental or embryological approach a more literal Lamarckian meaning.[44] But Grainger himself remained a conciliatory reformer, never moving as far or as fast as the ultras. For example, while he shifted some way toward an organizational explanation of vitality, he could never quite accept the austere mechanico-chemical reductionism of the older republicans.[45] From the first he was outflanked by his Webb Street colleagues Southwood Smith and John Elliotson. Southwood Smith battled privately with Bell over an organizational explanation of life, but it was the flamboyant Elliotson who remained "the strongest materialist" of his day.[46] And, unlike Lawrence, he continued under attack for his claim that medullary matter exudes thought as the liver does bile— for his no-hope "Spinozaism," which denied God and soul and took from the "thousands of the despised and the miserable" their "consoling" Christian hopes in a future recompense.[47]

Cheap education thus posed a moral, economic, and political threat. Coleridge and his medical disciples understandably despised these "lecture bazaars." They were breeding grounds of radical dissent, where the ethical naturalism of the old Jacobins met the developmental anatomies of the younger Wakleyans. The new medical peddlers had rudely set up as trading rivals, introducing the distasteful petit bourgeois practice of cost-cutting, unmindful of the old etiquette of the gentlemen surgeons. Like the merchant Nonconformists hammering on the town halls, the school proprietors stood outside the corporation doors, denouncing ancien régime monopolies protected by money and the law. In the wider view, the teachers were pushing for a political recognition commensurate with the growing "class" strength of their GP supporters.

44. *L* 1836–37, 1:21; Grant to C. C. Atkinson, 22 September 1837 (UCL CC 4166). "Conclusion of Dr. Grant's Lectures at the Hunterian School of Medicine," *LMSJ* 1836, 10:481–82.

45. Grainger 1829:8–9; cf. 1842–43b.

46. Morgan 1882:325.

47. Robertson 1835–36:205, 256–57; Elliotson 1835:39, 1831–32:289–90; Parsons 1832–36, 3:77. Jacyna 1983b:313–17 on materialist theories and republican strategies.

The Gerrard Street Teachers

It is my duty, as it is that of every lecturer, not only to give my pupils
full measure of scientific instruction, but also to open their eyes to
the real state of the medical profession.

—George Dermott opening his 1834 session[48]

George Dermott's school is important from our perspective, first because
its teachers had an interest in comparative anatomy, but also because we
can understand why they found particular approaches to this science attractive. Only one of the school's three teachers is well known: the phrenologist John Epps, whose *Diary* was published posthumously. The other
two, the corpulent Irishman Michael Ryan and Dermott himself, remain
shadowy figures.[49] Dermott was a political bruiser, rough and ready: a
frequenter of taverns for intellectual stimulation—a man who would ply
his students with punch, then bail them out after they had been arrested
staggering home. Conservative fears for the moral welfare of medical students, driven by excessive work "into the theatres and saloons to seek the
company of harlots and drunkards" (see fig. 4.3), were not allayed by the
likes of Dermott.[50] Nor did his manner help: he ended his introductory
lectures by threatening to thrash any consultant who should claim more
than his share of cadavers—to the undoubted delight of the reprobates of
his class. He was an uncompromising physical-force radical, temperamentally far removed from the piously Quakerish Epps, the epitome of a
moral-force temperance reformer.

As Brookes's protégé and colecturer, Dermott had hoped to acquire the
Blenheim Street premises. Unable to do so, he moved in 1829 to a school
attached to the Westminster Dispensary, in Gerrard Street, Soho, where
he stayed until 1837.[51] He was headstrong and impulsive, although no one
could turn out students faster or with greater success (the result of dis-
pensing with the prosecutor and dissecting and lecturing simultaneously).

48. Dermott 1835c:364.
49. To the extent that even Ryan's dates are wrongly listed in many sources, while Clarke
(1874a:126) and the *Medical Times* (1847, 16:618–20) obiturist disagree on Dermott's origins.
50. McMenemey 1966:138–39, 145; Dermott 1833:19–22. Clarke 1874a:128–32 on Dermott's drinking and teaching habits; Cope 1966:105 on the damage the students did to the
dispensary. Newman 1957:41–45 describes student lowlife in the period.
51. Before this (c. 1825–29) he taught at a school in Windmill Street. After 1837 he
lectured at his house in Charlotte Street, until it was condemned in 1845. See *Lancet* course
listings for each September; also *MT* 1847, 16:618; and Clarke 1874a:128–29. Loudon 1981
analyzes dispensary teaching generally.

Fig. 4.3. *Punch* took a dim view of the London student, particularly that drinking, smoking, debauched "son of the scalpel," the medical student. (From *Punch*, 1842, 2:149)

He was an earthy, plain-spoken son of a Methodist minister, with a connectional hatred of robed dons and plumed aristocrats, the targets of his merciless lampoons in class. He saw politics as an integral part of medical education, and he arrived for lectures armed with bundles of parliamentary petitions for his pupils to sign. There was a gruffly heroic quality about Dermott. It was he, for example, who braved the flames as the Commons burned in 1834 to rescue Warburton's committee records. He emerged after five hours of fire fighting, "burnt, bruised, drenched with wet, as black as Diabolus," having thrown the bundled papers to safety in the street, only to be arrested as a looter. Such heroism in aid of a "reforming government" had Wakley crowing with delight, the more so because Wakley thought the fire itself a blessing, clearing room for a new building with large public galleries more suited to an open ministry.[52]

Epps adopted a more Quakerish moral tone. Brought up in wealthy Sevenoaks, he early rebelled against his father's strict Calvinist teachings, with their gloomy emphasis on the devil and damnation. He disputed that God was vengeful and that only the elect would be saved, although he never lost the Calvinist belief in a "lawful" necessity. He overcame an infidel crisis, during which he delved into Tom Paine and Voltaire, and found in meliorist tenets and Arminian doctrine (the anti-Calvinistic faith that anyone could be saved) a new meaning to the Bible and his family's antislavery beliefs. While a student at Edinburgh University (1823–27), he joined the Scottish Baptists, attracted to them because they were at once "Scriptural and primitive," that is, they had a democratic congregational organization. He had been a phrenologist since his first exposure to the doctrine in 1821–22, and he joined the Phrenological Society in Edinburgh, where he was befriended by George Combe. In the late 1820s Epps integrated Combe's ideas on the "organs" of the mind with this "primitive Christianity," as he called it.[53] The outcome, a form of phrenological Christianity, was solidly Calvinistic. Indeed, Epps never wavered in his Calvinistic belief that man was "necessarily evil" and could do nothing to save himself. But to be "born again," man had only to believe the testimony of God, and phrenology for Epps showed that the faculties of Benevolence, Veneration, and Conscientiousness were "busily engaged" in every person in bringing about this end.

When Epps joined Dermott's London School he was an outdoor

52. Dermott 1834–35:160–62.
53. E. Epps 1875:354, 120–21, also 42–45, 60, 70, 114–67, 171. Epps 1836:17, 54–74 on necessity, evil, and phrenological salvation. Phrenology's wider social role among the Edinburgh petit bourgeoisie is well covered in Cooter 1984 and Shapin 1975, 1979a, 1979b. Murphy 1955 considers meliorism and the growing rejection of eternal damnation, and Moore 1989a its impact on one evolutionist, Charles Darwin.

preacher and phrenological lecturer. He remained a Baptist until 1831. That year he became the director of the Royal Jennerian and London Vaccine Institution, in succession to the Quaker democrat John Walker. Walker had been in Paris with Tom Paine during the violent days of the revolution (in 1797), although as a pacifist he baulked at Paine's dream of a tricolor flying over the Tower of London. Epps shared Walker's attitude toward Paine, praising his *Rights of Man* as "the best political essay in the English language," while abhorring his anti-Christian *Age of Reason*. "The one seems as a gift from Heaven," Epps wrote, "the other, as a pest from Hell."[54] Epps began moving closer to the Friends himself. Like Lawrence and other republicans attracted to the Quaker cause, he supported Quakers served with summonses for refusing to pay the church rates. He also learned from them the morality of resistance.

The *London Medical and Surgical Journal* virtually became the Gerrard Street house organ. Epps had helped found it in 1828, and Ryan took over the following year as sole editor. It lobbied for a living wage for the GP, looked after the private teacher's business interests, and sought medicine's professional parity with the law and Church. Like the *Lancet*, it insisted on the members' rights to elect the councils of the Royal Colleges, in whose grasp so many preferments lay.[55] Gerrard Street was a microcosm of medical radicalism, with its complex cross-threading of social, medical, and religious issues. Its teachers differed in their campaign styles. Epps had the organizational flair. He helped run the National Political Union, attended the Radical Club to consider the radical M.P.s' options, set up an ad hoc advisory group to assist Warburton's committee, and later worked with officials of the Anti-Corn Law League.[56] Ryan at the start was the most moderate of the three teachers. Unlike the conspicuously pious Epps with his Quaker's low-brimmed hat or the rough-hewn Dermott, Ryan was fastidiously attired in white choker and gold chain, and he sported an equally unruffled reformism. He was never an ultra-radical. While he denounced the "insolent aristocracy" and spoke eloquently on "one faculty" resolutions, he could never reconcile himself to ranking medical novitiates "with the *dignitaries* of the profession"[57] or countenance giving apprentice-trained GPs the title of doctor.

54. J. Epps 1832:27, 133–35, 141; *MCR* 1831, 15:26–35. E. Epps 1875:192–93; and on Quakers, 210ff. Also Clarke 1874a:139–40. Up to fifty thousand summonses were issued by the clergy in 1833 in an attempt to extract tithe money from resisting Nonconformists (Halévy 1950:150; Cowherd 1956:155).

55. "State of the Medical Schools in London," *LMSJ* 1833, 2:310–13; "Insufficient Remuneration of Medical Practitioners," *LMSJ* 1835–36, 8:56.

56. E. Epps 1875:194, 214, 221, 289.

57. "Medical Journalism," *LMSJ* 1834, 5:55–57; also 1833–34, 4:570–75; *L.* 1833–34, 1:366; Clarke 1874a:134.

Alone of the three, Dermott urged physical force and direct action. His recruitment of poor students and his medical pugilism alienated the moderate reformers (who hardly welcomed his offer to enter the ring with anyone opposed to democracy or the *concours* "and fight it out").[58] Feelings ran high against him, even, for example, at the Westminster Medical Society, where he was blackballed in 1834 for his abusive radicalism. Being neither a "pretended gentleman nor a pretended surgeon," he was scorned by the elite and excluded from a fellowship when the RCS was finally rechartered in 1843.[59] More than the others, Dermott actively politicized his students. He warned them of the "vineyard of corruption" they would have to pick their way through as GPs or teachers, confronted by hospital nepotism, corporation lies, and the corruption of official science. They would have to solicit patronage to obtain posts, rather than win them fairly. And in those uncharitable "arenas of warfare," the dispensaries and hospitals, they would have to watch corrupt governors giving their servants and friends preferential treatment at the expense of "the really indigent sick."[60] Dermott's blustery threats might have distinguished him from the urbane Ryan and Epps, but they too had their brushes with the law. Epps was cited in court for practicing without a license, and Ryan ran into financial difficulties. He was imprisoned and tried in the Insolvent Debtors' Court in 1836 after being sued by the *Journal*'s printer. Despite his acquittal, the *Journal* collapsed shortly afterward.[61]

Ryan's radicalization proceeded apace. When his *Journal* went weekly in 1832 it toed a moderate line. It saw itself steering a course between the Scylla of Wakleyism and Charybdis of reaction—castigating Wakley's "coarse and violent" language which appealed to the "lowest of human passions," while still storming the "brazen walls of unjust monopolizing corporations."[62] But Ryan found it remarkably difficult to plot a straight course and became exasperated with the prevarications of the corporations. By 1833 he was convinced that they were incapable of self-reform and that "nothing is now gained but by AGITATION."[63] The colleges had to be amalgamated into a single elected faculty or, failing wedlock, side-stepped altogether. (He never condoned talk of their destruction.) The

---

58. Dermott 1833:15, 1835c:365, 1835a:662–63.
59. "Mr. Dermott," *MT* 1847, 16:619; blackballing: *L* 1833–34, 1:646, 839.
60. Dermott 1835b:20–21.
61. *LMSJ* 1836–37, 10:335–40; E. Epps 1875:182, 313.
62. *LMSJ* 1832, 1:17–18, 211, 411–18, 503–4.
63. "Reform, Agitation, House of Commons," *LMSJ* 1833, 3:596; "What Is to Be Done with the Medical Corporations?" *LMSJ* 1834–35, 6:409–11. All three teachers helped organize the London College of Medicine.

reformed Parliament was expected to act quickly. But the medical radicals were impatient, and in 1833 Ryan was already accusing the "leading political gladiators" of failing to march out and meet their foes.[64] The teachers began casting around for their own M.P., someone who could represent their interests. Wakley was, if not a perfect choice, the best under the circumstances. He too was complaining that the medical colleges had their Tory champions while the GPs lacked their "orator" in the House.[65] He had fought the Finsbury seat as early as 1832, pledged to corporate and church reform, abolition of tithes, removal of religious disabilities, and the imposition of a wealth tax (see fig. 4.4). Epps had worked for his return, placing a giant placard in his front garden at election time. Ryan too urged a full turnout for Wakley, not necessarily agreeing with all his politics but "convinced that his exposures of the defects and abuses in our profession before Parliament would speedily effect their removal."[66] By 1834 the *Journal* was leaning Wakley's way: applauding the *Lancet* for rousing the surgeons—even if Wakley's ribaldry did leave him sounding like some "hot-brained, half-civilized, American editor."[67] But even these swipes were friendly asides by 1835 (the year Wakley entered Parliament, at his third attempt), and an impenitent Ryan now announced that "we glory in following the Lancet," "sincerely venerate" it "as the honest, fearless, and powerful advocate of the rights of the profession."[68] For a time the radical-Dissenting alliance worked, as in the country. Of course there remained differences between the sacrilegious, secular *Lancet* and the piously Nonconformist *Journal*. And the uneasy coalition always stood at the mercy of volatile temperaments—how much so becoming apparent when a belligerent Dermott severed all relations with Wakley after an apparent *Lancet* snub in 1835.[69] But the temporary alliance, based on

64. "Reform! Reform! Reform!" *LMSJ* 1833, 3:372–74; also 340–41, 596–98.

65. "Representation of Medical Men," *L* 1829–30, 1:750; Wakley 1834–35. Wakley was backed for the Finsbury seat by Cobbett, Hume, and the former tailor Francis Place (a onetime Jacobin with exceptional organizational flair who now sported a more genteel Benthamism). He campaigned against the Whig antilabor legislation, for instance the Poor Law Amendment Act (1834), newspaper stamp duty (which was designed to cripple the working-class press), and the Metropolitan Police Bill (1839), which extended police search powers. He was in favor of universal suffrage, secret ballots, and the abolition of property qualifications for M.P.s, and he supported the Chartists.

66. "Medical Reform," *LMSJ* 1833, 2:115–18; "Why Should Not the Medical Profession Be Represented in Parliament?" ibid., 597–98; Clarke 1874a:139; J. Epps 1875:235, 283, 314. The *Gazette* laughed at reports of Epps advising his phrenological "class of *grooms*" to vote for Wakley on account of the shape of his head: "Epps on Heads," *MG* 1834–35, 16:20–21.

67. "The Weekly Medical Press," *LMSJ* 1834–35, 6:442.

68. "A Kick for the Green Lizard," *LMSJ* 1835, 7:346–47.

69. *L* 1835–36, 1:212–13, 261–62; *MT* 1847, 16:619.

THE TORY PEACOCKS AND THE FINSBURY DAW.

Fig. 4.4. Wakley, Finsbury's radical M.P. from 1835, caricatured as a jackdaw plucking fine Tory feathers on issues such as the ballot, Charter, and New Poor Law. (From *Punch*, 1841, 1:139)

common professional needs, allowed an exchange of ideas, and the doctrines of a self-animating nature soon spread through the Dissenting bastions in medicine.

Gerrard Street was no Paleyan world of static creation. Nor could it be with liberal Anglican divines such as Buckland, Whewell, and the geologist Adam Sedgwick in the Church universities employing these doctrines to justify God's civil order—to provide a moral sanction and a physical bedrock for the existing political structure. In paternalist Anglican society the individual was stripped of sovereignty. He had no inalienable democratic rights; so the command for social change could not come "from below." Nature was no different. Divine intervention—outside "Creative Interference" Buckland called it[70]—operated whenever organic changes were needed: the abrupt appearances of fossil animals in the geological record were proof of this. For Oxbridge Anglicans nature was still an absolute monarchy. Sedgwick's Creator operated a spiritual close-borough, holding Personal control over "natural" appointments and organic changes. God was not a "uniform and quiescent" Legislator, but a careful meddler, "an active and anticipating intelligence,"[71] whose immediate attention to each mollusk and man obviated any need for self-developing species. There was no self-engendering "push" from below.

The institutions under Anglican patronage fostered this establishment ideal. The British Association for the Advancement of Science welcomed only medical moderates; nothing was heard of Lamarckism or the materialist anatomies here, nor was phrenology considered anything but the work of "crazy humourists."[72] Sedgwick used the opportunity provided by its countrywide meetings to praise the wisdom of the natural and social order—delivering antileveling scientific sermons to the black-faced "rabble" in the host towns. He preached to the Newcastle colliers on the providential "economy of the coal-field" and their own beneficial "relations to the coal-owners and capitalists," conflating the geological, economic, and moral orders to underpin the existing class divisions.[73] Pulpits throughout the country echoed these sentiments. A widely distributed tract demanded that "the Chartist leaders preach and teach the doctrine of 'equality'; but we have no such doctrine taught in the book of Nature or in the Book of God."[74] For their part, the London teachers accused of favoring tradesmen and indoctrinating their charges with medical materialism and radical propaganda had no truck with the BAAS.[75] The radical Dissenters, refusing to accept any supernatural sanction for the Anglican status quo, supported the "dark" sciences so despised by Sedgwick, brandishing the works of Geoffroy, Lamarck, Combe, Lawrence, Elliotson, and others as symbols of a defiant and democratic Dissent.

Epps was the only Gerrard Street phrenologist, although his col-

70. Buckland 1837, 1:586.

71. Sedgwick 1834b:315.

72. J. W. Clark and Hughes 1890, 1:265; Morrell and Thackray 1981:276–81, 287–90.

73. J. W. Clark and Hughes 1890, 1:515–16, 2:46–47; Morrell and Thackray 1981:31–32, 127.

74. Quoted in Vidler 1974:95.

75. Grant to Mantell, 16 July 1850 (ATL GM 83, folder 44).

leagues acknowledged the value of Combe's *Constitution of Man*. Phrenology was peddled as a rank-breaker, sold for its potential "to render a merchant or manufacturer a *gentleman*" through the cultivation of the correct faculties.[76] It was always a moderate reformist doctrine, too much so for many. Because the mental faculties—the "organs" of love, hate, greed, and so on—were anatomically determined it was assailed by socialists and libertarians alike. The old Jacobin William Godwin, with his Enlightenment faith in education and suspicion of hereditary "dispositions," considered many of the so-called "organs"—those showing propensities to rob or murder—"a libel upon our common nature."[77] He recognized phrenology as a sort of anatomical Calvinism. Were the faculties really inborn—beyond the reach of culture and education—they would burden man with an "intolerable chain" throughout life. Some did try to mate the science to a more environmentalist base and answer Godwin. John Elliotson—president of the Phrenological Society and (from 1831) professor of medicine at the London University—integrated it into his atheistic and reductionist physiology. He replied to Godwin, saying that phrenology indicated "only the general strength of each propensity, leaving its course to be explained by external circumstances."[78] But this still left the hereditarian base intact and its relationship to more subversive philosophies like Lamarckism and socialism problematic.

The apparent contradictions are even more evident in Epps. He accepted man's inescapably evil nature, yet saw in phrenology and "Primitive Christianity" salvation for all. And while he believed that man was stuck in his evil ways, Epps still threw himself into almost every reform cause. His "necessitarian optimism" was typical of many reformers of the period, reacting against their strict Calvinist upbringings.[79] More important from our perspective, these "neo-Calvinists" believed in a necessitarian nature, one of inexorable cause and effect, stripped of spiritual meaning and from which God stood apart. By the late 1820s the inrush of French and German anatomies began to push phrenologists into new directions. Epps himself exploited the Continental zoologies to demystify creation and explain human complexity naturally, cobbling together his science from Lamarck, the higher anatomist Gustav Carus, and Franz Joseph Gall, phrenology's founder. Lamarck's *Philosophie zoologique* equipped Epps to tackle life's progressive "transmutations" (though whether he understood these as Lamarck had is a moot point). Others

76. "Education—London University," *Phrenol. J*. 1825, 2:441–42.

77. William Godwin, *Thoughts on Man* (1831), extracted in *MG* 1831, 8:235–36; Marshall 1984: 369. Cooter 1984: 224ff. discusses phrenology and socialism fully.

78. Elliotson 1831–32:362.

79. Peel 1971:102–11; Cooter 1984:105.

used Tiedemann's comparative studies of the fetal brain and Serres's embryology to illustrate the natural and "gradual evolution" of the phrenological faculties.[80] Transmutation was given a more precise meaning when medical phrenologists translated Tiedemann's *Systematic Treatise on Comparative Physiology* (1834), which advocated the progressive generation of species under modifying environmental influences—a theory its translators endorsed, but which was denounced by one infuriated critic as a "godless, self-existing, self-destroying, senseless, aimless crotchet."[81] Many phrenologists were evidently eclectic, welding a variety of radical sciences; particularly in Combe's circle, it was not uncommon to find that phrenologists were also transmutationists. So when all is said and done, Epps's higher zoology ended up little different from that of his antiphrenological friends.

The serialist anatomy Ryan chose to promote through the *Journal* was again characteristic. He ran translations of Blainville's lectures. Grant's "successful labours" were applauded (with the *Journal* now backing the *Lancet*, Ryan too dubbed Grant "the English Cuvier").[82] But the real physiological tour de force for Ryan was John Fletcher's higher anatomy. Fletcher's 1834 course of "erudite, comprehensive, and incomparable lectures" at Edinburgh's Argyll Street school was published in the *Journal*, and his *Rudiments of Physiology* was rated above all the "national and foreign works of modern times."[83] It was a rich and powerful work, on a par with the best Germany or France had to offer (indeed, it was much like them in tone and content). But Ryan's panegyric testifies also to his eagerness to promote this kind of science—at once antivitalistic and developmental, with its organic unity and self-sufficiency underwritten by an irrevocable law. *Rudiments* was a systematic exposition of the new Continental anatomy, in which even Lamarck's transformism was treated with sensitivity (if finally rejected), and Geoffroy's homologies rather than Bell's beneficent adaptations were proclaimed as heralding a glorious new era.[84]

But Fletcher's rejection of transmutation was not enough to stop phre-

80. Sandwith 1827:493. J. Epps (1829:40, 43, 151, 155) talks (like the higher anatomists) of the cerebral ganglia and shields in squids and insects foreshadowing vertebrate cranial development, and of a uniform transcendental flow undermining Cuvier's discrete *embranchements*.

81. "On the Physiology of Man," *MCR* 1839, 30:452; Tiedemann 1834:13–15. One of Tiedemann's translators, James Gully, was to become Darwin's physician (Moore 1989).

82. "Solly on the Human Brain," *LMSJ* 1836, 10:250; "Physiology," *LMSJ* 1830, 4:231–33; "Professor de Blainville's Physiological Lectures," *LMSJ* 1833, 2:566–69, 587–88.

83. "Dr. Fletcher's Rudiments of Physiology," *LMSJ* 1835–36, 8:756–58; J. Fletcher 1834–36.

84. J. Fletcher 1835–37, 1:12, 13–16, 36–38.

nologists from bastardizing his doctrines and turning them into evidence for evolution. By the late 1830s higher anatomy was seeping out of the private medical schools and into popular culture. Fletcher's and Tiedemann's books were well read, and phrenologists were finding a new interest in Geoffroy. Another of Combe's adjutants in Edinburgh was the publisher and self-professed "essayist of the middle class" Robert Chambers, whose Edinburgh presses specialized in self-help manuals, anthologies, and, increasingly, popular science. Secord has shown that up to the time of the Reform Bill Chambers had been a Tory-Presbyterian, but his churchgoing gave way as his popular journals were attacked from the pulpit and he came under the sway of the phrenologists (whose artisan and shopkeeper audiences were the same as those for his magazine and miscellanies). Already in the mid-1830s he was deriding "those dogs of the clergy" and warming to progress, transmutation, and reform—and he was intent on creating a true "people's" science.[85] He had read Fletcher's *Rudiments* and other transcendentalist medical works; he knew of Geoffroy's and Tiedemann's ideas about the progressive animal series and its recapitulation in the fetus. And he was quite aware of the reformist credo that law and progress reigned in nature. By 1839 his conversion was complete, and he secreted himself away to synthesize these views in a popular evolutionary book with a disarming Providential gloss, *Vestiges of the Natural History of Creation* (1844).

In *Vestiges* Chambers argued that progress and "development" had been inevitable—from the coalescence of planets out of swirling nebulae, through the "chemico-electric" generation of the first living globules and the subsequent ascent of life, to the perfection of man. All of this had been achieved by natural law, which Chambers carefuly couched in terms of a Divine edict. Law, he argued, was the means by which God foreordained events. He did not need to step in personally each time a new mollusk or monad was needed; it would be demeaning to imagine that He did. To explain the steady advance of fossil life Chambers modified current Geoffroyan theories to produce the idea of a "higher generative law." Homologies, monsters (including Geoffroy's artificially produced ones), rudimentary organs, and the fetal recapitulation of the ancestral series "clearly show how all the various organic forms of the world are bound up in one— how a fundamental unity pervades and embraces them all, collecting them, from the humblest lichen up to the highest mammifer, in one system, the whole creation of which must have depended upon one law or decree of the Almighty." It was the idea of recapitulation that gave Chambers his clue to the "generative" law responsible for the birth of higher

85. Secord 1989.

species. If monsters are arrested at some lower level, and this is caused by the fetus stopping short in its development, then new species must be produced by the opposite—a prolongation of gestation. Catalyzed by changing "external conditions," he said, an animal mother-to-be will extend her gestation period beyond the normal and produce a more advanced offspring. And as the species progresses up the scale, so the one below rises up to take its place.

Whether the whole of any species was at once translated forward, or only a few parents were employed to give birth to the new type, must remain undetermined; but, supposing that the former was the case, we must presume that the moves along the line or lines were simultaneous, so that the place vacated by one species was immediately taken by the next in succession, and so on back to the first, for the supply of which the formation of a new germinal vesicle out of inorganic matter was alone necessary. Thus, the production of new forms, as shewn in the pages of the geological record, has never been anything more than a new stage of progress in gestation, an event as simply natural, and attended as little by any circumstances of a wonderful or startling kind, as the silent advance of an ordinary mother from one week to another of her pregnancy.[86]

Chambers published Vestiges anonymously (which added to the frisson of excitement when it appeared). He knew exactly how to market the book, sending preview copies to all the opinion makers. Of course it caused a furor among the old university elite, but it did notch up unprecedented high street sales, especially in London, where the rationalists loved it. The science was secondhand, errors were legion, and much of the subject matter (phrenology, the nebular hypothesis, transmutation) came from the intellectual fringes. But (and this was galling to his critics) it was beguilingly "got up" and clearly aimed at a popular audience. Sub judice subjects in medicine—fetal recapitulation and unity of plan—were being peddled to a scientifically illiterate class. Worse, in a paternalistic, male-orientated society, Chambers was giving indelicate topics such as pregnancy, abortion, and monstrosities (previously the province of medical men) a plebeian scientific importance. The trouble was that women— and not only emancipated socialist women—were actually enjoying the book; it was reaching a huge audience. This absolutely enraged critics such as Sedgwick, who warned "our glorious maidens and matrons" against soiling "their fingers with the dirty knife of the anatomist."[87] Sedgwick's fears were being realized: transmutation was leaving the shabby medical schools and entering the middle-class parlors.

86. Chambers 1844:224, also 197–98, 205, 214.
87. Sedgwick 1845:3; Ruse 1979:131; Desmond 1987:109.

This kind of developmental zoology was anathema to the divines because it evoked natural self-sufficiency, subverted orthodox creationist tenets, and put Paley's Deity out of reach. Sedgwick detested Geoffroy's "dark school" and greeted *Vestiges* with the apocalyptic cry:

I can see nothing but ruin and confusion in such a creed. . . . If current in society it will undermine the whole moral and social fabric, and inevitably will bring discord and deadly mischief in its train; and on this account also (having a belief in the harmony of nature and in an overruling Providence) I believe it utterly untrue.[88]

Charles Gillispie once noted that Sedgwick acted as if the Anglican state would collapse if Paley's natural theology were rejected, that is, if one could no longer infer Creative interference from the adaptations of animals. The radical Dissenters too were aware that Sedgwick's God was supposed to intervene actively to maintain the natural and social status quo. Their denial of "design" and advocacy of a lawful universe was calculated to destroy this image of active maintenance and support one of a self-sustaining progressive nature. This caused deep Anglican concern. But the severity of the clergy's reaction also reflects the fact that this Dissenting strategy was itself part of the campaign for disestablishment. Even Sedgwick's antitransmutatory polemics were sometimes couched in the kind of language used in the disestablishment debate. Take his metaphors of sexual profligacy and civil disorder. He execrated *Vestiges'* phrenology, spontaneous generation, and transmutation as an "unlawful marriage . . . breeding a deformed progeny" and threatened to stamp on "the head of the filthy abortion."[89] Such radical monstrosities threatened to corrupt "our glorious maidens" and through them poison the wellsprings of Victorian society. This was not solely a play on Chambers's fetal mechanism for evolution. It was also a reversal of the Dissenters' sexual rhetoric. Pamphleteers such as Epps had long exploited Paine's polemic, his denunciation of the Church's "adulterous connection" with the state.[90] Epps abominated the "fornicating" Church, indicting it for the "filthy crime" of adultery with aristocratic government, which had left it with the "stigma of disgrace" to degenerate into "an inflictor of misery . . . a destroyer of moral principle . . . a subverter of the good order of society." This sexual-social juxtaposition and the countercharges of infidelity (meaning both adultery and irreligion) characterized the angry exchanges over disestablishment. Epps in 1834 saw the Church continuing to "commit fornica-

88. Quoted in Gillispie 1959:169–70; J. W. Clark and Hughes 1890, 2:86.
89. Napier 1879:492; J. W. Clark and Hughes 1890, 1:2, 83.
90. Paine 1819:6.

tion, until the Dissenters tear her, by civil legislative enactments, from her ILLICIT EMBRACE, in which she has been playing the harlot" with civil government.[91]

The polemical language slipped over so easily into the debate on Geoffroy's and Lamarck's laws of form and development because they were perceived as integral to radical aims. Naturalism and disestablishment had become Dissenting bedfellows (to continue the metaphor), producing a disfigured progeny of deism and transmutation. For Epps, nature naturalized meant the Church disestablished. The Almighty had instituted self-adjusting physical and moral laws at creation, revealing them directly to Everyman through Nature and Revelation. Between man and his Maker, therefore, the state had no right to interpose priests, any more than it had the right to enforce Anglican creeds. Put baldly, Sedgwick's strategy for preserving Anglican power depended on his evidence for a providential superintendence of nature, whereas radical Dissenters favored more emancipating forms of knowledge—of a self-regulating nature mediated by a universal priesthood of all believers. They were attempting to divorce the Church as firmly from natural science as from the state. The scientific fears of the liberal Anglicans thus had a strong social basis. In the mid-1830s Nonconformists, now about four million strong, were intent on emptying the Church coffers of tithe money and breaking the Anglican monopolies on the rituals of life and death. Despite Church reforms and the introduction of civil registration for births, marriages and deaths in 1836, Dissenting and secular agitation continued. As the radicals pressed for disunion, they began to undermine Sedgwick's providentialism, redefining science and reinterpreting nature, creating a cultural atmosphere in which the *Vestiges* could become a best-seller.

This disestablishment mentality explains the license with which the radicals attacked Oxbridge claims to professional leadership. The London teachers talked of theirs too as a "ministry" with Supreme sanction.[92] As medical "ministers" they now demanded professional parity with the clergy (equal pay, power, and representation). No phrenologist himself, Ryan nonetheless recommended Combe's *Constitution of Man* because it allowed the physician to colonize increasing areas of Establishment domain. Combe had shown that the mental faculties were an intimate part of human organization, allowing the doctor to extend his control over both body and mind. Ryan's claim that "nothing can separate the study of [man's] physical and moral states" would have horrified Sedgwick, who

91. J. Epps 1834:3.
92. Ryan 1836–37:411.

was to savage Chambers on this very point.[93] This attempt to trench on the Anglican clergy's domain was typical of the new medical priesthood. Ryan told students that "neither in mental nor moral attributes," in benevolence, self-sacrifice, or "virtuous nobility of character," "does the profession of physic yield to that of theology." Dermott was as uncompromising in his professional demands as in his mental materialism. Having a profound contempt for the cloth, he warned the meddling divines to steer clear of physiology and leave speculation on mental functioning to the medical specialist.[94] The province of mind and soul became such a source of animated debate among the medical sectaries in 1830 that one disgruntled observer was "afraid that, between 'medical physics,' and 'medical metaphysics,' *common physic*, by which so many of us 'live, and move, and have our being,' will fall to the ground!"[95] But the very strength of this debate and support for the new anatomies explains the clergy's fears. Heretical sciences were being linked to democracy and disestablishment, and the radicals were plotting the "ruin and confusion" of the Anglican order by undermining its supernatural foundations.

The attempts to kick away the natural theological crutch supporting the Church and corporations are most evident in Dermott's blustery *Discussion on the Organic Materiality of the Mind* (1830), a work even Ryan considered practically blasphemous. The book shows how inextricable serial anatomy, mental materialism, and proprietorial attacks on priestcraft had become in ultraradical thought. Dermott was never backward in proclaiming his views. Charged by an irate parson with being a rank materialist after submitting a paper on the brain to the *Lancet*, he replied brusquely, "I am, as it concerns the mind; and what of that? does it necessarily follow that I deny the existence of a soul, or that I should believe the soul is a material principle? I answer, I am a downright materialist, as it concerns the mind . . . I am a *physiological* but not a *theological* materialist."[96]

In Dermott's model the brain was a continuous elliptical loop of cerebral fibers. Here the incoming sensations are assimilated into perceptions and judgments by ganglionic processes (or "digested," on analogy with the

93. Ibid., 411, 415–16. Sedgwick 1845:3. Ryan (1829) shows his antagonism to phrenology.

94. Dermott 1828–29, 1:40.

95. "Seat of the Soul," *MCR* 1830, 12:461.

96. Dermott 1830:43. For an example of theological materialism, see Jacyna's discussion of Thomas Laycock's "mortalism." Laycock, a Methodist London University medical student in 1833–35, applied Hall's reflex arc to the brain and considered the mind a function of neural organization. According to his mortalist creed, the soul was no less material—it perished with the body to await "physical resurrection" in the afterlife (Jacyna 1981:116; Cope 1965:175).

stomach), after which the volitions are dispatched through the outgoing voluntary nerves. Thoughts are material products, and mental functioning the cerebral equivalent of digestion. Comparative anatomy provided the "incontrovertible facts" to support this doctrine. It demonstrated a fixed-size relationship between an animal's cerebral faculties and the external senses supplying the undigested "impressions." On descending the scale, both sense organs and brain diminished—a relationship that can be traced to the level of the lowly polyps with their diffuse "granulated" nerves. Dermott believed that "*all these animals* have, more or less, a *mind*,"[97] something proved by their educability. The training of dogs, horses, and elephants, for example, capitalized on their abilities to retain "material" memories. Between man and animals there was only a quantitative mental difference. So-called rationality was no evidence of any qualitative barrier. True, man might possess a soul, but this imponderable entity was incapable of interacting with matter and remained totally "dormant" while the body lived.

Dermott denied that we can have any conception of these souls, "nor can any human language convey to the mind any just idea as to their nature." Others agreed: as Elliotson said, quoting the philosopher John Locke, these were "purely matters of faith, with which *reason had nothing directly to do*."[98] But just as this "dormant" spirit world was quite unknowable, so the known world was quite unspiritual: "let divines and philosophers say what they please," Dermott pontificated, "we . . . never shall gather any more, by researches into human knowledge, than that the works of nature are wonderful, and carried on by a concatenation of cause and effect." At a stroke he negated the religious edifice supporting priestly claims to social and scientific hegemony. The clerical naturalists were unjustified in setting themselves up as official interpreters of the attributes of God deduced from nature. The Bible and revelation, not parsons and natural theology, are our true guides—and the Scriptures are accessible to every man without clerical mediation. Revelation must be taken directly on trust, for it would be presumptuous of us "to seek out . . . sinister evidences [in nature], with the view of supporting the imagined deficiencies of [the Bible]."[99] By depriving nature of spiritual meaning, he was denying the sacredness of existing natural knowledge, the basis of natural theology, and the special status of the Oxbridge scientific

97. Dermott 1828–29, 1:41; 1833:11–12. On animal minds: J. Fletcher 1835–37, 3:93–95; Elliotson 1835:3–5, 39.

98. Elliotson 1835:43.

99. Dermott 1828–29, 1:42. The debate stirred up by Dermott was capitalized on by the working-class atheist and publisher of Paine and d'Holbach, Richard Carlile (Richardson 1987:94).

priesthood. Epps, too, praised Elliotson for exorcising physiology, spurning the superstitious divines who would "clothe palpable facts and sensible manifestations with a spiritual halo; as if truth could be more sacred when thus surrounded, than when clothed in her simple beauty."[100] This spiritual disrobing had a powerful meaning for the radical sects. It left morals as natural as matter; even the conscience in Dermott's view was a cultural artifact and capable of improvement. The mental materialists could therefore lay claim to the mind, breaking the clerisy's hold on morality, just as the democratic sects were shaking its grip on the sacraments. In the larger view, this naturalization of the spirit world was an attempt to transfer the old religious authority vested in the Established Church to the new Dissenting interpreters of nature. As the Nonconformist traders began to acquire their political voice in society, so their medical spokesmen were legitimizing the switch from an Anglican "theodicy" to a radical-Dissenting one.

Dermott was charged by one clergyman with rendering "Bibles, revelations, ministers, and religion as totally useless in this present world,"[101] and this was true as far as the ministers and creeds went. Like his colleagues he was constitutionally suspicious of organized religion. Epps analyzed the organ of veneration to illustrate the number of Christian sects that were capable of nothing but false devotion. But the extent of Dermott's materialism and dismissal of spirits dismayed even his friends, who now accused him of ignoring Christ's death to save man's soul and of producing a work "highly heterodox and injurious, nay subversive of the purity and simplicity of the Christian religion."[102] Notwithstanding this, his secular science promised the sort of social reformation so eagerly awaited by all the teachers. Disestablishing the Church was the crucial step toward dismantling the networks of social and professional privilege and unshackling the sects. For a democrat like Dermott, with his radical faith in education, competition, and change, this destruction of Old Corruption would usher in an age of unprecedented medical and social progress. In materialist physiology Dermott had found final deliverance: the means of wresting science from Anglican control and "knocking down *the golden bar of exclusion*."[103] This was the social rub, the imperative of

100. J. Epps 1828:100, 1836:3, 23. Moore 1985b, 1986a for a broader discussion on the naturalization of the spiritual world and consequent transference of religious authority, especially in the later decades.

101. On the charges and countercharges: L 1828–29, 1:582, 620; 2:230, 326, 454, 494, 625. Dermott's hatred of the clergy is evident from his rejoinder (1828–29, 2:230–32).

102. "Mr. Dermott on the Immateriality of the Soul," LMSJ 1831, 7:128; also 1833–34, 4:819–23. On Epps's sectarian use of phrenology: LMSJ 1829, 2:88–90.

103. Dermott 1833:20–21; Moore 1986a on the "new reformation" metaphor.

open competition which in the 1830s consumed the Nonconformist mind. He advised his students in 1834 to observe above all "political honesty . . . which is every thing that favours that sacred trinity, *industry, talent, and truth.*" For a radical in the 1830s this was the "moral basis of all true religion."[104]

Comparative Anatomy and the Anticruelty Ethic

A physiologist . . . scoops out the brain of a cat, and proceeds to record the effects of this operation on the *vital* phenomena—of what? *a dead cat.* Would he not be better occupied in observing the modifications of function which accompany the gradually curtailed development of the brain, from man down to the *mollusca?* In this series we find parts of the organ disappearing, and, if we go yet lower, to the *radiata* we find the whole disappearing, but all without *mutilation.*

—Michael Ryan in 1835, promoting the new serial anatomy as an alternative to vivisection[105]

The Benthamism that swept through the medical schools also reinforced the ethical awareness of the chapel anatomists. Like Bentham, the Gerrard Street teachers passionately opposed vivisection and animal mutilation. But it was not so much their alliance with the secular Benthamites on this point as their cooperation with the conservative evangelicals that I now discuss. For this anticruelty agitation brought Wilberforce's saints close to Dermott's sinners, and by selling the new philosophical zoology as a substitute for vivisection the Soho teachers could attempt to align these conservative religious reformers behind the new science.

To understand how far attitudes toward animal creation reflected religious differences, consider the problem of death. A Guy's Hospital surgeon told his students in 1825 that "the death of an animal is a very different thing from that of a man. To an animal, death is an eternal sleep; to man, it is the commencement of a new and untried state of existence."[106] But some Christian sects accorded brutes quite a different status. John Wesley in particular had extended the benefits of restitution to the whole of suffering creation, and with the rapid expansion of Methodism through 1840 the doctrine of animal immortality reached sizable audiences. Whatever else of Wesley's he rejected, Dermott believed that "animals partook

---

104. Dermott 1835c:367; 1828–29, 1:42–43.
105. "Importance of Comparative Anatomy," *LMSJ* 1835–36, 8:184–85.
106. Blundell 1825–26:116.

of the fall of Adam along with man."[107] The spread of the doctrine was helped, no doubt, by the fact that the philanthropic Society for the Prevention of Cruelty to Animals was selling Wesley's lectures for distribution at threepence a hundred.[108] But animal immortality was only one of a number of doctrines that underwrote the anticruelty agitation. The movement was also fueled by the growing meliorism and anti-Calvinist belief that all could be saved. For example, John Epps, as averse to slavery as to a salvation restricted to the chosen few, fashioned a religious egalitarianism whose ethical benefits extended far beyond enslaved men to the forlorn beasts. He had not merely come "to regard the poor Indian slave as my brother," but actually to "consider all creatures as being equally important in the scale of creation as myself."[109] (Charles Darwin's abolitionist views and evolutionary revelations led to his similar exclamation, as we see later.) Epps grew steadily more sensitive to animal pain and came to embrace Christ as the redeemer of all life. For him, as for Wesley, "the whole creation travaileth and groaneth," and in old age he was comforted that his dogs would rejoin him in a future existence.

The liberal Anglicans by contrast found Wesley's groaning creation distasteful. Oxbridge divines did discuss animal suffering and death, showing that they were no less affected by the climate of concern. But their defense of Creative beneficence led them to look through Paley's rose-colored spectacles at a "happy world" teeming "with delighted existence."[110] To some Dissenters it seemed as if liberal Anglicans were turning somersaults to defend the indefensible: a Paleyite Creation that buzzed with contented life, a world in which pain increased pleasure and inequality was the essence of harmony. At Oxford William Buckland, speculating on the Divine skill that had gone into fashioning predators' teeth, concluded that such exquisite weapons speeded the death of the ill and infirm, increasing the aggregate health and happiness of the animals they preyed on. That this presented difficulties was obvious from the response of his grave young admirer, Richard Owen. Owen reduced the situation to its bleakest, picturing the unseen and unrelenting destruction in the seas, as voracious predators with "adamantine jaws" grazed the verdant fields of fixed, "helpless, yet presumed Sentient" polyps. "The Mythic Story of Andromeda chained to a rock to await in shrinking terror

---

107. Dermott 1828–29, 1:41. But he baulked at bestowing a soul on "a spider, a lizard, a snake, or a cockle" since this would necessitate giving them "a bible—a revelation—ministers—religions—and a future state," which would have been an "unmerciful extravagance in the Creator." Stevenson 1956:149–51 more generally on Methodism and physiology.

108. *Eleventh Annual Report*, SPCA, 1837, p. 149.

109. E. Epps 1875:61, 71, 107.

110. Paley 1830, 4:317.

the approach of the Sea-Monster" was nothing compared to the "ceaseless carnage" of the sessile polyps. Owen found this impossible to reconcile "with the dispensation of a Creation founded on Benevolence" and candidly shifted to an automaton theory of invertebrate life, leaving insects and polyps in a state resembling man's "during Somnambulism or in a Dream." He limited pain and consciousness to the higher zoological reaches, restricting the worst "warfare & destruction" to a shockproof, unaware world.[111]

These attempts to diminish pain and preserve a contented Paleyite nature served to highlight the different messages extracted from nature by many Methodists and neo-Calvinists. The oppressed medical sects could hardly sanction clerical claims that this was the best of all possible natural and social worlds. For Dermott and Epps the animals shared in man's Fall. All life "groaneth," struggling for salvation, afflicted by pain and deformity. It was an image based on a quite different reading of the Bible. Suffering implied consciousness, and the radicals accepted that animals as low as the polyp "are as really endowed with mind,—with a consciousness of personality, with feelings, desires, and will,—as man." Elliotson's belief that even the zoophytes approaching "vegetable simplicity" possessed "consciousness and perception, and volition" was founded in his faith (shared with all higher anatomists) that a granular nervous system would eventually be detected in these microscopic "brutes."[112] This sensitivity to consciousness and pain also tended to have more personal repercussions. For Epps's group, the question often became one of the moral justification of killing for food or vivisecting for medical ends. It is perhaps glib and unfair to contrast Buckland's gastronomic views of creation—his notorious predilection for potted ostrich or crocodile at breakfast (a propensity for treating nature as a table d'hôte which he passed on to his son Frank)—with Epps's anguish over the suffering of a single animal "painfully put to death" for our food.[113] Nevertheless, worry over the slaughter of animals seems to have been a more frequent feature of private-school thought, and vegetarianism was sporadic among the sects. (Although it still savored of Oriental extremism; it was Hindu philosophy, for example, that enabled fellow phrenologist and physician Thomas Forster—a Pythagorean with an otherwise impeccable evangelical family tree—to refuse

111. R. Owen, "Hunterian Lectures on the Nervous System 1842: Lecture 1," 5 April 1842, ff. 9–10 (BMNH MN 1842–48).

112. Elliotson 1835:4, 32: J. Epps 1828:118.

113. E. Epps 1875:470, 558, 561, 582–83. Cf. Bompas 1888:69; Gordon 1894:104–5; Ritvo 1987:238 for Buckland's attempt, starting with "elephant trunk soup," to eat his way through the animal kingdom. On Forster see Stevenson 1956:132–33, 151; and vegetarianism, J. Turner 1980:41–43.

flesh.) So to a large degree the contented Paleyans and frustrated neo-Calvinists and Methodists adopted contrasting ethical attitudes, justifying their social actions according to rival interpretations of the Bible and animal creation.

The private teachers' response to the evangelical missions is more complex. After looking at Dermott and Epps, the religious differences of the wealthy, predominantly Church of England evangelicals are immediately apparent. While outsiders such as Epps abominated the "fornicating" Church, the Anglican evangelicals actually worked from within for its spiritual regeneration. Social differences were equally striking. These evangelicals were part of the respectable elite worried by working-class godlessness and crime, and, among their many organizations to combat this, the Society for the Prevention of Cruelty to Animals was set up in 1824 to police the "brutal" poor addicted to blood sports. The evangelicals offered the merchants, bankers, and magistrates, and still more their wives, moral certitude in an uncertain age. Whereas the medical Methodists and Baptists supported lay democracy and physiological materialism, the evangelicals reacted to rationalism and the dislocating effects of industrialization by looking inward, to the nation's morals, urging philanthropy and charity rather than democracy. Indeed, William Wilberforce tied Jacobinism to impiety and justified the government's repressive measures during the Regency, encouraging the prosecution of sellers of seditious literature. The salient feature of these evangelical missions was their obsession with working-class mores. Worried by urban crime and vice, they closed street fairs, gambling houses, dance halls, and, through the SPCA, London's cockpits and sporting dens. The SPCA, like its prototype the Society for the Suppression of Vice, used paid agents to pry into plebeian morality and used the courts to uplift it. The SPCA's "ultimate aim" in attempting to stop working-class barbarity toward animals was to "civilize manners, and hence make the masses more receptive to religious instruction,"[114] for which reason it was cordially hated by Cobbett. The society's patrons were bent on tempering the violence of the urban poor before it was manifested in more revolutionary ways—attempting to inculcate in the members of the lower orders "a degree of moral feeling which would compel them to think and act like those of a superior class."[115]

Whatever the medical Dissenters' antipathy to the Church (regenerated or otherwise), they were equally anxious to stop the cruelty and

114. B. Harrison 1967:100; Ritvo 1987:chap. 3. On the evangelicals see Bradley 1976:103, 111; Brown 1961:chap. 9. The society obtained its Royal Charter in 1840.

115. J. Turner 1980:55.

blood sports, if for different reasons. Because of their distinctive understanding of animal existence—as conscious, suffering, even immortal—
they backed the society's moves to curb the cockfights and bull runs. They
broke decisively with the secular radicals on this score. The parliamentary
radicals objected to the SPCA's methods and aims. They deplored its use
of paid informants and resisted any attempt to extend police powers. The
SPCA collaborated with the Metropolitan Police, rewarded its constables,
hired retiring "peelers" as inspectors, and sponsored bills that, it was
feared, would increase the scope for the "mischievous class of inform-
ers."[116] The radical M.P.'s main complaint was that those convicted were
invariably working men (mostly hackney coachmen and cart drivers fined
for beating their horses, although cases also involved cat skinning, dog
mutilation, bull baiting, and so on).[117] They were well aware that the society's inspectors and spies were vilified by workmen. Street fights frequently broke out as the dens were cleared, leading to injuries and even
to the murder of an inspector after one Hanworth cockfight in 1838.[118] The
radical M.P.s therefore voted against an SPCA-backed bill in 1835,
drafted to prevent cruelty on cattle drives and make it illegal to keep pits
for animal baiting or cockfights within five miles of Temple Bar. They argued that it would restrict the recreation of the poor and increase the
scope for police-spy activities. In contrast, Brougham and his followers
welcomed the bill, as did Nonconformists such as Brookes's pupil William
Youatt (honorary veterinary surgeon to the society), for whom the "cause
of suffering animals is a sacred one."[119] But this support was not unequivocal. Many Whigs and Dissenters were angered that the SPCA, hand in
glove with the squirearchy, was tacitly condoning stag hunting and grouse
shooting—leaving "the amusements of the higher classes, untouched"

116. *Hansard*, 1835, 29:537–38. *Ninth Annual Report*, SPCA, 1835, p. 6; *Eleventh Annual Report*, SPCA, 1837, p. 9; B. Harrison 1973:787 for the propolice attitude. On the
resistance to granting police wider powers, see Wakley's speech against the Metropolitan
Police Bill (*Hansard* 1839, 47:1291–92). Clarke (1874a:99–100) considered Peel's new policemen either "fools" (the country recruits) or "knaves" (the Bow Street hard types).

117. *Sixth Report*, SPCA, 1832, pp. 28–35; *Eleventh Annual Report*, SPCA, 1837, pp. 9,
76–88.

118. B. Harrison 1973:789, 800, 1967:116–17. On the 1835 bill: Moss 1961:49; Fairholme
and Pain 1924:71; *Times*, 8 October 1835; *RSPCA Records*, 1831–41, 4:56 (held in the society's offices); *Hansard* 1835, 29:538.

119. *Tenth Annual Report*, SPCA, 1836, p. 43, also pp. 5, 47–48. Youatt was editor of the
Veterinarian, and from 1833 the medical superintendent at the Zoological Society (Desmond
1985a:231). He lectured at the university in 1831–35, justifying veterinary science on the
Benthamite grounds that man's moral duty was to minimize the pain suffered by his animal
servants, although he also pointed out how much of the national wealth was squandered
through poor livestock care (Youatt 1831–32:80–82).

while curbing those of the "humbler classes."[120] They were adamant that the law should be stretched to include the equally bloody avocations of the rich. But the society's noble patrons, no strangers to the grouse moors themselves, saw the restriction of such gentlemanly activities as totally irrelevant to the society's tacit aim, the pacification of the poor.

The SPCA became very fashionable in the later 1830s. It was even powerful enough to have the troops called out in 1838 to stop London's bull running in Stamford. Nor did its inspectors patrol the slum districts alone. Splinter organizations proliferated early in the decade. There appeared the Association for Promoting Rational Humanity Towards the Animal Creation (f. 1831), which urged the substitution of education for prosecution; the more extreme Animals' Friend Society, founded in 1832 by Thomas Forster and the Jewish philanthropist Lewis Gompertz (who actually refused milk or carriage rides); even by 1835 a Ladies' Association for the More Effectual Suppression of Cruelty to Animals.[121] The grubby dens and dog-skinning barrow boys had offended more than the genteel parish ladies. The widespread concern explains the societies' broad-based support. Many Dissenting teachers backed the 1835 bill; some even sent in their guinea subscriptions to the SPCA. Youatt, Dermott, and Peter Mark Roget were all paid up, as was Bentham (who believed that the state should take action to protect animals), while the Congregationalist professor of mind at London University, Rev. John Hoppus, supported the Rational Humanity campaigners.[122] So a number of disparate groups collected under the animal welfare banner in the 1830s, and they agreed on the need for protective legislation, if for different social and religious reasons. In return for supporting the cause, the Gerrard Street activists could now canvass this wider constituency to gain backing for their own antivivisectionist strategies.

The general resurgence of interest in the moral issue of pain in nature was reflected in medicine, where it centered on experimentation on living animals. The "physiological butchery" carried out by vivisectionists was an evocative issue. Zealots such as Gompertz fulminated against the medical men who dissected "Dumb Animals Alive." He even advertised in the *Morning Herald* for information that could lead to the conviction of the "delinquents" involved. One angry surgeon replied that "Nature refuses to unveil many of her secrets to us by gentle means; that such se-

120. *Hansard* 1835, 29:538.

121. *RSPCA Records*, 1830–46, 5:1–14, 22–65, 66, 108–10. On Gompertz: J. Turner 1980:41–43.

122. *RSPCA Records*, 1830–46, 5:20. *Sixth Report*, SPCA, 1832, p. 26 for Bentham's and Roget's guinea subscriptions. *Fourteenth Annual Report*, RSPCA, 1840, p. 112; *Fifteenth Annual Report*, RSPCA, 1841, p. 156 for Dermott's.

crets are only to be wrung from her."[123] But this kind of response was atypical. Vivisection equally upset many medical Benthamites and Dissenters, for whom care was a sacred duty and wasteful sacrifice a sin against Creation. Like Youatt, himself a veterinary Benthamite, many actively supported the SPCA's manifesto declaration that "however justifiable it may be to conduct certain experiments of a painful nature . . . with a view to determine some important question in science, not otherwise attainable, yet all must agree 'that Providence cannot intend that the secrets of Nature should be discovered by means of cruelty.'"[124] This clause had been inserted in response to François Magendie's experiments in France on the spinal nerves of puppies in 1822, which had so horrified the British (and led to calls in the House for his expulsion when he crossed the Channel to repeat them). The popular revulsion was shared by many medical reformers. Knox had a "natural horror" of experiments on living animals and detested "the aimless probings and torturings practised by Magendie and his disciples."[125] Fletcher claimed that he had never exposed "a suffering animal even to students of medicine . . . for the purpose of elucidating any point of physiology," and he believed that vivisection was far less necessary for physiology than was supposed—although he would damn those who killed for sport before he condemned the surgeons who vivisected to improve man's lot.[126]

Vivisection became a heated issue in the mid-1830s. The earl of Carnarvon drew sustained applause during his 1837 address to the SPCA for equating street-urchin dog-skinners with those "barbarous" surgeons who mutilated living animals, describing "these scientific speculators in blood and torture" as more degraded "in point of civilization and Christianity than the benighted savages of Scythia." He feared that students were being habituated to the horrors in "these charnel houses" (the medical schools). Individual discretion was being so grossly abused, he insisted, that the law must step in to "protect and avenge."[127] The rhetoric was powerful but tarred all with the same brush. Youatt refused to condone the experimentation but exonerated British surgeons, laying the blame squarely on the Continentals. The *British and Foreign Medical Review* agreed that "there is in this country a very strong and, we think, a very

123. Chippendale 1838–39:357–58; "butchery," French 1975:28.
124. Fairholme and Pain 1924:191. On Magendie: Genty 1935:1271; French 1975:19–20; Gallistel 1981:357–60; Cranefield 1974. Not all of Methodist origins abhorred vivisection; Marshall Hall (Manuel 1987) was a striking exception.
125. Lonsdale 1870:215, 238.
126. J. Fletcher 1836:10–11.
127. Carnarvon 1837:19–21. Youatt's response: *Eleventh Annual Report*, SPCA, 1837, p. 49.

proper feeling against the indiscriminate employment of experiment
which characterizes some of the continental schools; and the objection
applies, not only to the cruelty, but to the worthlessness of the practice."
Magendie's experiments were considered scientifically indefensible because they were haphazard, wasteful, and the results inconclusive. For
the *Review* the "severest experiments" were only justifiable when carried
out to ascertain some "definite and important point from which improvement in practice may reasonably be expected to result." Elsewhere it suggested that experiments should be cautiously used only to verify opinions
"founded upon other grounds."[128] This was a common proviso—even the
vivisectionist Charles Bell supported it. It was however Grainger's ampli-
fication that pointed to the unique private-school solution. He protested
that "if vivisections are not performed in entire subordination to comprehensive views of the general laws of organisation," not designed and
tested in accord with the deductions of human anatomy, then they were
"much more likely to retard than to promote" physiology.[129] By mid-
decade the radical Dissenters had already taken the moral high ground,
pushing this program of subordination to an extreme and offering their
comparative anatomy as an effective total alternative to vivisection. Such
an idea could be sold not merely to fellow anatomists, but to the growing
evangelical and Dissenting antivivisectionist camp, thus attracting a new
audience for the new anatomy.

The Gerrard Street teachers were active animal welfare lobbyists. It
was Dermott in 1833 who alerted the SPCA inspectors to cases of the
Newport Market butchers dismembering sheep alive. He also testified in
the ensuing court case, obtaining a conviction and a ten-shilling fine.[130]
He told his students that zoology,

revealing to us the numerous senses existing in animals, and their high developed
state, teachers us an important Christian lesson. It shows us the degree of pain
which those we call brute animals are susceptible of; and this conviction brings
disgrace and shame upon us as a nation, that such monstrous cruelties should be
practised as are committed daily upon quadrupeds by the English—cruelties that
the Brahmins would shrink from with abhorrence. I allude more particularly to
the brutal manner in which we have the animals slaughtered for our daily suste-
nance. It is astonishing that persons knowing how cruelly animals are butchered
and skinned alive, can feed with satisfaction, even to a state of satiety. It is a national sin that seems to give our profession of Christianity the lie; and it will be a
great national sin until a legislative enactment is established controlling and su-

128. "German School of Physiology," *BFMR* 1838, 5:99; "Macauley's *Essay on Cruelty to
Animals," BFMR* 1839, 8:356–57.

129. Grainger 1842–43c:173.

130. *Seventh Annual Report,* SPCA, 1833, pp. 26–27, 42.

perintending the manner in which animals shall be butchered, and such an enactment as shall do honour to an enlightened nation professing Christianity.[131]

These teachers were thus predisposed toward anatomical approaches that circumvented the dissection of living animals. Having adopted a developmental zoology, an anti-Cuvierian lineal model in which organ growth could be traced from monad to man, they pointed out that this new science presented us with a living "history" of the human frame—a visual "explanation" of origin and structure. Ryan stated controversially that "nearly all that is satisfactorily known in physiology has been derived from an extended survey of the various forms of organized beings," arguing that physiology always had been subordinate to comparative anatomy. The need to stop any repeat of Magendie's mutilations and to protect laboratory life had become another reason for increasing the status of higher anatomy, moving it from a "collateral" position to center stage. Life's scale presented what Ryan called a natural series of "experiments" which obviated the need for any disruption of living tissues. Nature provides her own "dissections," in the sense that the structural and functional history of an organ in a higher animal can be read from the zoological scale. The liver, for instance, could be traced from its complex ducted form in man through a sequence of simpler states to its granular origins in invertebrates. The history of this complex structure could be ascertained while leaving the study animal "entire and healthy." Since

we have reason to be daily more distrustful of the results of direct experiment on the living organism, and further encouraged to trace the connexion of certain functions with certain organs, by marking the increased perfection of the former, invariably accompanying the fuller expansion of the latter, and *vice versa* in a descending scale, the system of animated nature, if properly viewed, presents a grand series of those experiments which physiologists ineffectually institute by the barbarous mutilation of living animals.[132]

So the "proper" approach, again, came from a study of higher zoology. Belief in the homology of organs and unity of the series also convinced the young Unitarian William Carpenter that studies of organ "histories" could replace vivisection—that we could make "observation a substitute for experiment." The results would be more certain and free investigators from the moral guilt "every humane mind must feel to the infliction of unnecessary tortures upon beings endowed with sensations as acute as our own."[133] In fatally mutilating animals "we are studying death, not life,"

131. Dermott 1835c:363.
132. "Importance of Comparative Anatomy," *LMSJ* 1835–36, 8:184–85.
133. W. B. Carpenter 1838c:341.

said Ryan, and no physiological investigation can be as effective as "following each organ or system from its most rudimentary form to its highest perfection."[134] So the Dissenting acclamation for zoologists such as Fletcher, Grant, and Knox testified to both the political and the ethical value of the new science. By concentrating on the ethical side, the school proprietors could now attempt to carry the religious conservatives with them. They could draw on the wider antivivisectionist coalition in support of their demand for the centrality of the new anatomy in physiological science. Tellingly, the SPCA, which had deplored the physiologists' "butchery," welcomed comparative anatomy's advance as a sign of the "spirit of the times."[135] The new image of higher anatomy as humane, safe, and sure had a strong appeal in an age worried that man's works were daubing blood on God's Word.

This state of affairs also helps explain the relegation of experimental physiology in early Victorian England to what Gerald Geison calls a "stagnant backwater."[136] Geison suggests that students taught Paley's natural theology in the hospital schools believed that they could deduce physiological function from anatomical structure. In his view, therefore, the rise of an independent physiology had to await anatomy's "emancipation" from the fetters of Paleyism some decades later. But if we look beyond Bell and the Paleyan consultants to the larger enterprise of London medicine we get a more complicated picture. In the Dissenting schools catering to the rambunctious "third estate," Paleyism had already been largely superseded by this Continental anatomy in the 1830s. In these schools, antivivisectionist sentiments and the deployment of higher anatomy combined to preserve the emphasis on structure. There was not so much a physiological impoverishment as a specialized comparative approach to meet local needs. Indeed L. S. Jacyna has shown that, far from the schools' unphysiological science being sterile, it was given an unexpectedly "creative and dynamic" aspect by the new philosophical anatomists.[137] The proprietors used their comparative anatomy to explain the origin of organic structures and proclaimed that the science was capable of generating higher laws to rival those of Newtonian physics. Control over such prestige-enhancing laws in an age of social change was crucial to the low-status teachers facing the might of the corporations. It brought them the help of the Benthamites. And, by turning the new comparative anatomy into an ethically acceptable alternative to vivisection, the Christian radicals were now beginning to attract a wider scientific and religious audience.

134. "Importance of Comparative Anatomy," *LMSJ* 1835–36, 8:184–85.
135. *Tenth Annual Report*, SPCA, 1836, p. 31.
136. Geison 1978:3; cf. Mazumdar 1983.
137. Jacyna 1984a:48.

5

Accommodation and Domestication: Dealing with Geoffroy's Anatomy

It is well that there should be in physiological science, as elsewhere, a conservative section, who may restrain the movement party from advancing with unsafe rapidity.

—The moderate *British and Foreign Medical Review* playing the Tories off against the radicals before steering higher anatomy in its own direction[1]

Ryan's *Journal* was not alone in its leftward drift in the early 1830s. The same trend occurred in the medical quarterlies, which were respectably reformist but equally frustrated by the colleges' intransigence. The *Medico-Chirurgical Review* (*MCR*), snarling in 1830 at the "lying Lancet"[2]—and observing that Wakley's tone owed more to his hatred of monopolists than "love of right and justice"—was thundering its own warning to the colleges within three years. By 1833 many moderates, including Wakley's erstwhile enemies on the *MCR*, had edged into an uneasy alliance with the ultraradicals. The *MCR*'s founder-editor James Johnson, a former naval flag-surgeon and author of Grand Tour guides, was himself now pleading guilty to "levelling upwards."[3] These years, then, saw a short-lived coalition between militants, GP teachers, and moderate reformers, and their combined strength was evident in their defeat of the filibustering Tories in the Westminster Medical Society's long-running debate on "one faculty" in 1833–34.[4]

But the alliance remained fragile. It centered on little more than a common commitment to break the existing monopolies. Even then, the

1. "German School of Physiology," *BFMR* 1838, 5:100.
2. "Hunterian Oration. By Mr. Guthrie," *MCR* 1830, 12:528; see also 455.
3. "Medical Statistics and Reform," *MCR* 1833–34, 20:567–71. By 1833 Johnson's "open hostility" to the *Lancet* had ceased, whereupon the *Medical Gazette* stepped up its pressure "censuring and carping at him," especially for his own "levelling sentiments" at the Westminster Medical Society: "Medical Reform," *MCR* 1833–34, 20:284. On Johnson's life: "Biographical Memoir of Dr. James Johnson," *MCR* 1839, 30:637–43.
4. For Johnson here: "Westminster Medical Society," *MCR* 1833–34, 20:185–87, also 284–85; *L* 1833–34, 1:466–69 for his leveling upward. For the *MG*'s attack on this: "The

professionals barred from positions of power and the radicals fighting for members' rights had different reasons for wanting the corporation privileges destroyed. Johnson's "levelling upwards" was a far cry from Dermott's leveling downward, and designed to satisfy a different section of the community. In such an alliance, where plans for the future medical government varied between the factions, joint action was necessarily restricted to common iconoclastic declarations. Johnson, for instance, could rival Wakley in slamming the "tyrannical" RCP monopolists, complaining that they recognized only Oxbridge degrees because these were supposed to be evidence of a moral education. "Holy St. Francis! is it come to this?" he cried:

Have the *medical* Fellows of Oxford and Cambridge monopolized all the morality and religion of the profession, as well as all the snug appointments belonging to the Corporations! Must we poor licentiates be damned in the next world for deficiency of religion—while we are degraded in this world, for want of that morality which is concentrated in Oxford and Cambridge![5]

This outburst was not aimed at the universities so much as the RCP's Anglican exclusivity. Indeed, Johnson, a Derry-born Protestant, deplored Wakley's vilification of the traditional seats of learning. He came close to approving their professional and classical curricula, and he actually sent one of his sons to Cambridge where he became a fellow. Yet Johnson despised the Pall Mall Tories, who dealt in discriminatory legislation, restricting the fellowship and electoral power to Oxbridge graduates.[6]

But whatever the differences, the coalition opened up channels for the dissemination of the radical anatomies through the liberal medical press. This was not a passive diffusion, however, made easier by common commitments; it involved an active uptake on the part of the moderates. The new doctrines were as valuable to them as to Wakleyans for iconoclastic purposes.

One Faculty," *MG* 1833–34, 13:402–5; and on Johnson's cease-fire with the *Lancet*, ibid., 529–39, 564–67, 601–2. While Johnson now called himself a radical reformer, he distinguished Wakley as an ultraradical.

  5. "Vindiciae Medicae," *MCR* 1834, 21:113–14.

  6. "Medical Reform," *MCR* 1833, 18:582–83, 19:568; "Biographical Memoir," *MCR* 1839, 30:643.

Morphology Enters the Moderate Quarterlies

Of two out of three English works, which are at present in the hands
of the student of comparative anatomy and physiology, we know that
the authors are largely indebted to [Grant] for their materials.

—*British and Foreign Medical Review*, 1842[7]

Here I want to sketch the political ideologies of the reviews, and relate
these to the growing liberal interest in higher anatomy as the decade progressed. In medicine, it was Johnson's *Medico-Chirurgical Review* and
John Forbes's *British and Foreign Medical Review* (*BFMR*) that catered to
a wealthier Dissenting clientele (hence their expense: six shillings a number). The reviews demanded a legislative curb on "aristocratic" privilege
and sought to shift power toward a new type of medical professional: often
London University-educated, industrious, Dissenting, concerned to in-
tegrate science and medicine into the Benthamite program for social welfare and legislative reform. These reviews promoted a naturalistic anatomy, with—as Carpenter put it—"law and order" replacing Divine fiat.
As a result, they commended the better parts of Combes's phrenological
program and Chambers's cosmogony of lawful progression. Indeed, they
were conspicuous in praising the *Vestiges*, delighting in the fact that it
would pique the "bigotted saints" who still believed in the Six Days. The
MCR admitted that the *Vestiges'* "doctrines have come out a century before their time," while Carpenter in the *BFMR* praised the "beautiful"
book for its ennobling conception of natural law as a divine "predetermi-
nation based on perfect knowledge."[8] (The fact that neither reviewer was
entirely happy with animal transmutation was almost an afterthought.) It
is not surprising, then, with the proprietors holding a naturalistic ideology of science, that both reviews greeted Geoffroy's lawful unity of structure enthusiastically.

Unlike the levelers, the reviews did not want the Royal Colleges abolished so much as legislatively checked by a new medical "Commons." In
other words, medicine was to move from absolute to constitutional monarchy, in which the licentiates could elect members to some sort of new
"Lower House" with powers of veto and ratification. In 1830 Johnson ha-

---

7. "Dr. Grant's *Outlines of Comparative Anatomy*," *BFMR* 1842, 13:218. The *Review* was
undoubtedly referring to Roget and Anderson or Carpenter.

8. W. B. Carpenter 1845a:155, 160, 167; "Vestiges of the Natural History of Creation,"
MCR 1845, 1:147, 157. "Combe's Phrenology," *MCR* 1831, 14:321–22; "Phrenology," *BFMR*
1840, 9:190–210.

rangued the Tory leader of the "Upper House," Sir Henry Halford, president of the RCP, urging him to sanction just such a lower chamber:

Has the constitution of England been weakened by an admixture of aristocracy and democracy? What would the House of Lords be, without that of the Commons? Why should not the LICENTIATES of the College have some tie or connexion with it. . . . The immense body of Licentiates labour under the same disabilities as the Catholics lately did . . . they must consider themselves as degraded outcasts from the College.[9]

This, remember, was at the time of the Reform Bill agitation, which sought to promote greater power sharing between landed wealth and new trading capital (much of it in the merchant Dissenters' hands). Johnson's was not a call for demolition of the RCP any more than the respectable reformers in the country wanted the abolition of the Upper House. The gutting of the real House of Lords in 1834 did not elicit the hallelujahs from the wealthy that it did from the watching crowd.[10] Moderates lobbied for an ordered extension of the medical franchise and the establishment of a licentiates' "Commons." But no more—both reviews barred access to "the multitudes of labourers who rush towards the harvest of medical practice."[11] They urged a uniform liberal education to "Scotch" standards, to be made fairly expensive to allow only the wealthy to specialize. Ostensibly, a uniform fixed-cost education would reduce discord, alleviate overcrowding, and raise standards. But with the working classes priced out and hospital nepotism crippled by educational equality, the wealthier middle classes would obviously have a head start in this professional free market. Johnson's "democracy of medicine"[12] was a middle-class meritocracy, his specialist training a means of upholding the "inequality of *real rank*." In short, it favored the middle-class professional expert. This is what he meant by "levelling upwards": providing a superior uniform education, but also rewarding wealthy talent with titles as a stimulus to drive the specialists above the common run. Given that the *BFMR* was also attempting to enshrine middle-class privileges in law and suggesting that the London University be granted educational powers over the whole profession and benefits beyond those enjoyed by both the GP teachers and hospital consultants, one can see how profoundly the reviews differed in their class interest from Ryan's and Wakley's radical weeklies.

9. "College of Physicians," *MCR* 1830, 12:521.

10. The young Thomas Laycock, watching the fire, was appalled at the mob's glee and "execrable language" as the House of Lords burned (Cope 1965:172–73).

11. "Tax on Entering the Profession," *MCR* 1830, 13:440–41.

12. "Medical Reform," *MCR* 1833–34, 20:281, also 567–71.

As the Reform crisis peaked in 1831, the *MCR* warned the "medical MAGNATES" that if liberal calls were not heeded, radical passions would erupt into "physical force"; better concessions now than an English July Revolution.[13] But there was no medical Reform Bill. And despite persistent rumors in 1833 that the RCP was about to undergo reorganization, wiser heads knew that "Kings and corporations" never surrender their privileges voluntarily. Now the *MCR* too organized parliamentary petitions and demanded a Royal Commission to investigate the colleges' discrimination and rotten-borough proceedings. Tempers frayed as still no word issued from Pall Mall:

Like the Jews of the Temple, nothing could persuade these ELECT that their sacred, time-hallowed, king-blessed, monk-erected edifice was vulnerable in any quarter, fallible in any point, or capable of improvement, even in the most trivial particular! All was Millennium with the College—they were the Heaven-born Esculapii of England—who sucked in medical science, and practical experience, with their Greek and Latin on the Cam and the Isis! The reign of Tory corruption and Corporation monopoly was never to have an end.[14]

So Johnson too threw his *Review* behind the call for "a thorough—indeed *radical* reform." Only it was to be orderly, legal, and benefit not Dermott's chimney sweeps or Wakley's republican rabble, but the disaffected non-Oxbridge professionals excluded from the corporate power structure.[15]

The infectious radicalism and fluctuating alliances helped the new morphologies spread right across the reform board. In the medical societies Paris-trained Geoffroyans rubbed shoulders with these angry moderates busily petitioning Parliament. The new anatomies thrived under these conditions. For the five years or so following his 1833 *Lancet* lectures on comparative anatomy, Grant was lionized by the left; now moderates too praised his course as the first systematic exposition of higher anatomy in the country.[16] The *BFMR* considered the course trenchant, original, and unblemished by the "pompous dogmatism of some foreign *savans*." Most of all it attracted that key word *philosophical*. It was sanctioned as a sign of professional competence, its superiority lying in its reduction of the arcane facts of morphology to a set of "general laws,"[17] which indicated a legislated organic progress. The *MCR* greeted the first numbers of Grant's

13. "Medical Reform," *MCR* 1831, 14:573–74.
14. "Medical Reform," *MCR* 1833, 19:567.
15. He did urge the GPs to combine, but, because this was to break the "thraldom of *trade*" in which they were held by the Apothecaries' Company, he expected the result to be nothing like workers' trade unions so much as the gentlemanly PMSA pressure group: "Medical Reform," *MCR* 1833–34, 20:279.
16. "Dr. Grant's *Outlines of Comparative Anatomy*," *BFMR* 1842, 13:218.
17. "Outlines of Comparative Anatomy," *MCR* 1835, 23:376–79.

cheap, paperback *Outlines of Comparative Anatomy* in 1835 as marking "an era in the history of anatomy and physiology in this country." The book was superior to the older dry compilations; it would also rid English physiology of the disfiguring vitalism that had long vanished on the Continent but still persisted among Hunter's heirs at the College of Surgeons. The *MCR* bemoaned the poor press that philosophical anatomy received in England. Its tenets had "too often met with the ridicule" of the professors in the hospitals, corporations, and old universities as a result of their "ignorance of the facts upon which the science rests." Grant and the growing "philosophical school" were "to dissipate this ignorance," overturning the scientific and moral claims of the corporation monopolists.

The *Review* supported Grant's Geoffroyan theory of vertebral elements and invertebrate-vertebrate analogies with but few cavils. It went on, in a sympathetic (if belated) review of Geoffroy himself in 1837, to praise the university's lead in philosophical anatomy, while raising Grant to the head of the handful of morphologists at home (including Knox and Owen) competent to tackle the intricacies of French "embryological anatomy." With the *MCR* no longer at loggerheads with the *Lancet*, it encouraged further patronage by Wakley, suggesting that he follow up the comparative anatomy course by publishing Grant's lectures on human physiology. The professor's "accuracy," "intimate acquaintance" with human and comparative anatomy, and "the philosophical cast of his mind" show that there is no "man in this country better fitted to do justice to the subject."[18]

The quarterlies in 1837–39 devoted a series of major reviews to the Continental comparative anatomy of Geoffroy, Serres, Meckel, and Tiedemann. The new morphological concepts were extensively discussed: Geoffroy's homologies, unity of composition (the "great principle" reigning "over the whole of zoological science"), recapitulation, Serres's centripetal development (the idea that the embryological growth of each organ starts at the periphery and works toward the center), and the "constant and precise rules" governing fetal retardation and the appearance of monsters (teratology, as this part of the science was now being called by Geoffroy's son Isidore). All tended to prove that the "elements" of animal structure were "everywhere identical [and] disposed according to invariable rules."[19] Commentators were not slow to see the value of this kind of science. We know that radicals were attracted to a progressive Geoffroyan nature, in which "the multitude of beings which compose the animal series" were treated as "the innumerable parts of one immense whole."[20]

18. "On Philosophical Anatomy," *MCR* 1837, 27:85, 87, 106.
19. "Saint Hilaire on Teratology," *BFMR* 1839, 8:4–7.
20. Ibid., 5.

Even if more moderate men were to subject the doctrines to a continual critique and revision in the 1840s, in the blustery 1830s they too were often prepared to use naked Geoffroyism and threats of "physical force" against the placemen: anything to wring concessions. Continental anatomy was useful in its leveling guise for undercutting the "magnates'" natural theology, and with it their moral and scientific leadership. The *BFMR*'s finger wagged at the entrenched medical baronets. Their Cuvierian anatomy, in which each animal type was "peculiar" and to be explained in unique functional terms, could give them only "partial views [of nature]"; it reduced zoology "to the sterile observation of facts, without reciprocal connexion, rational analogies, or possible consequences."[21] The Oxbridge school "does not penetrate beneath the surface." Restricting its donnish study to a functional explanation of form, it had abdicated its philosophic leadership to the new morphologists. Whatever their flaws, the higher anatomists had at least uncovered the lawful "*process* of change" in nature.[22] This is what attracted liberals: the binding legalistic requirements which the science entailed. Nature was governed by "invariable rules," which left it subject to a progressive lawful change. Continental anatomy might have "failed to realize its magnificent anticipations"—and conservative constraints had wisely impeded the radicals' rush toward pantheistic excess—but liberals now saluted the new anatomists for having brought nature, man, and mind under the control of the laws of progress and development. They had undercut Cuvierian teleology, the mainstay of the ancient universities and corporations, and ushered in a lawfully constrained, morphologically based science, in which the capricious Paleyite monarch was ousted by a more culturally appropriate Divine legislator.

By the late 1830s, when these reviews appeared, power was beginning to tilt toward the new class of medical specialists outside the Oxbridge orbit—to the sons of the Dissenters now taking their civic seats. The lawful morphology was welcomed by these reformers because it helped dislodge the old Anglican elites, outlawing their regal Paleyism with a piece of stern Calvinist legislation. Even so, some moderates still sought to modify Geoffroy's doctrines, to take off their radical edge. They also worried about the destructives' leveling tendency, which threatened perennial turmoil and the very stability of the new specialist class. A study of individual cases shows the subtlety and variation of this modification process. Since Unitarians were so influential in bringing a Calvinist ethos into science, I start with William Benjamin Carpenter and fellow Bristolian,

21. Ibid.
22. "German School of Physiology," *BFMR* 1838, 5:86, 89, 100.

physiologist, and Benthamite Thomas Southwood Smith: Carpenter because he strongly endorsed *BFMR* social policy and trimmed the new morphology accordingly, and Southwood Smith because he is indispensable to understanding the Unitarian origins of Carpenter's scientific Calvinism.

Divine Government, Unitarian Physiology, and Popular Sovereignty

Many writers . . . seem in general to think that the Deity bears *no* relation whatever to space; that, in fact, he is actually present *nowhere*, and that of course it is only in a figurative sense that he is omnipresent. But surely it is more just to conceive of him as *really* pervading all space, as *actually* present in every part of the universe.

—Unitarian minister, sanitary reformer, and physiologist
Thomas Southwood Smith on the *Divine Government*[23]

Unitarians were influential in the intellectual culture of the merchant and industrial regions. In the large towns, leading Unitarian families invested money in the literary and philosophical societies, Mechanics' Institutes, and liberal press. Unitarians had fiercely resisted the evangelical revival and carried Enlightenment rationalism into nineteenth-century science and theology. They rejected Christ's divinity and the supernatural structure of Christianity, and dismissed such orthodox doctrines as original sin, the immaculate conception, and the existence of hell. Southwood Smith's move to Unitarianism was typical. He had studied for the ministry at the local Baptist Academy in the commercial seaport of Bristol. But like many other Calvinists at the time he rejected the doctrine of election and denied that perpetual damnation awaited the impenitent and ignorant. For this his family cast him out, and he lost his Academy benefaction. He was welcomed into the Unitarian fold, and from the Lewin's Mead minister in Bristol, John Prior Estlin, he learned the "universality of Divine benevolence"[24] and accepted a moral government impartial in its distribution of Divine favors. Smith's move was common; Unitarianism had a strong Calvinist core, and it had a particular appeal to radical Baptists and freethinking Christians in their ethical flight from orthodoxy (especially in the decades of Tory oppression following the French Revolution).[25] The

23. T. S. Smith 1866:7.
24. Ibid., 117. Poynter 1962:382–84; Lewes 1898:7–12.
25. Seed 1986:112–15, 1982:3–13.

Unitarian community funded Smith's medical studies and ministry at Edinburgh (1812–16). It was in this period of struggle and tragedy (his young wife had died of fever in 1812) that the *Divine Government* was written. It was a powerful work: a sweeping condemnation of the malevolent doctrine of eternal damnation and a commitment to social change. Published in 1816, it sold well in the Regency. As a meliorist tour de force which claimed "divine authority for a deterministic 'Law of Progress,' "[26] it captured a reformist standard in the Peterloo period and ran to four editions during the next decade.

Progress, melioration, and law characterized Smith's theology, just as they did his physiology. In *Divine Government* he argued that divine omnipresence implied God's real, pervasive presence. This he equated with naturalism, accepting that God's operation gave nature its strict, predetermined course, which we could interpret in terms of cause and effect.[27] In other words, this immanent activity demanded a lawful cosmogony: since the animal and moral worlds were subject to God's ordinances, secondary causes were manifestations of His actions. There was no interfering "outside" Deity, no arbitrary providence to give nature the appearance of whim. Smith accepted a Calvinist predeterminism at both the genetic level (animals act according to the "settled principles of their nature") and the environmental: a man might act according to his free will, but God determines the circumstances and therefore foreordains the choice. It was a harsh naturalistic Calvinism, but it begat a benign reformism. With God devoted to the reclamation of character and to an increase of happiness, social progress and the alleviation of suffering were inevitable.

The rational Dissenters' campaign for religious liberties (Unitarianism itself had technically been a penal offence until 1813) pushed many like Smith into the Benthamite camp. In London in the 1820s he became Bentham's physician and his collaborator on the Constitutional Code, and he moved to the center of the city's Benthamite community based around the new *Westminster Review* (f. 1824).[28] But the optimism of the *Divine*

26. Poynter 1962:384. Studies of Unitarian science, theology, and politics have usually concentrated on Joseph Priestley in the late eighteenth century, especially on his rational Dissent and philosophy of matter and spirit (McEvoy and McGuire 1975; Wilde 1982:109–10, 126; Schaffer 1984), although Raymond and Pickstone (1986) cover Unitarian science teaching in the following century.

27. T. S. Smith 1866:6, 7, 11, 15, 16–17, 134–35. Raymond and Pickstone (1986:155) give another view of the way science for the Priestleyans led through the alleviation of suffering to moral progress.

28. Smith had authored probably fourteen articles in the *Westminster Review* by 1830. Houghton 1979 looks at the *Review*'s Unitarian contingent. For Smith's collaboration with Bentham: T. S. Smith and Bentham 1829; Smith 1832.

Fig. 5.1. Southwood Smith was involved in many urban improvement projects and campaigned for decent city dwellings for the poor. Here (seated, far left) he is attending a meeting of the Health of Towns Association, listening to the marquis of Normanby speak. (From *Illustrated London News*, 1847, 11:393; courtesy Wellcome Institute Library, London)

Government began to fade as he started working at the London Fever Hospital and experienced the poorer quarters of the East End (he also served in the Eastern Dispensary and the Jews' Hospital in Whitechapel). He now campaigned in the *Westminster Review* for health, sanitation, and educational reforms (see fig. 5.1). And with that young "briefless barrister" Edwin Chadwick,[29] recruited into Bentham's circle in 1829, he investigated child factory labor, laying the basis for the 1833 bill banning the employment of children under nine.

The creed of immanent Being encouraged Smith to develop a compatible physiological reductionism. Like other radical teachers, he rejected the liberal Anglican conception of vital powers and Creative interference. With matter animated directly by the Divine Presence, life could be defined physiologically by the phenomena accompanying organization. He announced in the *Westminster Review* in 1827 that life was characterized by a "combination" of properties and that the physiologist's task was simply to enumerate them. As he put it: "Life depends on certain conditions; these conditions depend on certain arrangements of material substances; such arrangements of material substances constitute organization; organization is thus an essential condition of life."[30] This as usual proved upsetting to the Anglicans. But it was also too extreme for the

29. Poynter 1962:381–82, 387–89; Lewes 1898:53; Halévy 1950:100, 114–15.
30. T. S. Smith 1827b:208, 212.

older Whigs and Broughamite educationalists intent on "diffusing" suitable science to the lower orders. Smith was now working alongside these men. For him, "redeeming the people from a degraded condition [was] a duty,"[31] and he believed that education could help the more motivated wage-earners escape from poverty and slum life. Because of this he backed the Society for the Diffusion of Useful Knowledge, sitting on the original committee. But, being on the Unitarian left, outside the Whig circle, he began diffusing a physiology that outraged many. Roget had been asked to write a book entitled *Animal Physiology* for the society; as he was unable to do so, the title passed to Smith in 1829.[32] The SDUK's referees—Bell, Roget, and the London University medical professor John Conolly—were all appalled at the result. On seeing the manuscript, Bell sent a terse note to the SDUK secretary, demanding that the society dissociate itself from Smith's "opinions about *Life*."

Life is nothing but organisation in action . . . you ought to know is an opinion highly objected to by many good men and is certainly incorrect. This is the line of argument which Mr. Lawrence followed, borrowed from the French, & is objected to by the Physiologists of this Country.

There appears to me no Call upon the Society to propagate these opinions. Were they correct I would disregard the Consequences, but they are incorrect and offensive to many.[33]

They certainly were to the moderate Broughamites and Paleyites. Bell's own gentlemanly Whiggism was much closer to the *Edinburgh Review* line. He hated radical extremism, saw the Reform Bill agitation doing "no earthly good," and—like Conolly—resigned his London University post during the student riots. He had no doubts about his own rank as the "captain of anatomists," and public confirmation came in 1831 with his knighthood shortly after the Whigs took power.[34] His own SDUK contribution, *Animal Mechanics*, toed a totally different line. It was the most design-orientated of the tracts, sold thirty thousand copies by 1833, and was praised by Brougham as the most original exposition of Paleyism to date. Before Bell's SDUK book, Brougham believed, the antimaterialistic physiologies had been "managed" so badly that "it is hard to say if scoffers did not, upon the whole, gain more than worshippers."[35] Bell had

31. T. S. Smith 1866:104.
32. Roget to T. Coates, 10 January 1828 [1829?], 8 June 1829 (UCL SDUK).
33. Bell to T. Coates, 2 September 1829 (UCL SDUK).
34. G. J. Bell 1870:173–74, 251, 318, 324.
35. Brougham 1827:519–20, 1835:2, where he rated Bell his favorite "fellow-labourer." Brougham and Bell (1836) for their cooperative writing. But through all Bell remained sensitive to the problem of reducing Paley to the hackneyed: G. J. Bell 1870:339, also 295, 302, 314–15, 317. On the print run, Grobel 1932, 3:681. See also Hays 1964.

changed all that. Because his book had relied heavily on mechanical analogies to prove intelligent design in nature (he saw skull tectonics better Gothic architecture, and spinal design improve on mizen masts), he was very careful to dissociate himself from any implied mechanism at a deeper level. Mechanical relations were unknown in the "finer textures of the body." Life in fact preceded organization and existed already "in simple and uniform substances, where there is neither construction nor relation." It was "an endowment, not resulting from organization . . . but, on the contrary, producing it."[36] The radicals' doctrine was stood on its head: the "living principle" was a beneficent endowment; in Bell's view it was a Divine power of arrangement.

Conolly shared Bell's caution. He deplored the "extreme admiration of the French schools" shown by some English teachers.[37] He also warned his students off Wakleyan politics lest youthful imprudence cause them to sacrifice "probity and honour." In the same way he abhorred Lawrence's scientific pyrotechnics, telling his pupils that they lived "when not *knowledge* alone, but *character* is power; when knowledge without character can procure no more than temporary and very transient pre-eminence."[38] Now the character of gentlemanly medicine was again under threat. Like Bell, Conolly refused to condone Smith's physiology. He urged Smith to amend his inflammatory text, the more insistently because the SDUK aimed at the working-class consumer,[39] and the extent to which radical artisans had plundered Lawrence's own "blasphemous" book was well known. A society set up to beat the street traders by supplying cheap stabilizing literature now stood in peril of fueling the agitation by sanctioning the very physiology it was trying to stamp out. The SDUK's ideals

36. C. Bell 1827–29:6–8, 10, 33, 44–50; G. J. Bell 1870:295. Looking at Bell and Smith, one understands why the SDUK deliberately refrained from republishing Paley in order to prevent a sectarian split (Brougham 1835:2).

37. Conolly (1828:475), championing Bell over Magendie. The same fear of the French existed among the zoologists at the time (Desmond 1985a:174–75).

38. Quoted in Wakley's antagonistic editorial: "The London University," *L* 1828–29, 1:50–52. J. Clark 1869:3–8; Bellot 1929:156–58. Conolly (1831) resigned from London University during the riots in 1831 with a plea to his students for order. He was succeeded at the university by Elliotson.

39. This belief informed the referees' actions in general; thus they advised authors that technical terms should be intelligible to laborers: e.g., Roget to T. Coates, 28 August 1829; also J. Conolly's letters of 21 December 1830, 18 November 1831 (UCL SDUK). This target class—the more literate artisans—is also discussed in Shapin and Barnes 1977. On the other hand, Bell (G. J. Bell 1870:295) and Brougham (1828:154ff.) did not doubt that the SDUK tracts would also reach a higher-class audience. And one has only to see Darwin's sisters "swearing" by the SDUK's *Penny Magazine* to realize how deeply these works actually penetrated wealthy Whig society (F. Burkhardt and Smith 1985–86, 1:284, 299).

were ill served by teaching life's innate sovereignty; a republican street literature juxtaposing a reductionist physiology and fierce democracy proved this. Organic parts had to be shown in interdependence, harmoniously functioning through the controlling agencies of Divine command. Nature was not a series of atomistic, self-controlled operations. Life was an irreducible property, whose power derived from above. Bell's and Conolly's fears were not groundless; Smith's physiology was indeed to be appropriated, like Lawrence's, by the pauper presses promoting atheism, socialism, and working-class suffrage.[40]

Conolly complained to the SDUK Committee more than once of Smith's "complete and almost contemptuous disregard" of the criticisms.[41] His and Bell's reports did eventually elicit a conciliatory note. Smith told the SDUK secretary that he was "not only *desirous* but *anxious* to obviate all possibility of misconstruction," believing that his "very guarded expressions" in the offending passages had been misunderstood. Of course the problem was not his reductionism being misconstrued, but his mooting it at all. He explained that, "far from wishing to countenance" subversive materialist doctrines, his "express object was to guard against them by a studied correctness & precision of language."[42] The first sixpenny part of *Animal Physiology* in 1829 remained unexpurgated. The offending passage stood: on life's "inseparable relation" to organization, and physiology being a study of "organization when in action."[43] He repeated that life was not a power, or a "real and distinct agent," indeed that the word "life" only had "*scientific* meaning" in the sense of an observable phenomenon. In fact at the time of Bell's criticism Smith had sent the secretary a new explanatory paragraph, intended to clear up the matter. Even if it was not too late for the printer, it was too brazen to be printed. In it he exposed the differences between his and Bell's physiologies:

The term life . . . has been used in two different senses, the one denoting certain phenomena cognisable by the senses; the other the presumed cause of these phenomena. Much confusion has been introduced into physiology by employing the term in the latter of the two senses. The existence or non-existence of the supposed cause of life as a principle single, undivided & distinct from organization has acquired an undue degree of importance from its being erroneously imagined to involve the truth & certainty of our hopes of a future state of conscious being. It must be evident however that these hopes rest upon our knowledge of the will of the Deity; that the subtle essence & the gross material are alike capable

40. Desmond 1987:101, n. 95.
41. Conolly to T. Coates, 26 February [n. y.], 28 January 1828 [1829?] (UCL SDUK).
42. Smith to T. Coates, 29 September 1829 (UCL SDUK).
43. T. S. Smith 1829–30:1, 2.

of being destroyed or perpetuated by that power which first breathed into man's nostrils the breath of life & that the alarm which has been felt upon the subject is therefore altogether unfounded.[44]

Carpenter was to make the same point, that vital and spiritual causes had been confounded, resulting in a fear that to relegate one was to relinquish the other.[45] In a sense, of course, Unitarians were equally conflating the issues: for them, the indwelling spirit meant that life could be defined by observable organizational phenomena. This definition already subsumed, as it were, the Divine dimension. The upshot of all this was that in Smith's theodicy, with its rationale for disestablishment, civil liberties, and spiritual sovereignty, there was no need for vital powers emanating from God, interpretable through an official priesthood. Nor was talk of eternal torment from the pulpit any use in controlling the working classes. With all matter alive to the Presence, an immanent spiritual guidance ensured that all men would be saved, according to the true meaning of Divine benevolence. Obedience came not from threats of Divine retribution but through education and democratic reform. Thus Anglican state hegemony was illegitimate theologically, pernicious socially, and unfounded physiologically. It is no coincidence that as the SDUK moderates condemned the spiritual values of Smith's physiology, so cautious council members at the London University refused to condone its moral consequences. The evangelical Zachary Macaulay, alarmed at the heterodoxy of *Divine Government,* overrode the council's declared nondenominational policy, defeated the Benthamites, and blocked Smith's application for the first chair of moral philosophy.[46]

Only a year after the *Divine Government* was published, Coleridge (no friend any more to the Unitarians) complained that the street "ruffians" were using this kind of atomistic, self-empowered physiology to give a spurious scientific respectability to their struggle for democracy. He was convinced that it was poisoning the minds of the rabble. True, Smith was a democrat, but he put quite a different gloss on the matter. The Benthamites wanted a centralized parliamentary meritocracy, a government by specialists, with the local legislative assemblies popularly mandated. Smith accepted that the working classes had to be helped to organize in industry, to exploit knowledge resources, and to improve their living con-

44. Smith to T. Coates, 29 September 1829 (UCL SDUK).

45. W. B. Carpenter 1839–47:144.

46. Morgan 1882:373; Bellot 1929:59, 108. Notice however that the council was not averse to using Southwood Smith to get itself out of a scrape: he was invited to complete the physiology course after Bell resigned mid-term in 1830: Smith to Brougham, 26 November 1830 (UCL HB 14346). Edinburgh Town Council refused to consider Carpenter's application for a chair in 1842 on the same grounds—that he was a Unitarian (W. B. Carpenter 1888:31).

ditions. They had to be assisted in their means of "acquiring an honest independence, of qualifying themselves for the possession & exercise of the elective franchise."[47] Chartist insurrections were symptomatic of a failure to implement a full Benthamite program, as he explained in his plea for a commutation of the death sentence on John Frost and the leaders of the Welsh Chartist uprising in 1839. Smith condemned the armed rebellion, but in his view it underlined the need for more educational enterprises and better living conditions. For the Unitarian Benthamites, civilizing the lower orders was to be achieved using books rather than sabers, housing rather than gallows. Pauperization, ignorance, and lack of proper representation were not conducive to a peaceful redress of grievances.

Smith's rational Dissent was accompanied by more than physiological reductionism. His demand in the *Divine Government* for the removal of the state-fostered evils of poverty and exploitation was based on a faith in social and organic mobility. The links between the moral and organic aspects in Smith's philosophy show quite clearly the social basis of his science of nature's development. This becomes evident as we look at the way he naturalized the social ideals of striving and progress.

To justify the universal restoration of sinners to a state of purity, Unitarians were forced to tackle the problem of evil. Smith rejected one possible utilitarian explanation—that evil is suffered by the few for the good of the many. To him this was incompatible with the idea of Divine justice. Since God was omnipresent in nature, the sin and evil committed by men exercising their "free will" must have been foreseen by God and a painful corrective automatically built into the system.[48] Thus free will was compatible with a predestined, regular, law-bound cosmos: all willed disorder engendered an environmental corrective, leaving society harmonious, progressive, and responsive. The Poor House, poverty, and the industrial ills were just such social correctives, indicating that the state had taken a wrong turn. Now the radical Dissenter's duty was to expose the cause of this suffering and to reform society in order to bring it back in line with God's original intent. Malthusian population pressures should not lead automatically to the Poor House; they could be counteracted by rapid educational and technological advances. But reformers also had to awaken the lower classes "to the need for self-development,"[49] and thus the imperative of providing good, cheap education to enable the industrious to

47. T. S. Smith, "Memorial to Lord Normanby" (BL Add. MS 44,919, f. 131). This is his plea for Frost's life.
48. T. S. Smith 1866:23–39, 66–72.
49. Ibid., 99–102, also 76. T. S. Smith 1830:280–81.

escape the poverty trap. As medical and manufacturing entrepreneurs,[50] Unitarians were underwriting free-market mobility and a natural inequality arising from competition. Commercial striving, not gentrified indolence, was conducive to the "highest happiness." Man was designed to strive in order to "form and prove" his character; the most natural society was competitive, capitalist, and progressive—with the stimulus for development being the desire to escape "poverty, dependence, and servitude."[51] In other words, the social evils were part of a benevolent dispensation and intended to sustain a self-reforming, progressive society.

Others making similar capitalist demands had by 1831 already accepting a self-reforming, transforming nature. As we have seen, the commercial tree grower Patrick Matthew argued that in nature the selection of "hardier, more robust, better suited to circumstances individuals" kept the species at its competitive best, but that in society the process was being circumvented by the existence of aristocratic privilege. The "law of entail, necessary to hereditary nobility," he warned, "is an outrage on this law of nature which she will not pass unavenged." In the wild, species had a "self-regulating adaptive disposition," honed by competition; Malthusian superfecundity and the ensuing selection caused them to undergo "new diverging ramifications," a self-transformation. Noblemen and their priestly supporters, in disobeying nature's law, in upholding the inheritance of rank and position, were blocking this progress in human society. The situation could be overcome by opening up trade fully to mercantile incentive and competition—in short, by establishing a democratic, capitalist society free of protectionism and tariffs. A "new state of things is near at hand," he announced, a time when the "merchant and manufacturer will no longer be . . . harassed"; they will be in control.[52]

The link between capitalist self-government and organic mobility is even clearer in the *Divine Government*. According to Smith, just as escape from the poverty trap is the raison d'être of social mobility through competition, so nature's inexorable ascent enabled animals to escape the "evil of imperfection." Nature and society were congruent. Both were improving and progressive: the moral evils of society were benevolent dispensations favoring working-class improvement, while in the animal kingdom the inferior organisms triumphantly progressed to escape their lowly station. Nature was dynamic, with all creatures "continually advancing

50. Holt (1938:36–68) and Seed (1982, 1986) discuss industrialization and the Unitarian families in the cotton, engineering, chemical, and pottery trades.

51. T. S. Smith 1866:78–79.

52. Matthew's work reproduced in Dempster 1983:98, 99, 100, 107; Wells 1973:229, 323–29.

from one degree of knowledge, perfection, and happiness to another."[53] The natural and social were therefore "inseparably connected," subject to the same Divine dispensation, driven to escape the same evil, subject to the same everlasting ascent. Smith spoke as a social Lamarckian when he wrote that "all reasonable beings, however inferior the condition in which they commence their existence, are destined to rise higher and higher in endless progression, and to contribute to their own advancement." The belief more obviously legitimated the removal of civil, religious, and trading restraints (to allow the realization of God-given potential and ensure a stable society) and made health, education, and medical reform inevitable. But it also explains why the Unitarian intelligentsia—Smith, Carpenter, we might include Charles Darwin (who had attended the Shrewsbury congregation as a boy, and whose mother and wife were Wedgwood Unitarians)[54]—had little trouble accepting evolution. Even if they disputed Chambers's idea of Divine law as a kind of regal edict, Unitarian doctors could still accommodate the Vestiges' upward-sweeping development,[55] unlike most Anglicans, whose moral authority rested on state privileges, eternal retribution, a social hierarchy, and static Creation.

The sciences that underwrote evolutionary and democratic progress were Oxbridge targets because they threatened Anglican privilege and raised the specter of Dissenting hegemony and even working-class emancipation. But Unitarian interests were tied to an industrializing, changing Britain, and it was no coincidence that the new physiologies took hold as the Nonconformist manufacturers and professionals began making political headway. Wellington's bluster about the hated Reform Bill switching power from the Anglican landowners "to another class of Society, the shopkeepers, being Dissenters from the Church, many of them Socinians [Unitarians] [and] atheists,"[56] contained a germ of truth. As power began shifting toward medical Dissent, so the new physiological sciences became more conspicuous. The civic pockets in which the improving anatomies flourished—in the Nonconformist schools and new university, in the

53. T. S. Smith 1866:65, 66, 104.
54. Barlow 1958:33; Moore 1989a on the Unitarian ambiance of the Darwin family home.
55. Unlike some Unitarian theologians. But notice that when Vestiges was panned, for example by the Christian Reformer, which had a conservative scriptural bias (McLachlan 1934:186), it was because the reviewer—taking a Humean view of "law" as a constant order of events—could not abide Chambers's theologically "dangerous" concept of law-as-logos, as "something which is separate from the Deity himself": "Vestiges of the Natural History of Creation," Christian Reformer 1845, 1:34–39. So on this point the Reformer actually stood on the same ground as Smith and Carpenter. Sympathetic Unitarians ignored Chambers's own meaning and rendered Vestigean "law" more immanent (W. B. Carpenter 1845a).
56. Holt 1938:132; Cowherd 1956;79; Seed 1986:111, 1982:6.

Benthamite bureaucracies, and among radical journalists—confirm that a manufacturing, utilitarian, radical Dissent was the best carrier.

W. B. Carpenter and Lawful Morphology

We cannot . . . see any ground for the indignation with which Mr. W. regards the speculations of M. Geoffroy St. Hilaire, on account of their neglect of what he seems to consider the *end* of physiological research.

—Young William Carpenter defending Geoffroy against Whewell's criticism that he did not explain structures in terms of their end function[57]

Southwood Smith's moral government testified eloquently to his training at Lewin's Mead chapel. Here Estlin, ministering to Bristol's Unitarian merchants, preached the doctrine of universal salvation.[58] In 1817 Estlin was succeeded by Lant Carpenter, a controversial and extremely influential preacher. Carpenter inspired a rising generation of Unitarian intellectuals, including James Martineau and the *Westminster Review*'s John Bowring. Like Smith, Carpenter argued the absurdity of God's suspending cause and effect, when this was the actual manifestation of His action in nature (thus ruling out liberal Anglican conceptions of "Creative Interference"). Carpenter also preached the exclusivity of "physical causes" in nature, and his theology of immanent action sustained an equally thorough mental determinism. The "operation of divine agency," he believed, resulted in "all the properties (including the *powers*) of the mind" being "subject to laws, in the same manner as the properties of matter," which made the mind at once natural and susceptible to God's "immediate influence."[59]

Lant's austere young son William was to embrace this naturalistic cosmogony as a matter of course. With his father preaching at Lewin's Mead, William was at the center of the patronage web linking the city's leading Unitarian families—those who had paid Southwood Smith's tuition fees a few years earlier. The young Carpenter was himself encouraged to take up medicine and was apprenticed to the eye surgeon John Bishop Estlin (the late minister's son), whom he accompanied to the West Indies in 1833. Carpenter's reformism was typical of the Unitarian elite's. Like Est-

57. W. B. Carpenter 1838c:339.

58. Estlin 1813. W. B. Carpenter 1888:39–40 for his belief in universal restoration. Neve 1983 on the merchant context of Bristol science.

59. L. Carpenter 1822:57–59. Fees at Carpenter's school were a hundred guineas a year, which gives an idea of the wealth of his congregation (McLachlan 1934:113).

lin he became a passionate temperance campaigner, and with Smith he encouraged the study of human physiology as a necessary adjunct of the public health movement.[60] His deterministic theology also left him holding a physiology much like Smith's. He eschewed vital agents and saw matter endowed with the properties through which the Creator effected his purpose in nature. For Carpenter in the 1830s, physiology as much as physics was a science of material causation. Gravity was an effect of the properties of matter as life was of tissues.[61]

How he then applied this Unitarian understanding is interesting. Being a generation younger than Smith, he was primarily interested in the new French morphology imported into Edinburgh and London during his student years. And not only into the capitals—Tory Bristol too was subject to the same trends. Bristol in 1830 was a declining Atlantic port, with a wealthy merchant elite and a querulous working class. (Carpenter witnessed the riot in 1831 when the Bishop's Palace was burned down by a crowd furious at the Lords for blocking the Reform Bill.) Through conservative Whigs like the Rev. William Conybeare, the town fathers retained intellectual links with Oxford, and as governors of the Bristol Institution they sponsored an anti-Lamarckian science teaching Paleyite subservience.[62] But by the late 1820s the old anatomy took to the defensive as the new morphology entered the city. First the young physician Henry Riley returned from Paris, impeccably French-dressed and mannered, and enamored of the new Geoffroyan anatomy. Then in 1831 the suave Edinburgh-trained John Symonds arrived, a rich Whig reformer intent on voting in Reform whatever the costs, and full of Knoxian notions.[63] The two subjects that would prove so important in Carpenter's professional life—physiology and forensic medicine—were taught by Riley and Symonds in the new medical school in Old Park in 1832. (Carpenter was to take over Symonds's summer course in 1836 and lecture jointly with Riley in 1839.) The Bristol Institution was already beginning to reflect these changes by 1831. Here the young Carpenter (see fig. 5.2) may have heard Riley's technically proficient, probing expositions of Geoffroy's

60. W. B. Carpenter 1843, 1:1–9. W. James 1855:14 and W. B. Carpenter 1851 on temperance.

61. W. B. Carpenter 1839–47:142, 150, 1838b:331–33, 348–53; Jacyna 1981:114. Jacyna (1984b:43–44) shows that Carpenter's attempt to give physiology and physics "comparable epistemological status" was also a device to enhance the prestige of a low-status discipline. It is less surprising, knowing his theological background, that Carpenter should later have taken up study of the correlation of forces (V. M. D. Hall 1979; R. Smith 1977).

62. Neve 1983:187–90. Conybeare 1835–36:6–7 and G. T. Clark 1835–36:21–22, 39 for the Bristol elite's attacks on Geoffroy and Lamarck.

63. Symonds 1871:xi. On Riley: Prichard 1894:4–9; M. A. Taylor and Torrens 1986:140.

Figure 5.2. William Benjamin Carpenter, an adept systematizer of the newer biological sciences. By T. H. Maguire, 1850. (Courtesy Wellcome Institute Library, London)

unity of structure between 1831 and 1833: courses delivered to packed houses and receiving detailed press coverage.[64]

In 1834–35, after returning from the Caribbean island of St. Vincent, Carpenter attended Grant's course in London. This made a strong impact,

---

64. "Bristol Institution. Lectures on Anatomy by Dr. Riley," *Bristol Mercury*, 15 March 1831, gives an inaccurate transcription of the first three 1831 lectures. For a full transcription

and he always "looked back with peculiar interest" on it for "the mental quickening and special love of the subject which it aroused within him." [65] The young Unitarian appreciated the unifying power of Geoffroy's principles taught in Gower Street. He immediately penned a paper for the Bristol Institution's house organ, the lackluster *West of England Journal of Science and Literature*, arguing for the functional unity of respiratory organs and developmental unity of vertebrate lungs and gills. [66] The twenty-two year old was already immersed in the works of Grant, Carus, Roget, and Tiedemann, reading about unity of plan and structural reducibility, and refining the writing skills that were to make him the foremost popularizer of biological principles before Herbert Spencer.

The powerful hold of the new morphology over Carpenter was apparent during his stay in Edinburgh (1835–39). The Athens of the age might have been declining, but higher anatomy was still dominant in the extramural schools. Carpenter's intent was to reduce physiology to a set of naturalistic laws proclaiming Divine omnipresence, and he began to conceive his role as a systematizer, tilling the works of others and cropping the physiological principles from which to forge expansive new "generalizations." (Carpenter and Spencer gave an unexpected new meaning to the contemporary title "cultivator of science.") Men such as Fletcher and Carpenter were quite conscious of the need to retail their new science, to get popularizations before a book-buying public. Partly this was to counteract the "inflammatory political trash" streaming off the working-class presses (much of which found its way onto tradesmen's shelves). [67] But there was also a need to displace unsatisfactory works such as Perceval Lord's *Popular Physiology*, published in 1834 under the auspices of the Society for the Promotion of Christian Knowledge. Geoffroy's morphological laws provided perfect material for Carpenter's mill grinding, but the new science was to be subtly refined to make it attractive to a much wider medical audience. Carpenter at first explored the idea of unity of function. At Edinburgh's Royal Medical Society in 1837 he used Geoffroy's principles to justify the analogy of wings and gills in aerial and aquatic

of the two subsequent years' courses: "Bristol Institution—Lectures on Erpetology" (13 lectures, April-May 1832), unidentified newspaper cuttings (Bristol Corporation Archives Office: Richard Smith Biographical Memoirs, 13: ff. 734–54); and the April 1833 course "On Comparative Anatomy and the Philosophy of Zoology" (ff. 720–31).

65. W. B. Carpenter 1888:10. He was awarded Grant's certificate of merit in 1835: *Distribution of the Prizes*, College Collection, UCL.

66. W. B. Carpenter 1835–36:221, 228, 286–87.

67. J. Fletcher 1836:18. Lord (1834:27) toed a Bell-like line, teaching that life was itself the cause of organization. Carpenter's *Principles of General and Comparative Physiology* was incomparably superior to Lord's book. It was also suitable as a classroom text, which Lord's was not.

insects. At this time he still accepted that each natural group "passes by almost imperceptible gradations into every adjoining one." And he acknowledged a structural unity that transcended Cuvier's divisions, like Geoffroy and Grant taking as his prime example the "very gradual transition" from mollusks to fishes. Despite Cuvier's warning that "an impassable gulf" separates these groups, "more extended researches" had shown that the cephalopod's nervous, skeletal, and locomotor systems were almost "on a level with those of the lowest cartilaginous fishes," while its "circulating apparatus" was "strikingly intermediate between that of the mollusca in general and that of fishes."[68] He too now claimed that the cartilages protecting the cephalopod's cephalic ganglia "obviously foreshadow the *neuro-skeleton* of Vertebrata."[69] And he sought precise analogies between the hagfish's brain and the squid's cephalic and suboesophageal ganglia. He believed that the vertebrate brain was constructed from invertebrate components, with added cerebral and cerebellar lobes. Indeed this is what was expected from Lamarck's "beautiful" classification, in which the mollusks and articulates rise from a common radiate base and progress in parallel toward the vertebrate level. Carpenter was now able to pinpoint the cepalopod and insect components which went to make the primitive vertebrate brain.

Other factors reinforce the view that Carpenter at this time was closely following the program laid down by the Gower Street Geoffroyans. Grant in the early 1830s was training students to apply his friend Marshall Hall's reflex theory to the insects, and Carpenter took up the challenge. The sensory and motor ganglia of mollusks and articulates were the subject of his prize thesis at Edinburgh in 1839 and his Bristol lectures in 1840. He extended the work of Grant and his skilled working-class protégé George Newport (studying the motor nerve tracts in insects) and confirmed the existence of Hall's reflex arc in invertebrates. Like Grant and Grainger, he held that vertebrate and articulate nervous structures were "conformable": they contained analogous segmental reflex systems (responsible for instinctive acts), as well as similar cephalic masses where the "*sensations* can be felt." And although these systems were differentially developed to leave insects more instinctive, there was still a mental continuum between insects, mollusks, and vertebrates.[70] His 1839 Edinburgh thesis

68. W. B. Carpenter 1837:97–98.

69. W. B. Carpenter 1839:44–45; 1840–41, 28:57–58; 1837:98.

70. W. B. Carpenter 1840–41, 27:939–40, 944; 1888:26. This taxonomic continuum in the "reasoning" faculties extended finally to man, whose mind differed only in degree from that of the higher animals (W. B. Carpenter 1843, 2:541–42). Jacyna (1981:112–13, 1984a) gives an excellent account of the romantic physiologists' understanding of the nervous system.

was by his own admission a compilation, but it clinched the case for many reviewers, and one journal at least promptly dropped its opposition to Hall's reflex arc in insects as a result.[71]

But revealing the connecting passages between *embranchements* was, in his view, different from accepting that "the whole animal kingdom is formed upon the same type."[72] On this point he was already moving away from the Geoffroyans by 1837. One reason was his interest in the new embryological ideas being imported from the Continent. Martin Barry— who had trained under Tiedemann in Heidelberg—began to publicize Karl Ernst von Baer's nonrecapitulatory embryology in 1836. (In this, embryos of different animal types were seen to diverge away from a similar-looking initial germ, rather than all to climb the same ladder of development. So, according to von Baer, a human embryo never passes through, or recapitulates, any states corresponding to adult mollusks, echinoderms, or insects.) Barry's imports ultimately undermined the recapitulatory axioms supporting an extreme unity of composition. He stripped away the embryological basis for believing in the homologies of, say, mollusks and fishes. Carpenter came to accept that the new embryology was incompatible with the notion of a single animal type, one model of which all actual animals were variants. And he began to doubt that "the transitory states [passed through by the embryos] of the higher animals furnish exact representations of the permanent forms of the lower."[73] But the doctrine of divergence away from the common germ toward an archetype characteristic for each division took time to catch on. It was also to find different uses among the rival groups. In Owen's Tory-Anglican hands it would quickly undermine the foundations of the despised mollusk-fish bridge built by Geoffroy's disciples. But not having Owen's anti-Lamarckian ax to grind, Carpenter's assimilation took longer and tended to be less destructive. During this time his Grantian model—with its connected *embranchements*—yielded only slowly. In 1841 Carpenter was still promoting Geoffroy's laws, a "single scale," transitional types, and von Baer's embryology—the concepts all apparently nestling together comfortably.[74] He came to a clearer understanding of the respective arche-

71. "Dr. Carpenter *on the Physiology of the Nervous System*," BFMR 1839, 8:511.

72. W. B. Carpenter 1837:99, 1841b:191.

73. W. B. Carpenter 1837:99, 1841b:196–97; Barry 1837a, 1837b. Ospovat 1976 and E. Richards 1987 on von Baer's embryology in Britain.

74. Jacyna 1984a: 59; Desmond 1985b:46–49. In *Principles* Carpenter (1841b:191) still believed that one division would show an "approximative tendency" to another; thus he talked of the nervous and osteological systems of the cuttlefish as the "rudiment" of vertebrate organization. As a result of von Baer's introduction (among other things), the later numbers of Grant's *Outlines* were beginning to look out of date as they came off the press in 1841: "Dr. Grant's *Outlines*," BFMR 1842, 13:217–18.

types only later in the 1840s. But even here we must tread carefully, for Carpenter's appreciation contrasted with Owen's. Carpenter, whose Unitarianism left him unperturbed by the kind of progressive lawful creation fashionable in London's rationalist circles in the forties,[75] differed socially, theologically, and scientifically from an Oxbridge-supported conservative such as Owen. Owen's archetypes were Platonic ideals and used for Coleridgean ends at the unreformed College of Surgeons, but Unitarians hated the "half-crazed" Coleridge and his idealism. They could no more accept an archetype as a Platonic ideal than the notion of law as a Divine logos interposed between the Creator and nature.[76] Carpenter was to interpret the archetypes materialistically, as immanent in nature. Indeed, later egged on by the brash young T. H. Huxley, he became a merciless critic of Owen's idealizations. Ultimately, Carpenter put von Baer's embryology to use in his different social program. It served as a guide to the true morphological laws through which the Divine expressed itself in Nature—laws of strategic importance in his Dissenting opposition to the Oxbridge clerisy.

How this program functioned we can see from his press articles, especially for the *British and Foreign Medical Review*. This was a journal resisting demands for a medical revolution and urging steady educational changes as the basis for a redress of professional grievances.[77] Carpenter was championed by the *BFMR* moderates. Circulating in-house was a belief that "almost in his pupilage" (he was twenty-six in 1839), he had "rivalled, if not outstripped, the most learned of his contemporaries."[78] His *Principles of General and Comparative Physiology* (begun in 1835, published in 1839) was hailed as a rich mix of Fletcher's philosophy and Müller's physiology. It equaled the *Review*'s "most sanguine expectations."[79] Recruited as a writer for the *Review* soon after its launch in 1836, Carpenter established himself here as a major medical and educational critic, eventually becoming its editor in 1847.

75. W. B. Carpenter 1845a; Desmond 1982:29ff.; Secord 1989. E. Richards 1987 for Owen's view of the *Vestiges*' embryology. In later years Carpenter was actually to help patch up the *Vestiges* (Chambers 1884:xxv). Yeo 1984 also discusses Chambers's audience.

76. Holt 1938:343; "Mr. Green on Vital Dynamics," *BFMR* 1840, 10:545–47. Desmond 1982:16–17, 37–38, 41, 92–93, 212–13, n. 39 documents Carpenter's later attacks on Owen.

77. "Medical Intelligence," *BFMR* 1839, 8:300.

78. "Dr. Carpenter," *BFMR* 1839, 8:507. Conolly was the *Review*'s original coeditor, but he retired, leaving it totally in Forbes's hands. The *BFMR*'s reformism was more compatible with, say, the PMSA's (with which Conolly was also associated) than with the trenchant BMA's.

79. "Mr. Carpenter's *Principles of General and Comparative Physiology*," *BFMR* 1839, 7:168–69. Carpenter (1838a:100) was distinctly impressed by Fletcher's *Rudiments*.

As the first Unitarian mayors took their municipal seats in the industrial towns, so Carpenter proposed ways of strengthening the professional hand of the merchants' sons in medicine. Hs saw the need for a "liberal" preliminary education for all medical men, with foreign languages and sciences substituted for the "exclusive" Oxbridge system of classical study.[80] Like all reformers he accepted the need for a standardized scientific training (modeled on that in Gower Street) and the removal of religious disabilities. But his was no Wakleyan voice. In opposition to the levelers, he proposed a subsequent "subdivision of medical *practice*" along the lines of the existing estates (surgeon, physician, GP, midwife); indeed his omnipresent metaphor of von Baerian progress from general education to medical specialization completely undermined Wakley's one-faculty call.[81] Carpenter's aim was to break the Oxbridge monopoly on position and place, and promote trained specialists to the higher ranks and income bracket. Rank for Carpenter equaled level of technical expertise, not inherited status. He was angling for legal reforms in order to redistribute power and create a new wealthy elite.

He held no destructive brief, nor was he promoting the independent teachers (who were training GPs to minister to the tradespeople and poor). The best evidence of this is that, in opposition to Wakley and the private school proprietors, he wanted London University's privileges increased. He suggested that the university should be amalgamated with the rechartered medical corporations (which would have given it inordinate power over the profession, decisively breaking the Oxbridge grip).[82] His *Review* was conspicuous among medical journals for its policy in this respect. Editorials praised the university's "complete" approach to medical education and suggested that it should be copied countrywide—indeed that similar "liberal" institutions should be founded right across the empire. It should be granted "higher powers and privileges" and be made the sole examining *and* licensing authority for England. Had this been

80. W. B. Carpenter 1840a:177. Following the 1835 Municipal Corporations Act, Unitarian mayors were elected in Manchester, Liverpool, Leicester, Bolton, Derby, Leeds, and Birmingham (Holt 1938:23, 217–41; Fraser 1979).

81. W. B. Carpenter 1840a:200, also 1840b:411–12.

82. W. B. Carpenter 1840a:203. Increasing the university's privileges was condemned by those both left and right of the *Review*. Wakleyans objected to any monopoly on principle—the private teachers because it threatened their trade, while the hospital consultants saw it undermining their established privileges. Even the *MCR*—drifting leftward but still sympathetic to many of the *BFMR*'s goals—was appalled at the prospect of the university getting a monopoly on degrees. Such an "odious mark of favouritism" would have been reprehensible. Johnson joined his "ultra-liberal contemporaries" in denouncing this kind of "invidious distinction": "Metropolitan University," *MCR* 1835–36, 24:597–98.

effected, it would have curbed not only the corporations' power, but also
the private schools'. A leader suggested that those receiving the bachelor's degree might be licensed as GPs, while others continuing to specialize for doctorates would constitute the "higher ranks":

> This arrangement, while recognizing the unity of the profession and the right of
> all to practise any of its branches, would still admit of those divisions into classes
> and those differences of rank, the convenience and utility of which have been
> sanctioned by experience. The equality of all the members of the profession would
> only be broken by the legitimate claims of superior talent, higher studies, or more
> enlarged experience. [83]

Such an extension of university privileges would have proved catastrophic
for the corporations: place and rank would no longer have been in their
gift, and with their loss of licensing power would have gone their ability
to control the profession through the regulation of diplomas. But shifting
power from the corporations, Church, and Cambridge to the new university would have suited the Dissenting nouveau riche. The private teachers and poor-practice GPs knew that the shift would replace nepotism
with new wealth: it favored the merchants' and professionals' sons by giving status to those who could afford to specialize longer. In the long run,
Carpenter and the quarterlies were restructuring rather than relegating
rank to give bourgeois wealth the educational edge.

Science was important in Carpenter's strategy to claw more power for
the metropolitan professionals. A good science-based education was already proving itself in London; as he noted, the Gower Street medical
school was outstripping its hospital and Oxbridge rivals, showing what a
combination of secularism and middle-class talent could do. And having
slipped its Oxbridge yoke, science was acquiring a new meaning in the
hands of the Benthamite "expert." Carpenter's own philosophy speeded
this process, for it undermined the Anglicans' approach to nature and
their claim to intellectual authority based on it. As a Unitarian, he was
concerned with material causation. Because God had impressed matter
with its properties in the Beginning, Carpenter felt free to interpret cause
and effect, and therefore natural law, as a sign of His action and intent.
This led Carpenter to deny that God could interrupt Nature (God could
hardly interrupt Himself). So all talk of miraculous interference on the
earth—or of fresh Creations of animals and plants at the beginning of each
geological epoch—was nonsensical. A conservative Cambridge divine
such as the mineralogist William Whewell might have sanctioned the miraculous introduction of mice and moles on high philosophic grounds, but

83. "Medical Intelligence," *BFMR* 1839, 8:299–303.

this only exposed the poverty of his philosophy. Nature was uniform. And with life subject to the same invariant cause and effect as the planets, it became the rightful province of the new biological specialist.

Another of Carpenter's differences with the Anglicans emerged when he reviewed Whewell's three-volume *History of the Inductive Sciences* in 1838. Carpenter told the physicist John Herschel that Whewell's "comprehensive mind had failed to appreciate the true import of the data" of comparative anatomy, no less than the "true mode of reasoning from them."[84] Carpenter was particularly galled by Whewell's chapter entitled "Doctrine of Final Causes in Physiology." Here, Whewell had restated Cuvier's axiom that the final purpose or function of an organ was the explanation of its structure. And he censured Geoffroy's followers for rejecting this doctrine of final causes, led on by their "false philosophy."[85] Geoffroy's theory of a unity of plan throughout nature was "utterly erroneous," insofar as it denied "an intelligible scheme and discoverable end, in the organization of animals." The doctrine of final causes was a guiding principle; to ignore it was a "mischievous error." It had been responsible for almost all the major discoveries in physiology. Without it, Cuvier could not have penetrated the mysteries of the past, or reconstructed extinct animals from scanty fossils.

Whewell would have hated Carpenter's ideal syllabus of university studies, which included the kind of higher anatomy taught in Gower Street. But Carpenter just as firmly deplored Whewell's narrow emphasis on individual adaptations and his attempts to stop all investigation at this point. The philosopher's job was to discover the "general laws" that transcended final causes and to comprehend the overall plan of nature. This Geoffroy had attempted, and Whewell's "indignation" at Geoffroy's work roused Carpenter to a spirited defense of his "higher" laws of animal structure. They offered a more exalted conception of Omnipotence and revealed the "vastness of that designing Mind, which, in originally ordaining them, could produce such harmony and adaptation."[86] Carpenter's *Principles* promoted this grander theology. Here he complained again of the extreme difficulty of proving a Designing Creator "from individual cases of adaptation of means to ends." But he believed that on witnessing the conformity of animals "to one comprehensive plan, and trac[ing] this throughout the extinct as well as the living beings of each type," everyone would admit that "such a plan *could* have originated no where but in Infi-

84. Carpenter to Herschel, 29 November 1839 (RS JH). Holt (1938:343) mentions other Unitarians attacking Whewell and Cambridge science.
85. Whewell 1837, 3:464.
86. W. B. Carpenter 1838c:338–39; Ospovat 1981:12, 1978:38–39. W. B. Carpenter 1840a:183 on his projected university syllabus.

nite Wisdom, and could have been executed by none but Infinite Power."[87]

Geoffroy's anatomy was incorporated into Carpenter's strategy. From it, he could point to a higher unity of plan and purpose. Rational Dissenters had uncovered a more magnificent design than the one conceived by the dons. All the time, though, the more radical connotations of Geoffroyism were being stripped away; eventually the emphasis was no longer on an extreme leveling unity, embracing life from the lowest worm to the highest human, but on a set of standard types of animals, each with its own regulatory plan.[88] Yet, however domesticated, Geoffroy's science still generated impassioned feelings. On top of this, there was an inevitable backlash against any naturalistic strategy that served bourgeois ends to the detriment of Oxbridge interests. This explains the polarized reactions to the *Principles*. Traditionalists objected to its lawful tone. Carpenter was adamant that "one simple law" had been impressed on matter by "Almighty *fiat*" at the Creation, to ensure the "general uniformity" of the cosmos, control the emergence of the planets, and "people all these worlds with living beings." The cosmos was regulated by God by means of this original "law." To say otherwise, to imagine that this "plan of the Universe, once established with a definite end, could require alteration," was "to deny the perfection of the Divine attributes."[89] Because the act of promoting organic laws rather than Divine interruptions and final causes was so contentious, he expediently dedicated the volume to Herschel (himself Cambridge educated). He told the natural philosopher that the book emphasized "the *laws* which modern [physiological] researches" have unfolded and that it illustrated "them by a comprehensive survey of the structure and function of *all* Classes of living beings."[90]

Those ideologically in line applauded *Principles*. The reform press welcomed it, and the Edinburgh and London professors adopted it as a textbook.[91] Carpenter was encouraged by this to drop his practice and devote

87. W. B. Carpenter 1841b:192, cf. 560.

88. By the time Carpenter (1843:52) popularized the type system, based on the dissection of the dog, lobster, slug, and starfish, he was effectively exploiting the idea of four plans.

89. W. B. Carpenter 1840c:2–3. This higher teleology brought Carpenter close to extreme latitudinarians such as the Oxford professor of geometry Baden Powell, whom he had met by 1838: Carpenter to Herschel, 24 July 1838 (RS JH). Carpenter (1838d:548–49) praised Powell's *Connexion of Natural and Divine Truth* (1838:150–56) for similarly exposing the "antiquated prejudices" of the Paleyites. See also Carpenter to Herschel, 29 November 1839 (RS JH); Desmond 1982:44–46; Corsi 1988:200, 264–65, 273–74.

90. Carpenter to Herschel, 24 July 1838 (RS JH). This letter also mentions Powell's support for his book.

91. It was recommended in Edinburgh by professors William Alison (medicine), Robert

himself full time to physiology. Still, he informed Herschel, "a heavy pecuniary sacrifice" was involved. Money was always tight, and as a family man in the 1840s (his first child was born in 1841) he remained financially straitened;[92] hence his sensitivity to those reviews accusing him of irreligion. In an age when a savant without a personal fortune had to have several livings, a bad press threatened to block future openings. One critic did indeed declare that Carpenter's view of nature, running its course "without requiring the continued superintendence of the Creator," rendered him "unfit for the duties of a Public Instructor." But Carpenter denied that this was his view—or that he was guilty of materialism. He repeated that the "properties first impressed upon matter [do not] *of themselves* continue its action," because these "impressed" laws of physics and life are "nothing more . . . than a simple expression of the *mode* in which the Creator is constantly operating on organic matter, or on organised structures."[93] It was this sort of attack that prejudiced his chances of getting a job, as he told Herschel. Accusations of impiety had nearly lost him his newly acquired lecturership at the conservative Bristol College, and he was forced to scratch around in 1840, collecting testimonies to counter the calumnies, clear his name, and certify his design arguments free from "the doctrines of a dangerous tendency."[94]

So promoting this Unitarian naturalism in Tory Bristol had its drawbacks. Carpenter had to maneuver adroitly to remain financially buoyant and prevent civic propriety from taking fright.[95] His career was eventually secured in London, where his physiology was well received—not only by the medical reformers, but by the rationalist intelligentsia generally. In 1844 he was appointed Fullerian professor at the Royal Institution. By now the bastions of hospital privilege had been breached by the new man,

---

Jameson (natural history), and Robert Graham (botany): Carpenter to Herschel, 8 February 1840 (RS JH); and by William Sharpey in London (1840–41:142).

92. As late as 1851 Carpenter was forced to decline a place on P. B. Ayre's committee collecting money for the impoverished Grant, apologizing that he was overloaded with work trying to make a living out of physiology and that he could only spare two guineas out of his limited income: Carpenter to Ayres, 12 June 1851 (WI). In London Carpenter alternated with Grant for the few paying positions, following him and Rymer Jones as Fullerian professor at the Royal Institution in 1844 and preceding him as Swineyan lecturer on geology at the British Museum in 1848.

93. W. B. Carpenter 1840c:2, 3.

94. Carpenter to Herschel, 8 February 1840 (RS JH).

95. In 1842 he tried (unsuccessfully) to get Herschel himself to review the second edition of *Principles* in the *Quarterly Review* as this "would have a most valuable influence on my future career" by silencing conservative critics: Carpenter to Herschel, 21 January 1842 (RS JH).

and he joined Hall and Grainger at the reorganized St. Thomas's Hospital.[96] Here, and later as registrar of the University of London, he continued to churn out books promoting temperance, tolerance, and a biology based on "*Law* and *Order.*"[97] His prolific output in the 1840s was rivaled only by that of his Methodist admirer Herbert Spencer a generation later—a man who, with the Huxleys, Tyndalls, Lewes, and Martineaus, was to continue Carpenter's bourgeois assault on science and society.

Peter Mark Roget and the Whig Compromise

Unlike the Benthamites, the older patrician Whigs never cared for the new scientific naturalism or for the rational Dissenters' innovative design. They now attempted to domesticate Geoffroyism in a distinct way, alloying it with Paleyan natural theology in order to make it palatable to genteel Bridgewater tastes. A study of the wealthy Whig Peter Mark Roget shows the political rent that this action caused and its institutional reverberations.

Roget, later of *Thesaurus* fame, was a medical practitioner, scientific manager, and popular physiologist. At Edinburgh University he was a contemporary of that group of brilliant Whigs shortly to make their mark as advocates, statesmen, *Edinburgh* reviewers, and educationalists, including Brougham, the Horners, and Henry Petty-Fitzmaurice (later the third marquis of Lansdowne). He was supported, educated, and initiated into Whig society by his uncle, the law reformer Samuel Romilly. Romilly patronage cushioned Roget in a world of wealthy Whiggism—antiradical to be sure (the Whig-turned-Tory James Scarlett was another Romilly protégé and visitor to the Roget home), but one aware of the need for a stabilizing reform. Roget's upbringing was sheltered: the young man emerged prim, overprotected, and easily shocked by the fashionable freethinking radicalism of the early decades. In 1800 Romilly arranged for him to assist Bentham in planning an experimental ice-filled "frigidarium" (to preserve fruit out of season),[98] but Roget quickly abandoned Bentham,

96. Parsons 1932–36, 3:92–93; *L* 1844, 2:8. He succeeded South, Solly, and C. L. Meryon here as lecturer on comparative anatomy.

97. W. B. Carpenter 1843, 2:viii. In zoology too he saw the task as exposing the "prevailing uniformity" in the animal kingdom, denying—like other Dissenting reformers before him (for example, J. E. Bicheno)—that naming was anything other than a taxonomic means to that end (1845b, 1:1–5). Cf. Bicheno, discussed in Desmond 1985a:164–66.

98. Roget to Bentham, 9 September 1800 (BL Add. MS 33,453, f. 409); 17 September 1800 (f. 417); 5 October 1800 (f. 421). Emblen 1970:49.

appalled by his unbelief. On the other hand, he carefully cultivated the patronage of the Whig noblemen who shared Romilly's gentler reformism. Through Romilly, Roget became physician to Lord Lansdowne and secretary to Lord Howick (who as Earl Grey was to see the Reform Bill through Parliament). Romilly committed suicide in 1818 (Southwood Smith delivered the funeral oration), and Lansdowne's son, the third marquis, stepped in as Roget's benefactor. When the Whigs finally took office in 1830, Lansdowne House was transformed from an Opposition retreat into the great "ministerial salon,"[99] putting Roget at the very center of landed Whig power.

The noble lords did not neglect the government of science. With Lansdowne president of the Zoological Society (1827–31),[100] Statistical Society (1834), and a governor of the Royal Institution (1811–36), Roget was eased into the highest echelons of metropolitan scientific management. He had also been groomed to assume responsibility in the Whig educational empire. Romilly had installed him in a £1,500 house in Russell Square, and in 1809 Roget taught animal physiology at the new Russell Institution endowed by Scarlett, Romilly, and Francis Horner. This Whig group also sponsored the new Northern Dispensary, a charity serving the Camden Town poor, where Roget was appointed physician. (Bell, another favored client of the "Scotch" Whigs, was offered the post of surgeon.) Roget began lecturing at the Royal Institution in 1812 and contributing to the *Edinburgh Review*. In the 1820s he served on the SDUK Committee, overseeing its *Penny Cyclopaedia*, writing his own treatises, and becoming a resident referee.[101]

Roget later acquired the image of a courtly Whig placeman. But he was a brilliant mediator in his early career, bringing about sensible negotiated settlements in age of constitutional crisis (see fig. 5.3). True, he had always been cautious, as reflected in his choice of medical dining clubs. He was active in the Medical and Chirurgical Society (f. 1805) and became its president in 1829. Respectable and rigidly graded, this society steadfastly barred GPs from office and eschewed the kind of impromptu discussions on medical government which enlivened the rival Westminster. It locked its doors on the *Lancet* and only lifted reporting restrictions in 1836 (even then sending abstracts to the *Gazette* first to beat Wakley's men). Inevitably the society was overtaken by events and tended to the corporations' elitism—not surprisingly, for its personnel increasingly overlapped with

99. Halévy 1950:13.

100. Desmond 1985a:226. ZS MC 1: ff. 1, 15; 2: ff. 103, 147. Berman 1978:109, 123.

101. See the letters from Roget to T. Coates (UCL SDUK). On the Northern Dispensary: G. J. Bell 1870:160; Emblen 1970:107.

Figure 5.3. Peter Mark Roget, well connected to the leading Whig families and a subtle mediator in scientific society. By J. Cochran. (Courtesy Wellcome Institute Library, London)

Figure 5.4. A meeting of the Royal Society in Somerset House. Roget was appointed one of the two secretaries in 1827 and served in this post for twenty-one years. By Melville, after Fairholt. (Collection of the author)

those of the Royal Colleges and hospitals, to an extent that Wakley indicted its "exalted members" in 1841 as *particeps criminis* in the efforts to block reform.[102]

Locking the doors proved less practicable at the Royal Society (see fig. 5.4). In 1827, the year Lansdowne crossed the House to form a coalition with the Tory prime minister George Canning, Roget accepted the post of secretary to the unreformed society. Its president, the Tory M.P. Davies Gilbert, attempting to stem the reform tide in Somerset House no less than in his Cornwall constituency, was under attack from both Charles Babbage's "philosophical" faction—gentlemen wanting a more strictly scientific society—and from the ubiquitous medical radicals. Roget's talents for negotiation proved indispensable. With Whig credentials and landed contacts, he could moderate between Babbage's party and the Tory stalwarts, and at the same time liaise with Broughamite bodies such

102. "Royal Medical and Chirurgical Society," *L* 1841–42, 1:265–67; Clarke 1874a:215–21.

as the SDUK.[103] His political acuity in managing the factions was prover-
bial, and it served him well so long as all sides abided by traditional paternalist principles (of course, the Wakleyans would not). He has been
painted a conservative manager in 1830, but this gives a false picture,
despite his stonewalling of reforms a decade later. He was a constitutional
"Royalist" practicing Grey's compromise politics. When the duke of Sussex (George III's son) became president of the Royal Society in 1830, the
vocational scientists, having failed to elect one of their own, John Herschel, began to boycott meetings.[104] Roget welcomed Sussex's "spirit of
conciliation" and his attempt to placate Babbage's faction "who now assume so hostile an attitude, & threaten to secede from the Society". He
actually doubted Sussex's chances, but only because he believed that the
"imbecile Council" of old hardliners would attempt to "paralyse our efforts." And he threatened to resign if they did.[105] So he was no diehard.
But then neither was he a medical reactionary. He might have infuriated
GPs by accepting a fellowship in the College of Physicians in 1831, but
once in he seconded Elliotson's doomed motion to abolish the Oxbridge
fellowship restriction, which would have left all licentiates of five years'
standing eligible.[106] He was, in the end, what Bentham once dubbed
Romilly—a "mere Whig," practicing a self-protective sort of reform for
party ends.

　　　He was widely outflanked by such gentlemen radicals as Elliotson and
Lawrence. In fact his antireductionist philosophy fashioned after the Peterloo period was very much a reaction to their self-sufficient science, and
it played a leading part in the physiological reconsecration of paternalist
values. The Aldersgate Street school, one of the earliest to establish physiology classes, engaged Roget in 1826 to teach human and comparative
physiology. Here he criticized physiologists for showing "too great an ea-

　　　103. Roget to T. Coates, 18 April 1829, 8 May 1829 (UCL SDUK). On his mediation
between Babbage's faction and the executive: Roget to Swainson, 5 December 1830, 2 March
1831, 12 March 1831 (LS WS). MacLeod 1983:63; Emblen 1970:202–12; M. B. Hall
1984:56–57.
　　　104. Roget to Swainson, 2 March 1831 (LS WS). The ultraradicals were of course no
friends of the "Royalist" party supporting Sussex's presidency. "Royalty and science!" sneered
Wakley. "How the terms assimilate! As harmoniously as poison and antidote, ignorance and
knowledge": "The Royal Society, L 1832–33, 2:245–46.
　　　105. Roget to Swainson, 5 December 1830 (LS WS). On Sussex's attempt to conciliate,
unsuccessfully inviting the abstaining "reformers, both moderate & radical" to join a committee to revise the statutes, see Roget to Swainson, 12 March 1831 (LS WS).
　　　106. "Reform in the Royal College of Physicians," LMSJ 1836, 9:153–54; G. Clark 1964–
72, 2:663, 682–89. On the fury at his fellowship: LMSJ 1836–37, 10:704. Reformers condemned the RCP Council's "Machiavelian" divide-and-subdue policy of raising certain
wealthier licentiates to the medical peerage: MCR 1833, 18:582.

gerness to attempt the reduction of all phenomena to a single principle, or law of life."[107] With Lawrence teaching surgery at the school that year and stirring passions at the medico-democratic rallies in the Freemasons' Tavern, this attack on physico-chemical reductionism could only have appeared as a shot across Lawrence's bow. Roget saw the "new world" of life transcend the operation of "mineral" laws. In organic nature

a number of new and subtle agencies are at work; and a totally different class of phenomena make their appearance. These phenomena are not capable of being explained simply by the laws of Mechanism or of Chemisty; they are of too complicated a character to admit of being reduced by inductive reasoning to one single principle, in the same way in which the movements of the celestial bodies are now reducible to the single law of gravitation: they imply the operation of a number of principles quite distinct from those which govern inorganic matter.[108]

Roget was too astute not to sense the vacuousness of some of Paley's teleological reasoning. Nonetheless he saw the "new principles" exhibited by life "strongly and indelibly impressed with the character of INTENTION" and bearing the stamp of "intelligence and of power" beyond human comprehension. In 1826 this belief in irreducible physiological principles pointing to Superior purpose characteristically demarcated the Whig elite from the democratic reductionists. The response to Roget predictably followed party lines. Benthamites yawned; Southwood Smith noted Roget's faith in the ruling "general powers" but argued an opposing thesis in the *Westminster Review.* He interpreted the latest microscopical work (some of it his own) as proving that all life, from the "meanest plant" to man, was composed of the same elementary "globules"—and he inferred from electrical experiments on albumen that these "globules" were ultimately produced by physical rather than vital laws. With no vitalizing agents from preexistent life apparently present at their birth, these globules defied Roget's antireductionist logic.[109] But for moderates life attended on Divine pleasure, not galvanic shocks. Bell insisted that the theatrical use of electricity to make hanged felons move again or to create microscopic cells fooled nobody—it held out no promise of the key to life.[110] Roget's attitude placed him much closer to the older Paleyites of the Brougham-Bell school. For him animal design argued for a prescient Power who could accomplish His "distant purposes" through an "immense chain of causes and effects." Divine instrumentality remained the key. Change was

107. Roget 1826:18. Lawrence's Freemasons' Tavern speech is reported in *L* 1826, 9:725–30.
108. Roget 1826:14–15, 19, 20, 22; cf. 1838, 1:102ff.
109. T. S. Smith 1827a:439, 443, 1827b. Jacyna 1984b:21.
110. C. Bell 1827–29:49.

preordained, the command coming from the Divine Administrator. Like all of the patrician Whigs, he hated Lamarckism and saw the divine command structure strip organisms of their self-developing powers; as the authority and power of the base collapsed, so must an "extravagant" science characterized by "spontaneous" species elevation and improvement.[111]

Roget took increasingly prestigious posts in the government of science. In 1834 he accepted the first Fullerian chair of physiology at the Royal Institution (where a popular physiologist was required to draw the crowds—and Roget had long trod the boards here). He was also perfectly placed at the Royal Society to be slated to write one of the Bridgewater Treatises (the nominations being in Gilbert's hands). Historians have considered this series the epitome of Paley's natural theology, and unarguably the contributions of Bell and Buckland praised Paley at the expense of the new morphology. But not so Roget. He adapted Grey's parliamentary strategy and opportunely set about harnessing the radical morphology, repackaging it to pick up the widest possible vote.

Roget had actually taken no more interest in the anti-Cuvierian morphology during the 1820s than Grey had in returning to reform before the king's call. Roget's own command came from Canterbury by way of Gilbert. The Bridgewater brief was to provide a stabilizing physiology—to chant "the hymn which the living world, in the rich drapery of its loveliness, raises to its sovereign king."[112] But Roget was not prepared simply to restate old Paleyan values in an era of mounting class hostility. He was now past fifty and conscious of the need to brush up on modern trends. This was evident from his letters of the period. For instance, asked by the publishing entrepreneur Dionysius Lardner in 1829 to supply two volumes on physiology and comparative anatomy for his Cabinet Cyclopaedia, Roget apologized about the time needed to catch up with the latest work, and he pulled out of the project after wrangling over copyright control.[113] However, given the £1,000 Bridgewater inducement, he did make the effort. In April 1832 he paid £3 and joined Grant's zoology class. That fall he sat the full comparative anatomy course of fifty-eight lectures (joining William Farr, C. A. Tulk, and George Newport),[114] and he remained for the opening zoology lectures of 1833, boning up on the latest Continental approaches. For his benefit Grant delved into more esoteric matter

111. Roget 1826:89–90, 102; 1838, 1:143.
112. Brewster 1834:146.
113. Roget to D. Lardner, 18 November 1828, 19 March 1829 (WI). Hays 1981 on Lardner.
114. Grant 1846. Grant of course dissented from Roget's antireductionism; indeed, in his copy of Roget's *Introductory Lecture* (1826) Grant put a question mark in the column against

after class, providing fuller explanations and references to sources. Throughout the year Roget continued writing his massive two-volume *Animal and Vegetable Physiology Considered with Reference to Natural Theology*, negotiating a deadline extension for the manuscript from December 1832 until March 1833. The quarter-of-a-million word text was finally ready for the printer by April and awaited only the engravings.[115] The impact of the new anatomy was immediately apparent. The *Physiology* of the title gave no clue to the book's new-found morphological theme. Its core—a descriptive and functional anatomy of the animal orders—was sandwiched between tentative transcendental explanations, and the whole then encased in Paleyite wrappers. The result was that only the opening and closing words really fulfilled the spirit of his Paleyite charge. That the adaptations locking each animal into its niche implied a Designer he took for granted; it was his connecting of this with bolder concepts of organic unity that provided the novelty. This was compromise politics working its way to the very heart of science. As a "ministerialist," Roget was diluting Geoffroyism and mixing it with doctrines palatable to wider Whig tastes.

In 1826 Roget had seen species exhibit a bewildering complexity and had clustered them into a "complicated net-work" to deny Lamarck his transforming chain.[116] But by 1832 Roget had recognized the potential in the theory of morphological unity for reconceptualizing natural theology. Thus *Animal and Vegetable Physiology* was unlike anything he had written before[117] and unlike any other Bridgewater Treatise. He now denied that organic variation was "indiscriminately followed"; it was "circumscribed within certain limits, and controlled by another law . . . that of *conformity to a definite type*." All existing forms were therefore "as so many separate copies" of a "certain ideal model." Nor was this an endorsement of Cuvier's discrete *embranchements*, for he continued:

To regard any of the beings in the creation as isolated from the rest, would be to take a very narrow and false view of their condition; for all are connected by mu-

---

the statement that life's "phenomena are not <u>capable</u> of being explained simply by the laws of Mechanism or of Chemistry" (p. 14, Grant's underscoring). Grant was of course making a directly contrary statement in his own lectures. This divergence and the flattery of Roget's attendance might have made Grant the more eager to explain his views.

115. While writing, Roget constantly discussed publishers, profits, and deadlines with William Buckland, himself a Bridgewater author. Roget to Buckland, 29 and 30 August, 9 and 16 October 1832, and 5 April 1833 (announcing his completion) (UMO WB).

116. Roget 1826:11, 88–90; 1838, 1:144–45.

117. In 1838 Roget told Macvey Napier, wanting to reprint his entry "Physiology" (originally published in 1824) in the new edition of the *Encyclopaedia Britannica*, that it was now

tual relations. Even among the leading types which represent the great divisions of the animal kingdom we may trace several points of resemblance, which show them to be parts of one general plan.[118]

Gently, in deference to the superimposed Paleyite adaptations, Roget bent the morphological model into a less menacing shape, making it the basis for a new unity of Divine plan. Like the radicals, he placed intermediate species between the *embranchements*, recognizing that "the steps of gradation by which one type passes into another, are so numerous and so regular, as to preclude the possibility of drawing a decided line of demarcation." He too imagined that this "law of *Gradation*" was a consequence of the unity of composition; it had to be so because the latter left all "the races of animated beings . . . members of one family." Again, with Geoffroy's supporters, he took the mollusk-fish bridge as his paradigm: the cuttlefish's cartilages were analogous to the spinal column, and he projected an "easy" transition to the cartilaginous fishes.[119] He went on to endorse Geoffroy's theory of vertebral elements and (more cautiously) discussed the vertebral skull. He might have lacked Grant's assurance, Fletcher's brilliance, or Knox's bluster, but he had still shifted significantly from Bell's out-and-out anti-Geoffroyism. Not, of course, that he came any closer to endorsing a naturalistic explanation for the graduated variation. Despite his lukewarm acceptance now of successive creation, according to which "the standard types have arisen the one from the other," he refused to picture this progressive development as the result of "simple laws," let alone condone Lamarck's "presumptuous reveries."[120] But this could not disguise the fact that the book contained a reverential reworking of the new morphology. From radical doctrines, Roget had extracted a higher design, envisaging an organic ground plan which testified to the unitary Cause of all created existence.

Radicals were appalled by the apparent piracy as much as the anatomical castration. To Wakley's followers it simply confirmed a long-standing prejudice: that the Whigs, so high principled in private, were untrustworthy in public.[121] Ryan's *Journal* upbraided the "pseudo fellow" for forgetting to mention that his "Bridgewater Gleanings" were largely gleaned in

---

"quite obsolete" and required rewriting: Roget to Napier, 19 February 1838 (BL Add. MS 34,618, f. 565).

118. Roget 1834, 1:48, 49, 51–52. On the tensions within Roget's natural theology: Yeo 1986:268, 273; also Rehbock 1983:57–59.

119. Roget 1834, 1:268, 263n., 338–96, 407, 52; 2:627–29.

120. Ibid., 2:637–38.

121. *L* 1831–32, 1:840.

Gower Street. *Lancet* letter writers dubbed him Grant's "plagiarist."[122]
Grant was quietly hopping. Even while proof sheets were circulating, he
began agitating against this "improper use" of his lectures.[123] But as a case
of plagiarism it was never simple. The later publication in the *Lancet* of
the Grant-Roget letters showed the complexity of the issue. Roget's
"facts" and their morphological framework were said to have been pirated;
Grant claimed that even the book's illustrations of vertebral elements and
cranial vertebrae came straight from his blackboard. But beating the radicals at their own game, Roget insisted that scientific facts become the
"property of those to whom they are communicated." Grant was selling
his wares in a lecture hall, and Roget, having dipped into his pocket, demanded the right to make "whatever use" he pleased of knowledge thus
imparted by "a public professor."[124] Grant was on a slippery slope. Piracy
claims fell badly from a radical's lips. Hadn't Wakley pirated the monopolists' lectures, selling them at sixpence a time in the *Lancet?*—justifying
this democratizing of knowledge on the ground that the best should be
freely available to all. Roget's act also proved the difficulty for a professor
who fails to publish of protecting his "property"—his rights to personally
discovered "facts." Was there, as it ironically fell to the *Gazette* to argue,
an "intellectual patent," and did this take effect before or after publication?[125] Or were property rights waived, as Roget now claimed, when
these new truths were placed on the open market? The treatment meted
out to the monopolists was rebounding on the moralistic radicals. Grant's
financial plight undoubtedly increased the bitterness. Roget was wealthy;
he had reached almost forty before Romilly's quarterly allowance had
stopped, and his fellowship of the College of Physicians would have swollen his already lucrative practice. Now he was to receive the Bridgewater
£1,000 plus royalties. It must have rankled. Nor would Grant have rel-
ished seeing his naturalistic lectures thrust into a theological context and
sanctioned by the eccentric earl's executors in Canterbury.

This attempt to hold some sort of copyright control over lecture material proved a perennial problem. Grant incorporated his discoveries into
his lectures rather than publishing them. As a result his protégés, in reworking the ground, ran the occupational risk of a priority dispute.

122. TJ [Thomas Wharton Jones?], *L* 1836–37, 1:624; *LMSJ* 1836–37, 10:844.

123. "Letter from Dr. Roget," *L* 1846, 1:420; Grant 1846; and Roget's rejoinder, *L* 1846,
1:482–83; Godlee 1921:101.

124. *L* 1846, 1:483, also 446.

125. "The Rival Discoverers," *MG* 1837–38, 21:906. Property rights on transcendent
"facts" were again an issue a decade later when Edward Forbes fought off charges of plagiarism leveled by Hewett Watson (Rehbock 1983:176–84, esp. 180).

George Newport's case is the most revealing, partly for its rancor, but also because he was implicated in the radicals' indictment of Roget. Newport had been an apprenticed wheelwright and amateur entomologist before entering London University in January 1832. A sympathetic Marshall Hall acted as patron, financing, feeding, clothing, and housing the financially distressed student.[126] At Hall's request, Grant inducted Newport free into his classes, also touched by his "adverse circumstances" and "humble occupation," and he persuaded other professors to do likewise. In the 1832 session Grant first announced the motor function of the abdominal nerve columns in articulates (Hall corroborated Grant's claim, although the technical issue is far from clear-cut).[127] When Newport, in a series of Royal Society papers on the articulate nervous system, seemingly plagiarized Grant's work and then added insult to injury by accepting the society's Royal Medal in 1836—a "philosophical Order of Knighthood" signaling establishment approval, in Wakley's words—a bitter dispute broke out. It was the more acrimonious for Hall's and Grant's feelings of personal betrayal. Hall insensitively denounced Newport's science, manners, and sharp financial practice. He also agitated on Grant's behalf (not that Grant was incapable of giving the "ungrateful parasite" his due) and began proceedings at the Royal Society to reconsider Newport's award.[128] Because Newport and Roget suffered similar accusations of plagiarism from the same radical source, and because Newport had also helped correct the proofs of Roget's *Physiology* after Grant refused to have anything to do with it, radical suspicions that the "knighthood" was the payoff from Roget's society were impossible to allay.

This belief that Newport's medal was a kickback for his help in polishing up Roget's "pirated" Bridgewater refused to go away.[129] Wakley's friends were notoriously good haters and treated Roget with unmitigated scorn for the next fifteen years. As a patrician Whig holding court at the Royal Society, he was a visible reminder of the anachronistic ideals of salon politics. The presses indulged in a war of attrition, for he met a constant hail of criticism from the GPs' journals. Public awards, "like a woman's honour," were expected to be "above all suspicion."[130] Charges and

126. M. Hall 1837–38; Newport 1837–38a; Grant 1837–38a.

127. "Dr. Marshall Hall on the Nervous System," *L* 1837–38, 1:650.

128. Grant 1837–38b. "Parallel Passages by Marshall Hall, M.D.," *L* 1837–38, 2:17. The "knighthood" metaphor: "The Royal Society," *L* 1846, 1:635.

129. *L* 1836–37, 1:624; 1846, 1:391, 499–501, 634–36. Hall to Herschel, 1 October 1839 (RS JH). On these medals generally, see MacLeod 1971.

130. "Award of the Royal Medal to Mr. Newport," *L* 1836–37, 1:715–16. The complete correspondence is: *L* 1836–37, 1:624, 656, 715–16, 799–800; 1837–38, 1:715, 746–49, 812–17, 897–900, 950–52; 1837–38, 2:17–18.

rebuttals concerning "Roget's Medal" filled the letters pages of the *Lancet* for fourteen months in 1837–38. Eventually the clamor forced the council to establish a Physiological Committee to assess this award and look at research generally. But the radicals' smears alienated many moderates. Grant and Hall's gutter tactics—indelicately questioning Newport's morality and financial practice, accusing him of sucking at a savant's brains and purse—were judged beyond the bounds of acceptable taste. The *British and Foreign Medical Review* drew back at this display of "low and evil passions."[131] For once it backed the *Gazette*, which condemned Grant and Hall's muckraking and claimed that their expletives "shewed that they were tolerably versed in other terms besides those of science." The *Gazette* did not scruple to use radical tactics in portraying Newport as the sympathetic underdog, persecuted by powerful and monopolizing teachers. It also allowed him the use of its pages to fight on, to prove at the end of the day that Grant had mistaken blood vessels for motor nerves.[132]

But the radicals had grass-roots support in the unions and schools, and with Roget's courtier ways so loathed, their criticisms continued to be widely echoed. On top of the medal issue in 1837 came Roget's attempts to retain the Fullerian professorship beyond three years. He announced that John Fuller, whose deed had established the chair, had actually intended him to have it in perpetuity. But the managers upheld the triennial nature of the appointment, which Ryan considered a blessing given the "great dissatisfaction universally expressed" at Roget's lecturing performance,[133] and Grant took over the chair. Nothing smacked so much of Old Corruption in Roget's machinations. By now the destructives were ruthlessly targeting timeservers who shunned competition and treated official posts as life-tenured party gifts. They were a visible reminder of the leisured days of gentlemen's dining clubs and had no place in the new era of competition and accountability. The radicals' long practice of damning the surgical self-elect meant that they had no trouble switching targets to Roget and the Royal Society. And not only the Royal Society: Roget's sponsors controlled a considerable number of posts. There were mixed feelings, for example, when the Whig governors gave him a seat on the senate of the newly chartered University of London in 1836. Ryan's *Journal*, unlike some of its correspondents, tried to be charitable, even if it ended up damning him:

131. "The Rival Discoverers," *BFMR* 1838, 5:621–22. Grant was also criticized for ignoring Newport in his *Outlines: MCR* 1835, 23:384–87. M. B. Hall 1984:68 on the setting up of the "Physiology Committee."

132. "The Rival Discoverers," *MG* 1837–38, 21:903–6, also 930–32, 985–86. Newport 1837–38b.

133. "Royal Institution," *LMSJ* 1837, 11:408–9.

whatever errors he may have committed when compiling his works, in not acknowledging the sources from which he derived his information, and however much he may have mixed himself up with the disreputable intrigues which have for some years past disgraced the Council of the Royal Society, there is something amiable and flexible in his nature which we do hope and trust will enable him to adapt himself and be moulded into that form which will render his services useful and acceptable to the New University. Let it . . . be recollected [though], that he is one of those licentiates [of the RCP] who deserted the cause of his fraternity, and accepted the fellowship on terms which many others of a higher tone of mind would not stoop to receive![134]

The consequences of this cultural politicking for Hall and Grant were enormous. Hall's new paper on the reflex system in 1837 was, depending on one's source, either "blackballed" by the Royal Society (the radical claim) or "withdrawn" by the author (Roget's official communiqué).[135] Galling as this was, the situation was exacerbated by the second secretary returning Hall's manuscript embellished by a "lampoon." Scribbled next to a description of some tortoise experiments was the query, pinpointing the real gastronomic nature of a country gentleman's interest: "Would they live after they had been made soup of?" This at least dispelled the myth that regenerate Christians lacked a sense of humor, for the sender was the evangelical John Children. Being himself a notorious dealer in patronage and doyen of the old guard, he had long been the subject of radical venom. Children's squib backfired, inflaming yet more passions: "The miserable scoffer," raged Wakley, "will live in this vile anecdote when nothing of him" remains; "the name will be gibbeted to the deed."[136] Hall's radicalization now became more acute as he moved to the center of the British Medical Association. In 1839 he made a direct appeal on his and Grant's behalf to the "acknowledged head of British Science," John Herschel, sending him an eleven-page list of charges drawn up against Roget and the Council of the Royal Society.[137] But it was to little avail. By now Grant and Hall had disqualified themselves from any Royal Society post, at least while Roget's faction remained in office.

134. "Liberality of the Government," *LMSJ* 1836–37, 10:704; cf. a reader's response, p. 844. Roget was appointed the examiner in physiology and comparative anatomy by the senate in 1839 and was thus empowered to test Grant's students. The irony of the situation—Roget having audited the course himself to pep up his knowledge—was probably not lost. Grant's protégé P. B. Ayers (1841–42) complained after his own failure to win a medal that the senate was negligent in its physiology department.
135. "History of the Royal Medal," *L* 1846, 1:391–93; Roget 1846.
136. "Proceedings of the Royal Society," *L* 1846, 1:418–19.
137. Hall to Herschel, 1 October 1839 (RS JH).

By 1840 the scientific elite's assimilation of radical doctrines, like the government's, was at an end. Nonetheless its appropriation of Geoffroy's science had been an important step. Through Roget the Royal Society—the upper tier of English science—had been able to absorb a sanitized morphology, dissociated from any leveling Lamarckian laws. We turn now to the London acolytes of the Oxbridge divines, where we see this modification process continued to the extent that the anatomical doctrines could be used against the very destructives who had first imported them.

6

Science under Siege: Forging
an Idealist Comparative Anatomy at
the College of Surgeons

> If . . . past science is viewed as an activity which was socially orga-
> nized and countenanced, we may expect it to have shown sensitivity
> in various degrees to some of the diverse elements in its social en-
> vironment. This claim should receive particular justification from
> the study of the scientific activity of individuals who were associated
> with institutions in which sciences were taught, especially during a
> time of persistent stress and repeated crisis.
>
> —J. B. Morrell on the *Theophobia Gallica* of scientific Tories in the
> aftermath of the French Revolution[1]

No scientific institutions suffered such persistent stress and crisis as the
medical corporations in the 1830s. The sergeant-surgeons were assailed
in the press, investigated by parliamentary committees, and involved in
interminable legal disputes to preserve their privileges. Here, as nowhere else, we should be able to plot the conservative response to the
medical unrest: to the unionization, one-faculty calls, and new professional demands on education—but most of all, to the radical sciences promoted in the university, private schools, and free press.

Here the Lamarckian and Geoffroyan anatomies nestled comfortably
alongside democratic and Dissenting demands. Radical groups had embraced a welter of rank-breaking sciences, including atheistic forms of
evolutionary development, materialist mental physiologies, and reductionist comparative anatomies. All presupposed an atomistic self-
sufficient nature; all returned sovereignty to the individual and sanctioned development from below. Jacyna has shown how the medical
republicans in Regency London drew heavily on a monistic physiology;
how they believed that matter itself possessed "endogenous" powers,

---

1. Morrell 1971b:43.

making it the sole cause of life and mentality.[2] In the 1830s agitators were still exploiting this sort of self-animating nature. Infidel science and seditious politics were mixing freely in the fiercer medical unions, anatomy schools, and secular press. The medical gentry were too aware that a similar philosophic malaise had led to the Terror to ignore this dissident literature. Coleridge told the prime minister, Lord Liverpool, in 1817 that the new crop of foul fruits being harvested in Britain—treason, blasphemy, and riots—was the result of a diseased "speculative science" spawned by the French Revolution.[3] Nor were Tories slow to accuse the radicals of inciting the working classes. The artisan presses were pirating medical republican works, forcing them to serve more insurgent ends. Illegal penny prints harped on the fatal consequences of an atheistic self-developing nature for the authority of kings and priests. Imprisoned deists adopted d'Holbach's ultramaterialistic *System of Nature* as their bible; inflammatory street tracts promoted a godless evolutionism; and Lamarck was put to subversive use by cooperators and atheists.[4] Hence Coleridge's wrath—his belief that insurrection was the rotten fruit of an unsound philosophy. Cultural determinism and unaided progression had become militant creeds, and in the filthy bazaars of science "Ouran Outang theology" was sullying "the Book of Genesis."[5]

The onus lay squarely on the medical patriots to provide a remedy. It was after all the radical manufactories—the London University and private schools—that were churning out the higher anatomies and Lamarckian biologies. The challenge for corporation conservatives was to fashion an equally sophisticated rival, a biology with the same morphological sweep, but which would return sovereignty to the Godhead and authority to the traditional elite. This was the task ahead as the grave young Anglican Richard Owen settled in at the College of Surgeons. Owen was patronized and petted by the college councillors. In politics he was a Peelite. (Peelites were moderate Tories—or, as they were shortly be be called, Conservatives. They followed Sir Robert Peel, leader of the party from the early 1830s, rather than the Iron Duke of Wellington. Although Peel originally opposed the founding of the London University, Catholic emancipation, the Reform Bill, and so on, he eventually accepted the Whig reforms of the 1830s as a fait accompli and left them intact.) The political climate in which Owen's patrons fought to preserve their power goes a long way to explain the moral force of his rival science. Owen was ensconced in a corporation where a conservative social philosophy was

2. Jacyna 1983b:313.
3. Griggs 1956–71, 4:758.
4. Desmond 1987.
5. Coleridge 1972:51–52.

used to justify existing privileges, where traditional values provided a moral arbiter for science, and where the democratic sciences were despised as criminal and unpatriotic. He was to become Britain's leading biologist, assuming Grant's mantle as the "English Cuvier" in the late 1830s. His case shows how crucial contemporary political factors were in determining the rejection of a science of "self-developing energies."[6] It is not only the contents of Owen's anatomy that can be pinned to a political backcloth. Job restructuring was an integral part of the council's response to external threats—part of its strategy to fend off radical attacks. Owen's college appointments themselves were in a real sense political.

The intensity of these attacks was increasing as Owen joined the college in 1827. Wakley was stepping up his dogged, grinding campaign against the college's "self-perpetuating, tyrannical council" laboring in "its sordid vocation."[7] By 1833 all the reforming editors had swung his way, and both Ryan's *Journal* and Johnson's *Review* were drawing council blood. "We have thus a triple alliance against us," moaned the *Gazette* in 1834, with all the publishers "engaged in the plan of levelling—radical abolition against rational reform."[8] Month in, month out, the tirade against the council continued. The journals campaigned ceaselessly for curbs on executive privilege, for rank-and-file rights, and for increased access to college lectures, library, and museum. They demanded suffrage and an end to council self-election as a check to nepotism. They insisted that certificates from the private schools be recognized, not just those issued in the councillors' own hospitals. At the same time they derided the surgeons' unsophisticated science, pointing out that the jingoistic gentlemen in their Lincoln's Inn enclave seemed quite ignorant of foreign thought. "When we peruse the published lectures" of the hospital teachers, wrote a critic, "we blush at their puerilities and shallowness which characterize them, and for the want of reference to the exact state of science in other countries. We look in vain for the opinions of celebrated foreigners, and are almost disposed to imagine that we reside in one of the Lilliputian islands, or in the celestial empire, beyond the precincts of which all mankind are supposed to be fools."[9] By this time the radical papers were reaching large audiences. Wakley was selling four thousand *Lancets* a week and enjoying huge public support judging by the collections to pay his court expenses (he fought ten law suits in ten years). Even the respectable *Gazette* was itself becoming worried by the extent of corporation intransigence and cautiously calling for a "reformation in the

6. R. Owen 1841c:202.
7. "Address," *L* 1830–31, 1:4.
8. "Illustrations of Consistency," *MG* 1833–34, 13–529.
9. "State of the Medical Schools in London," *LMSJ* 1833, 2:311.

medical aristocracy"—for the admission of Dissenting fellows into the RCP and a stronger "bond" between the RCS Council and commonalty. Only the gesture of self-reform could preserve the corporation privileges, which were in imminent danger of being lost through council obstinacy.[10]

Of all the radical approaches, Wakley's was still distinguished by its vulgarity. He used caricature and character assassination, cudgeling councillors and delivering the heaviest blows to the college presidents: to Owen's patrons such as John Abernethy (president, 1826), and George IV's surgeons, Sir Astley Cooper (1827) and Sir Anthony Carlisle (1828) (see fig. 6.1). All of these gentlemen execrated the "reptile press." Abernethy fought to stop Wakley from pirating his college lectures, only to lose the court battle. Cooper suffered persistent criticisms over his nepotism. Carlisle was a despised hardliner and derided as a "conceited, crotchety, knight." Such crude abuse naturally galled the senior surgeons. Carlisle did not relish being called an old "crustaceous philosopher" and was undoubtedly referring to Wakley when he complained to Owen of the "active malevolence of ignorant savages."[11] Carlisle was a College hawk, utterly opposed to any democratization of the council elections. He was not a great surgeon. He had gained his Royal Academy chair through carefully cultivated connections, while his knighthood was for services as surgeon extraordinary to the Prince Regent. In dotage he was taunted for his incompetence at the Westminster Hospital. The *Lancet* campaigned to have him removed. In 1839 Wakley's reporter was set upon by Carlisle's students at the hospital for publishing a transcript of the surgeon's insensitive remarks made during rounds and was only saved by the intervention of some coalmen in the street.[12] There was never any love lost between the court surgeons and disaffected democrats, and in Carlisle's case hatreds festered to the extent that personal violence became a real danger.

10. "Reform," *MG* 1832–33, 11:487; also 1831, 8:56–60; 1831–32, 10:394–96; and its strongest call, 1834–35, 15:561–64.

11. R. S. Owen 1894, 1:86. *Report SCME, Pt.2,* 141 and Cope 1959:60–61 on Carlisle's hardline attitude; "crustaceous": *L* 1832–33, 1:154; "conceited": *L* 1830–31, 2:662. On Abernethy: *L* 1828–29, 1:1–7. Wakley's pirating was a "democratic" measure, i.e., designed to deal a blow to RCS exclusivity and to make the best lectures accessible to everyone in the profession. For the lecturers it was not so much a financial threat—students still had to pay £5 to attend their courses in order to obtain certificates—but simply galling to find copies available at sixpence a week on the bookstands.

12. Clarke 1874a:283–94. The offending report was "Specimens of Clinical Instruction," *L* 1839–40, 1:93. Other blunders were reported in "Sir A. Carlisle," ibid., 23. Cole 1952 on Carlisle. Granville was Carlisle's pupil in 1813, but had to move out of Carlisle's house after three months because of his intolerable personal quirks (P. B. Granville 1874, 1:326–27).

The Political Nature of Owen's Appointment

He has rendered himself one of the very first persons in his own
branch of science in the kingdom; and, I am proud to say, that he
has been educated in the College of Surgeons, and that it is owing
to us that there is such a man.

—College president G. J. Guthrie, praising Richard Owen before
Warburton's committee in 1834[13]

The political situation was fraught as the twenty-year-old Owen moved to
the capital in 1825 (see fig. 6.2). Owen was an odd, diffident man. The
affable conservator of the Hunterian Museum, William Clift, found him
"sober and sedate very far beyond any young man I ever knew."[14] Contemporaries offered wildly differing assessments of his character. Even a
sympathizer such as Edward Forbes saw him guarded by two spiritual
policemen, one good and the other evil. To political friends he was charm
itself, but to foes he was ungainly, obstinate, and argumentative.[15] Everyone however acknowledged his one great asset: his propensity for work.
Owen had originally intended to study at Edinburgh University, where
he had gone in 1824 from his native Lancaster. But after only two terms
he took John Barclay's advice and transferred to Abernethy's class at St.
Bartholomew's Hospital. In Owen's quaintly deficient *Life*, little beyond
the skeletal facts of his subsequent rise are recorded. To make sense of it
we need to restore the social flesh, the machinations and political jockey-
ing, and examine the direction and impact of his science. We have seen
that the eminent surgeons were suffering smear campaigns and attacks on
their nepotism and privilege. It was to these gentlemen surgeons that
Owen owed his appointments. Indeed, his recruitment into the College
of Surgeons in the first place was part of the council's strategy to overcome
one of the reformers' most damaging charges.

The radicals in the mid-1820s echoed the frustration of members trying
to gain access to the college library and museum, or to see John Hunter's
manuscripts and preparations. Hunter had died in 1793, the most famous
surgeon-anatomist in London. Although from a poor background, he had
risen to become surgeon extraordinary to the king and had amassed a
huge museum with over ten thousand anatomical preparations. These and

13. *Report SCME, Pt. 2*, 47.
14. Clift to J. Hodgson, 7 January 1830 (BMNH RO 8: f. 113).
15. "Ungainly": Flower 1894:xiii; "spiritual policemen": Forbes to Huxley, 2 December
1852 (IC THH), quoted in Desmond 1982:26; "charming": Pollock 1887, 1:274.

Fig. 6.1. Sir Anthony Carlisle, infuriated at being taunted by Wakley's men, lashed out at the "savages" of the reptile press. By H. Robinson, 1841, after M. A. Shee. (Collection of the author)

his extensive folios of unpublished manuscripts (many dealing with comparative anatomy) had been purchased by the government for £15,000 for the nation. They were entrusted to the college in 1799, on condition that catalogs be drawn up, a yearly lecture series instituted, and the museum kept open for certain hours.[16] These conditions had not been adhered to. The Hunterian Museum in the mid-1820s was only open twice a week, and then only for four months of the year. The library in 1827 was still not open at all to the public, causing critics to accuse the council of keeping the nation's treasures for its own use. The fellows "take our money, give us ex post facto laws, lock up our property, insult us with mock orations, live at our expense," raged Wardrop in 1825—yet the rank and file is as "entitled to the museum, and the property of the college, as any member of the court":

some years ago Sir William Blizard promised that the library should soon be opened. We are still outside the door however, and a part of the bust of Sir William, like the molten calf of the Israelities, may go down our throats, before we shall see a book, especially the Hunterian manuscripts.[17]

Wakley, Lawrence, Hume, and twelve hundred others met in the Freemasons' Tavern in 1826 to lobby Parliament over this state of affairs. Warburton took up the case inside the Commons, his petition in 1827 finally spurring the House to demand financial accounts from the college. Faced by what amounted to a censure, the president, Abernethy, announced that the reason for the members' nonadmittance was the absence of a catalog, the destruction of many of Hunter's manuscripts, and the poor condition of the library. He then hired his Barts protégé Owen in 1827 at £30 a quarter to catalog the collection, creating his assistant's job in the museum as part of an overall strategy to preempt further criticism. At the same time, the council hastily prepared the library for opening to the members, the first being admitted in January 1828.

Owen was subsequently groomed by a succession of eminent surgeons. Carlisle in particular saw in the dedicated, moralistic Owen an image of his younger self. He poured out help and hints, talked over his own failed aspirations as a comparative anatomist, and believed Owen perfectly qualified "to fulfil the hopes & wants of the College and to pitch the Bar of Physiology a throw forward."[18] But the "wants of the College" were defiantly conservative, and the throw was to be in a decided direction. The paternalistic interest of the older Paleyite surgeons was complemented by the more dynamic initiation into Coleridge's social thought by Joseph Green. In short, Owen was inducted into an intellectual priest-

16. Cope 1959:23–25; Cross 1981:15.

17. *L* 1825, 7:247; Sprigge 1899:174; Brook 1945:79, 80, 83, 84; Dobson 1954:74–76. Once inside the museum, one delegate in 1826 said, visitors were shunted around as though at the wild beast show at Exeter 'Change: *L* 1826, 9:735–36.

18. Carlisle to Owen, 26 February 1835 (BMNH RO). On Carlisle's help and recollections, Carlisle to Owen 12 March 1834 (RCS 275.h.15 [10]).

Fig. 6.2. The young Richard Owen. (Reproduced by kind permission of the president and council of the Royal College of Surgeons of England)

hood of professionals united to the gentry. He was "raised" to abhor the "ignorant savages" of the press and their iniquitous attacks on the institution. From the first he espoused a social philosophy that was as inimical to Lamarckism as to Wakleyan democracy.

Rather than abate, radical attacks during the years of Owen's assistant conservatorship intensified dramatically. The private teachers slammed the college's discriminatory legislation. The university posed a new moral and social threat, with Grant in his 1833 address to the medical school denouncing the privileges of the Royal Colleges as "ruinous" for the country.[19] Reformers welcomed the setting up of Warburton's Select Committee in 1834, and a succession of witnesses damned Owen's college, their testimonies being reproduced, dissected, and embellished by Wakley. Warburton's radical directive was to press the college elite for clear answers on the methods of council selection, and its powers of patronage and preferment. This he did to great effect. Stalwarts such as Carlisle were pushed into defending the status quo. Guthrie agreed that the council's self-selection procedure was atypical for a corporation but saw no alternative. Hardly any councillors in fact followed Bell in urging a "large" body of electors among the membership or the admission of GPs to the council.[20] Warburton's *Report* in 1834 was embarrassing to the college: uncomfortable council witnesses, eminent men unused to public self-justification and rough cross-questioning, too often gave disingenuous responses to questions probing their self-interest. It was against this backdrop of agitation for fundamental reforms that the council's flurry of activity makes sense: its rebuilding of the museum and library to accommodate the expanding collections (1834–37) (see fig. 6.3), production of expensive catalogs, extended opening hours, and installation of a Hunterian professor.

The council was eager to publicize its achievements before the Select Committee. Guthrie informed Warburton that museum upkeep alone cost the college £2,000–3,000 a year, and Carlisle revealed that £600–700 had been spent on producing just one illustrated catalog, with a future volume expected to cost £1,000.[21] Others lavished praise on the college for its "religious care" of Hunter's collection. Not merely care but expansion, for half of the 7,833 specimens on show had been added since Hunter's day. Hunterian orators too harped on the council's expenditure on the museum and "liberal" policy toward visitors.[22] Owen's role was crucial in the council's strategy to disarm critics, for he was actually to do the lion's share of the cataloging. He too boasted before a select committee in 1836 of the council's no-expense-spared attitude in making the Hunterian

19. Grant 1833b:19.

20. *Report SCME, Pt. 2*, 2, 130, 141, 152.

21. Ibid., 46, 140.

22. W. Lawrence 1834:22–23. Jacyna 1983a:94–96 deals with the self-serving nature of the Hunterian Orations—how the orators put Hunter's "icon" to social and political use, paying homage yearly to the supremacy of his surgical knowledge in order to secure their gentlemanly status and social parity with the physicians.

Fig. 6.3. Rebuilding the museum and library of the College of Surgeons in 1834. Many saw the sudden flurry of council activity as a response to probing parliamentary questions and the widespread criticism of the college. By George Scharf, July 1834. (Reproduced by kind permission of the president and council of the Royal College of Surgeons of England)

Museum "as generally and extensively useful as possible."[23] By then all but two of the catalogs had been printed, along with a number of scientific monographs—including his own *Memoir on the Pearly Nautilus*—at the college's expense. Owen's unflinching support at a time of crisis had important career implications. Guthrie praised him unstintingly before Warburton's committee, announcing that the success in identifying and describing Hunter's preparations had "been owing in a great degree to the labour and assiduity of Mr. Owen." The fact that Owen was becoming one of the leading biologists in Britain Guthrie put down to his education at the college. "I trust we shall raise up a great many others, in a similar manner. I believe we are the only institution in the empire which does devote its money in this particular matter." Owen had become the darling of the institution. Guthrie admitted that, although Owen was only paid

23. *Report SCBM*, 44, 46.

£300 a year, his "assistance alone is worth 500*l*. a-year: indeed I scarcely know the limit that ought to be placed to his salary. The Hunterian Museum of comparative anatomy is more valuable [as a consequence of his labors], in a practical point of view, than that of the *Jardin des Plantes* at Paris."[24]

There was another reason why Guthrie should have rated Owen's work so highly—indeed, why Owen's cataloging was so difficult. It had been known since 1824 (at least to the board of curators) that Hunter's brother-in-law and executor Sir Everard Home had actually burned most of Hunter's manuscripts.[25] What had not been publicly appreciated until Warburton's committee sat in 1834 was the extent of the destruction, or the degree to which Home had been systematically cannibalizing the papers for thirty years. Home by now was dead. In the Regency he had been an ambitious courtier. He had become a friend of the Prince Regent and had been knighted for his services to the duke of Cumberland (the king's son). But (according to radical tittle-tattle) he had taken to drink, "never went to bed sober," and had died "a regular sot" in 1832.[26] Now, before Warburton's committee, Hunter's former amanuensis William Clift revealed the magnitude of the damage. Home had removed literally a cartload of Hunter's papers to his house in 1800, ostensibly to draw up the catalogs. But while the council continually requested these, Home had used the manuscripts instead to compose his hundred or so papers for the Royal Society's *Philosophical Transactions*, burning the scripts in 1823 shortly after finishing the proofs of his *Lectures on Comparative Anatomy*.[27]

The extent of Home's duplicity has been keenly debated. Home always maintained that he burned the papers in accordance with Hunter's dying wishes. Jane Oppenheimer, the most sympathetic interpreter, notes that Hunter was himself cautious about publishing. She points out that the manuscripts remained imperfect on his death; Home was expected to complete and publish them, and he evidently did so with a clear conscience, seeing his work as a faithful continuation of Hunter's own. He burned the originals not so much to cover his tracks, but because his thirty-year labor was over. Few historians, however, take such a charitable view. Nor did many contemporaries. Even some sergeant-surgeons,

24. *Report SCME, Pt. 2*, 46; "Mr. Guthrie on the Museum and the Apothecaries' Act," *L* 1834–35, 2:379.

25. Negus 1966:16ff. Attendants at the 1826 meeting in the Freemasons' Tavern were told of Home's arson (an act difficult to cover up; he had practically set fire to his house in the process, and the fire brigade had had to be called): *L* 1826, 9:735–36.

26. Dobson 1954:91.

27. *Report SCME, Pt. 2*, 61–68; "Destruction of John Hunter's Manuscripts," *L* 1834–35, 2:238; "The Hunterian Museum," ibid., 471–76; Dobson 1954:59, 62; Oppenheimer 1946:71.

generally circumspect on the question, pointed an accusing finger. Guthrie grudgingly admitted that Home had made use of the papers "and did not wish the record to remain behind."[28] Because Clift's was the most caustic criticism inside the college, Oppenheimer has looked for some deeper motive for his vindictiveness, suggesting that he had always been envious of Home's success and station. This is at least plausible: Clift, the humble miller's son risen into the comfortable middle classes and aspiring to send his son to Charterhouse, could have resented the court surgeon's fashionable life. But it is unlikely. The evidence tells of friendship and respect for his patron before 1823. Home had actually augmented Clift's meager salary out of his own pocket, and Clift had christened his own son William Home.

Far more convincing is Jessie Dobson's reconstruction. Being familiar with Clift's manuscripts and marginalia, she has been able to illustrate the plagiarism precisely. She is in "no doubt that Home abstracted practically the whole of Hunter's unpublished work" and attempted to cover up the theft. Clift's anger after 1823 stemmed from his loyalty to Hunter (who had first taken him on as a draughtsman). Nor need we turn the evidence inside out to understand the resentment. After all, Clift—fiercely protective of Hunter's reputation—had visited Home shortly before the destruction only to find Hunter's manuscripts being used as toilet paper.[29] But the rights and wrongs are not really the point. It was Clift's traumatic, tearful revelations before Warburton's committee that galvanized the medical world. These public disclosures increased the council's discomfort and provided grist for the radical mill. Democrats were appalled by the council's lackadaisical attitude not only in failing to censure Home publicly, but in actually allowing him to remain a museum trustee until his death. Wakley claimed that Home and the other councillors were on the same "Hail fellow, well-met" terms after the event as before.[30] The radicals were able to make great play of the surgeons' cavalier attitude toward national property and their disregard for the law in cases involving one of their own.

An explanation of the council's actions was expected from Sir Everard's pupil Benjamin Brodie in his 1837 Hunterian Oration (the first since Warburton's *Report*—the museum having been closed since 1834 for rebuilding). But, like Guthrie and Cooper, Brodie refused to condemn Home,

28. *Report SCME, Pt.2*, 45; Oppenheimer 1946:39–71.

29. Dobson 1954:59–62, 69.

30. "Mr. Clift's Evidence," *L* 1834–35, 2:488–89. This was true of Cooper, Brodie, and Guthrie (Oppenheimer 1946:41–43). But on learning of the deed, the Board of Curators in 1824 was angry and heaped "the severest Censures" on Home in an effort to wrench the remaining manuscripts from him (Dobson 1954:63–67).

infuriating radicals by actually speaking warmly of him. Ryan's *Journal* immediately charged the trustees with "a gross dereliction of their public duty" in allowing the "literary plunderer" to walk roughshod over them, "wickedly" destroying property left in their charge.[31] The council's subsequent failure to certify which of Home's papers were genuine led the *Journal* to accuse "those connected with the CRIMINAL" of complicity—of attempts "to stifle all *investigations*" and cover up the vandalism.

The situation was particularly embarrassing because the council had been charged by the government with looking after Hunter's collection as a national treasure. In fact the prestige of the Hunterian Museum rested to a great extent on Hunter's fame. Clift reckoned that nine or ten folio volumes and perhaps thirty papers had been burnt—nine-tenths of all the manuscripts. The loss was "irreparable" and made the cataloging extremely difficult. (Even with Clift's help, Home had only managed to produce one twenty-four-page "numerical" catalog, and that practically useless.)[32] All of Hunter's specimens, having been shorn of their accompanying notes, had to be redescribed by Owen. His laborious work in the one area where radical censure was concentrated made him indispensable to a council intent on defusing the situation. By 1836 Owen was able to tell a new parliamentary committee that seven catalogs—six years in the making, and listing some eight thousand specimens, or two-thirds of the collection—had been printed. Some catalogs were huge. The "Physiological Series of Comparative Anatomy," for example, comprised three volumes.[33] Thus while he had not been able to undo the damage done by Home, he had shown the council's intentions to make good (and, by his own estimate, he had quadrupled the value of the collection). The council was naturally eager to publicize Owen's labor. By reworking the old material, Guthrie claimed, Owen had made immense advances, outshining all his predecessors, Cuvier as much as Hunter. Guthrie pointed to Owen's proof of the nonlarval nature of the college's bottled *Proteus* (a permanently gilled cave amphibian)—a specimen Cuvier had traveled expressly to see. It was this kind of accuracy which gave the catalogs their scientific worth. Owen's value was not lost on a deflated and besieged council.

31. "The Hunterian Oration," *LMSJ* 1836–37, 10:774–75, 800–802; "Sir Benjamin Brodie and the Hunterian Manuscripts," ibid., 865–67. Others noted the irony that it had been Home himself who had founded the Hunterian Oration: "Sir B. Brodie's Hunterian Oration," *BFMR* 1837, 4:189–91.

32. *Report SCME, Pt.2*, 43; Dobson 1954:68.

33. *Report SCBM*, 44, 46. W. Lawrence (1834:24) reported that the catalogs in print comprised some 1,015 pages. Guthrie's testimony: *Report SCME, Pt.2*, 46. The physiological series by 1840 consisted of five catalogs (Rupke 1985:241). On Owen's quadrupling the value of the collection: Owen to Whewell, 22 January (TCL WW Add. MS a.210[63]).

Although Owen's catalogs eased the situation, criticisms continued through mid-decade. Ryan in 1835 admitted that the council, "thanks to the Lancet," was now improving conditions, and that Clift and Owen were "polite and obliging." The major rebuilding work was finished by 1835. But opening hours remained restricted, which was inconvenient for GPs, and Ryan urged that the museum be opened all day and every day.[34] In 1836, in another of the "bit-by-bit reforms," the library was opened for three hours on three weeknights. But the struggle to gain these concessions left considerable bitterness. One less-charitable reviewer took the opportunity of Tiedemann's visit to London in October 1835 to twit the curators:

he will probably expect to find the guardians of the invaluable museum not only deeply informed . . . but zealous to diffuse such information through the medical community. He will be disappointed; these dog-in-a-manger rogues, who have stolen John Hunter's museum, take good care to keep the profession excluded from all effectual access to it.[35]

The fate of comparative anatomy was intertwined with these political events. There was a widespread feeling that it would be recognized in the one-faculty reorganization of the teaching curricula following the publication of Warburton's report. This made rescuing Hunter's collection "from the fraudulent gripe of its present self-constituted guardians" all the more urgent. If comparative anatomy was to receive prominence in the reorganized profession, then the museum must be made accessible to everybody: midwives, apothecaries, GPs, and all. Also, given this projected status-rise for comparative anatomy, the radicals thought it essential to establish a chair associated with the collection in the democratized College of Surgeons. A GPs' spokesman announced in 1835 that

the establishment of a chair of comparative anatomy, in connexion with the Museum, would be attended with great advantages, and, in the event of one faculty being immediately instituted, we have little doubt that such a professorship would forthwith be created. When the general intelligence of the profession is brought to bear on its institutions, the wise men now in office will soon see how much the *subordinates* are in advance of them, and how rapidly the genuine light of science has been diffusing itself through the medical community for some years past, while *they* have been supinely slumbering in the unwholesome shade of corporate indolence.[36]

34. "The Hunterian Museum," *LMSJ* 1835, 7:536–37; "Re-Opening of the Library of the College of Surgeons," *LMSJ* 1836, 9:867–69.

35. "Professor Tiedemann in London," *LMSJ* 1835–36, 8:379.

36. "Importance of Comparative Anatomy," *LMSJ* 1835–36, 8:184–85. Cf. R. S. Owen 1894, 1:61.

But the council surgeons had been carefully grooming Owen. His allegiance was unquestionable and his capacity for work extraordinary; most of all he had proved himself ideologically sound. The college had, as Guthrie said, "raised" their man. Owen was by now closely integrated into the college network. He had lived in Clift's house until 1832 and had, after a difficult affair, married his daughter Caroline in 1835, after being provided with premises over the museum.[37] He was everything to everybody: Abernethy's favorite, Carlisle's reincarnation, Guthrie's corporate savior, and Green's Coleridgean heir. He had shown himself able to better the French on their own ground, thus stifling the radical taunts of incompetence in foreign science. He was also closely in touch with the medical officers in the hospitals. Indeed, he had himself been appointed the lecturer on comparative anatomy at St. Bartholomew's in 1834 and had joined Abernethy's other protégé there in petitioning Parliament against the granting of a charter to London University.[38]

The radicals were nonplussed to learn in April 1836 that the suggestion of a Hunterian chair had been taken up, but that the surgeons had placed their own man in the job. Prof. Owen was to reaffirm the corporation's control over comparative anatomy and uphold the interests of the hospital staff. Perhaps the title of Professor gives an unfair impression of his new status, for, as Guthrie reminded him, he was still expected to be deferential to the gentlemen of the council, "whilst you shall be engaged to the best of your ability in carrying out their wishes."[39] Ignoring the radicals' one-faculty demands and Geoffroyan program at the university, Brodie proclaimed during his 1837 Hunterian Oration that Owen,

devoting himself entirely to anatomy and physiology, will prove, as I venture to predict, a more efficient professor than any of us, who have preceded him. Thus there will be established, by means of this great museum and the lectures, a school of what may be called the "science of life," such as has never existed in this metropolis before.[40]

According to Brodie, all the hospital students wanting to "earn for themselves pre-eminence" would now be able to study "the phenomena and

37. On their emotionally fraught affair, and the difficulties of residing under the same roof: Caroline Amelia Clift to R. Owen, 1 May 1832 (BL Add. MS 39,955, f. 212); C. A. Clift to Caroline Harriet Clift, 16 May 1832 (BL Add. MS 39,955, f. 218).

38. Thornton 1974:61–63, also 58. Owen failed to obtain a staff post at Bart's (lacking the necessary apprenticeship), so he himself could not have been totally happy with the existing privilege system (Thornton 1974:265).

39. Guthrie to Owen, 10 February 1842 (BMNH RO).

40. "Sir Benjamin Brodie's Hunterian Oration," *MG* 1836–37, 19–972; "Sir B. Brodie's Hunterian Oration," *BFMR* 1837, 4:189–91. R. S. Owen 1894, 1:95–96, and 61 on the first mooting of the idea of a chair in 1832.

laws of life generally" at the college. Of course, such a school is precisely what the university had aimed at. The corporation was belatedly attempting to regain the initiative by setting up a rival program, equally comprehensive but more conservative. Through the influence of the new "liberal" councillors, men such as Brodie (promoted to the council in 1829), Benjamin Travers (1830), and Joseph Green (1835), an attempt was finally being made to redirect comparative anatomy. As a young Peelite, Owen was at ease with the new men. He was well suited to advance the corporation's interests while developing a new science to substitute for the radical nostrums.

Contemporary observers were in no doubt that the rapid modernization of the college in the 1830s was a response to the widespread criticisms and threat from the university—that the extension, renovation, longer opening hours, expensive catalogs, elevation of Owen, and institution of yearly Hunterian lectures were strategically necessary if it were to regain the initiative and develop its own school of the "science of life." Wakley saw the council's activity—hectically refurbishing even while Warburton's committee was sitting—as a "scandalous device" to stave off parliamentary demands for its reorganization.[41] If so, the ploy worked. Travelogues marveled at the new ninety-foot museum, officially reopened in the presence of Wellington, Peel, and five hundred guests in February 1837 (see fig. 6.4). It was a "magnificent place," lit by high recessed alcove windows, with three-story walls and a Doric pillar base. The facelift and restocking expenses—still nearly £3,000 a year in 1841—allowed the college to display the richest collection of comparative preparations and skeletons in the metropolis. The museum "possesses almost everything the imagination of man can conceive of that can be useful or necessary for the study of physical life," wrote a visitor; it was as if "the whole earth has been ransacked to enrich its stores."[42] The piecemeal reforms had removed the sting from the radical criticisms, and all but the hard-core critics were now won over. In May 1837 the formerly hostile *Medical and Surgical Journal* was "quite unprepared, notwithstanding the investigations of Mr. Warburton's Committee, for the gigantic strides in liberality" shown by the council in throwing open the museum and establishing a reading room for the "Profession at large"—where, it added devilishly, GPs could "ponder over the luminous contributions which are so abundant in the pages of the *Lancet* and the *Medical and Surgical Journal.*"[43]

41. "Proposed Outlay of College Money," *L* 1833–34, 1:830–32.
42. C. Knight 1841–44, 3:200–203; Dobson 1954:99–108.
43. "Working of the Improvements of the Royal College of Surgeons," *LMSJ* 1837, 11:174–76.

Opening hours were ample, the rooms well lit and ventilated; the exhibits were displayed to advantage, and the officers helpful. The grievances in this area at least were being met. And in 1839 moderates went on to applaud the "progressives" on the council, men "who can read the signs of the times," for instituting two studentships in human and comparative anatomy.[44]

Only the doctrinaire radicals in the university, private schools, press, and unions (typified by Grant, Dermott, Wakley, and Webster) refused to see the council's actions as a sign of serious ideological softening. The piecemeal reforms were denounced as cosmetic, a sham. Wakley stuck to his hard line, deploring the state of society

which allows various sets of mercenary, goose-brained monopolists and charlatans, to usurp the highest privileges. . . . This is the canker-worm which eats into the heart of the medical body. The "improvements" which have been effected in our colleges we regard as *nought*. It is not sufficient to tell us that the Council . . . [is] now more liberal, and that the insults which are offered to the commonalty have become less numerous. . . . It is not sufficient to inform us that the museum of the College is better kept, and more freely open, and that a catalogue of the preparations, after twenty-six years of labour, has at last been produced. . . . A thousand acknowledgments such as these, if they were all of admitted truth, could not still the voice of one sensible medical practitioner who . . . [demands] a thorough change in the statutes and charters which concern medical education and practice in this country.[45]

As an astute strategist, Wakley was well aware that each concession actually made it harder to justify the complete overthrow of the autocratic system. His hardline in the face of minor reforms meant that the appreciation of Owen remained split along party lines well into the 1840s. Green portrayed him in 1840 as "the able vindicator of Hunter's fame" and asserted that the establishment of the chair "is calculated to form a glorious epoch in the annals of science, reflecting honor alike on this College and on the country."[46] It was hardly a view shared by the radicals. Wakley deplored his elevation. He pointed out the "impropriety" of giving Owen new "discretionary" powers, instancing his opposition to Grant's reelec-

44. "Studentships in Anatomy in the Royal College of Surgeons of London," *MCR* 1839, 31:284–85. These were worth £100 and tenable for three years. Three studentships were offered in 1842: *MCR* 1842, 36:299.

45. *L* 1838–39, 1:2–3. Cope 1959:49. Wakley consistently challenged the sincerity of "the '*liberal*' party in the College," attributing disingenuous motives to those proposing reform, e.g., in "Messrs. Guthrie, Blizard, Lynn, and Lawrence Hold up Their Soiled Hands for Reform," *L* 1831–32, 1:122–23.

46. Green 1840:29.

Fig. 6.4. The magnificently refurbished Hunterian Museum, reopened in 1837. By T. H. Shepherd. (Reproduced by kind permission of the president and council of the Royal College of Surgeons of England)

tion at the Zoological Society.[47] Grant himself continued to berate the self-elect, whose traffic in diplomas had for decades led them to ignore the unprofitable zoological sciences. For him, comparative anatomy had failed at the college, and his shafts struck the new Hunterian professor

47. "Re-Opening of the College of Surgeons," L 1863–37, 1:766–77.

hard. Grant considered the Hunterian Museum's authority actually the greatest "impediment" to the progress of comparative anatomy in the country, with any "affected encouragement . . . more calculated to insult than to promote" the science. The museum's very existence was detrimental to the "diffusion of knowledge," being designed to obstruct the formation of collections in the universities and private schools.[48] These politically motivated attacks continued to reflect the deep divide between the university/private schools and the gentlemanly elect. During the depression of 1837–42, as agitation for suffrage in the country turned to violence, so ultraradical opposition to the college had grown more extreme. A romantic conservative such as Owen, educated in the ways of the National Church, would have seen the criminality in these rabble-rousing attempts to overturn the accepted order. The pernicious assaults on his superiors incensed the young Peelite with social pretensions, making a scientific response morally and professionally expedient.

On the Need for a New Conservative Philosophy

Her MAJESTY's Ministers are responsible to Parliament; the Legislature to the constituency; town-councils to the rate-payers; guardians to parishes; the councils of scientific societies to the members at large, who elect them annually; but the Council of the London College of Surgeons remains an irresponsible, unreformed monstrosity in the midst of English institutions—an antediluvian relic of all, in human institutions, that is most despotic and revolting, iniquitous and insulting, on the face of the earth. According to the law of correlation, Mr. OWEN, treading in the steps of CUVIER, though not with the assured step of our great GRANT, divined the whole structure of an animal from the inspection of a single part. The same law will apply to political structures; and when we put this single fact—"irresponsibility in the management of a public institution"—before the eyes of the student, it gives him the key to the elaborate evils which have marked the course and history of the "Twenty-one" [councillors] of the College of Surgeons. This is the frightful "claw;" he will readily conceive what was the rotatory power and agility of the grasping arm; how the teeth were sharp, and set like scissors, to cut and tear; how keen the stomach was in digestion, and how

48. Grant 1841:49–51.

ruthless and fierce the instincts which governed the whole organisation.

—Wakley using Cuvier's methods to ascertain the evil nature of the RCS
beast. This spoof was based on Cuvier's claim that he could reconstruct an
entire animal from a single bone. [49]

Given the new needs, Owen's response to the assaults would make little
concession to the old conservative standbys in science. Vital fluids, Paleyism, crude empiricism, even aspects of Cuvier's functional anatomy—all
were showing signs of cracking under the radical pressure. The medical
hacks, intent on breaking into the surgeons' "celestial empire," gleefully
pointed up their "puerilities." By now too, numerous rival sciences were
providing stiff competition: Wakleyism had its scientific cutting edge in
Grant's Lamarckism; the medical Benthamites promoted philosophical anatomy; while the democratic Methodists and Dissenters brought their own
sectarian understanding to bear on the new naturalistic zoologies. Coleridge saw the need for a strong combative Anglican science—a science to
reassert the traditional values of the gentry and professional clerisy.

Owen's education too left him aware that the traditional sciences were
inadequate to meet the threat. When he entered Barts, his elderly patron
John Abernethy was still reeling from his "hot controversy" with Lawrence over the basis of life, mind, and morality. [50] Abernethy himself had
argued that a vital principle animates tissues, vivifying them. He had also
raised mind over neural matter, suggesting by analogy that the soul
equally stood apart from the body. These were all doctrines he—and Regency Tories—believed would enforce traditional loyalties and deference. But Lawrence had laughed this mystical force out of court. Life was
a function of organization, not of peculiar vital powers, nor were these
needed to prove the soul's existence or enforce social obligations. He
looked to the rigors of an atomistic physics to shield biologists from such
"metaphysical chimeras." He believed that liberation from this reactionary science was just part of a secular revolution that would lead to the
"complete emancipation of the mind, the destruction of all creeds and
articles of faith, and the establishment of full freedom of opinion and belief." [51] It was these social implications that brought the debate to Coleridge's attention. Coleridge, like Abernethy, abominated the atomistic
and atheistic excesses of Lawrence's republicanism. He more than any
other sensed that Lawrence's sort of science was at back of the godless,

49. "The College Conversazioni," L 1841–42, 2:246.
50. Griggs 1956–71, 4:928.
51. W. Lawrence 1844:2–9, 55, 68, 1816:166–77; D'Oyly 1819:3.

seditious evils of the age. Coleridge complained to Green, his confidant and amanuensis, that

a system of Materialism, in which Organization stands first, whether composed by Nature or God, & Life &c as its *results*; (even as the sound is the result of a Bell)— such a system would, doubtless, remove great part of the terrors which the Soul makes out of itself; but then it removes the Soul too, or rather precludes it.[52]

Coleridge wearied of the radical successes in medicine and society. He delved deep into medical philosophy and kept close links with the sergeant-surgeons. Given his sympathy for Green and Abernethy, it is obvious that his scientific delegation to the National Church, mooted in *On the Constitution of the Church and State* (1830), would have included a contingent from Lincoln's Inn Fields. Abernethy had attended Coleridge's lectures at the Crown and Anchor in 1818 and had quoted them against Lawrence. "By their fruits should trees be known" was Abernethy's Coleridgean maxim, intelligible to Lincoln's Inn listeners in the troubled Regency period. He conceded that while godless materialism in Catholic France might have broken the yoke of "superstition and bigotry," it had nevertheless led to terror on the streets and the Napoleonic Wars, and he apocalyptically repeated Coleridge's words: "There can be no sincere cosmopolitan, who is not also a patriot."[53] Lawrence's materialism was a political crime: in Coleridgean circles, it was tantamount to treason against the state. But when it came down to it, Coleridge actually had no greater love of Abernethy's explanation of life. After the turn-of-the-century discovery that electricity could dissociate salts, direct chemical operations, and make muscles twitch, Abernethy had begun to identify the electrical fluid with the vital principle itself. The vital agent became what he called a "subtle, mobile, invisible substance superadded to . . . animal matter," making it uncertain now whether it was really a power or a form of "mobile matter." To Coleridge—hating any offshoot of the "mechanico-corpuscular" philosophy—this smacked of just another sort of materialist explanation. He clearly hoped to see Lawrence defeated on quite different grounds, and he tried "to insinuate into Mr Abernethy" more philosophic means of repulsing "the attacks of Lawrence, and the Materialists."[54] This was the crux: not that Lawrence was wrong and Ab-

52. Griggs 1956–71, 5:47.

53. Abernethy 1825: 65, 68. W. Lawrence (1844:4) also talked of his supposed "treason against society." Abernethy's 1819 oration was delivered to an audience that included Coleridge (Coleridge 1949:24–25, 28–29). Abernethy was quoting from Coleridge (1949:236n., 422, n. 9).

54. Griggs 1956–71, 4:809, 5:49–50; Levere 1981:46. On Abernethy's vital electrical fluid: Goodfield-Toulmin 1969:294–96.

ernethy right, but that Lawrence was wrong and the old guard had nothing credible to put in his place. Coleridge shared Abernethy's view that
life does not proceed from organization. But attributing life to quasi-
electrical fluids was no solution to the problem.

These imponderable fluids were to be replaced by Coleridge's brand of
German nature philosophy. Coleridge eschewed all forms of atomistic explanation, whether derived from conservative Newtonian matter-theory
(of inert atoms in a world upheld by God) or the radical Enlightenment
philosophy of active, self-empowered atoms—the basis of the extremists'
evolutionary doctrines (and an atheistic dogma, Coleridge noted, common among the "Scotch Physicians"). Coleridge's rival nature had a strong
"organic" or dynamic aspect; he pictured nature "unfolding" under the
influence of "Constructive Powers." On occasions he actually called the
natural laws themselves the "Constructive Powers, excited in Matter by
the influence of God's Spirit." Elsewhere he seemed to suggest that the
laws—which as expressions of Divine Will were antecedent to matter—
only described the "inner necessity" of this unfolding. In effect, each of
nature's "minute elements" was "a living germ in which the present in-
volves the future"; that potential is realized by productive powers reflecting God's Will, and its unfolding course is described by what we call laws.
Although very close to German *Naturphilosophie*, Coleridge was always
concerned to distance himself from Schelling's pantheism—from the idea
of Will as nature. He was not always successful, but this concern was inherited by his disciples Green and Owen. Whatever Coleridge's precise
view, the "philosophic Naturalist" could now account for nature's development by invoking the ideal powers of Divine Thought, rather than by
resorting to the deist's shuffling atoms with their self-existent energies.[55]

What eventually cowed Lawrence in the debate was a mixture of career
threats and professional inducements. While historians have shown how
an evolutionist such as Darwin was to learn the lessons of Lawrence's persecution,[56] it is what happened afterward that is important when considering Owen's perceptions. For Lawrence's situation changed dramatically
after 1828 as he reaped the rewards of political compromise. The Chancery court case and threat to his hospital post might have silenced him.
But inducements did their work too: Lawrence actually accepted a council
seat in September 1828, without, Wakley snarled, any recourse to his
much-vaunted plebiscite. Wakley viewed his apostasy with "intense re-
pugnance." It became even more painful in 1831 because Lawrence was

55. Levere 1981:106; quotes above from 60, 79, 98–100. McFarland 1969: chap. 3 considers Coleridge's flirtatious skirting of pantheism.
56. Gruber and Barrett 1974:204–5.

now mixing "in friendly combination with those very men who are *prose-cuting criminally* his former friends." Wakley depicted it as an act of polit-ical "suicide" and predicted that his practice would crash as he lost the confidence of the profession.[57] Lawrence's tone moderated. He no longer attended public meetings, and as a reluctant witness before Warburton's committee he spoke against summer certificates and total suffrage and even commended the council for having removed the cause of his earlier grievances. Foreigners expecting a fiery demagogue were now surprised to find him docile and "dull." "He appears to have allowed himself to be frightened," wrote one, "and is now merely a practising surgeon, who keeps his Sunday in the old English fashion, and has let physiology and psychology alone for the present."[58] Educated to rank and privilege, he was caught between the iniquities of Old Corruption and the career dan-gers of a republican remedy. Radicals recalled his "quiet sneering smile" (see fig. 3.3), which they came to see concealing a disingenuous nature.[59] But democrats in general practice and in the press could afford to be un-compromising with so much less to lose. As a hospital surgeon Lawrence appreciated the value of political expediency, and he continued to func-tion as a moderate reformer on the council. As Owen joined the college, Lawrence's case left no one in any doubt of the value of compromise.

Other conservative doctrines besides those of vital fluids were being reappraised. By the 1820s Paley's "tricksy sophistry"—as Coleridge called it—was coming under the stern eye of the Romantics. To Coleridge Pa-ley's natural theology was one of the more insidious fruits of the "mechan-ico-corpuscular philosophy." Having made a monarch of the senses, Lock-ean philosophy had duped naturalists into concentrating on the attributes rather than the "Personal Being of God." Coleridge feared "the prevailing taste for books of Natural Theology, Physico-Theology, Demonstrations of God from Nature, Evidences of Christianity, and the like." "*Evidences* of Christianity!" he wrote, "I am weary of the word. Make a man feel the *want* of it; rouse him, if you can, to the self-knowledge of his *need* of it"— let his love lead him to God, not some "tricksy" argument from nature.[60] While Paleyism still figured prominently in the works of the older sur-geons, by the 1820s it had begun receding from Coleridgean medicine. It was not only that it made a poor foundation for belief. More practically it was not substantial enough to meet the new radical threat. Bell's case

57. "Mr. Lawrence," *L* 1830–31, 2:533–36. In terms of seniority Lawrence was the next in line, but his appointment split the council, and he scraped in by a majority of one: *Report SCME, Pt.2*, 172.

58. Carus 1846:88; Neuburger 1953, 2:266.

59. Clarke 1874a:20.

60. Coleridge 1913:168, 271–72; Garland 1980:60ff.

shows the demands now being made of it. Admittedly Bell was not a typical figure in RCS affairs, and he never aspired to the "surgical Woolsack" (the presidency).[61] But he spoke the ultrateleological language of the college oligarchs such as Abernethy, Carlisle, and Cooper. His RCS lectures in 1832–33 highlighted the inadequacy of his gentlemanly approach. Like his books they were models of lucidity: simple, direct, "delightful."[62] His rhetoric was Cuvierian. With form and function strictly correlated, he said, one bone, indeed "one joint," holds the key to the entire animal. Like Cuvier, Bell could build an entire anteater from a single bone by a series of logical steps. All adaptations for him pointed to creative wisdom. Some even showed "a prospective design," an anticipating intelligence— exhibited for example in those organs developed in the larva which were destined to function in the metamorphosed and dramatically dissimilar adult. Yet this marked the theoretical limit of his explanation of structure: he went no further than design, except to deny any role for Geoffroy's morphological laws. His papers—as both Knox and Fletcher complained—lacked the subtlety and excitement of the new French works. His approach appealed to a gentleman's taste, not to the new student's professional needs. Hence Bell's ousting from the university, itself a sign that design was no longer de rigueur in a Benthamite education.

Geoffroy's doctrines were known to be conquering radical medicine.[63] As a result, Bell was forced more and more to thrust his gentleman's anatomy into competition with the French imports, pushing it into work for which it was ill equipped. In his lectures to the surgeons he attacked Lamarck's environmental explanation of shape and structure. He urged that the *"prospective* design" of fossil animals militated against their having been fashioned by contemporary environmental changes, whatever certain comparative anatomists "of high character" insisted to the contrary.[64] In his Bridgewater Treatise he denied that Geoffroy's homologies could justify equating the fish's operculum with the mammalian ear bones. These were differently adapted bones, each explicable by the "function which is to be performed." He detested "the cold and inanimate influence" of homologies and transmutation, which threaten "to extinguish all feeling of dependence in our minds, and all emotions of gratitude [to the 'intelligent, designing, and benevolent Being']."[65] But all Bell could offer in exchange was "prospective design." It is ironic that at the moment

61. Gordon-Taylor and Walls 1958:173.

62. "Sir Charles Bell's Bridgewater Treatise," *MG* 1833–34, 13:253.

63. E.g., R. Owen (1839b:46) acknowledges the wide acceptance of Geoffroy's vertebral theory.

64. C. Bell 1833–34:281–82.

65. C. Bell 1833:139, 144.

when old Tory wealth was relaunching Paleyism on a popular footing in
the Bridgewater books, the doctrine was faltering in medicine. Paleyism
and imponderable fluids were being seen as passé—as ineffectual against
the radical anatomies. Bell's lectures in 1832–33 were the last in the RCS
conceived within an exclusively Paleyan frame. College tyros, aware of
Paley's weakness, were already seeking a stronger antidote to the radical
imports. The need was for a new philosophy that relied on neither naïve
arguments from adaptations nor the existence of subtle fluids to prove a
higher power—a biology sophisticated enough to meet the rival Geoffroyan anatomies on their own ground.

Cautious Reform and Coleridgean Romanticism: J. H. Green and Owen's Philosophical Apprenticeship

> [There is no] power in the lower to become, or to assume the rank
> and privileges of, the higher, upon any such fanciful scheme as that
> proposed for the invertebrated animals by that laborious and other-
> wise meritorious naturalist, Lamarck.
>
> —Owen's mentor J. H. Green, qua comparative anatomist, opposing self-
> development and arguing for a "higher power" in whose thoughts life's
> "generic" thread is located[66]

> The colleges of learning . . . work by descent; they are to be the
> suns of the system . . . and not mere mirrors, reflecting only the
> light they had been previously bestowed; and their characteristic
> form, from the very beginning, is by *appointment*—appointment by
> a higher, in contra-distinction from *election* by a supposed lower, or
> equal.
>
> —J. H. Green, qua antidemocrat, arguing that power must be dele-
> gated from above and cannot be mandated from below[67]

The leading Coleridgean inside the college was Joseph Henry Green (see
fig. 6.5), and it is his conservative romanticism that concerns us in connection with Owen. The son of a wealthy merchant, Green had completed
his education in Germany. On his return he was apprenticed to his uncle
Henry Cline, surgeon at St. Thomas's, becoming a surgeon there himself
in 1820 on the death of Cline's son. Green was skillful and above all humane. Wakley, trained at St. Thomas's, never ceased to commend his

66. Green 1840:108.
67. Green 1831:16ff.; "Distinction without Separation," *MG* 1831, 8:215.

Fig. 6.5. Coleridge's leading medical disciple, Joseph Henry Green, German educated and the first to lecture on transcendental zoology at the College of Surgeons. By J. H. Lynch, after G. F. Tenniswood. (Courtesy Wellcome Institute Library, London)

"natural goodness of heart,"[68] even while hating his paternalism and laughing at his labyrinthine language and bombastic idealism.

In 1817 Coleridge had visited Green's house to meet the German philosopher Ludwig Tieck. (It was Tieck's recommendation which enabled Green to continue his study of philosophy in Berlin that year.) Green and Coleridge shared a powerful Germanizing interest, and they formed a lasting friendship. Green attended Coleridge's Thursday class and became his medical disciple, introducing Coleridge to the College of Surgeons' library and guiding him through its zoological works. Coleridge, for his part, included Green's poems in his own collections. Green may also have helped Coleridge revise his manuscript "Theory of Life" while preparing his own college lectures in 1824. Certainly at this time, Coleridge alludes to their "confab" on the German transcendental zoologist Lorenz Oken, and they talked about the succession of life, and the "*generic* in Thought" as contrasted to "*Genetic* in Nature."[69]

In 1824 Green was appointed to the anatomy chair at the College of Surgeons and delivered his first twelve lectures on comparative anatomy. These lectures, Coleridge observed, "have deservedly attracted much attention."[70] The course extended over four years (1824–27), and Owen attended for three, providing the accompanying dissections in 1826 and 1827." "For the first time in England," Owen later recalled, "the comparative Anatomy of the whole Animal Kingdom was described" and illustrated by colored diagrams. Rather than take a Paleyite view, Green promoted Coleridge's philosophy and German *Naturphilosophie*. He based his course on the concept of an ideal "Unity" underlying all creation, using the Dresden professor C. G. Carus's *Comparative Anatomy* as a textbook. So Owen, during his first years in London, was exposed to "the dawning philosophy of Anatomy in Germany, rather than the teleology which Abernethy and Carlisle had previously given as Hunterian."[71] When Carus

68. "Mr. Green's Views on Surgical Reform," *L* 1830–31, 2:569; also 1832–33, 1:127. Wakley claimed that Green was the only teacher he learned anything worthwhile from at St. Thomas's: "Alleged Hauteur of Mr. Green to His Pupils," *L* 1829–30, 2:96. Green's infuriatingly eliptical language was one of his less happy debts to the *Naturphilosophen*. It elicited raucous laughter: *LMSJ* 1834, 5:284–85; *L* 1833–34, 2:52 ("the queerest of all comical writers"); *L* 1830–31, 2:598–602; and *L* 1832–33, 1:127: "away he shoots like a sky-rocket—Whiz—pop—diddle—diddle—diddle—confusion—*darkness*." Green unfortunately led Owen into the same linguistic limbo.

69. Griggs 1956–71, 5:372. Levere 1981:44–45. Details of Green's life are taken from J. Simon's memoir in Green 1865.

70. Griggs 1956–71, 5:369–70.

71. R. Owen in Green 1865:xiv. Green echoed the *Naturphilosophen* who had followed Schelling's thought into zoology, particularly Oken, Johann Spix, and Carus (whose *Comparative Anatomy* was translated into English in 1827). These transcendental zoologists took a

visited London some years later (as the king of Saxony's physician), he was received "with visible pleasure" at the college. Owen's admiration of Carus was reciprocated. "Owen pleases me thoroughly," Carus wrote; he is "a sensible, able man—deeply versed in what is old, and ready for the reception of what is new."[72] Indeed, with Peelite perspicacity Owen recognized that there was no going back—that the higher anatomy was established, but that reinterpreted in ideal terms it could provide a conservative bulwark against the godless radicals.

Green's was an effusive, devout, antiatomistic reverencing of German philosophy. His aim was to reconcile "the study of Nature with the requirements of our moral being, and [to connect] science,—which even as the noblest offspring of our intellect is but a fragment of our humanity,— with the philosophy of Coleridge."[73] His pious *Naturphilosophie* was the first heard within the college walls. His Coleridgean romanticism was a hymn to that Divine "Intelligence whose thoughts are acts." It was also a philosophic prescription to seek the enduring Ideas behind nature's facade: the "Exemplars" and "Archetypes" symbolic of "that Mind which is the identity of truth and reality." The prescription Owen learned well— "to discover the laws which give permanence and regularity, to discern the eternal Ideas, which are the regulating types and standards." With German nature-philosophers, Green portrayed nature advancing in perfection toward mind and individuality, "the final purpose of Divine Law." Life was striving toward consciousness, ascending to the human level where "nature must not only feel,—she must know—her own being," become, in short, an image of the "invisible Supreme."

Green's idealized understanding of nature's "advancing perfection" was the antithesis of the materialists'. Owen left telegraphic notes covering the course in 1827, which record Green's classroom denunciation of Lamarckian "genetic" continuity. In his *Naturphilosophie* the "generic" thread running through life was a Divine projection. There was no actual transmutatory continuity, no detestable "Ouran Outang theology." Owen

---

special interest in serial homologies, that is, the repetition of homologous parts in the same animal. They approached the body as a repeated number of vertebral segments, seeing the skull itself as composed of fused vertebrae. Consequently the jaws were the homologies of the vertebral processes of the spine and even of the limbs. Green (1840:57) took the view that "all the varied organic forms are but modifications of but one simple primary form, and that, for instance, the osseous system in every part, and in its most complicated total result, is but the repetition of a simple vertebra." Owen was to criticize the excesses of German zoology, but he was still deeply indebted to this approach. He too interpreted the arms as the massively developed processes of the first cranial vertebra (see chap. 8).

72. Carus 1846:60.
73. Green 1840:xx, xxiv–viii, 41–43.

noted that Green "considers nature as a series of evolutions—not under the idea that the lower can assume the characters of the higher—Lamarks [*sic*]."[74] The last lecture of Green's course was published in his *Vital Dynamics*, the Hunterian Oration for 1840. Here we find his original statement. He conceived nature

as a series of evolutions from the lowest to the highest. Not, allow me to remind you, in supposing that there is any power in the lower to become, or to assume the rank and privileges of, the higher, upon any such fanciful scheme as that proposed for the invertebrated animals by that laborious and otherwise meritorious naturalist, Lamarck,—a scheme in which the ground and cause is everywhere meaner and feebler than the effect, and in which blindness is made the source of light, and ignorance would be the parent of mind and thought;—but in assuming that the ascent is the indication of a law, and the manifestation of a higher power acting in and by nature.[75]

Knowing the beleaguered corporation context, and also Green's hatred of atomism and haughtiness toward the lower orders, we are in a position to relate his rejection of life's spontaneous ascent to his Coleridgean social strategy. Harmonious class relations, in Coleridge's view, depended on sound scientific foundations, which made the eradication of any Lamarckian tendency toward self-development imperative. Coleridge and Green's desire to reduce all biological and social "knowledges into harmony" had a serious political intent.[76] They were naturalizing a patrician style of government using a physiological science which testified to the supremacy of higher powers. To connect "natural history with political history" Coleridge employed a "moral *copula*." Like the *Naturphilosophen*, he envisaged society in organic terms, stratified into classes and orders, each functioning in strict relation to the others. The state's "integral parts, classes, or orders are so balanced, or interdependent," he wrote, "as to constitute, more or less, a moral unit, an organic whole."[77] As Ben Knights says, since society was not "a mere conglomeration of self-fulfilling individuals," disruption of the body politic by democratic action would lead to cultural decay,[78] as disruption of the body anatomic would lead to death. This

74. R. Owen, "Notes and Annotations: Prof. Green's Lectures on Zoology," f. 131 (RCS 275.b.21). Compare Coleridge on the "ascending series of distinct evolutions" (Levere 1981:136).

75. Green 1840:99–135, esp. 108. Green was fully aware of Lamarck's taxonomic merits, having structured his own "general view of the Gradual ascent of the Animal Kingdom from the most simple . . . [after] the two-fold style of Lamarck," that is, accepting his two invertebrate streams: J. H. Green's 1824 lectures, f. 32 (RCS MS 67.b.11).

76. Coleridge 1917:157.

77. Coleridge 1972:91; Harris 1969:221.

78. Knights 1978:65.

metaphoric stretching from nature to society was commonplace. At times the body was conceptualized like so many government ministries. Carlisle spoke of "the governing and directing powers which order all the offices of living organisms."[79] But something more than a shared metaphor is involved in Coleridgean philosophy. Here bodily organs and stratified society actually depend on the same irradiating power—ultimately on the one Divine Presence—for their harmonious functioning. David Bloor talks of the "spiritual resonance" between Coleridge's idealized nature and patrician culture.[80] For conservatives, the Logos actually penetrated society and passed on through nature. It is one and the same chain of command, establishing order and authority in society, Church, and nature. Coleridge himself saw an "antecedent Unity" as the very "cause and condition" of the material and social atoms. The "One universal Presence" unites the community of nature with the "Community of Persons."[81]

So there was a clear political ring to Green's actions and an obvious target group. Radicals marching under tricolor flags and massing outside the corporations based their cultural relativism and emergent morality on the precepts of a self-empowered nature. Their rival organic-social metaphors allowed them to draw quite different conclusions from biological theory. As physiologists they explained higher organic wholes by their physico-chemical constituents, while as democrats they saw political constituents—voters—empowered to sanction a higher authority: a delegate to act in their interest. This democratic legitimation became a prime Tory target in the aftermath of the French Revolution. At the height of Lord Liverpool's crackdown on sedition in the Regency, Coleridge wrote the prime minister, laying the blame for the evils of society squarely on the "speculative science" of the republicans: on their "Physiology" of "pure fiction," of self-energizing or "Demiurgic atoms . . . that are the stuff, the tools, & the workmen of the material Universe." This atomistic physiology—"out of the pale of which there is no salvation for reason"—bore a "powerful, tho' most often indirect influence" on the "Taste and Character, the whole tone of Manners and Feeling, and above all the Religious (at least the Theological) and the Political tendencies of the public mind." Coleridge was adamant that the sedition, blasphemy and deism so troubling to the authorities were the upshot of this venomous philosophy. It was hardly "a sport of chance" that the language of the "mechanic philosophy" was a "fac-simile" of the rhetoric of the French democrats. For the "army of Ruffians" following d'Holbach, "an Atom is an Atom . . . and by

79. Carlisle 1826:35.
80. Bloor 1983:615; Jacyna 1983b:325–26.
81. Coleridge 1913:40–41.

the pure Attribute of his atomy has an equal right with all other Atoms to be constituent & Demiurgic on all occasions." For the republican rabble,

the independent atoms of the state of nature cluster round a common centre and *make* a convention, and that convention *makes* a constitution of Government; then the makers and the made make a contract, which ensures to the former a right of breaking it whenever it shall seem good to them, and assigns to the govern'd an indefeasible sovereignty over their Governors.

A physics of atomic sovereignty and physiology of organizational supremacy underlay the cooperators' belief that collectively they were the power of the state—that their democratic conventions and trades unions were valid. Yet "neither historically nor morally," said Coleridge, "neither in right nor in fact, have men made the state." On the contrary, "the state & that alone makes them men." The State was antecedent to man as the Logos was antecedent to life. Until, he told Liverpool, the gentry and clergy were grounded in a strong counterphilosophy, "all the Sunday and National schools in the world will not preclude Schism in the lower & middle Classes."[82]

The solution was a program of vocational training. Coleridge proposed the establishment of a nationally organized teaching clerisy to enforce the moral union between science and state. It was to be a gentlemanly coalition of professionals from the church, law, and medicine. Only with such an intellectual spearhead would it be possible to regain the cultural initiative. In his view

a permanent, nationalized, learned order, a national clerisy or church, is an essential element of a rightly constituted nation, without which it wants the best security alike for its permanence and its progression; and for which neither tract societies nor conventicles, nor Lancasterian schools, nor mechanics' institutions, nor lecture-bazaars under the absurd name of universities [i.e., the London University], nor all of these collectively, can be a substitute.[83]

Coleridgeans had been mortified at the radical demands: demands for democracy, disestablishment, popular education, corporation takeover, land taxes, and abolition of the tithe, all of which would have hit the gentry, clergy, and professional elite. The lecture hagglers and science diffusers had inflamed labor relations. And by breaking ranks with the upper classes, some medical man had made the situation worse. Lawrence's "treason" was shared by all the medical radicals; they were traitors to the state (the intellectual clerisy, their own class). The democrat intellectuals were the real enemy within. The clerisy was now to counteract their social

82. Griggs 1956–71, 4:758–62; Knights 1978:38.
83. Coleridge 1972:53.

Lamarckian subversion by inculcating principles of religion, responsibility, and sobriety into the masses. The theologians, lawyers, and savants—to be educated as a class at the ancient seminaries—were to strengthen the moral union between philosophy and polity. Bloor characterizes Coleridge's goal as connecting the *is* of nature with the *ought* of politics.[84] The moral strength of political conservatism was to lie in natural truth. But it was not to be the radicals' truth. Coleridge was using a different physiology to extract his political precepts. His nature was a subject state. An idealist physiology shifted the emphasis from base nature back to the Godhead, pulling the "ontological rug from beneath the feet of those who would set themselves in opposition to spiritual authority."[85] It undermined the Lamarckian justification for fierce democracy, the sort stemming from the university and workers' coops. Corporation Anglicans were now advocating a hierarchic natural and social order dependent on God's ordinances and stripped of all "self-developing energies." Neither brute matter nor the populace could order themselves—all progress was dependent on the "higher powers" and subject to Divine law.

This was the thrust of Green's "Vital Dynamics." It left no room for spontaneous self-development in nature or radical self-reliance in society. The postulates of emergent evolution and democratic change violated Divine and constitutional law and, as with all such infractions, threatened atheism and mob rule. The electorate could not mandate a higher authority. The masses could not authorize constitutional change. Authority could only come from a sovereign body: the king in the case of corporate reform, the Supreme Legislator in the case of life's "series of evolutions." Coleridge's physiological dynamics gave more than metaphoric significance to Green's "descensive" spiral of power.[86] It ultimately justified advancement through autocratic appointment. The "meaner and feebler" could only be hoisted up by their beneficent betters.

Even the sympathetic *Gazette* commented on the "aristocratic caste" of Green's mind.[87] For the democrats, who rejected this downward delegation of power, such idealism was denounced as an unprogressive, unproductive prop to Church and state power. The Unitarians detested Coleridge's social blueprint. Lant Carpenter's successor in Bristol, the Rev. George Armstrong, hated "the instinctive, transcendental and whatnot German school of moral and metaphysical philosophy—the spawn of Kant's misunderstood speculations—the dreams of the half-crazed Cole-

84. Bloor 1983:614.
85. Bloor 1981:208.
86. Green 1831. Wakley lost no opportunity to ridicule Green's "descensive" logic; "Mr. Green's Sky-Rocket Lecture," *L* 1832–33, 1:151–55.
87. "Distinction without Separation," *MG* 1831, 8:312.

ridge, and the inane fancy of the Hares, Sterlings, Whewells" at Cambridge.[88] W. B. Carpenter's mouthpiece, the *British and Foreign Medical
Review*, simply dismissed Green's mystifications as a Platonic irrelevance
to science.[89] These Dissenters, with their commitment to lay democracy
and self-government, were reading quite different texts:

> The essence of their Nonconforming zeal was the denial of divinely ordained or-
> ders in society, an attitude which the Quakers symbolized by their refusal to doff
> their hats in the House of Lords and to bend their knee to the King. While the
> Established clergymen were preaching on the scriptural text, "be subject unto
> higher powers . . . for the powers that be are ordained of God," the Dissenting
> preachers were exhorting their flocks to be . . . superior to, and independent of,
> the magistrate in moral and religious affairs. Because Dissenting parents had been
> denied a proper place in local and national government, their children learned
> liberal political action about the Lord's Table.[90]

Green's talk of the romantic unfolding of society under the "pre-disposing
power of a Divine providence" infuriated the radicals.[91] For Green, all
aspects of society were subject to this Divine superintendence. Wakley
shook with anger at Green's discussion of the "predetermined order and
providence in the successive evolutions of the professions"—the ordained
appearance during history of the legal, clerical, and medical professions—
seeing it as an invidious attempt to legitimize the unreformed medical
order. This "stupid species of cant" would only bolster corruption and bar
competition, giving placemen a free run. Even putting law and divinity
on a par with medicine was wretched in Wakley's eyes. The Inquisition
bore eloquent testimony to the providential origins of divinity. And what
was law but "the great chain used by tyrants" and slavers. Indeed, "What
has so long deprived Mr. GREEN's own countrymen of their rights?
Law."[92]

The providential continuity between natural and civil history provided
the basis of Green's own reform call. In a series of political pamphlets he
articulated the feelings of the younger councillors, those who believed
that to prevent a catastrophe the council must make adjustments. It was
an essentially conservative reform call—an attempt to disarm critics while
leaving the concentration of power fundamentally unchanged. In 1831, as
reform fever swept the country, he addressed *Distinction without Sepa-*

88. Quoted in Holt 1938:343. On William Whewell, Julius Hare, John Sterling, and the
other Cambridge Platonists: S. F. Cannon 1978:47–51; Garland 1980:65ff.; Preyer 1981.

89. The review hated the notion of separately existing laws and Platonic ideas: "Mr.
Green on Vital Dynamics," *BFMR* 1840, 10:545–47.

90. Cowherd 1956:66, quoting Romans 13.1.

91. Green 1832:3.

92. "Mr. Green's Sky-Rocket Lecture," *L* 1832–33, 1:151–55.

ration to the president. Here he argued against electoral representation. This was acceptable for guilds and companies in "political, municipal, or commercial life," where bodies were elected to protect the members' common interest, but it was inappropriate for scientific institutions. Here royal appointment was preferable, that is, selection by a sovereign above party-professional interests, by a monarch who was impartial in his dealings with the medical estates. "From a source like this, and from no lower or narrower, can be derived or conferred the authority and the duty of superintending . . . the interest of any particular class of men, for the *public* good, for the weal of the *nation*."[93] Only by such a mechanism could those suited to promote the general good be appointed—those whose advancement might otherwise be "frustrated by popular elections."

Conservatives in the country feared that total suffrage would lead to the successive annihilation of the Tory party, Church, and Lords. Green mirrored these fears inside the college. He was acutely aware that GPs vastly outnumbered pure surgeons, who risked being electorally routed. He knew of the GPs' dissatisfaction and their susceptibility to the demagogy of "the turbulent few" intent on running off with the college funds. But he insisted that reducing the higher grades to plebeian level "would infallibly vulgarize the profession." The GPs must be excluded from the administration. They were ill suited as "guardians of professional honour"—unlike the surgeons, whose disinterested cultivation of science, "liberalising" views, and "intercourse with those most influential in rank and talent, render it more likely that they should take enlarged views."[94] And anyway, with the elite surgeons exalted by the nobility and feted for their scientific successes, the lowly would automatically "partake of the honour" and be elevated by the experience. Aristocratic suns were to bathe the third estate in a sublimely reflected light.

Owen's promotion at the college coincided with that of Green's liberals. He was no Carlislean hawk, having far more sympathy for Brodie and Green's political ideals. These gentlemen continued to argue through the 1830s that medical professionals should receive their moral, religious, and scientific education with the gentry in Oxbridge or the London hospitals, believing that this alone would enable them to aspire to the ranks of the "learned class" "forming the moral strength of the country." In 1834 Green proposed the appointment of a National Council of Medicine to oversee the training of the various grades, each of which was to cater to its client class and be educated accordingly. Behind such educational ideals were

93. Green 1831:16, 30, 1834:36. Unlike the radicals' "twopenny trash," Green's were expensive pamphlets; his *Suggestions Respecting the Intended Plan of Medical Reform* (1834) sold for two shillings.
94. Green 1831:32–33, 1834:26.

strict financial factors. The consultants were to minister to the wealthy gentry who could afford to pay more and demanded better service. Hence the pure surgeons and physicians were to receive an extra three years' training, be better qualified, earn more, and occupy the executive posts in medical society.[95] GPs heckled Green for his attempt to isolate "our order," accusing him of cheapskate motives in limiting the practitioners' education. They responded with a barrage of counterproposals: abolition of all nominal distinctions, uniform rather than class-based education, and all honors to be the reward of "superior talent."[96] But Green's suggestions were acceptable to the new councillors, and he received Brodie's support.

Brodie had taught comparative anatomy at the college before Green (in 1820–21) and had a similar paternalistic cast of mind. He too denounced utopian democratic and educational schemes, justifying a hierarchical "division of labour" on the grounds that the scientific advances were made by surgeons, not practitioners. But he too saw scope for educational improvement. He agreed that the religious restrictions in the RCP were iniquitous and that examiners' appointments in his own college should be based on ability rather than seniority. Like Green he urged the establishment of a regulatory body, which could override council interests where they clashed with those of the profession. But again, it was not to be a body responsible to an electorate. Brodie was appalled by Warburton's ill-fated bill for the "Registration of Medical Practitioners," which proposed that all registered practitioners should become the constituency for a new medical parliament. As he protested in the *Quarterly,*

it must be almost unnecessary to point out the classes of persons of whom we may expect these parliaments to consist: we must not look among them for those who love the tranquil pursuit of science—who pass their days and nights in accumulating knowledge for future use; nor for those who by their labours have already earned the good opinion of the public . . . but rather for the vain and the idle—for those who hanker after a noisy notoriety, and have abundance of leisure because they have no professional employment.[97]

Like Green, he remained suspicious of the rank and file and determined to keep power in traditional hands. Thus his solution to corporation self-interest was the establishment of a Board of Control which could ratify

95. Green 1834:1–2, 5, 10.
96. "Mr. Green's Suggestions for New Modelling the Medical Profession," *LMSJ* 1834, 5:315–16; also 343–45. The *Gazette* (1833–34, 14:25) agreed with Green that a policy of uniform education was "dangerous."
97. Brodie 1840:58, 68, 74–78. On Brodie's Whiggism: Lefanu 1964; J. F. Clarke 1874a:387, 515.

the council's actions—a nonpartisan board that derived its authority from the Crown rather than from the electorate or home secretary.

Aware of the need for a conserving reform (to preempt more drastic measures), Green urged token concessions to the GPs and a "closer union" between the governing and lower classes generally. He suggested modifying the charter to allow the GPs a certain advisory capacity. He proposed a two-tier house, consisting of a Supreme Council and General Council (both appointed), with one-third of the General councillors (twenty in number) being GPs, to be nominated after stringent examination. But the lower chamber was only to be a kind of committee, able to advise but not to legislate itself.

With their "bold nervous" criticisms of the existing system, Green's pamphlets generated considerable interest. Reactions across the political spectrum varied predictably. The *Gazette* praised his repudiation of the "universal-suffrage" nonsense and saw no need for any concessions to the "riotous few."[98] The *Medico-Chirurgical*—putting itself forward as the sensible center—hailed Green as a fellow "liberal reformer" and urged that his reform package be acted on immediately, even though it would "be considered *revolutionary* by the ultra-aristocracy" and inadequate by the democrats.[99] But radical GPs found Green's tone offensive and demanded total destruction of the "close-borough system." Their alternative was to place the colleges within the chartered London University or, as in Paris, within a national institute. The *Lancet* rejected Green's proposal outright, noting that the outnumbered GPs stood no chance of carrying reforms. Wakley, always curiously sympathetic to his old teacher, portrayed Green as institutionally enslaved, a "giant manacled by the Lilliputians," fettered by privilege and educated into "sentimentality, folly, and obscurity" in Germany.[100] He urged Green to break ranks and join the reformers in toppling a corrupt system before it was too late. For, as Ryan's *Journal* put it, "those who live by monopoly and corruption, will, like the fallen and despicable Tories," reap the reward.[101]

Where political goals were compatible, there Coleridgean philosophy had a pronounced impact, and this was particularly true of the King's Col-

98. "Distinction without Separation," *MG* 1831, 8:215–16.

99. "Mr. Green on Medical Reform," *MCR* 1831, 15:161–71.

100. "Mr. Green's Views on Surgical Reform," *L* 1830–31, 2:569, also 280–82, 658–63; 1832–33, 1:127. Wakley (*L* 1833–34, 2:695) deplored Green's "unsurpassed insolence" to those he considered to be of inferior rank (e.g., the apothecaries). For his part, Green, as Astley Cooper's godson, cosigned an order banning Wakley from St. Thomas's Hospital after his attack on the Cooper clique's nepotism (Brook 1945:44).

101. "Mr. Green on Surgical Reform," *LMSJ* 1831, 4:468–72.

lege in the Strand (f. 1828). Strong links were forged between the medical corporations, ancient universities, and King's. Indeed the latter stood preeminent among the Oxbridge "preparing Schools," the one most frequently attended by future Cambridge wranglers.[102] King's carried the bishops' blessings (the bishops who were about to vote in a block against the Reform Bill). It was to reaffirm Anglican standards and meet the threat from the "godless college" (see fig. 6.6).[103] Wakley, with typical Gilrayish crudity, saw the "dirtiest portion of the law-church-and-King faction" building the Strand College to belch out black fumes and blot out the brilliant light from Gower Street.[104] The bishops' insistence on Anglican exclusivity and their rejection of the best Nonconformist medical minds caused dismay in the press. The moderate *Medico-Chirurgical Review* denounced the "sapient big-wigs" for blowing on "the dying embers of bigotry," crying "shame on those who vainly attempt to roll back the tide of religious liberty from the nineteenth to the ninth century!"[105] Unlike the university, where a governing body of Benthamite lawyers and merchants did the hiring, King's medical professors were recommended by the corporation strongmen: Cooper, Brodie, and the high Tory physician Henry Halford.[106] They appointed Green professor of surgery in 1830. His address to the medical school in 1832 was the antithesis of Grant's in Gower Street the following year. Grant demanded democracy and an end to the despotic corporations. Green summoned support for the national clerisy. He launched into a panegyric on the nobility of the professions and the moral hegemony of Oxbridge institutions. He deplored class conflict and "turbulent" innovations, and praised those "seminaries of learning" where professionals and gentry could be united to form a national class of "guardians, and extenders of civilization."[107] King's and Oxbridge alumni, from their common training, were to be the self-constituted custodians of science, morals, and manners. His Coleridgean

102. Green 1832:36; Becher 1984.

103. An expression often used, e.g., Forbes to Owen, 2 November 1846 (BMNH Ro).

104. "Origin of King's College," *L* 1832–33, 1:124.

105. "Metropolitan University," *MCR* 1835–36, 24:597; "Prosperity of King's College," *LMSJ* 1836, 9:569–71. Not all non-Anglicans were barred. The Scottish minister James Rennie claimed that his best-selling books had "obtained" the chair of natural history at King's for him, "while I was not only personally unknown to [the council and bishops] but *belonging to a different church*." Rennie to Swainson, 17 May [n.y.] (LS WS).

106. Also the college consulted the Council of the College of Physicians about medical courses, and the RCP tried to get King's exclusively granted a charter to confer degrees in medicine (G. Clarke 1964–72, 2:693). On Coleridge's support for the professors: Gold 1973:25–29.

107. Green 1832:iii, 34, 36, 41. "Professor Green's Introductory Lecture," *MG* 1832–33, 11:30–31.

Fig. 6.6. A spoof of the vested interests behind the Anglican King's College in London. Brougham (pronounced "broom," which is what he is wielding), Bentham, and the London University radicals, standing for "Sense and Science," are outweighed by the King's-supporting Tory bishops, who have thrown "Money and Interest" in the balance. Among the churchmen is the *Quarterly* reviewer (Rev. George D'Oyly) who six years earlier had slated William Lawrence for the irreligious tone of his lectures. (Courtesy The Library, University College London)

message was wrapped in transcendental prose, a pious mystification the radicals saw designed to please the "political archbishops."[108] The listening prelates were of course disgusted at the godless utilitarianism of Gower Street, hence Green went to great lengths to emphasize the medical alternative—providing a model of elite professionalism that guarded against Benthamite excesses and guaranteed "the precious birthright of an *English gentleman."* At this point radical tempers snapped. "The King's College in Strand-lane, to be named in the same sentence in connexion with OXFORD and CAMBRIDGE! *King's College,* founded for the perpetuity of prejudice, in honour of toryism and bigotry! Fah, how sickening it is!"[109]

Owen's rise was not an isolated event. The same political winds swept a number of his protégés into King's posts. His "old Pupil" and friend Thomas Rymer Jones took the new chair of comparative anatomy there in 1836. The Joneses were to enjoy half a century of "uninterrupted and unabated friendship" with the Owens,[110] cementing the family ties by christening one of their children Owen. Jones's *General Outline of the Animal Kingdom* (1838–41) was an expensive, descriptive work dedicated to Owen. It promoted his terminology and was a standard conservative text for over a decade, competing with Carpenter's more adventurously theoretical *Principles.* Another King's appointee was Arthur Farre. Farre (like Jones) had been coached by Owen at Bart's, becoming his successor there as lecturer on comparative anatomy in 1835 (when he also followed up Owen's work on parasitic worms). He was a model product of the Oxbridge network: educated at Caius College, appointed professor of obstetric medicine at King's College, London, in 1842 and a fellow of the College of Physicians a year later. He had a practice among "the very highest ranks" and went on to achieve great eminence as an obstetrician, attending at the birth of Queen Victoria's children.[111] Opponents saw King's now carry the "fondest hopes of the *aristocrats* in science."[112] Science was indeed to be recast from an idealist mold here. We have only to look at

108. "Origin of King's College," *L* 1832–33, 1:124.

109. "Mr. Green's Sky-Rocket Lecture," *L* 1832–33, 1:153.

110. Rymer Jones to Owen, 14 February 1879 (BMNH RO 16: f. 277, also f. 285). At King's he founded a museum of comparative anatomy (Letter Book 1834–, f. 226, King's College Archives). Jones promoted Owen's nomenclature in his *General Outline of the Animal Kingdom* (1841:8, 10). The costliness of this textbook is mentioned in "Jones's Manual of Comparative Anatomy," *BFMR* 1841, 12:218.

111. "Dr. Arthur Farre," *Proc. Roy. Soc.* 1889, 46:iv. Farre (1837) did seminal work on the Ciliobranchiata (a group comprising bryozoa, certain polyps, and the polyzoa). He lectured at Bart's on comparative anatomy from 1836 to 1839 or 1840.

112. "London Medical Education," *LMSJ* 1836–37, 10:445. On Forbes: Mills 1984:377; Rehbock 1983:103–13; J. Browne 1983:149–55.

Green's addresses, or the adulation of Owen, or at Edward Forbes's talk of genera as Platonic ideals. So the conservative alliance—corporations, Cambridge, and King's College—experienced a broad shift in the mid-1830s with alert young Coleridgeans taking the initiative. This intellectual network now provided the channels through which Owen's pioneering morphology was to be promoted as the new conservative standard.

To conclude, then, Owen was well schooled in the ideals of the medical clerisy. He was "raised" by romantics who saw the trade in base physiologies and "plebification" of science threaten to wreck the medical priesthood's mission—to preserve gentrified power and still the lower orders.[113] Green was a chapter-and-verse Coleridgean who looked on the *Constitution* as a safeguard against fierce democracy and an antidote to the necessitarian, utilitarian, industrial decay of traditional religious culture. Green promoted Owen's career, acting as literary midwife, submitting his papers to the Royal Society, and recommending him for its fellowship in 1834. Owen was immensely impressed with Green ("that noble and great intellect")[114] and followed his college patron in detesting Lamarckism as a beastly contamination of the Divine thought—as a cause of profligacy, immorality, and democratic unrest. "Ouran Outang theology" had become the prime conservative target, one Owen lined up squarely in his sights. He was now to reconstitute comparative anatomy along romantic lines, to strip nature and society of their innate powers and return these to the Godhead. Since it was the teachers in the "lecture bazaars" who promoted the materialistic sciences, it was they who would take the brunt as Owen started his reconstruction.

113. Coleridge 1972:53.
114. Owen, in Green 1865:xiv.

7

Engaging the Lamarckians

Owen and the medical and legal gentlemen of Lincoln's Inn shared a set of political and moral assumptions. One of them, paramount in Green's philosophy, was that nature and commonalty were divested of innate powers: monads and masses could only advance under the divine influence. As Green's intellectual heir, Owen was committed to pointing up the absurdity of the radicals' self-sufficient nature. During the 1830s he increasingly aligned himself with the Oxbridge dons, ever sensitive to the Lamarckian-radical threat. For the Anglican divines, Lamarckism denied man an immortal soul; it also fueled the secularists' demands for disestablishment and democracy, and was turning ignorant men toward action to achieve betterment in this life without waiting for the next. Working in a corporation under siege, patronized by the Oxbridge clergy, Peelite judiciary, and surgeon-baronets, Owen was now to strengthen the liberal Anglican view of nature. In this view, man was modified to become "the seat and instrument of a rational and responsible soul"[1]—and "responsible" meant socially accountable in this life and morally accountable in the next.

Making Contact with Lamarckism

Saturday . . . In the Evening at Cuvier's. Mad. Cuv. & Madlle Duvaucel [Cuvier's step-daughter] to both gave Mr and Mrs Clifts regards on which they returned kind regards stayed till 11. Cuvier shook hands at going away—Had a long convers. about Orang with him. He said he had never dissected a Chimpanzee, was going to write upon Sternum contra Geoffroy. Dr Grant introduced me to Fred. Cuvier.

—Owen's notebook, Paris, 20 August 1831.[2]

Owen had been exposed to Lamarckism before his first trip to Paris in 1831. He had attended John Barclay's Edinburgh classes in 1824–25, just as they were being taken over by Knox and Grant; even in London he

1. R. Owen 1835a:343.
2. R. Owen, Notebook 5, entry for 20 August 1831 (BMNH).

learned his invertebrate zoology partly from Scottish sources, dissecting specimens using Grant's work as a guide.[3] He also read Lamarck's *Histoire naturelle des animaux sans vertèbres* to aid his compilation of the Hunterian catalogs.[4] But, as Pietro Corsi says, there was no substitute for actually meeting the deists in Paris to appreciate the strength of their position. Lamarck might have died a blind octogenerian in 1829 and have been ceremoniously interred by Cuvier with a disparaging *éloge,* but transformism was no straw man in Paris around 1830. Owen was to become personally acquainted with its advocates. This familiarity is important to establish, for reading Lamarck's books alone could not have brought home to him the viability of transformism. He had to see for himself how it was being applied.

Cuvier opportunely slipped into England at the time of the July Revolution, while the tricolor fluttered atop Notre Dame.[5] He was escorted round the College of Surgeons by Owen, the only French-speaking member on hand who was familiar with the preparations. Cuvier's return invitation led to Owen making his first trip to the Jardin des Plantes in July 1831. It would have been surprising if Owen's contact with the Parisian savants had not influenced him profoundly. He spent a month attending lectures and visiting the museum and menagerie, although of course theaters and restaurants were not ignored, and he attended at least two of Cuvier's Saturday soirées. Owen presented Cuvier with copies of the Hunterian catalogs and saw Geoffroy, Blainville, and others lecturing.[6] What he thought of the recent riots or the rampant anticlericism which had fueled the July Revolution and left two thousand dead on the streets we do not know. He could hardly have missed the burned-out shell of the archbishop's palace next to Notre Dame, the result of an attack in February. A grave church-goer, already a recipient of Cuvier's patronage, he could not have been blind to the republican mood.

But we can glimpse the way in which he was introduced to the contentious issues in French comparative anatomy. Owen recorded in his pocket book that he stayed at the same hotel as Grant, who was summering in Paris as usual. It shows the two men becoming increasingly friendly: on at least five occasions they breakfasted or dined together. Grant regaled Owen with accounts of his "wanderings" across Europe and accompanied him to lectures, pointing out who was old-hat and who up-to-date. Discussions obviously ranged from the sublime to the ridiculous, and given

3. Sir Richard Owen Scientific Notes, c. 1828–32, f. 38 (BL Add. MS 34,406).

4. R. Owen, "Books Referred to for Natural History" (RCS 275h.3.5). Corsi 1978:241.

5. For an eyewitness account of the flag and fighting: Philip Taylor to Richard Taylor, 9 August 1830 (Taylor Papers, St. Bride Printing Library, London). Also Cobban 1981:74–101.

6. R. S. Owen 1894, 1:50, 51–58. Limoges 1980 on the Muséum.

Grant's funny streak it is not surprising that they indulged in long conver-
sations "de omnibus rebus anatomico physiologico-mathematico-
nonsensicology."[7] Grant was familiar with the Muséum and its professors,
and he appears to have taken Owen in hand—introducing him to savants
such as Frédéric Cuvier (Georges's brother), warning him off Fleurens's
superficial lectures, and so forth. Grant had an unrestrained enthusiasm
for Lamarck (as Darwin had discovered five years earlier). And with the
Académie debates between Cuvier and Geoffroy barely a year old, and
Geoffroy currently working on teleosaur transmutation, it is likely that
Grant also praised his friend Geoffroy and his paleontological endeavors.
Certainly Owen records buying the "Philos. Zoologique" (presumably
Geoffroy's newly published Principes de philosophie zoologique) which he
read avidly.[8] So Owen was made intimately aware of the Academy clash
over the homologies of mollusks and fishes. And being introduced to the
debate (and some of the debaters) by a partisan, he was well placed to
judge Geoffroy's position and the strength of Lamarckism.

By all accounts Owen was given a radical's-eye view of that cluster of
concepts so hated by Cuvier—Lamarckism, serialism, and unity of composition. He found Geoffroy, Blainville, and Grant all supporting a unity
of structure that cut across Cuvier's discrete embranchements, while
Grant and Geoffroy went further to accept the reproductive continuity of
all life. The trip obviously had a dramatic impact on the young anatomist.
On his return, he began musing in his notebook on the term Nature, like
Coleridge and Green castigating the pantheists who would use it to exclude an external moral power.[9] He also had the opportunity to dissect a
new mollusk, a Polynesian nautilus, presented to the college by his former assistant George Bennett on his return from Australia in July 1831.
And in his Pearly Nautilus memoir in 1832 Owen openly attacked Geoffroy's Principes for portraying a unity between mollusks and vertebrates
and an "unbroken series" in nature.[10] This marked the beginning of his
campaign against radical notions of unity of composition. Over the next
few years Owen was to disentangle the logical complex that caused Cuvier
such consternation: self-development, unity across divisions, recapitulation, and the unilinear series—theories finding their way into the radical
anatomy schools at home. He was to restrict the scope of Geoffroy's unity,
allowing homologies to function only within Cuvierian embranchements,

7. R. Owen, Notebook 5, entries for 10, 11, 12, 19, 21 August 1831 (BMNH).

8. Ibid., entry for 17 August 1831; Geoffroy 1830, 1833 (read from 4 October 1830 to 29
August 1831). Appel 1987:130–35, 155, 166–67 discusses Geoffroy's Principes and transformist paleontology in this period.

9. R. Owen, Notebook 7, f. 64 (BMNH).

10. R. Owen 1832b:1.

thereby denying that all life was tied into a threaded sequence. Owen was as unprepared as Cuvier to accept homologies between vertebrates and invertebrates because a series from monad to man had clear transformist ramifications. But he did accept a unity within divisions, and he criticized Cuvier for his failure to go at least this far. This itself made Owen suspect in some eyes. He therefore had to be seen to be purging zoology of all its Lamarckian connotations before he could proceed to modify the prevailing philosophical anatomy.

Through the decade 1831–41 Owen can be seen identifying with increasing accuracy the weak points of the serial-transformist target. As he tackled successive issues, he also built up his contacts in the conservative medical, legal, and scientific community. This guaranteed him a large sympathetic audience, sickened by this irreligious foreign science. His confidence in turn was boosted, enabling him after mid-decade (when the radical impetus was itself slowing) to switch from attacking the Parisian pantheists to engaging the London democrats directly. The shifting scientific ground on which the debate took place allows us to break the story into discrete episodes. But all show how single-mindedly Owen devoted himself to abolishing the central Lamarckian tenet—the serial continuity of life. First, in 1832–34, he tackled the supposedly transitional nature of monotremes. Then in 1835 his study of chimpanzee development was designed to distance ape from man. In 1838–39 he undermined radical interpretations of one of the most celebrated fossils, the Stonesfield "opossum." And in 1841 he demolished the paleontological argument for inexorable ascent and promoted an image of "punctuated" progression to break Lamarck's iron law.

The Generation of Monotremes

In the 1820s English anatomists generally classified the Australian *Ornithorhynchus* (duck-billed platypus) and *Echidna* (spiny anteater) as lowly mammals, usually—following Cuvier—as edentates (a group embracing the sloths, armadillos, and anteaters). But from the colonies there came persistent rumors of egg laying, and many naturalists became openly skeptical of Cuvier's classification.[11] These reports were also used by French transformists to justify separating the platypuses into a transitional class. However, Lamarck's suggestion that this should lie midway between birds and mammals was already looking untenable by the 1820s; Knox, for one, found nothing birdlike in the male platypus shipped to

11. Traill told Swainson, 20 January 1829 (LS WS), "I hope Cuvier in his new edition will make a new *class* for the Ornithorhynchus and Echidna which are now duly ascertained to be oviparious animals; a sort of connecting link between Mammifera & Aves."

Jameson's museum by the governor of New South Wales.[12] On the other hand, Lamarck's belief that they laid eggs and lacked mammary glands was shared by Geoffroy. He told Grant in 1829 that, from the genital organs, the platypus looked more like a "true *oviparous reptile*."[13] In print, however, he admitted that warm blood and the respiratory system made it "necessary to see them as an essentially new type," and placed the "monotremes" (his word) in a fifth vertebrate class between mammals and reptiles. Animals of this class were characterized by a common cloacal opening, oviparity (egg laying), and lack of mammary glands. But in 1830 the last two points remained highly contentious. Geoffroy printed a letter from Grant describing two cylindrical eggs reputedly from an *Ornithorhynchus* nest, but Grant himself was cautious, conceding that the eggs resembled a lizard's or snake's. Another problem was that these particular eggs were actually larger than the platypus's pelvis; this was to force Geoffroy into some dubious ad hoc reasoning to explain how they could have been laid.

With the increase in military activity in the Australian colony, firsthand reports now began to reach London on a regular basis. In 1832 Lieutenant Maule of the New South Wales Garrison sent an account of the platypus's lifestyle to the Zoological Society. He had actually set out to test the colonists' belief that "the female *Platypus* lays eggs and suckles its young." He confirmed that the animals lived in burrows on riverbanks, and on digging into the nests he found egg shells. Females that had been shot were also found to contain eggs. But while this seemed to confirm Geoffroy's view, other observations were more problematic. Maule captured a female with two young, and when she died and was skinned, milk was seen quite clearly oozing through her stomach fur.[14]

This was the position in 1832 when Owen first engaged Geoffroy; and we will see how each sifted the evidence, emphasizing some aspects and suppressing others. Take the question of oviparity first. Geoffroy was prepared to defend this egg laying, even if it meant proposing a peculiar type of reproduction, for example, allowing the huge egg to develop rapidly in the cloaca before being laid. His problem, because he thought the egg was so large, was to explain how it could pass the restricted pelvis; he was not prepared to sacrifice the principle of oviparity itself. Owen by contrast was convinced that the platypus was a mammal, however extraordinary.

12. Knox 1823:172; Lamarck 1809, 1:145–46, 342.

13. Grant 1830b:149–51; Geoffroy 1829:158.

14. *Proceedings of the Committee of Science and Correspondence of the Zoological Society of London* 1832, 2:145–46; also 1830, 1:149–50. Milligan 1838, Appleyard and Manford 1980 on the Swan River colony, first settled in 1829; and Whitley 1975 on Australian natural history.

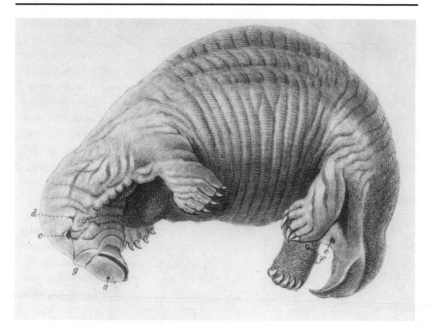

Fig. 7.1. The larger of Maule's nestling platypuses. Owen dissected it to find coagulated milk in the stomach. (From Owen 1835b, pl. 32)

So he was unreceptive to any hint of oviparity or incubation, and endorsed only the contrary facts in Maule's letter. In his first paper, Owen accepted the observation of milk secretion but claimed that the dissected-out eggs were of little value because Maule had failed to state where he found them (whether in the ovary, oviduct, or cloaca). Nor did egg shells in the nest prove anything because these could have been expelled covered in salts during the birth. (Owen actually believed that the young platypus hatched in the oviduct and therefore that monotremes were ovoviviparous.) Again, he enthusiastically noted the presence of "coagulated milk" in the nestlings' stomachs,[15] which proved the correctness of Maule's sighting of milk on the mother's fur (see fig. 7.1). Yet he refused to admit the importance of an egg tooth which he himself detected on the bill of the smaller specimen.

So the value Owen set on Maule's observations depended on whether or not they corroborated his preconceived opinions. Owen went on to devise a set of anatomical and physiological arguments between 1832 and 1834 which were designed to disprove Maule's inference of oviparity and

15. R. Owen 1835b:225, 1832c:534.

incubation. He pointed to the lack of any shell-secreting membranes, to the narrowness of the pelvis, preventing a large egg from being laid, to the lack of sufficient yolk to enable the embryo to survive incubation inside an egg, and to the presence of mammary glands, which in mammals substitute for egg yolk and render incubation superfluous.[16] He attempted to construct a watertight case founded on the anatomy of preserved specimens to discredit Maule's on-site observations. He did not convince everyone. John Marshall of the Military Museum in Chatham considered Owen's "Paper War" with Geoffroy a perilous affair. His museum, he said, possessed a platypus with eggs in the oviduct.[17] Geoffroy quickly learned of this and wrote for details. Owen visited the museum and observed three well-developed ovisacs, but still in the ovary. He therefore informed Geoffroy that this specimen offered no proof of oviparity.

I have emphasized Owen's devaluation of conflicting evidence in order to highlight his unshakable faith in the mammalian nature of monotremes. He had elevated this belief beyond the reach of empirical refutation. He was not consciously distorting—selection is always part and parcel of normal evaluation. Nor did he imagine himself having an ideological ax to grind. Although he accused Geoffroy of prejudging the case, he declared himself "in no way biassed" by his belief in "the mammiferous nature of the Ornithorhynchus."[18] But we have gained an unfair impression of Owen's position by concentrating on eggs and incubation, which after all supported Geoffroy's case. Owen's real strength lay in his elegant demonstration of the existence of mammary glands in monotremes, and on this subject he was able to push Geoffroy onto the defensive. As early as 1824 Meckel had detected tiny glands composed of tubular tissue in the platypus, which he interpreted as mammary. While acknowledging the monotremes' reptilian affinities, Meckel nonetheless agreed with Cuvier that they were edentatelike mammals. Geoffroy himself had detected this gland in monotremes but reported that it possessed none of the characteristics of a marsupial mammary gland. The tissue was different, nipples were lacking, and the glands were smaller than Meckel suggested;[19] he maintained that they were either aquatic lubricating glands or similar to the scent glands in shrews, which also follow the phases of sexual development.

Owen's papers were designed to sustain Meckel's interpretation. But

16. R. Owen 1834:563–64. The illogicality of a mammal with mammary glands laying eggs was widely appreciated: "On the History of the Ornithorynchus," *L* 1827, 12:170–71.

17. Marshall to Owen, 7 April 1833 (BMNH RO); Geoffroy to Clift, 9 May 1833 (BMNH RO 23: f. 42). R. S. Owen 1894, 1:81–82; R. Owen 1834:557.

18. R. Owen 1834:555–56.

19. Geoffroy 1826.

since Owen also maintained that the outcome of the debate over the glands would decide the "true affinities of the Monotremata,"[20] he was prepared to let his evidence support much broader conclusions. Owen discredited Geoffroy's alternatives and demonstrated the milk-secreting function by an ingenious comparative study of five adult females. His procedure was to dissect and measure the gland in each, as well as the uterus, to assess the state of egg development. This enabled him to relate gland size to the ovarian cycle. He showed that the glands (as one would expect of milk glands) were full size "*after* gestation," that is, when the ovaries were already shrinking, having released the eggs. He simultaneously eliminated Geoffroy's counterproposal. When the eggs were mature, the glands had only just begun developing. So they were not scent glands, secreting "an odoriferous substance attractive of the male," which at this time should have been functioning maximally.[21] Owen's strategy was perfectly executed. He displayed a mastery of difficult dissection techniques (perfected through his work on the Zoological Garden's corpses), allied to a clear conception of the points to be proved. Finally, he clinched his case by dissecting out similar glands in the echidna, thus proving that the monotremata as a group shared with all mammals "the characteristic function of lactation."[22]

Geoffroy (see fig. 7.2) corresponded increasingly with the Zoological Society in 1833 and 1834, but his prevarications and position shifts showed his difficulty. In February 1833 he insisted that, because the platypus's urinogenital system is reptilian, it could only produce eggs. And he now suggested that the gland in question might secrete a lime compound to harden the shell. Or, after reading Maule's letter in March, he speculated that it could be a mucus-producing structure. In either case, he pleaded for "further examination"; better this than to adopt a complacent attitude and accept the beast's "normality, founded on strained and mistaken relations."[23] Such tactlessness did more harm than good, and Owen gave his speculations short shrift. The "Paper War" ran on through July 1833, with Geoffroy branding Owen a reactionary (being hidebound by "the rules of *the past*") while placing himself on the side of "progress." He shifted ground again, to make this a "Monotrematic" gland sui generis, and a justification in itself for taxonomic uniqueness. Owen countered Geoffroy point by point—dismissing his final argument that "conglomerate" mammary glands produce milk, therefore the simple

20. R. Owen 1832c:517.
21. Ibid., 531.
22. R. Owen 1832a:180.
23. See the letters from Geoffroy in *Proc. Comm. Sci. Corres. Zool. Soc.* 1832, 2:28–30, and *Proc. Zool. Soc.* 1833, 1:15.

caeca in monotremes must have another function—by pointing out that whales too possess simple caeca. Geoffroy's ad hoc shuffling now reached a climax, and he concluded not that the monotremes should be returned to the mammals, but that the whales should also be removed from them.[24]

This provides one of the most graphic examples of piecemeal retreat and desperate maneuvering. Above all it shows the tenacity with which Geoffroy clung to his oviparous theory because it fitted so well with his view of serial development and transitional types. He forfeited support because of his lack of subtlety, and he was forced to concede early in 1834 that the "monotrematic" secretion in porpoises really was milk.[25] Owen's success reflects his astute judgment of the form that a convincing refutation must take. He often began his papers by suggesting that Geoffroy's new class must stand or fall with the verdict on this gland; thus he introduced into the initial equation the elements that would allow him to draw anti-Geoffroyan conclusions. Owen's arguments convinced potential critics on both sides of the Channel. Blainville accepted that monotremes were a unique sort of mammal. Grant, lecturing on monotremes in March 1834, ignored the debate, and for that matter Owen, and simply credited Meckel and Geoffroy with elucidating the structure of these glands. Grant said that the platypus's primitive traits should be "viewed as marks of inferiority generally," rather than as indicating its special affinity with the birds or reptiles. He agreed that the monotremes resembled edentates, but thought that the "low condition of their generative system" warranted their separation into a new order.[26] Owen—acknowledged or not—had convinced the opposition of the monotreme's mammalian nature.

Owen had achieved his main aim: to throw doubt on the monotremes' egg laying and therefore on their intermediate nature. This victory over France's leading morphologist was acknowledged by the jingoistic elite entrenched at the Zoological Society, obsessed by scoring points against "la grand nation."[27] For his work on marsupials and monotremes, Owen was also elected a fellow of the Royal Society in May 1834, supported by the college surgeons and gentlemen zoologists, both celebrating the end of France's hegemony in natural history. Another factor besides zoological acumen contributed to his success. This was the wealth of his material,

24. Letter from Geoffroy, *Proc. Zool. Soc.* 1833, 1:94.

25. Ibid., 1834, 2:26–27.

26. Grant 1833–34, 2:1, 3, 4; and Broderip and Owen's comment on this (1851–52:377). On Blainville: *Proc. Zool. Soc.* 1833, 1:30. In 1833 Blainville gave them distinct status as "ornithodelphs," on a par with marsupials (didelphs) and placentals (monodelphs) (Appel 1980:312).

27. Desmond 1985a:174. R. S. Owen 1894, 1:80 on the nomination for the fellowship of the Royal Society.

Fig. 7.2. Geoffroy about age seventy (c. 1842). (Courtesy Wellcome Institute Library, London)

which reflected the superior resources and increasing colonial contacts of the College of Surgeons and Zoological Society. Monotreme specimens were reaching London in growing numbers as a result of Britain's military expansion in New South Wales and the new settlements along the Swan and Fish rivers. Like Herschel annexing the "Southern Skies," and Murchison laying territorial claim to the world's Silurian strata, Owen was gathering the southern fauna under the British flag, establishing himself as the leading exponent of Australian zoology.[28] The Zoological Society Council, with its imperial pretensions, made great play of Owen's success with the Australian animals and of foreign naturalists now being forced to look to England for guidance.[29] One reason he could cut so quickly through the Meckel-Geoffroy stalemate was that he could muster five female platypuses for comparison. He was actually in the position of being able to order the anatomical parts he needed. George Bennett, back from Australia, was present when Owen first dissected the *Ornithorhynchus* in 1832. On returning to the colony that year, Bennett was equipped by the college and carried Owen's list of requirements.[30] The first full crates arrived home in the summer of 1833; by July Owen was exhibiting Bennett's specimens at the Zoological Society and reading his descriptions of "the milk gland" into its *Proceedings* (see fig. 7.3). By May 1834 the number of specimens received had topped five hundred. Many were unique—including generative organs with small ova—in a "good state of preservation" and accompanied by accurate field notes.[31] Bennett was awarded the college's gold medal for his work. Being able to put an assistant so quickly into the field obviously gave Owen the advantage. It also reduced Geoffroy to requesting drawings and information from Grant, Clift, or the Zoological Society.[32] Geoffroy was sent one of Maule's nestlings, but generally the material remained the property of the college or Zoological Society—

28. MacLeod 1982:8; Secord 1982; Moyal 1975.

29. *Reports,* 1839 (ZS); Desmond 1985a:230.

30. R. Owen, "General Account of Specimens on Comp Anatomy and Natural History Collected and Presented to the Museum of the Royal College of Surgeons by George Bennett Esq MRCS FLS &c &c" (RCS Cabinet VIII [1] b.L); *Proc. Zool. Soc.* 1833, 1:82; G. Bennett 1835. Bennett went on to run Sydney Museum in 1835–41 (Kohlstedt 1983:2–3).

31. R. Owen, "General Account of Specimens," as note above. See the letters from Bennett to Owen, esp. 4 February 1833 (RCS); also those in BMNH RO 3: ff. 252–371, 4: ff. 1–54. The college was eagerly acquiring specimens at this time as the new museum was being planned. Hence their published guidelines: Royal College of Surgeons 1835.

32. Geoffroy to E. T. Bennett, 9 April 1834 (BMNH RO 23: f. 41); Geoffroy to Clift, 9 May 1833 (BMNH RO 23: f. 42). In the 1820s Southwood Smith had been Geoffroy's contact: "On the History of the Ornithorynchus," *L* 1827, 12:170–71. On Maule's nestling: ZS MC 4: f. 13.

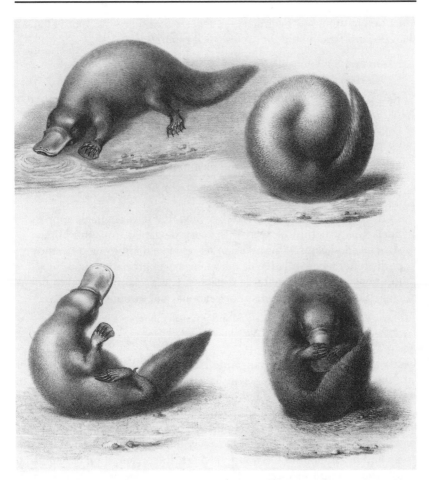

Fig. 7.3. Life sketches of the platypus that accompanied George Bennett's paper to the Zoological Society. (From G. Bennett 1835, pl. 34)

hence his difficulty in answering Owen with anything like enough detail, and his recourse to speculation and special pleading.

The question of transformism was not raised during the debate, although Owen was dealing with known transformists, and Geoffroy's new class had an obvious bearing on the taxonomic gradualism so essential to contemporary transformist theories. Indeed, the tenacity with which Geoffroy clung to his new class—and his ad hoc explanations of how a large egg could pass through a small pelvis, or why the abdomenal glands must be anything but mammary—suggests that there was more at stake than pedantic taxonomics or professional reputations.

The Proximity of Ape to Man

In a comparison of the frame and capabilities of man with those of the inferior animals . . . it will be found . . . that man is unquestionably endowed with [a perfect] structure . . . revealed in such a balanced relation of the parts to the whole as may best fit it for a being exercising intelligent choice, and destined for moral freedom.

—J. H. Green in his Hunterian Oration[33]

In tracing the successive stages by which the lower animals approximate the structure of *Man* . . . every deviation from the human structure indicates with precision its real peculiarities, and [a study of the differences will give us the] means of appreciating those modifications by which a material organism is especially adapted to become the seat and instrument of a rational and responsible soul.

—Owen on the spiritual and physical differences between
man and ape[34]

Faced with the specter of a transmuted human, Owen brought the subject into the open. Like so many scientific gentlemen he was acutely aware of the danger of brutalization, and by the mid-1830s he had made the morphological separation of man and ape a moral imperative. For Green, human self-awareness was the "final purpose of Divine Law." He taught that "below man the body may be said to constitute the animal, in him it is the organ and instrument of the mind; in short, the organization of man is no longer the mere perfecting, but the *apotheosis*, of the animal structure."[35] Owen was to provide an anatomical rationalization of this view. He might not have been able to prove that man's body fitted him for moral acts or that it was shaped to receive its spiritual host. But he did physically dissociate man and ape, leaving the moral consequences of the divide unstated but unmistakable.

The fear of bestialization was widespread in the 1830s. The Lamarckian threat to "human dignity" affected Charles Lyell as much as Owen; even the gentleman radical John Elliotson—though an extreme materialist—was shocked that some anatomists "perversely desirous of degrading man"

33. Green 1840:60.
34. R. Owen 1835a:343.
35. Green 1840:42.

should be pushing him so close to the apes.[36] Many of these fears stemmed from Lamarck's *Philosophie zoologique* (reissued in 1830), which contained a graphic scenario for the ape's transformation into man. Lamarck guessed that the chimpanzees, being forced to the ground, would lose their grasping big toes as they became used to walking. Once there they could command a distant view by standing erect, and doing so for generations would result in calf development. As these ground-dwellers ceased "using their jaws as weapons" and developed them for chewing, so "their snout would shorten" and the face would become flatter, resulting in a higher facial angle.[37] Lamarck's speculations had been given credibility in 1827 by the observations of Bory de Saint-Vincent. Bory was a leading anti-Cuvierian materialist who blended the best of Lamarck's philosophy with Geoffroy's higher anatomy. He actually surpassed Lamarck on the question of the ape's ability. Lamarck considered the ape very much man's inferior in intelligence, but Bory supported Tiedemann's conclusion that its brain was far superior to a monkey's and more like a man's.[38] For Bory this explained the orang's cultural adaptability. Enlightenment rationalists, believing that ideas and mental development were products of circumstance and sensory input, had even suggested that, given the advantages of civilization (and a sign language), the orangs might themselves be made into "little gentlemen."[39] But Academicians in the religiously conservative 1820s repudiated such speculative fictions. The cautious Frédéric Cuvier, long an expert on ape behavior, increasingly feared this encroachment on human uniqueness. Bory in 1827 attacked Cuvier outright for denying apes reason. He concluded that words were not proof of mental superiority since "idiots" often spoke distinctly. Speech was the feature distinguishing man from ape; although orangs cannot articulate words, Bory hazarded that given an adequate voice box a chimpanzee might still outshine a Hottentot.

Bory related the case of peasants in the Marensin canton who had actually acquired dextrous toes after climbing trees for generations to collect resin. They could write with them, yet flat feet and a "parallel" big toe were supposed to distinguish man from ape. He goaded Cuvier by asking whether the *résiniers* should not be classified with the monkeys. (Like Lamarck, he refused to discuss the soul because it lacked any anatomical features.) He concluded that only vanity drove us to ally orangs with the "stupid brutes," while elevating ourselves to a dignified position. This was the kind of flippancy that proved so upsetting to the English

36. Elliotson 1835:11.
37. Lamarck 1809, 1:349–57.
38. Bory 1827:266–67. Corsi 1978:228–29, 1988:230.
39. R. J. Richards 1982:276–78.

Anglicans. However imponderable, the soul was a Divine gift responsible for man's reason and dignity. To deny it, to convince man that he was just a better sort of brute, would be to unleash the forces of moral decay and social degeneration.

The evidence for a smoothly increasing facial angle from monkeys to white men was already contentious by the 1830s. Geoffroy's older series, from the flatter-faced *Troglodytes* (the young chimpanzee), through the *Pithecus* (orangutan), to the more brutal, big-jawed *Pongo* (the so-called Wurmb's Ape), was exploded when the last two were found to be merely age variants. By the later 1820s both Bory and Georges Cuvier realized that Wurmb's ape was the adult orangutan, classified separately from its young. But the series from orangutan, through *Troglodytes,* to humans held because the middle-rung chimpanzees were still only known from immature, flatter-faced specimens. Lamarck's explanation of this muzzle shortening had prompted quite different reactions in England. Charles Lyell, who had already adopted a nonprogressionist paleontology to avoid the Lamarckian snare, had no use for a scale from higher mammals to "apes with foreheads villanous low," and then to African and European men.[40] He was willing to concede that the European's capacious forehead might indicate "a large development of the intellectual faculties," but dismissed a parallel "scale of intelligence" in animals as nonsense. As befits a son of the Kinnordy gentry, he concluded that the ape's intelligence had been exaggerated "at the expense of the dog." The ranking of human races was also under attack as a justification for the "abominable traffic" in slaves. During his visit to London in 1835, Tiedemann caused a stir by claiming that the Negro's brain was neither smaller nor lighter than a European's, but in fact equal in intellectual and moral capacity. Black inferiority was a "prejudice" fostered by studies of slaves, who had been crushed and demoralized by "oppression and cruelty."[41] Tiedemann's findings caused controversy. But the antislaving lobby was pleased, Tiedemann having disproved the "dastardly allegation" that Negroes were "a degraded and inferior race."[42] So the whole question of the facial angle was topical and loaded in a complex way. Anti-Lamarckians such as Lyell were refusing to use it to scale up animals, and transformists such as Tiedemann were dismissing its importance among humans (and in the process denying that Ethiopians were a bridge between orangs and Europeans). Though a result of "hard" measurement, the concept of the facial

40. C. Lyell 1830–33, 2:60; Geoffroy 1812:87–89; Greene 1961:196–98.

41. "On the Brain of the Negro," *BFMR* 1837, 4:529–30; "Organization of the Brain in the Negro," *MCR* 1837–38, 28:249; Tiedemann 1836:520–26.

42. "Organization of the Brain in the Negro," *MCR* 1837–38, 28:249; Combe 1838:585–89.

angle was easily molded into ideological shape. And because the Parisians were still using it,[43] Owen was able to extract fresh anti-Lamarckian capital from his new analysis.

Owen's use of his ape material changed dramatically over the crucial 1830–35 period. He had dissected his first ape in 1830—a young male orangutan which had died three days after arriving in Bruton Street.[44] His ensuing paper was factual and unprovocative. Five years later matters were quite different. His discussion of chimpanzee osteology in 1835 was an ideological tour de force. In the intervening years, of course, he had visited Paris, debated with Geoffroy until the latter came to "lay down his arms,"[45] and watched the rise of the radical Dissenters and *Lancet* Lamarckians. Nor was he any longer the novice: institutionally settled, patronized by Coleridgeans, a newly elected fellow of the Royal Society, secure within the Tory "junto" at the Zoological Society—he was in a strong social position. Given the radicals' Francophilia, a new conservative paper couched in anti-Lamarckian terms could have a nationalistic appeal, as a repudiation of the secularism and materialism at root of France's continuing instability. It would also have a career payoff, appealing to the Coleridgean patriots and Peelite squirearchy at the Zoological Society—gentlemen who were already quizzing Owen over the intelligence of the zoo's new baby chimpanzee, acquired by the entrepreneurial council in 1835 (see fig. 7.4).[46]

Owen now studied ape skeletons to prove that the facial angle had lent undue support to "theories of animal development."[47] The novelty of his new paper lay in his description of the hitherto-unknown mature chimpanzee. By showing that it, no less than the adult orang, had a protruding snout and "bestial" physiognomy, he was finally able to knock the middle rung out of the orang-chimpanzee-human facial sequence. It was a superb strategy. In his notebook he had already worked out the best way to present it. He would prove that chimpanzees had been misunderstood because only the unrepresentative young had been studied. A human baby's cranium, he jotted, with its "disproportionate" brain and small jaws, gives a totally misleading view of man's "endowments." A statuary, for example, would see in it "the exaggerated proportion & facial angle of a demigod,"

43. E.g., Latreille 1825:43–44.

44. It died before it could be exhibited (*Reports*, 1830, ZS). R. Owen 1830:5, 9, 67–72.

45. Geoffroy to E. T. Bennett, 9 April 1834 (BMNH RO 23: f. 41).

46. Broderip to Owen, 20 October 1835 (BMNH RO), commenting that the chimp's "intelligence is quite marvellous." It was purchased at Bristol for £35 (ZS MC 4: ff. 241, 256). Youatt 1835–36a, 1835–36b on this chimp; and Ritvo 1987:30–39 on the worrying humanlike behavior and appearance of apes in general.

47. R. Owen 1835a:343.

Fig. 7.4. George Scharf, whose livelihood depended on selling prints (he sold this one in the zoo itself), in 1835 deliberately made the zoo's baby chimp tantalyzingly humanlike. But then even Broderip thought its intelligence "quite marvellous" and tackled Owen on the subject. (Courtesy Zoological Society of London)

while a phrenologist would "predict from its undue cerebral development the intellectual powers of an Aristotle or a Bacon."[48] Owen argued in print that chimpanzee endowments had been exaggerated for the same reason. Knowledge of the flat-faced, bigger-brained young had completely misled classifiers into imagining that "the transition from the *Monkey* to the *Man*" was "more gradual" than was really the case. True, man's resemblance to the infant chimpanzee was "startlingly close,"[49] but it was only by comparison to an adult ape that the real relationship would become apparent. No adult chimpanzees had reached Europe alive, and only immature animals were exhibited in London and Paris. But Owen had found an adult skeleton in a local surgeon's museum, and he was now able to publish the first description of the mature animal. He described how the young ape threw off its human mask (see figs. 7.5–7.7). Shedding its milk teeth, the jaws elongate, canines protrude, and biting muscles develop, anchored to the

48. R. Owen, Notebook 11 (1834–36), f. 87 (BMNH).
49. R. Owen 1835a:354, 343, 349.

massive brow ridges. Brain growth is relatively retarded, the "cranial box" remaining almost static as the jaws expand. These changes resulted in the adult dwarfing its brain, explaining why the "gentle manners of the young *Ape* rapidly give way to an unteachable obstinacy and untameable ferocity in the adult" (another hit at the notion of educability). So great was the facial change that anatomists might be forgiven for mistaking mother and offspring as separate species. The adult's skull was unhumanlike; it had expansive crests accommodating powerful chewing muscles and a prognathous face with protruding jaws. The "irrational ape" possessed doglike canines as "weapons of destruction," quite unlike "the master of the animal creation." From skull architecture alone, Owen could paint a bestial picture, emphasizing the taxonomic chasm between ape and man. By removing the artificial middle rung created for the immature *Troglodytes* he had made the last step to man morphologically impassable.

Owen then challenged Bory and Lamarck on other points. He proved the impossibility of an orang standing erect and being counted a man, showing that the flexor muscle, terminating in a single tendon on the big toe in man, ends in three tendons to the middle three toes in the orang. So the muscle that helps raise the heel in man enables the orang to grasp. Owen had turned Lamarck's sequence of civilizing steps from a behavioral possibility into a physical improbability. To imagine the tree-dwelling ape's transformation was simply anatomically naïve.

The importance of Owen's paper was recognized in Britain and abroad,[50] but it drew scorn from his enemies. In a bellicose outburst, Knox totally denied Owen's originality. He accused Owen of illustrating an orang that was clearly a composite—the trunk of one specimen and the head and arms of another "of probably an entirely different species."[51] He also cast doubt on the chimpanzee material, despite Owen's "pompous display of measurements and comparisons." But even where Owen's evidence was accepted, transformists put a different gloss on it. Geoffroy condeded that the childlike young ape grows into an adult of "revolting bestiality." He also recognized that the ape posed a threat to human dignity, but he warned against resorting to religion to stifle progressive opinion. He saw the evidence in quite another way. He likened the contrast between young and old apes to that between generically distinct dogs and bears, and marveled at finding such ontogenetic changes which "reveal the facts of a successive development in a single species."[52] It was as if, for

50. "Heusinger on the Skull of the Simia Satyrus," *BFMR* 1840, 10:251–52. According to this source, Blainville had criticisms of Owen's paper. If so, he did not express them in his lectures (Blainville 1839–40:216).

51. Knox 1839–40:290.

52. Geoffroy 1836b, 1836a; Gould 1977:353–55. Owen was quite able to hold his own

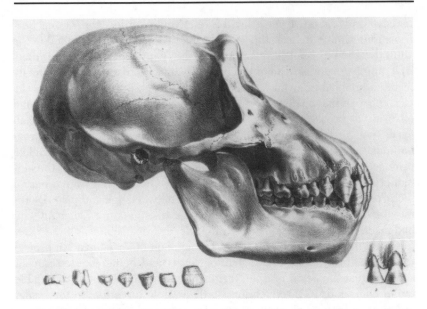

Fig. 7.5. Owen was to show that the maturing chimpanzee's brow ridges develop, its brain-box retards (relatively), and its jaws elongate into a pronounced muzzle, making the adult much less humanlike. (From Owen 1935a, pl. 51)

him, these changes were further proof of transmutation. Nothing better illustrates the degree to which the same facts could be given different meanings and accommodated in rival programs.

Owen and the Conservative Gentlemen of Geology

Owen's approach appealed to three groups sharing a liberal conservative outlook at this time: the Zoological Society gentry, the corporation execu-tives, and the Oxbridge elite of the Geological Society. Strong links ex-isted between these ruling groups. The 1830s, after all, were still a time when gentlemen of superior education (meaning an Oxford or Cambridge degree) were expected to take in trust the nation's moral development, and that included science. Owen's patrons such as the Old Etonian Sir Philip Egerton or Police Magistrate William Broderip (both accomplished dispensers of fossils and patronage) show this in full. Egerton had studied at Christ Church, Oxford, under the Rev. William Buckland (the univer-

against the French: see, in a different context, his (1839a) defense of his new species of orangutan, *S. morio;* and Owen to the secretary of the Académie des Sciences, 10 January 1839 (BL Add. MS 42,581, ff. 225–34).

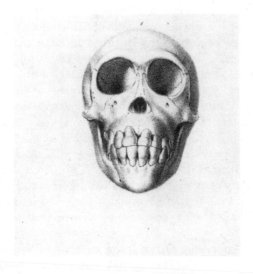

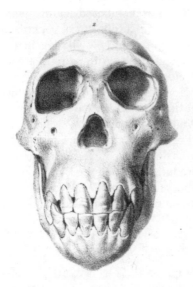

Fig. 7.6. Owen's comparison of the skulls of young and adult chimpanzees, showing the larger snout-to-cranium ratio of the latter. (From Owen 1935a, pl. 56)

sity reader in geology and canon of Christ Church). When Owen first met him, Egerton was a distinguished fossil-fish-collecting Tory M.P. He played host to the savants of the British Association at his Oulton Park estate in Cheshire, a "beautiful place" where Owen sometimes stayed.[53] As a gentleman of wealth and rank, he exercised control over a large area of London science. He was a trustee of the College of Surgeons as well as of the British Museum, a councillor of the Geological Society, and one of the active managers of the Zoological Society. Owen's defeat of Geoffroy and repudiation of Lamarckism coincided with the conservative victory at the Zoological Society's council elections of 1835. And as the political complexion in Bruton Street now changed, so Owen strengthened his grip on the society's material until, in 1840, Egerton and Lord Braybrook tipped him off privately that he would be "allowed to dissect whenever and whatever he liked" at the gardens and "have precedence over any other person."[54]

Broderip too supported Owen at the society (and was backed in turn when his own vice-presidential position came under democratic fire in 1835). Of all the gentlemen naturalists, Broderip was Owen's closest confidant. Although he is remembered for his chatty "Zoological Recreations," like Egerton he was a collector of scientific *objets*, owning the celebrated Stonesfield "opossum" jaws and keeping a conchological cabinet in his Lincoln's Inn chambers. With patrons of science such as Brougham and Peel in both Houses, such an avocation did no harm to his professional prospects. He applied to Brougham for a judgeship in 1831, telling Babbage that "all the time which has not been employed in my official duties and all the money I could spare has been cheerfully devoted to the advancement of [science]. Lord Brougham, perhaps, will not think the worse of me for such devotion coupled as it has been with a close attention to my office."[55] Broderip attended the lavish soirées which did so much to put science on the social map. He would arrive at Babbage's or Lord Northampton's armed with curios to astonish the socialities (and what bet-

53. Owen to Clift, 23 August [1848?] (BL Add. MS 39,955, f. 249); Egerton to Owen, 26 October 1840 (BMNH RO). R. S. Owen 1894, 1:141. Owen to Buckland, 12 December 1838 (UMO WB).

54. R. S. Owen 1894, 1:169; Desmond 1985a:241. I say privately because, despite Caroline Owen's diary note that this order was carried by the council, the minutes state only that Owen was to "have the earliest information of the death of any animal at the Gardens" (ZS MC 6: f. 309). However I am sure that Owen's allies made the fuller meaning clear to him. Braybrook (whom Owen had asked to be present to ensure the success of this resolution) was an active vice-president.

55. Broderip to Babbage, 25 March 1831 (BL Add. MS 37,185, f. 510). As a visitor to Britain observed (Dean 1981:121): "every ambitious young man studies geology; so members of Parliament are made, and churchmen"—although evidently not in this case judges.

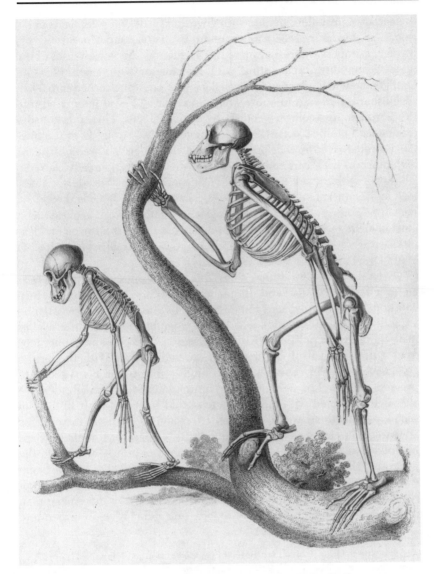

Fig. 7.7. Young and adult chimpanzee skeletons. (From Owen 1935a, pl. 48)

ter talking point than a sponge that could grow "as large as Cardinal Wolsey's hat").[56] And he would invite Egerton and his geological companion,

56. Broderip to Babbage, 18 March, 6 May, 7 May, 6 December 1842 (BL Add. MS 37,192, ff. 65, 77, 78, 210). On the soirées: Morrell 1976:137; Secord 1986a:123. The cold

the earl of Enniskillen, to dine with Owen on turtle and cold meats in his chambers. Owen by 1836 was attending Lyell's teas and Murchison's extravagant dinners. More and more, he frequented the town houses of the scientific nobility, breakfasting with Egerton or with Enniskillen (an Orangeman whose Fermanagh mansion, with one wing turned into a fossil museum, he came to know well). So Owen was well integrated into Broderip's polite geological coterie by the mid-1830s. This coterie had a pronounced political orientation. Its members had strong Oxford affiliations; Broderip's insistence on an Oxbridge education as a prerequisite for high office in the "learned professions or the state" had already caused a falling out with his self-made friend William Swainson.[57] Broderip, Egerton, and Enniskillen had all been Buckland's pupils at Oxford, and by mid-decade Owen too was working closely with the Oxford geologist. But most of all its members shared Church and king values. Egerton and Enniskillen were shire Tory M.P.s, and Owen was himself being invited to Peel's Drayton Park estate by 1839.

Owen proofread Broderip's penny pieces and praised him publicly as an excellent naturalist and "humane magistrate."[58] Broderip in turn checked Owen's papers and immortalized his client's Cuvierian feats: he argued—despite Blainville's and Knox's ridiculing of the similar claims made for Cuvier's powers—that Owen's reconstruction of an extinct moa from a broken shaft of femur was an astounding feat. With Broderip a *Quarterly* reviewer, Owen could arrange publicity for his works. He pressed his friend to publicize the Hunterian catalogs, and Broderip—although burdened with judicial duties—told Buckland in 1842 that "in such a case as this I would make an effort and if you think it would be advisable and can get Lockhart [the editor] to consent, I might perhaps be able to cook up a mess that though solid should be palatable to the general. I should take the Catalogue of the Mus. Coll. Reg. Chir. begin with John Hunter and end with Owen, the English Cuvier, who has already done enough for a long life; and [has gained] . . . more fame than many a first rate philosopher."[59] Broderip's extended reviews turned into

---

cuts: Broderip to Owen, 6 May 1841 (BMNH RO). On Egerton and Enniskillen (Lord Cole): K. W. James 1986; Desmond 1982:67–68; R. S. Owen 1894, 1:102–3, 113, 122, 141, 156, 161.

57. A customs clerk, naturalist-traveler, and author, but not Oxbridge educated (Desmond 1985a:170).

58. R. Owen 1848a:119. On the moa: Broderip to Buckland, 20 January 1843 (BL Add. MS 38,091, f. 193).

59. Broderip to Buckland, 14 January 1842 (BL Add. MS 40,500, f. 247). This project was not in fact tackled until much later: E. Richards 1987:141, n. 47; Broderip and Owen 1851–52; Rupke 1985:244.

partisan celebrations of Owen's labors, even if their cloying nature galled such young Turks as T. H. Huxley.

Through Broderip too we begin to see Owen's connections with the legal elite at the nearby Inns of Court. We know that the judiciary was to play a key role in Coleridge's clerisy, and Owen had begun developing ties with prominent Lincoln's Inn lawyers while he was still practicing in Cook's Court in the 1820s. Many of Owen's patrons in the 1830s held Peelite views. Broderip greatly admired the Tory leader, through whom he had obtained his appointment as magistrate to the Thames police court in 1822. He considered him "something more than a great man, he is a good one."[60] Owen gained the ear of a number of Peel's legal advisers, among them Frederick Pollock, with whom he was to form a lasting friendship. Pollock became king's counsel in 1827 and a Tory M.P. in 1831, and he was knighted in 1834 on becoming attorney general in Peel's first ministry. His son, who found Owen "a most wonderful" person and regularly attended his lectures, often sitting alongside Bishop Wilberforce, recalled the gloom in his father's circle at the prospect of reform. He remembered the talk about the "approaching destruction of everything after the Reform Bill was carried"—the fear that "the Church establishment and the House of Lords would go first, and the monarchy itself would soon follow."[61] These Tories were assiduous in countering the democratic threat wherever it arose. As counsels to the College of Surgeons, Pollock and James Scarlett had prosecuted Wakley for libel in 1828 and sought a criminal information against him for provoking a "riot" in the theater in 1831.[62] They literally stood on the opposite side of the dock from the Wakleyans. But it was not only in court that they faced the radicals. They defended the Royal Colleges at Westminster and challenged their critics. Pollock presented petitions from the RCS Council against the radical bills before Parliament, while it was Scarlett, for example, who confronted Epps for practicing without a license. These gentlemen, hated by the Wakleyans as enemies of the third estate, brought Owen to the very heart of Tory corporation politics.

60. Broderip to Buckland, 14 January 1842 (BL Add. MS 40,500, f. 247). On the appointment: Broderip to Peel, 19 March, 20 April, and 23 May 1822 (BL Add. MS 40,345, f. 241; 40,346, f.189; 40,347, f. 135). He was appointed by Lord Sidmouth, a high Tory hated by Regency radicals for his repressive measures (Inkster 1979).

61. Pollock 1887, 1:31–32, 273–74; R. S. Owen 1894, 1:42–43, 157. The lawyer David Pollock (Frederick's brother) gave Owen a letter of introduction on his trip to Paris (26 July 1831, RCS MS Cab. VIII [1] a75).

62. Clarke 1874a:38ff. "The Council of the College v. the Members," L 1830–31, 2:273–77; G. Clark 1964–72, 2:704. It was Pollock's house adjacent to the college that was bought and demolished by the college in 1834 to make way for the museum extension: "Proposed Outlay of College Money," L 1833–34, 1:830–32; E. Epps 1875:182.

Of all the coterie members, though, it was Buckland himself who gave Owen his most prestigious scientific endorsement. By 1833 Owen was already experimenting on nautilus flotation for Buckland. Two years later the Owens honeymooned at Oxford, and Buckland tapped Owen's brains on marsupial reproduction and the Stonesfield "opossum."[63] The Bucklands became family friends. Buckland would come down to attend Owen's Hunterian lectures, with his wife Mary taking the opportunity to visit Caroline Owen. He made a point of being present at the reading of Owen's papers and advised him on obtaining BAAS grants and the best publishing terms in London.[64] Their letters at this time show how broadly their scientific interests conspired, from agreeing on *Nautilus's* biology (of importance because Blainville was also working on the problem), to understanding the Stonesfield "opossum" and perfecting a paleontological strategy to defeat the transformists.

Transformist Fossil Zoology

The modern Pantheists . . . see nothing but absurdity in the supposition of a Great First Cause, they deem the belief in a Deity as unfit to be even entertained by their philosophy, and they substitute in his place that most extravagant of all suppositions, that most grovelling of all religions—the self-created, self-endowed, and self-creating powers of Nature.

—A medical reviewer abominating Tiedemann's transformism[65]

While Owen had tackled the issues of platypus eggs and the ape's cultural aspirations, transformists themselves were looking to the fossil record and the results of domestication for their main evidence. In his inaugural lecture (1828), Grant had outlined a program that stretched far beyond the safe empirical confines of conservative metropolitan zoology. The new science was to embrace "the origin and duration of entire species, and the causes which operate towards their increase or their gradual extinction . . . and the changes they undergo by the influence of climate, domestication, and other external circumstances."[66] Domestication was widely discussed in Paris and Edinburgh circles, and Fleming in 1822 considered

63. R. S. Owen 1894, 1:66, 90; Buckland to Owen, 25 January 1835 (BMNH RO). To the ambitious Mrs. Clift's delight, the honeymooning Owens met "the great *Lord Chief Justice*" at Buckland's: C. H. Clift to C. Owen, 22 July 1835 (BL Add. MS 39,955, f. 225).

64. Buckland to Owen, 24 February [1839] (RCS MS [1] a/19); on the nautilus, 9 March 1838 (RCS MS [1] a/11).

65. "On the Physiology of Man," *MCR* 1839, 30:450.

66. Grant 1829:6; cf. the London zoologists, Desmond 1985a:161ff.

that the changes recorded by horticulturalists and breeders provided the strongest evidence supporting the transformists' cause.[67] Grant in 1826 emphasized these changes to bolster his Lamarckian case. Shortly afterward, the Scottish arboriculturalist and advocate of free market forces, Patrick Matthew, argued that culturing exploited the "plastic quality" of life, and he explained the fossil ascent as an analogous "self-regulating" change, forced by varying circumstances and made possible because of life's natural variations and Malthusian fecundity.[68] By now the records of expatriate farmers were also being scrutinized, and Geoffroy took a keen interest in the reports of changes undergone by European livestock transported to South America.

But Grant's speculative interests had always lain primarily in fossil zoology, and he eventually acknowledged this as the "highest" department of biology.[69] Since life's "extreme branches only are visible on the surface," the fossil "roots" alone could provide an accurate chart of animal development.[70] This belief was reflected in the structure of his first university course, which ended in the late spring with a section on the "nature and origin of *Fossil Animals*." He discussed the order of their succession, the relations between living and extinct species, and contemporary global changes.[71] This became the foundation for his regular summer "Fossil Zoology" course, begun in 1831, in which he broached the "direct [i.e., natural] generation" of successive faunas and its paleoclimatic causes. He continued to visit the European collections yearly, adding to his dry muster-roll of fossils. The course remained uniquely Continental in content, with Grant characteristically championing the European savants at the expense of the English geologists. Events following Bell's resignation increased the student audience for these lectures. Despite the geological interests of the university's founders, the geology chair remained unfilled. The Yorkshire geologist John Phillips had toyed with applying in 1831; he even delivered a trial course, but being on the spot brought home to him the university's frightful financial condition and the interminable squabbling. The punctilious Leonard Horner had resigned, "fairly scared and

67. Fleming 1822, 1:27; Grant 1826b:298. R. J. Richards 1982:248.

68. Dempster 1983:106–7. Geoffroy 1828; Roulin 1829. Cuvier's heirs continued to play down the "superficial" changes caused by domestication (Flourens 1834–35:305–6).

69. Grant to J. Barlow, 3 May 1856 (RI General Archives, box 14, file 142); Desmond 1984c:407.

70. Grant 1833b:10.

71. Prospectus of Grant's first (1828) Comparative Anatomy and Zoology course in his hand (UCL CC 1179 [13]); printed version: item 4, p. 6, in *Grant on Zoological Subjects*, College Collection DG 76, UCL. See also Grant 1829:6; Grant to Horner, 8 April 1830 (UCL CC 139); 24 April 1830 (CC P145). This course changed little over the decades (Desmond 1984c).

worn out" by the Pattison affair, and he also warned off Phillips. With no geologist willing to take the gamble, Grant's fossil lectures in 1832 doubled for part of the geology course, with Edward Turner (mineralogy) and John Lindley (fossil botany) providing the rest. Fossils and rocks were borrowed from the Geological Society, and the impromptu geology course continued until Turner's death in 1837.[72]

Grant's contribution was defiantly Continental. His developmental metaphors closely resembled Tiedemann's. Both accepted that matter had the "power of acquiring, by degrees, different simple forms of living bodies"[73] and that these became progressively more complicated through environmental changes (the complex climatic shifts accompanying plane-tary cooling in Grant's case). Otherwise Grant's terminology was strictly Geoffroyan. As Geoffroy, while discussing the changes in extinct croco-diles, spoke of "la transmutation et la métamorphose des partes," so Grant in class talked of a slow metamorphosis of fossil animals to meet changing conditions.[74] Translations too were making this kind of Continental trans-formism better known. Tiedemann's *Treatise on Comparative Physiology* containing his *metamorphose-theorie* appeared in English in 1834. In it he argued for an emergent development and progressive complication of fossil organisms, based on the inherent "plastic power" of organic matter.

The point to be made here is that medicine at this time was so much more socially diverse than, say, geology. What with the pro-French atti-tudes at the Benthamite university and the down-market Dissenting schools with their heterodox materialism, Continental sympathies were far more pervasive, and even transformist works such as these could find supporters. Not that the respectable medical men liked these "godless, self-existing, self-destroying, self-contradicting, senseless" theories any more than did the gentlemen geologists. They deprecated talk of innate powers, lumping Tiedemann with Lawrence as a purveyor of "pernicious doctrines." Pantheists who rejected a vital controlling force and made Geoffroy's "unity" embrace the whole of evolving creation were dismissed as "lunatic" slaves "to superstition."[75] Given the social diversity within the huge, shambling world of medicine and the intellectual free-flow that still

72. Minutes of Council, 3: ff. 38, 155 (GS); Edmonds 1975. On Horner running scared: G. J. Bell 1870:317–18; Bellot 1929:212–13. Lindley was an authority on extinct plants, being coauthor of the three-volume *Fossil Flora of Great Britain* then in press.

73. Tiedemann 1834:15; Sloan 1985:75–80.

74. Grant 1833–34, 2:1001; Geoffroy 1825:151. At this time "metamorphosis" was pri-marily used in the sense of fetal germ development: e.g., Brewster 1834:143; Rush 1835:8; Carlisle 1826:37.

75. "On the Physiology of Man," *MCR* 1839, 30:452; Tiedemann 1834:13–15, 39–40; Schumacher 1973.

existed between it and the orderly, gentrified world of London geology, we can understand why there was a periodic clash of professional and social cultures—why sparks flew as the wealthy geologists rubbed up against the medical Geoffroyans. To an extent it was Owen, having come from a medical corporation himself, who now focused the geological mind. We know he was familiar with the medical transformists. (He probably met Tiedemann himself in 1835. At least Tiedemann, in London that year, asked to see Owen's brains—or rather, Owen was told by the secretary of the Zoological Society, not those "in your *cranium*, but such as have been removed from the *crania* of our animals.")[76] To understand how Owen chose to tackle the fossil evidence for the "metamorphosis" of species, we need to look at the Geological Society in some detail. We will then be able to see how the gentlemen responded to this foreign fossil zoology with its theory of a self-organizing nature.

Recent work on the Geological Society has concentrated on the small group of wealthy geologists who by the 1820s had usurped the place of the original mineralogically orientated founders. These gentlemen and their Oxbridge mentors now "dominated the meetings of the Society, dictated its social tone, and engineered key appointments such as the Presidency."[77] By the 1830s they had constituted themselves into a self-contained, "self-validating knowledge elite."[78] Martin Rudwick and James Secord have greatly increased our understanding of the elite's self-image and day-to-day geological activity. Most of the independent sporting gents could pursue geology in a way that was impossible for the society's medical backbenchers, tyrannized by their laissez-faire teaching trade. It is the relationship of these backbenchers to the star chamber that I now explore. The society was more than its productive elite; it was never a social monolith. True, contemporaries praised it as a stable meritocracy, free from the sort of corruption endemic to the Royal and Zoological societies. But there were strong incentives at the Geological Society to present just this public image of social cohesion. If we now focus on the radical democratic members, we glimpse another side of the society—one that saw the makings of a struggle against the dominant conservatives. The tension becomes clearer as we chronicle the attempts to marginalize the backbencher Grant and discredit his serial Lamarckism.

I talk of star chambers and backbenchers because the society's main room in Somerset House was set up like a learned commons, with

76. E. T. Bennett to Owen, n.d. [1835] (BMNH RO 3:f. 190).
77. Morrell 1976:139; Porter 1980:145–56; Weindling 1979; Secord 1986a:14–21; Rudwick 1985: chap. 2.
78. Porter 1978:810.

benches facing one another and the president holding the speaker's chair. As at Westminster, the geological ministerialists were part of a larger hegemonic elite. Buckland's friend and fellow "saurologist" Rev. William Conybeare spoke for this larger community in 1833 when he told his friends at the BAAS that the Whigs and Tories must unite against the radical unions. It was a feeling shared by most of the geological fellows; Charles Lyell, for example, inveighed against the "mob-rule which I see daily in the papers." As a result, conservative consensus politics was the order of the geological cabinet. A united front and a presentation of what appeared "indisputable factual knowledge"[79] was tacitly agreed by the society's managers. Rash theorizing was effectively banned; scripturalism, for example, was not tolerated, and while it was permissible to discuss Creation, one could not talk approvingly of transmutation. The main preoccupation of the group was with a stratigraphic science that spoke of "permanence, impersonality, and neutrality."[80] Permanence was important. As Secord says, a science that looked provisional threatened the elite's claim to be the guardian of the rock of knowledge. But more, as Morrell and Thackray show, this image of incontrovertible science was essential if the natural order were to be seen pointing toward God's immutable moral order.

And many saw it this way. The gentlemen were peeling away the strata to portray an ordered geological creation, pregnant with Divine intent. It was a sign of God's fixed design that appealed to the propertied class at a time of violent unrest. Hence Sedgwick on occasions read his sermon directly from these rocks, warning the laborers that their place was equally ordained in the social strata. Ultimately these field geologists were involved in a taxonomic enterprise. They were "taming the 'chaos' of the strata,"[81] imposing a Cambrian, Silurian, and Devonian order on the "chaotic" older rocks. Buckland talked evocatively of this as a kind of scientific enclosure act: just as the commons were fenced off, keeping the poor from grazing their stock on public land, so the gentlemen were roping off the "common field of geology," reserving the rocks for their own use (though he did not quite mean it like this).[82]

Secord observes that the gentlemen were not concerned with processes and causation so much as with mapping the strata. And Rudwick has shown how the elite's stratigraphic claims were then validated by means of group consensus. But a radical such as Grant stood defiantly

79. Rudwick 1985:25; Secord 1986a:22; K. Lyell 1881, 1:291–92. On Conybeare: Morrell and Thackray 1981:2; and Neve 1983 on the larger hegemonic elite.

80. Secord 1986a:317; Morrell and Thackray 1981:31.

81. Secord 1986a:4, 33–34, 315, 1982:413, 415, 1985:185–87.

82. Secord 1982:415.

outside this consensus. Both in his approach and conclusions he differed dramatically from the geologists. For them, "organisms were of primary interest as stratigraphical markers rather than as ancient forms of life."[83] For Grant, life and process were paramount; his major interest was the course of life through geological time and its explanation, not the entombing strata. His classification reflected this and departed radically from the gentlemanly norm. It was based on the major breakthroughs in the history of life. Thus his "Protozoic" period—as he taught in his later lectures—encompassed the spontaneous appearance of infusorial life and ended at the dawn of the terrestrial invertebrates (a wholly aquatic phase). His "Mesozoic" commenced with the development of the varied land-living worms, insects, and mollusks and finished with the emergence of the fishes. Finally his "Cainozoic" covered the rise of the vertebrates and will continue until all life is extinct and the planet barren.[84] This was an extraordinary leveling classification, making no concession to man, nor even the appearance of mammals, but broken solely by the invertebrates' success in conquering the land and by their development into vertebrate forms. Grant's classification signaled a history of development and achievement, but not from a homocentric perspective. It marked the stages of life's self-powered ascent.

The gentlemen would have seen this as idiosyncratic, irresponsible, and a violation of their stratigraphic canons. Their own classifications were now beginning to be exported worldwide. Secord has depicted British geology as an expansionist emblem for the Empire; a militarist such as Murchison, for instance, envisaged his Silurian nomenclature accompanying British trade goods to the four corners. More to the point, the clerisy's "internal" imperialism—its attempt to conquer the chaotic lower orders at home—was quite unlike the ultraradicals' program justifying a self-powered democratic ascent. Consequentially, the respective sciences embedded deep in these rival strategies were very different. This all helps to explain why Grant rarely referred to the Oxbridge dons and their London acolytes, preferring to cite French and German sources. Grant's case shows that the stratigraphic "norm" was not necessarily "natural." It demonstrates just how different a radical science based on fundamentally distinct conceptions of nature, causation, and history can be. Through Grant, the Wakleyans realized a non-Anglican paleozoology—one that did not acknowledge discrete steps recording acts of "Creative Interference," but a self-governing and sovereign ascent in which the achieve-

83. Secord 1985c:187.
84. R. E. Grant, Palaeozoology Lectures (BM Add. MS 31,197, ff. 23–25); Desmond 1984c:399–403 on his sources.

ments of the organism provided the criterion of classification. True, Grant was only a peripheral figure in geology. Still his deviant view is interesting for opening up an alternative geological reality with different social roots.

From the first, Grant had taken a lively interest in the Geological Society. With a telling slip, Horner had even introduced him to its leaders in 1829 as "our Professor of Comparative Anatomy & Geology."[85] He became a fellow in 1830 and anticipated a long-term commitment, taking out life membership in 1831 at some cost (£31.10s, or more than his previous year's earnings from comparative anatomy classes). He was elected to the council in 1832, and it is not hard to identify the reform vote he picked up.[86] That year, Warburton was on the council, Turner was secretary (and only the year before had dubbed Grant the "English Cuvier"), while the treasurer was his future backer at the Royal Society, the Unitarian mining entrepreneur John Taylor. Grant brought a succession of students and guests,[87] but he was never more than a minor figure here. He did however stand up to Owen, and it is their sharp exchanges that I shall focus on. The first of these clashes concerned the Stonesfield "opossum" jaws, in a debate that was pregnant with meaning for both the Lamarckian "progressives" and their conservative opponents.

The Debate over the Stonesfield "Opossum"

The final judgment of M. de Blainville met with approbation and support from the stricter systematists, since it harmonized with their preconceived opinions on the progressive appearance of organized forms on this planet.

—Owen explaining the opposition to a mammalian interpretation of the Stonesfield jaws[88]

In 1837 Owen was still largely indebted to Buckland for his reading of the fossil record, and by now the two men were working ever more closely to stem the radical tide. Buckland was well aware of the French situation.

85. Horner to G. B. Greenough, 22 February 1829 (Greenough Corres. GS).
86. Ordinary Minute Book 5, 1830–32, ff. 119, 371 (GS). His backers at the Royal Society in 1836 give an idea of the reform vote he was picking up at the Geological Society. They included Benthamites such as the London University treasurer William Tooke and professors Turner and Elliotson, medical corporation critics Robert Lee and John Bostock, as well as John Taylor: RS *Certificates 1830–1840*, VIII, 182. Crosland 1983:179–83 on entry into the Royal Society.
87. Turning up with them on fifty occasions during the 1830–45 period. For his early appearances: Ordinary Minute Book 4, 1828–30, ff. 14, 371; Book 5, 1830–32, ff. 74, 85, 119, 229, 273, 331 (GS).
88. R. Owen 1846a:35. Desmond 1984a for a synopsis of this section.

He had been sent Geoffroy's papers, and these had caused him privately to open a file on transmutation. In this he repeatedly turned over the evidence brought to support the "Absurd doctrines of Lamarck & Geoffroy": the recapitulation of animal stages—"like Shakespeare['s] 7 ages"— in the human fetus; the liassic mollusks developing "rudiments of Vertebrae"; and Geoffroy's experiments to deform chick embryos by altering the egg's environment, with its implication that "nature also has adopted that circuitous course instead of making each species at once for its destined office." Much of their evidence he caricatured, especially Lamarck's notion of "volition," according to which "Reptiles tired of Crawling at length [and] by the mere wishing to fly were converted into Birds." Being an expert on saurians, however, he was most preoccupied with Geoffroy's plesiosaur-to-crocodile transformation, and he reasoned that in Geoffroy's scheme the crocodiles in turn "must be the ancestors of man." Buckland never published his more sarcastic comments on Geoffroy's attempt to make "Man the Son of a Crocodile,"[89] but in 1836 he took a strongly anti-transformist line in his Bridgewater Treatise, *Geology and Mineralogy Considered with Reference to Natural Theology*. In this he confronted the serial transformists with examples of "*retrograde* development." He pointed out that the oldest fishes in the fossil record were heavily scaled and therefore "advanced," and were accordingly placed high up in the conventional classification. The same was true of the oldest cephalopods and crinoids: they were among the highest invertebrates. None of this squared with the Lamarckian ideal of advancing perfection. The early appearance of high-born animals could only be explained by "the direct interposition of repeated acts of Creation."

Owen, in his first Hunterian lectures (1837), simply borrowed these illustrations from Buckland, making the quiet observation that "the different organized forms which have succeeded each other do not display regularly successive stages of complication, or perfection of Structure."[90] The tameness of this statement contrasts with his dramatic application of the principle of retrogression in 1841. But by then he had made extensive contact with the fossil data, which convinced him that it could offer the clerisy a stronger antidote to the Lamarckians' inexorable ascent.

By then, too, he had had his first run-in with Grant at the Geological Society on the Stonesfield "opossum" issue. It was in this episode in 1838–39 that we first see Owen and Buckland working on a joint strategy to outmaneuver Grant, who was refusing to accept the tiny jaws from the

89. William Buckland, file of assorted notes marked "Species Change of Lamarck" (UMO WB); Rupke 1983:175.

90. R. Owen, "Nature and Character of Organized Beings," Hunterian Lecture 3, 11 May 1837, ff. 34–35 (BMNH MN 1828–41). Buckland 1837, 1:vii, 294–95, 312–13, 431.

(Jurassic) Stonesfield slate as mammalian. (By serialist criteria, that would have made them fossil anachronisms—mammals out of sequence, living in the Secondary age when climatic conditions were thought to have permitted only a reptilian grade of existence.) This section deals with the unfolding debate over these fossils, to show how the personal interactions, divergent ideologies, and local contingencies helped sustain the rival interpretations.

Coming from the Oxford Stonesfield slate, the jaws had first fallen into the possession of the university gentlemen. Broderip, as an Oriel undergraduate, had acquired two from an old stonemason in 1812, one of which he sold to Buckland. Because of the jaws' exceptional antiquity Buckland refrained from publishing until Cuvier had examined them. This he did on visiting Oxford in 1818, when he compared them to the jaws of the opossum *Didelphis*. Still Buckland did not announce the discovery until 1824, and then only mentioned it in passing in his paper on the newly unearthed Stonesfield "giant lizard" *Megalosaurus*. Blainvillean serialists realized the implications. The following year Constant Prévost, refusing to accept that a mammal could be so ancient, tried to reinterpret the Stonesfield slate as a younger, Tertiary deposit. Cuvier reconfirmed that Buckland's animal was opposumlike, but with a longer toothrow (it had ten "grinders"), and he called it *Didelphis Prevostii* (see fig. 7.8). If this English dating was accurate, then it was, he conceded, a "remarkable exception" to the rule that mammals were of Tertiary age.[91] Broderip meanwhile had mislaid his fossil and only recovered it and published a description in 1827. He noted that his jaw was "generically different" from Buckland's and, with only seven "grinders," was still more opposumlike.[92] He christened his specimen *Didelphis Bucklandi*. And he saw further evidence that these really were marsupials in the associated invertebrate fauna, which resembled that still surviving in the Australian colonies.

In the 1830s the leaders of the Geological Society all accepted Cuvier's diagnosis and an Oolitic (Jurassic) age for the embedding slate. All the same, Buckland was inconvenienced at first by an odd Oolitic mammal. In 1836 he still believed that global conditions, judging by the Secondary fauna, were somewhat ill suited to land mammals. On the other hand, Owen was shortly to speculate on the advanced respiratory mechanics and cardiovasculature of his new "dinosaurs" (which in his view pointed to "improving" environmental conditions in later Secondary times),[93] so he

91. Cuvier 1825b, 5:349; Prévost 1825; Buckland 1824:391; Broderip 1827. For overviews of the subject see R. Owen 1871; Blainville 1838a.

92. Broderip 1827:410–11.

93. Buckland 1837, 1:72; R. Owen 1841c:203–4.

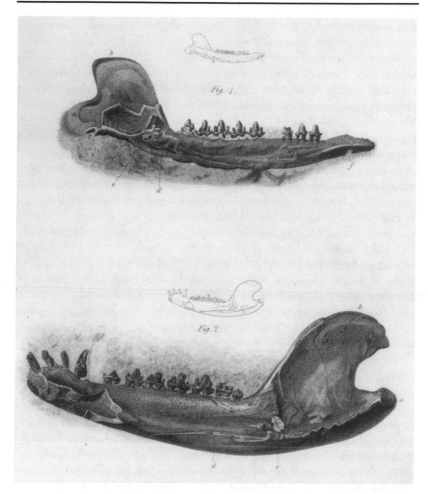

Fig. 7.8 Buckland's fossil "opossum" jaw (above), with the longer tooth row (*Didelphis* [*Thylacotherium*] *Prevostii*). Below is Broderip's jaw with only seven "grinders" (*Didelphis* [*Phascolotherium*] *Bucklandi*). (From Owen 1841b, pl. 6)

could more easily accommodate marsupial contemporaries, and Buckland too soon came round. But Grant's image of a lineal fossil ascent made him extremely suspicious. He undoubtedly drew support from Geoffroy's belief that the birth of mammals in the age of reptiles was unlikely on account of their respiratory needs. Anyway, Grant was the first to dissent from the marsupial diagnosis. In his university lectures in 1834, he asserted that Cuvier's Montmartre excavations had revealed opossums in

Eocene rocks, but that the older Stonesfield jaws had been "erroneously ascribed to the same animal."[94]

There followed five years of growing dissention among the anti-Cuvierians. Grant's opposition was widely reported in Paris, and according to Blainville, voicing his own doubts in 1838, Grant in class was now detailing his objection to the standard Cuvierian interpretation.[95] Grant, having examined the jaws, was always more forthright in his opposition than Blainville. In fact Blainville had equivocated at first, conceding that Buckland's jaw resembled a tree shrew's and Broderip's an opposum's, even though the teeth differed in number and shape. Also, only mammals were known to have incisors, canines, and complex-crowned molars like those in the tiny jaws. Then, searching the literature, Blainville managed to find a description of a reptile with complex teeth—the large Alabama saurian *Basilosaurus*, whose multicusped molars had been described by Richard Harlan, professor of comparative anatomy at the Philadelphia Museum. Blainville concluded that if the basilosaur were a reptile, then it was probably "an animal of the same kind as that found at Stonesfield."[96] He nonetheless coined the equivocal name *Amphitherium* for the Stonesfield jaws. If at length he was cautious, it was because he had not personally seen the fossils, although he too considered the presence of reptiles far more likely, given the age of the rocks.

Grant was Blainville's friend and frequent visitor. He immediately accepted the new name and continued to deny the opossum diagnosis. Late in 1838 he published an abstract of his dry Fossil Zoology course under the title *General View of the Characters and the Distribution of Extinct Animals*, separates of which were struck off by the publisher Bailliere and sold for 3s.6d. This tract caused immediate consternation among the Geological Society "saurologists." In it Grant appealed as usual to Geoffroy's unity of plan and made the fossil record one of continuous ascent. In three places he noted *Amphitherium*'s wrong identification and insisted that "no unequivocal skeleton of bird or quadruped" had been found in Oolitic rocks, making the oldest "authenicated" mammal still Cuvier's opossum from the Eocene gypsum.[97] Grant had now examined the four known jaws. He published a description, identifying their compound structure (a compound jaw is a reptilian feature), and agreed that Harlan's *Basilosaurus* was a "closely allied" genus.

He was attempting to rid the record of what he saw as fossil anachro-

94. Grant 1833–34, 2:72. Valenciennes 1838:573 points out that Grant was the first to dissent.
95. Blainville 1838b:730, 1838a:405, 416, 418.
96. Blainville 1838a:417, 1838b:736.
97. Grant 1839:7, 42–43, 54.

nisms. He had no time for a Creationist science which allowed species to pop up contingently, out of step with life's progressive ascent. In his lectures he continually reinterpreted such "anachronistic" fossils. He also tackled fossil footprints, like those found in New Red Sandstone (Triassic) rocks in Saxony (made by an unknown animal christened *Cheirotherium*)—the more earnestly because they were being used as evidence in the Stonesfield affair. Johann Kaup believed that marsupials had made these tracks. Buckland reported Kaup's opinion in his *Geology and Mineralogy;* although Buckland himself had identified tracks of similar age in Dumfries as those of a tortoise, he rather incautiously gave Kaup's view credence. Kaup, he said, believed that the Saxony footprints

may have been derived from some quadruped allied to the Marsupialia. The presence of two small fossil mammalia related to the Opossum, in the Oolitic formation of Stonesfield . . . are circumstances which give probability to such a conjecture. In the Kangaroo, the first toe of the fore-foot is set obliquely to the others, like a thumb [the Saxony prints were handlike], and the disproportion between the fore and hind foot is also very great.[98]

Triassic kangaroos were, if not fatal to Grant's ultraserialism, then a cause for concern. So on visiting Liverpool to lecture at the Mechanics' Institute in August 1838, he made a point of examining similar tracks uncovered the previous year in Stourton Hill Quarry, five miles from the city center. Grant's talk on the footprints was reported in the *Liverpool Mercury* and extracted (with only his opening remarks on the undeviating progression of life excluded) in the *Magazine of Natural History,* which had already run translations of Blainville's *Amphitherium* papers. The editor, in his introduction, made it plain that the question of the tracks bore very closely on "that of the 'supposed fossil didelphs.'"[99] Therefore it was essential to interpret them correctly. Grant urged caution in dealing with tracks of such antiquity, and he stressed the unlikelihood of their having been made by a mammal. Such false attributions were common enough: he had already diagnosed Lord Greenock's "wolf's" tooth from the New Red Sandstone as that of a fish, and had listed a catalog of similar "errors." He now deftly reinterpreted the tracks, fitting them into a normal crocodile sequence. He switched the right foot for the left so, instead of a marsupial thumb (on the right foot), this became the small toe (on the left). No longer would the animal have a unique marsupial gait, a mammalian

98. Buckland 1837, 1:265n. Buffetaut 1987:93–95 discusses the history of the various fossil trackways.

99. "Scientific Intelligence," *Mag. Nat. Hist.* 1839, 3:43. On the discovery of the footprints: "An Account of Footsteps of the Cheirotherium, and the other unknown animals lately discovered in the quarries of Storeton Hill," *Proc. Geol. Soc.* 1838, 3:12–14.

duck walk, crossing its own line at each step. Grant rendered the prints wholly unexceptional; they were exactly what we should have expected, had they been made by the teleosaurs common in the period. They offered no proof of the existence of New Red Sandstone mammals. Geologists had only considered them marsupials in the first place, he said, because of their mistaken belief that the Stonesfield jaws were the remains of opossums.[100]

Owen too had come to see *Cheirotherium* as a reptile. Buckland's promotion of Kaup's view was evidently proving embarrassing, for Owen wrote reassuringly to Buckland, admitting that they were getting

on the right scent to the true animal, which should certainly be a reptile if the thumb be a little finger. I cannot imagine how any reader of your B.T. could suppose that he was obtaining any thing else than Kaup's opinion through your translation and at that time it deserved undoubtedly every consideration.[101]

The trouble was, with the issue of tracks and jaws so entangled by the press, this discrediting of Kaup's view was seen to weaken the case for marsupials in Stonesfield rocks.

By now the jaws had acquired international notoriety. In the summer of 1838, with disaffection spreading to Paris and Germany, Buckland set out on a European tour, taking two jaws with him. Just missing Blainville in Paris, he left the jaws with the zoologist Achille Valenciennes for casting, with the copies to be presented to members of the Académie. Buckland's gamble paid off: a flurry of papers appeared in the *Comptes Rendus*. Valenciennes vindicated his late patron Cuvier. Disagreeing with Grant, he concluded that each jaw was composed of a single dentary, making it indisputably mammalian. What allowed Valenciennes to be so positive was his use not of the Virginian opossum for comparison, but of the South American mouse opossum which was closer in size and structure. He announced that the fossils represented a distinct genus of opossumlike marsupial and, finding nothing ambiguous about them, proposed substituting the name *Thylacotherium* for Blainville's *Amphitherium*.[102] This was gratifying to Buckland, though better was to come. In the *Comptes Rendus* Geoffroy himself admitted the marsupial diagnosis, although this was

100. Grant 1838:44, 46, 48; 1834.

101. Owen to Buckland, 12 December 1838 (UMO WB).

102. By the end of the year so many names were in use that everyone was confused. The situation was made worse by Owen (1841b:57) at first accepting the name *Thylacotherium*, then switching back to *Amphitherium* (1842:62). The situation did have its lighter side. The *Athenaeum* (1838:731, 747, 841) eagerly reported the seesawing fate of the tiny jaws in its "Weekly Gossip" column. To "avoid making an invidious selection of the different claimants to the right of christening," it renamed the beast *Botheratiotherium*. Blainville (1838b:735), whose English was presumably none too good, evidently missed the joke, and in the presti-

something of a Pyrrhic victory, for Geoffroy then turned round to assert that marsupials were not mammals at all but, like monotremes, a separate lower-ranking class.[103] Finally, in September 1838, Buckland took the fossils to Freiberg (where he joined up with Owen) to place the problem—and the jaws—before the congress of German naturalists, hoping for a decisive result.

But Grant proved more intractable, and because he was a serial transformist it became imperative to dispose of his views as publicly as possible. Buckland now determined to get the Grant-Blainville diagnosis "officially" discredited at home. He asked Owen to counter Blainville's criticisms in print, lending him the two original jaws for the purpose. Owen was the perfect choice. He had been elevated to the council of the Geological Society in 1838, and was acknowledged both there and in the Zoological Society as an expert on marsupial anatomy and, increasingly, vertebrate fossils. In the same year he had received the Geological Society's Wollaston medal for his work on Darwin's South American fossil rodent *Toxodon*. Grant of course was still powerful, having just taken the Fullerian chair at the Royal Institution (1837–40)—a post that Owen (the managers' first choice) had been forced by the RCS Council to turn down.[104] By 1838 the medical journals were themselves bracketing the two men together as the leading comparative anatomists in the city, although the radicals were still rooting for "our great GRANT."[105] On French science, Owen was close to the anti-Blainville faction in Paris. Many of Owen's correspondents now openly disparaged Blainville's work and communicated their frustrations to the College of Surgeons. The Irishman Joseph Pentland, who worked alongside Blainville in the Muséum, told Clift in 1832 that, despite succeeding to Cuvier's chair, Blainville was "too old, too idle, and too stubborn" to do any good for science, while Geoffroy was a "terrible *wrong-head.*"[106] Cuvier's sycophant Charles-Léopold Laurillard also complained to Owen of Blainville's unreasoning criticism of Cuvier and appalling lack of logic. Owen was himself aware of these

gious *Comptes Rendus* protested at this infraction of the zoological rules. Needless to say, English journals saw the funny side of Blainville's reply. Poking fun at the French was a common pastime: "Owen on Odontography," *BFMR* 1840, 10:211.

103. Geoffroy 1838:629–33; Appel 1987:184.

104. Managers Minutes, 1832–53, 8: ff. 307, 552 (RI). On Owen's GS medal: K. Lyell 1881, 2:37, 39; R. S. Owen 1894, 1:121–22.

105. "The College Conversazioni," *L* 1841–42, 2:246. The *MCR* considered that of the "very few" people in Britain qualified to write a textbook on "embryological anatomy," only Grant, Owen, and Knox stood out: "On Philosophical Anatomy," *MCR* 1837, 27:87, 106.

106. Pentland to Clift, 10 May 1833 (BMNH RO 21: f. 219); and on Blainville: Pentland to Clift, 5 November 1832, copy in Owen Notebook 9 (1832–33), f. 90 (BMNH).

shortcomings and alerted Whewell to Blainville's nitpicking and contempt for Cuvier's prowess in reconstructing fossil animals from a single bone.[107] So Owen was sensitive to the general social situation—to the Cuvierian interests in Paris, and to the radicals' investment in the "saurian hypothesis." And after the debacle for the radicals at the Zoological Society in 1835, and their failure on the British Museum Committee in 1836, he must have known that Grant had a lot riding on the outcome of the Stonesfield debate. The jaws could provide more than another nail in the Lamarckian coffin.

Owen now made a close study of all four jaws—those in the Ashmolean, Broderip's, now in the British Museum, and Colonel Sykes's in York Museum. In the first part of his paper to the geologists in November 1838 he concluded that these definitely were Secondary mammals, and he singled out the newly discovered numbat *Myrmecobius* as the closest living marsupial (see fig. 7.9). A specimen from the Swan River settlement in Australia had only recently been described by the curator of the Zoological Society's museum, George Waterhouse. Waterhouse had portrayed the long-jawed numbat as an insect-eating, shrewlike marsupial. With nine molars in each jaw, this animal "decisively" proved in Owen's view that the Stonesfield jaws belonged to true mammals.[108] But he encountered heavy resistance at Somerset House. His paper elicited "a protracted and brilliant discussion," according to the *Athenaeum*, and "the result was more favourable to the views of M. de Blainville than we were prepared to expect."[109] Grant's was undoubtedly the main voice raised in opposition. The other known skeptic, the Zoological Society secretary William Ogilby, was more undecided than opposed and was content to list the pros and cons of a marsupial relationship. Even then, he later told Owen (no doubt with a modicum of hindsight), he had expressed his anti-mammalian objections "more strongly than he intended."[110] So the main opposition rested with Grant, and its force stemmed from his commitment to the "'progressive' theory," as the *Athenaeum* dubbed it.

Buckland now busied himself canvassing influential parties to attend

107. Owen to Whewell, 11 February 1839 (TCL WW Add. MS a.210[55]); Broderip and Owen 1851–52:39; Laurillard to Owen, 12 October 1843 (BMNH).

108. R. Owen 1841b:57; Waterhouse 1841, read 13 December 1836. The importance of colonial imports in raising London comparative anatomy to Parisian standards cannot be overemphasized. Blainville, Valenciennes, etc., were still not using the numbat in their 1838–39 studies, relying, like Cuvier before them, on museum specimens of the opossum.

109. *Athenaeum*, 1838, no. 578:841. I assume that Grant was present. The minutes only record those occasions when a guest was introduced; hence he is recorded as signing in guests on 5 and 19 December, and 9 January 1839: Ordinary Minute Book 9 (GS).

110. R. Owen 1846a:37; Ogilby 1839:23.

Fig. 7.9. The numbat *Myrmecobius* from the Swan River settlement in western Australia. The shrewlike marsupial was first described by George Waterhouse in a paper to the Zoological Society in 1836. (From Waterhouse 1841, pl. 27)

the reading of the second part of Owen's paper on 5 December. He approached Lord Brougham, sending news of his own visits to footprint sites, and alerting him to the coming "paper by Mr Owen on Blainvilles Botheratiotherium," and suggesting that he might like to dine with "Sedgwick Darwin Greenough Murchison Lyell & some more of the élite of the Society" before the meeting.[111] Brougham had long patronized the gentlemen geologists (he had obtained Sedgwick a prebendary at Norwich in 1834) and corresponded with Buckland. As canon of Christ Church, Buckland had sought his lordship's aid in obtaining a second living and (unsuccessfully) the Radcliffe librarianship.[112] Buckland also acted as intelligence gatherer, sending paleontological papers and passing on news of discoveries. Like Sedgwick, he had corrected Brougham's own works on natural theology and persuaded him to subscribe to books such as Louis Agassiz's *Fossil Fishes*. Thus Buckland was in a good position to advise Brougham of important Geological Society debates. Actually the omniscient Brougham was *au fait* with the Stonesfield debate. His own *Dissertation*

111. Buckland to Brougham, 26 November 1838 (UCL HB 20,100).
112. Buckland to Brougham, 14 June 1832 (UCL HB 20,098); 28 January 1834 (HB 46,563); 2 February 1834 (HB 46,809). Sending intelligence: 26 November 1838 (HB 20,100, 20,101); 23 December 1838 (HB 20,104); 26 March 1839 (HB 20,166); 29 February 1835 (HB 20,099) on Agassiz's *Fossil Fishes*.

on Subjects of Science Connected with Natural Theology (1839) dealt extensively with the implications of "fossil osteology" for Paleyite religion.[113] But one can understand Buckland's eagerness to entice his lordship if a political coup was in the offing.

The second part of Owen's paper was not in fact ready by 5 December and he did not attend that week. After the meeting Buckland dropped him a letter:

> I will thank you to inform me as soon as possible whether you will have ready your paper on Broderip's Stonesfield jaw by the next meeting of Geol Soc on the 19 as I know Lord Brougham is interested about the Botheratiotherian & if your paper will be read I will invite him to the meeting. Your absence was felt at the meeting of the 5th when we had the Chirotherium footsteps which I believe after all to be Reptile. Sir P Egerton will have told you why. Dr Grant seemed disappointed that he could not differ from me. I think it desirable for the sake of everybody both in London & Paris to put the Marsupial Character of the Stonesfield beasts beyond all doubt as speedily as possible especially after what Grant has published in the Annuari of Bailliere [i.e., Grant's *General View*].[114]

Owen, staying at Egerton's country manor, had also glimpsed an advance copy of Grant's *General View*, with its evidence for a reptilian *Amphitherium*.[115] Owen's second installment was now scheduled for 19 December, and Buckland again invited Brougham "to witness our Skirmish," when "we hope to give the coup de grace to those who would make a Reptile of this highly organized, tho early Representative of the mammals."[116] So there is no doubt that Grant was the target and that "skirmishes" were anticipated and materialized, with the "elite" geologists ranged against him.

The second paper provided more evidence to back up the Owen-Buckland theory, and Broderip's shorter-jawed *Amphitherium Bucklandi* Owen now renamed *Phascolotherium*, removing Blainville's equivocal term. There remained the problem of the supposed reptilian traits of the American *Basilosaurus*, which supported the saurian hypothesis. At the meeting on 19 December, Buckland and Owen evidently conferred on this point, because Buckland wrote on 4 January 1839:

> Our last talk about old Nick has, as usual produced his Horns. The first thing I did on my return from the last meeting of the G.S. was to begin a paper to show that

113. Brougham 1839, 2:113–242.

114. Buckland to Owen, 11 December 1838 (RCS MS [1] a/6).

115. Owen to Buckland, 12 December 1838 (UMO WB): "I saw Baillieres Annual at Oulton: some pages were printed a little too soon."

116. Buckland to Brougham, 14 December 1838 (UCL HB 1957).

Basilosaurus was a true aquatic mammal, commencing with a statement of my reasons for entering on this, questioning the use lately made by Blainville & Grant of Dr Harlan's paper to support their notion of the Stonesfield Mammals being Reptiles.[117]

The Quaker Richard Harlan was a skillful paleontologist, if rash at times. He was ambitious, distributing copies of his *Medical and Physical Researches* (1835)—with its lengthy description of the *Basilosaurus*—to the London savants in order to promote his discoveries.[118] Early in January 1839, in the midst of the controversy, he arrived in London carrying basilosaur bones from the Alabama plantations. This gave metropolitan geologists an opportunity to examine the fossils firsthand. Buckland, presumably cued by Owen, had already made up his mind that the American "saurian" was in fact a fossil whale. He now pointed out to Owen the advantage in persuading Harlan himself to recant publicly, knowing that nothing would so decisively swing the vote. Failing that, Buckland offered a number of alternative strategies:

If the bones are on the table & I presume they will be some sort of Notice sh^d be read in order to draw attention to them, & get the fact recorded in the Report. The best thing w^d be 2 or 3 pages of recantation by himself, if you have convinced him of his error. The next best thing will be a short paper by you founded on the specimens you have examined, the 3d alternative will be a statement of my reasons for dissenting from his published paper. Till I hear from you I will make no further progress with what I have begun. I sh^d do little more than state in writing what I uttered in words at the meeting, when Stonesfield beasts were on the Tapes. We had better be guided in all this by Dr. H's own feelings. Do persuade him if you can to sign his own recantation it will be the most agreeable to his feelings, most honourable & most influential way of setting the matter right.[119]

Buckland's intention was transparent. It was to turn the basilosaur bones against those who had appealed to them in support of the "'progressive' theory" and to ensure maximum publicity for the rout. He added a postscript:

117. Buckland to Owen, 4 January 1838 [1839] (RCS MS [1] a/19). The "elite" accepted Owen's evidence for the fossil's marsupial nature as conclusive. Darwin wrote to Lyell on 14 September 1838: "I suppose Owen has pointed out to you the internal process in the Stonesfield jaws, which amongst Mammalia, is exclusively confined to the Marsupiata" (F. Burkhardt and Smith 1985–86, 2:106).

118. Gerstner 1970. Harlan also sent boxes of fossils to Murchison and had the imperial geologist elected to his own Geological Society of Pennsylvania: Harlan to Murchison, 18 May 1832, 11 April 1834 (Murchison Corres., GS). John Le Conte to Swainson, 11 May 1828 (LS WS), calls Harlan "very rash and inconsiderate." A. H. Harlan 1914:335.

119. Buckland to Owen, 4 January 1838 [1839] (RCS MS [1] a/19).

Some how or other we must contrive to anticipate on the 9th what will otherwise be done with the Basilosaurus at Paris & promote him to the Rank of a mammal & get the promotion gazetted in the Report of the Geol. Society.

The ruse evidently worked. Although Harlan visited the society as Grant's guest on 9 January, the statement he then read must have disappointed Owen's opponents. Harlan announced that he had originally believed the bones to be those of a marine carnivore, but then a study of the jaw had convinced him of their reptilian nature. Now he credited Owen with finally having solved the problem. Owen had persuaded Harlan to let him section and examine the teeth under the microscope at the College of Surgeons. And at the society he now followed Harlan with a paper on the *cetacean* nature of the bones and jaw, dismissing this last vestige of support for the "Saurian hypothesis."[120]

Formerly historians have seen this episode as unproblematic. Looking only to the published papers, they stripped away the social framework, ignored Grant's input,[121] and attributed Owen's success to his more "correct" identification. But correctness is an anachronistic evaluation. Take into account the cultural alignments and we can see that something more was at stake. At its starkest, the political protagonists were perceiving anatomical characteristics in divergent ways. But is this so surprising? As Jacyna reminds us, a microscope does not present a privileged close-up of reality so much as a set of images that await interpretation. The ability to interpret requires a period of training within a cultural tradition, and it is this educative process that supplies the social dimension to perception: social prestructuring allows meaning to be extracted from the magnified image.[122] The protagonists saw the jaws in ways that reflected their stand on serialist science. There was considerably more at stake than simply correct interpretation. Buckland's and Owen's machinations stemmed from their sensitivity to the Lamarckian-radical threat. As defenders of the moral values implicit in corporation-Anglican science, they were championing a mammalian *Amphitherium* for conservative ends. And their "victory" now meant that the marsupial diagnosis was embedded more firmly by the geologists in their Peelite strategy. While the Anglican gentlemen and divines ruled science, the diagnosis remained secure. Only when their polite, gentlemanly ethos gave way a generation later to

120. R. Owen 1841a:69; R. Harlan 1841:67–68, read 9 January 1839. As Grant's guest: Ordinary Minute Book 9, 9 January 1839 (GS).

121. Even though Owen, rehashing the Stonesfield debate in his books (1846a:37–42, 1871:13), made it quite plain that Grant was the leading protagonist.

122. Jacyna 1983c:76–80. Rachootin 1985:155–74 for an excellent study of how an anatomist such as Owen could "read" another fossil, the camel-like *Macrauchenia*, quite differently from the naturalist Charles Darwin.

a new bourgeois professional ethic in Britain and America were the Stonesfield animals again to be reconceptualized (as generalized, submarsupial mammals).[123]

Previously Owen has been portrayed as acting in "consultation with Harlan" in the interests of international cooperation.[124] But the letters suggest that Owen and Buckland conspired to present just such an image. Of course, Harlan's change of camps might well have been an attempt to gain greater recognition for American efforts. But Owen and Buckland's original intention was far from furthering diplomatic relations or creating a "transatlantic" science. They had contrived to present the society dissidents with a fait accompli—Harlan, the radicals' guest, recanting at his first appearance. It was less an exercise in international goodwill than an attempt to regain the coopted fossils and turn them to antiserialist ends. With science integrated into wider political strategies, it was essential to weaken the paleontological base of serial transmutation—to show that there could be no capitulation to secular extremists. Grant's role in the affair underlines the radical interest in retaining a progressive series governed by Lamarckian and Geoffroyan laws. Acknowledging his part and the contingent interests enables us to paint the episode onto our broader political canvas. The debate at this level was about science as an arbiter of authority. The clerisy needed success to vindicate its anti-Lamarckian approach and legitimate its control on science. Lamarckism with its democratic consequences had to be shown to be scientifically bankrupt. "Neutral" nature was being made to speak out against radicalism. And Buckland, intent on capturing hearts and minds, on "gazetting" Owen's triumph, made sure she was heard.

The journals in 1840 conceded that the marsupial character of the Stonesfield jaws was now "generally admitted."[125] Owen also claimed in the early 1840s that a nearly complete skeleton of the fossil whale (renamed *Zeuglodon*) discovered in Louisiana "fully confirmed" his earlier deductions.[126] How Grant reacted to Buckland's coup is difficult to say. Caroline Owen recorded in her diary on 10 January that "Dr. Grant was obliged to admit, in spite of his teeth, that they were mammalia and not saurians."[127] But, not being there, she was only repeating what Richard had told her. In truth Grant always doubted Owen's clinching arguments (for example, that the posterior end of the jaw was inflected inward, mak-

123. Desmond (1982:43–44, 200–201) takes up the story and looks at the later reinterpretation.
124. Gerstner 1970:147.
125. "Owen on Odontography," *BFMR* 1840, 10:211.
126. R. Owen 1840–45, 1:360.
127. R. S. Owen 1894, 1:152.

ing it characteristically marsupial), and he never accepted that these animals were marsupials. Even as late as 1853, in his Palaeozoology course, although he admitted that *Amphitherium* might have had mixed mammalian and reptilian features, he still saw its specifically marsupial character as "without anatomical proof."[128] Clearly, the cultural dimension of established facts meant that many outsiders simply could not accept them.

From now on there was open hostility between Owen and the radicals, with perennial accusations of literary and, on one occasion, almost literal theft. In 1839 he was already locked into dispute with the dentist and geologist Alexander Nasmyth over the discovery of dentine in the teeth. The argument was drawn out, debilitating, and public, with Nasmyth's priority upheld by the Grant-Wakley faction. One can understand Owen's resoluteness, because his alleged plagiarism of Nasmyth's microscopical work was as usual given political color by the *Lancet*. It placed Owen's action on a par with Home's, seeing his filching of the "toil-gotten facts" from the real "workers" in science as an illustration of BAAS corruption.[129] This petty vindictiveness was typical as relations soured; the clerisy closed ranks, and Buckland urged Owen to publish swiftly, "for there are living as well as fossil Sharks with prodigiously voracious teeth."[130]

The menacing sharks refused to move off. By 1841 Owen was mired in another dispute with Grant, this time over the ownership of a crate of zoological oddments shipped from Hobart by Grant's former pupil Edmund Hobson. While at the university in 1837, Hobson had donated skeletons sent by his brother in Australia to Grant's museum. But before sailing to Tasmania himself, he also questioned Owen on the local fauna and promised him specimens. Unaccountably, his first shipment to Grant in 1841 was readdressed en route and ended up at the College of Surgeons, where Owen naturally claimed it, and the dispute was only resolved when Grant produced witnesses who testified to a switching of the labels aboard ship.[131] Tempers flared yet again the following year, in an argument over

128. R. E. Grant, Palaeozoology Lectures, f. 185 (BL Add. MS 31,197). Geoffroy's attitude to the discovery of "marsupials" in strata so old was consistent with the line he had taken against Owen on the monotreme question. He simply relegated marsupials with monotremes to a lower submammalian level (Appel 1987:184).

129. "Mr. Nasmyth and Mr. Owen," *L* 1839–40, 2:376–78; also 1840–41, 2:841–43. Nasmyth 1839:xi; bound with a copy of this in the Smith Woodward Library (Zoology Library) at UCL are transcripts of relevant letters and Owen's rejoinders. This is possibly Owen's personal copy. See also Nasmyth 1841:iii–xvi for his treatment by the BAAS Council. Also Owen to Buckland, 12 December 1838 (UMO WB).

130. Buckland to Owen, 24 February [1839] (RCS [1] a/19).

131. See Grant's letters to C. C. Atkinson through April and May 1841, and Owen's responses (UCL CC). On Hobson's gift to London University: Grant to Atkinson, 19 April 1837 (UCL CC 3968). Hobson was enrolled in Grant's classes in 1836–38 and his gold medal

a fossil elephant, the Missourium, mounted in the Egyptian Hall in Piccadilly. This episode is of more interest in showing the practical problems that Grant, as a poorly paid lecturer, was now facing. When the skeleton arrived from America it caused laughter among the knowledgeable because it was so "miserably put together."[132] Owen diagnosed it as *Mastodon giganteum*, but Grant and Nasmyth considered it a different genus, *Tetracaulodon*. At the Geological Society in 1842 there were angry exchanges between Owen and Grant over the issue.[133] But the society was now very much Owen's terrain. Its managers laid the blame for this incessant bickering at the feet of menials envious of his achievements. Even Owen was adopting the supercilious attitude for which he was to become famous, telling Whewell that "in London there are never wanting men whose jealousy augments at each fresh evidence of successful labour in a contemporary."[134] The society published Owen's paper in its *Proceedings*, but only printed Grant's as a one-sided abstract. Yet during the months of the Missourium affair, Grant had completed an eight-part, two hundred–page monograph on the mastodons, discussing their development, structure, and dental changes and tackling the vexed question of the differences between the genera in great detail. Yet it proved "too long" to be read.[135] After the society refused it, publishers were probably wary of taking it, and Grant was in no position to print it himself. (The estimate for printing Owen's long memoir on the fossil ground sloth *Megatherium* a decade later was £700.) So at the very least, Grant's loss of society support and his teetering finances were affecting his ability to publicize his science to the extent that Owen now was.

Publicizing the Evidence against Continuous Ascent

See how the greatest—am I wrong in calling him so?—of the British disciples of Cuvier walks among the shattered remnants of former worlds, with order and arrangement in his train. Mark how, page after page, and specimen after specimen, the dislocated vertebrae

winner in 1837–38. But see also Hobson to Owen, n.d. (Cabinet VIII [1] a71, RCS). On Owen's acquisitiveness: Dobson 1954:112, 116; Woodward 1893; Desmond 1982:198; and M. Benton 1982.

132. Curwen 1940:150; K. Lyell 1881, 2:59.

133. Curwen 1940:159; Gerstner 1970:138–41. Grant introduced Albert Koch, Missourium's owner, to the GS: Ordinary Minute Book 10, 18 May 1842 (GS).

134. Owen to Whewell, 5 November 1842 (TCL WW Add. MS a.210[68]).

135. "Biographical Sketch," *L* 1850, 2:691; Grant 1842. On the £700 cost: Desmond 1982:28.

fall into their places,—and how the giants of former days assume their due lineaments . . . all alike bearing the indelible marks of adaptation to the modes of their forgotten existence, and pregnant with the proofs of wisdom and omnipotence in the common Creator.

—Lord Francis Egerton, testifying to the ideological correctness of
Owen's fossil reconstructions[136]

After Owen's Stonesfield victory, his funding was stepped up by the gentlemen geologists, who now controlled the coffers of the British Association for the Advancement of Science (f. 1831). The BAAS hierarchy largely overlapped with that of the Geological Society; it was again a comfortable coalition of gentry and prosperous middle classes. Morrell and Thackray picture the association as an incarnation of Coleridge's clerisy, seeing it promote the sort of science that could guarantee "God's order and rule."[137] In a turbulent age it was designed to foster "national identity, common commitments, and a continuous acceptance of the leadership claims of traditional rulers." To legitimize these claims, its wealthy managers projected an image of the unbiased observer—the "scientist," as Whewell now called him—who could appeal to neutral nature for evidence. And it was as a leading association "scientist" that Owen was henceforth to be cast.

The association's organizers distrusted radicals, disliked the artisans (whom Murchison tried to bar from meetings), and shunned disreputable sciences (phrenology, Lamarckism, Geoffroyism) because of their reformist pretensions. They encouraged responsible science, the sort that ensured the ideals of duty and subservience. Owen's productivity, Coleridgean standard bearing, and astute antitransformism met their criterion exactly. Nothing testifies so eloquently to the managers' expectations as their allocation of funds. The association's profits had soared by 1837 as a result of a huge rise in the membership, and Owen was perfectly placed to reap the reward. His grants—£618 in all between 1838 and 1845—placed him among the top recipients.[138] His payments were the more impressive when one considers that the major awards were usually for practical, naval, or geographical work. In cash terms, the managers put his science on a par with the imperial and navigational research that was of such commercial benefit to a colonial nation. But then Owen's anti-

136. F. Egerton 1842:xxxv.
137. Morrell and Thackray 1981:96, also 11, 31. This was dignified as the "Parliament of Science" in Traill 1837:xlii; MacLeod 1981:17–24.
138. Morrell and Thackray 1981:319, 551.

Lamarckian anatomy was essential to the gentlemen's internal imperialism—their strategy to conquer the new worlds at home, the godless urban tracts and growing regions of industrial Dissent. Owen was restoring the "shattered remnants" of older worlds and putting the *embrouillés* fossils into a new anti-Lamarckian order, undoing the leveling damage of the fiercer democrats. His science had a social, cultural, and nationalistic appeal to the clerisy: social because of his support for the traditional rulers; cultural for his Coleridgean idealism, which could secure the hegemony of the National Church; and patriotic because, among a jingoistic elite, he was establishing Britain's reputation in the eminently French preserve of comparative anatomy—not, like the secularists and Dissenters, by importing French philosophical zoology, but by pursuing an independent British path.

The managers were quite unabashed in their partiality. Murchison, awaiting Owen's second paper on British saurians in 1840, advised him "completely between ourselves to give me such a *report* of the state of that branch of our knowledge as you would read if you were General Secretary commenting on *Professor Owen's Memoir*. . . . My soul object is to give you the widest possible extension to your views and *wishes*." [139] This injection of funds and encouragement to thrust his personal views forward amounted to a vote of confidence. It was professionally expedient for Owen to present his most damning paleontological case against Lamarck, Geoffroy, and their British disciples from the floor of the "Parliament of Science." This is what he did, and the size of his funding was reflected by the length of his report on fossil reptiles, the two parts of which took a total of five hours to read.

Some who encouraged his excursion into paleontology had explicitly nationalistic motives. Landed members, who knew from their Grand Tours the state of the Continental museums, were concerned with sovereignty, with British priority in natural history. With priority came prestige; thus it helped to whip up patriotism at home, so important in an age clamoring about "Old Corruption." This nationalistic message was spelt out in the 1842 presidential address by another of Owen's admirers, Lord Francis Egerton. He saw scientific discoveries bestow privilege. They "elevate the country in which they originate in the scale of nations, and gratify the most reasonable feelings of national pride." [140] It was foreign competition that worried his namesake Sir Philip Egerton. Sir Philip entertained the visiting savants at his Cheshire manor during the Liverpool

139. Ibid., 217; *Report BAAS 1838*, xxviii; *Report BAAS 1839*, xv.
140. F. Egerton 1842:xxxvi; Morrell and Thackray 1981:385.

meeting in 1837 and advised Owen to draft a report in order to stake his claim to the British fossils, before they fell to the expansionist Germans. As he later explained to Owen:

I had just returned from the Continent, where I had an opportunity of seeing what von Meyer, . . . Münster and others were engaged upon, and so confident was I that a vast field for original [research] in this branch of Paleontology was offered in our Collections, at the same time so fearful of the harvest being gathered by a Foreigner, and so anxious that you, for whom I have so great a regard, and whose Talents and discrimination I considered so [supremely] fitted for the inquiry, should have the fruits, that I felt myself impelled to take this step, and to follow it up by the second application at Newcastle, to enable you to undertake the preliminaries for your task.[141]

Count Georg Münster was well known to these gentlemen. Murchison had described him to Sedgwick as "the prince of fine, honest-hearted, intelligent travelled Germans" with a cabinet that was "the most instructive in Europe."[142] Münster offered this collection for sale (perhaps this is what attracted Egerton to Bayreuth in 1837); a duplicate series, "nearly as good as the first," went on sale in 1839 (Sedgwick persuaded the trustees of the Woodwardian Museum in Cambridge to buy it). The BAAS executives would have been in no doubt about the potential threat from collectors such as Münster and Hermann von Meyer; hence they were keen to see Owen seize priority.

So there were nationalistic reasons that Owen's report was sponsored, but his own motives for beginning work on the saurians were probably different. He hinted at them in his first notebook jotting on fossils. In an isolated comment, entered sometime before November 1834, he wrote: "fact wholly at variance with every theory that would derive the race of Crocodiles from Ichthyosauri & Plesiosauri by any process of gradual transmutation or development."[143] This is a clear reference to Geoffroy's five "Mémoires sur de[s] grands sauriens" (1833). These memoirs had been read to the Academy in 1830–31. In the fourth, Geoffroy investigated the environmental factors influencing the ichthyosaur's transformation via a teleosaur into a crocodile. Geoffroy emphasized the intermediate nature of the teleosaur's palatine and skull-roof plates and, while discussing the mechanism of change, praised Lamarck's laws.[144] The last

141. Egerton to Owen, 26 October 1840 (BMNH RO); J. W. Clark and Hughes 1890, 1:490.

142. J. W. Clark and Hughes 1890, 2:18.

143. R. Owen, Notebook 11 (1834–36), f. 1 (BMNH).

144. Geoffroy 1833:77; Grant 1839:40–41. Owen's notes on Geoffroy's 1824 terminology of the crocodile cranial bones are in RCS Cab. II (14).

memoir was read in August 1831 when Owen and Grant were together in Paris. This kind of theorizing was not peculiar to France. As the *Lancet* finished printing Grant's lectures in September 1834, he too was seen to be discussing the "metamorphoses" of fossil animals, while in his fossil course he praised Geoffroy's views and discussed the teleosaur's relationships. Geoffroy continued to be interested in marine reptiles. In September 1836 he set off for London, bringing Grant a paper on the hyoid bones and planning to search for teleosaur paddles in the Oxford rocks.[145] He also continued to promote transformism. And in 1837, during a debate with Blainville over the gigantic Indian fossil *Sivatherium* (believed by Geoffroy to be an ancestor of the giraffe), he made his famous claim that the age of Cuvier was passing.

Late in 1837 Owen set out on his tour of the major fossil collections at home and abroad. He documented all the British saurians, described new species, and devised new criteria for classifying them. Some of the best fossils were actually close at hand. The spectacular Lias saurians of the Glastonbury scriptural geologist Thomas Hawkins—"the Elgin marbles of fossil zoology"—had only been unpacked at the British Museum in 1835.[146] Those fossils that Owen did not go to see came to him; Buckland plied him with plesiosaur paddles and more besides, carrying the fossils up to London, or having Enniskillen deliver them.[147] Owen read his first paper at Birmingham in August 1839. Here he described twelve new species of plesiosaurs and six new ichthyosaurs, which put him in a position to refute the transformists on stratigraphic grounds. By looking at his conclusions, to the work he expected his fossils to do, we can see the grasp he already had on the problem. He wound up in 1841 using the distribution of these reptiles to show that Geoffroy's anatomical sequence ran counter to nature. "*Ichthyosaurus, Plesiosaurus* and *Teleosaurus* are . . . genera which appeared contemporaneously," he said, "one neither preceded nor came after the other," a fact that was impossible to square with the "transmutation theory." Moreover the ichthyosaurs could be traced "generation

145. Geoffroy to Grant, 10 September 1836, "Biographical Sketch," *L* 1850, 2:691–92; Bourdier 1969:55. However, *L* 1835–36, 2:787, has Geoffroy coming over for information on orangs. Geoffroy 1837:77.

146. Officers Reports, 1834, 16: f. 3737 (6 November 1834); 1835, 17: f. 3819 (12 February 1835) (British Museum Archives). Hawkins 1834. Grant calls them the "Elgin marbles" in *Report SCBM*, 135. But there were complaints that Hawkins had reconstructed a number of missing bones, and the museum had bought the fossils unaware of this: Curwen 1940:111; *Mag. Nat. Hist.* 1840, 4:11–44. This "discovery of a rather vexatious nature" caused consternation in the British Museum ("Officers Reports," 1835, 17: ff. 3818–19). R. Owen 1839b:44 lists the museums he visited. His notes on enaliosaurs are in Notebooks 13–15 (1838–39) (BMNH).

147. Buckland to Owen, 9 March 1838 (RCS MS [1] a/11).

after generation" through the strata without showing any signs of change, and they disappeared in the Chalk as suddenly as they had appeared in the Lias. There was no succession. Only if the species were pulled out of their chronological order could a sequence be artificially constructed. Allowing that, it would be "as easy as seductive to speculate on the metamorphoses" of an ichthyosaur into a crocodile, and we could investigate the "physiological possibilities of such transmutations." But it cannot be allowed; the ichthyosaur-teleosaur sequence was unnatural, and the fossils spoke against Geoffroy and his disciples. Since Grant too had inferred an undeviating succession of fossil life, Owen finished by quoting his *Lancet* lectures to show how "the transmutation-theory" in England was also being "supported by palaeontology." [148] Superficially, fossils might appear to lend support to this Lamarckian hypothesis, "but of no stream of science is it more necessary," Owen warned, "to 'drink deep or taste not.'"

Owen used these stratigraphic ranges to prove that species remained stable for immense periods. But this kind of approach lacked the wider cultural impact of his coup de grace. At Plymouth in 1841 he introduced the dinosaur, built from Buckland's *Megalosaurus* and other fossils. Owen used body size and the configuration of the sacral vertebrae to distinguish the dinosaur from other reptiles, living or fossil. He also cleverly shortened the formerly lizard-shaped megalosaurs—"reduced the sesquipedalities of some of your old friends 'with tails as long as St. Martin's Steeple,'" Broderip laughed to Buckland[149]—to effect a gross change. The resulting dinosaurs were modeled along "pachydermal" lines. He gave them an advanced cardiovasculature and a mammalian stance. Owen now exploited his creation to argue that the unabated Lamarckian ascent envisaged by ultraradicals was imaginary. The "Reptilian organization culminated" in these dinosaurs, with their quasi-mammalian physiology, he argued. If they "were on the march of development to a higher type," they would have "given origin" to the mammals themselves. But the Stonesfield opossums showed that the mammals originated independently; indeed, the *Thylacotherium's* "abrupt appearance" was contemporaneous with that of "the most ancient Dinosaur," which was "inexplicable" by any known law. The dinosaurs did not move up to a higher type at all; they died out, and the reptiles "subsided" into a "swarm" of small lizards. This evidence of degeneration demolished the Lamarckians' case for an unaided and inevitable progress. Though there had been a general ascent from fish to man, the lack of exact succession and the intermittent re-

148. R. Owen 1841c:196–97, 198–99.
149. Broderip to Buckland, 14 January 1842 (BL Add. MS 40,500, f. 247).

trogression gave the lie to radical notions of nature's own "self-developing energies," pushing life blindly upward.[150]

Why Owen's Strategy Was a Conservative Success

[Owen's] grand conclusion, so essential to science and our knowledge of creation [is] that in all these creatures [fossil reptiles] the structure was peculiar to each, and adapted to the condition of the earth at the time; that there was no gradation or passage of one form into another, but that they were distinct instances of Creative Power, living proofs of a divine will, and the works of a divine hand, ever superintending and ruling the existence of our world.

—*Literary Gazette* on one moral to be drawn from Owen's work[151]

The advantage of Owen's punctuated progression was that it could serve any number of conservative masters. It permitted a "general ascent," proving that Man was the culmination of Creative effort. This could be interpreted in Coleridgean terms as an expression of Divine Will in Nature. Alternatively, many at Plymouth would have seen the unevenness of the record as proof of a contingently active Creator, a caring father concerned with fitting each organism to its niche. It was certainly important not to alienate the Paleyite natural theologians, for Owen was addressing clergymen such as Sedgwick, Buckland, and Conybeare, each in his "double capacity as divine and saurologist."[152] To understand the value of his science to them we need to appreciate their limited anti-Lamarckian options in 1841.

The urbane lawyer and geologist Charles Lyell had offered them one rather extreme strategy. His *Principles of Geology* (1830–33) also enshrined the social values accepted by the clergymen; it too was an attempt to stop the Lamarckians from hijacking fossil history. Lyell denied the validity of the "physiological laws" employed by transformists to explain changes induced by domestication; laws which, in the context of fossil life, were supposed to account for the generation of new species and even the "production of man."[153] But he went so far in the opposite direction as to propose a fundamentally nonprogressionist life history. Indeed, in his effort "to preserve man's special place in creation," his whole geology had

150. R. Owen 1841c:201–2. Desmond 1979.
151. "The Fossil Reptiles of England," *Literary Gazette*, 14 August 1841, no. 1282:512.
152. Whewell 1832:117.
153. Ibid., 110, 113–16.

become affected. Lyell's nondirectional fossil sequence itself required a "nonprogressive series of climatic conditions," which led him to deny the idea of a steadily declining central heat in the planet.[154]

Lyell's science was a singular reaction to the threat of bestialization. Like his teacher Buckland, he feared that, with the adoption of a brute ancestry, man would be "degraded from his high Estate."[155] This is a revealing metaphor, for the gentlemen using it regarded themselves as the guardians of morality and culture. In fact, as the wealthy intellectual classes, they were the "high Estate"; it was they who stood to be "degraded." Cultured man to them was everything, whereas to the artisan cynics he was "nothing," his body destined to rot "like a dunghill" without spiritual trace.[156] This attitude shocked the rulers, and they despaired of atheists who toyed with transmutation and threatened to drag humanity down to this gutter level. Lyell's science was an attempt to save man's "dignity." Historians have long recognized this, but have interpreted his use of the term in light of his individual religious beliefs. The context however suggests that it encompassed a wider social dimension. Of course religion (or the lack of it) was a worrying point. Lyell saw Lamarck's "worshipers" deliberately adopting transmutation in order to dispense with an interfering God.[157] But the conditions under which they embraced it also tell us a good deal. During his trips to France in 1828–29 and 1830, Lyell saw a number of influential savants treading a dangerous path. He met former collaborators of Lamarck's, current sympathizers of Bory's, and republican recruits, notably the mollusk specialist André de Férussac, who were now hoisting their own tricolor flags. Lyell himself witnessed fighting in Paris during the July Revolution. He had actually written home about the immense ages that would be required for "Ourang-Outangs to become men on Lamarckian principles," after watching jubilant sansculottes rampaging through the streets outside his house. He was equally worried by events at home. While on the Continent he kept a constant watch on the election turmoil and "mob-rule" in Britain.[158] So Lyell in revolutionary Paris was exposed to materialist philosophy, fossil progressionism, and Lamarckian biology. Back in London he now witnessed the same academic trends, as teachers such as Grant explained the progressionist record as the result of a cooling earth and transforming creation. Lyell, who always preferred the decorum of King's College to the

154. Ospovat 1977:318–21; P. Lawrence 1978.
155. W. Buckland, "Species Change of Lamarck" file (UMO WB).
156. Carlile 1821:98, 132; Desmond 1987:87.
157. C. Lyell 1830–33, 2:18; K. Lyell 1881, 1:260; Bartholomew 1973, 1979.
158. K. Lyell 1881, 1:291–92, 308, 363; Corsi 1978:227 on Lyell's contacts in Paris.

unruliness of Gower Street, was well aware of Grant's position at the university.[159] But the threat came much closer to home. By 1832 Grant had brought his fossil zoology into the very heart of the Geological Society, when he joined the council for a season. So the year that Lyell published his anti-Lamarckian volume, a transformist had actually penetrated the geological gents' sanctum sanctorum.

Lyell penned this volume of *Principles* during the Reform Bill crises of 1831–32. He composed it on his father's Scottish estate. But he remained "distracted by the disturbed state of politics,"[160] not surprisingly, for his sisters had brushed with roving gangs of "reformers" and (as daughters of the Tory gentry) feared being stoned. This intimidation of his family could only have increased Lyell's sensitivity. Like other elite geologists, he feared the social degradation threatened by the sea of hoveled poor. Just as the artisans' "levelling doctrines" were designed to smash the barriers "between rank and rank,"[161] so their Lamarckian science would reduce man to lowly parentage. The emphasis on human dignity that informed Lyell's social, political, and religious thought now entered *Principles* to meet this Lamarckian threat. In a telling passage he pictured a credulous naturalist being seduced by the idea that animals and man have "one common origin." The tyro falls for this "continuous and progressive scheme of development," with a consequence that "he renounces his belief in the high genealogy of his species" and takes radical action to perfect man in this life without waiting for the next.[162]

The ruling gentlemen had of course a rather literal insight into man's exalted birth; after all, history for them was a continuity of noble descent, not a chronicle of working-class emancipation. It was a view hardly shared by the artisan agitators spouting d'Holbach and Paine. They were using a social Lamarckian science of progressivism, materialism, and environmental determinism to underwrite the change to a democratic, cooperative society. In the "new moral world," land and wealth were to be taxed or confiscated, and the squires and priests forced to prove that they had something "worth exchanging for the products of labor," as the Red La-

159. Lyell was familiar with Jameson's journal, the main organ for Grant's papers in 1825–27. He knew Fleming, Grant's Scottish patron, and Horner, who was introducing Grant to Geological Society members in 1829 (and whose daughter Lyell proposed to in 1831). Lyell was an inveterate university-watcher and had declined the geology chair in 1829. He was also aware of Grant's inability to attract a large audience at London University at this time (K. Lyell 1881, 1:176, 257, 397).

160. Wilson 1972:320; cf. Rudwick 1975:234ff.

161. As fellow Whig Adam Sedgwick said (J. W. Clark and Hughes 1890, 2:47).

162. C. Lyell 1830–33, 2:20–21.

marckian William Thompson put it.[163] The radical image was one of escape from humble origins—the reverse of the squires' insistence on man's high-born status. Lyell's passage seems clearly aimed at this radical rhetoric, with its promise of heaven on earth when rank and privilege were abolished. The peddlers of the deistic and Lamarckian sciences were undermining the Church, demoralizing society, and arming the artisans with intellectual weapons. His message was one of caution: the middle-class deists and corporation haters must beware of jettisoning man's noble pedigree and going over to the socialists and democrats. Utopian promises were no substitute for divine redemption, and sinister social consequences would follow the rejection of mankind's "high estate."

Though all the gentlemen were alive to the danger, most geologists remained opposed to Lyell's cosmogony. It seemed to defeat the purpose, for being nonprogressionist, it was essentially non-Christian, providing no reflection of a directional sacred history, stretching from the first Day of Creation to the Last Day of Judgment. Faced with Lyell's *Principles*, most clerical geologists followed Sedgwick's lead: they restated their belief "that the approach to the present system of things" had been progressive—and that a directional fossil history afforded proof that man was in God's mind at the outset. However much Lyell had "abused" the idea of a general ascent of life, Sedgwick still considered it sound and founded on "toilsome induction."[164] It was not that Lyell's colleagues failed to appreciate his anti-Lamarckian ploy, but that they refused to use such a "rash and unphilosophical" device to defeat the transformists. In some ways, the *Principles of Geology* actually increased their quandary. Like Buckland in his Bridgewater Treatise, caught between refuting Lyell's steady-state geology, yet offering nothing to encourage the progressive transformist, Sedgwick dismissed the "phrensied dream" of the Lamarckians, but hardly had kinder words for Lyell.[165] These liberal Anglicans were obviously receptive to any anti-Lamarckian strategy that could meet their progressionist specifications. They were all increasingly worried by events. Whewell believed that "many of the geologists of France" were turning to transmutation. Sedgwick saw Geoffroy's "dark school" gaining ground in England itself, depriving physiology "of its beauty and meaning."[166] Owen's patrons, Sir Philip Egerton and Sir Harry Inglis, had al-

163. W. Thompson 1824:238. Thompson (1826:250–54; Desmond 1987:94) used Lamarckism explicitly to legitimate a cooperative society, female emancipation, and an equal education program.

164. Sedgwick 1834a:207, 1834b:305–6. Buckland 1837, 1:44, 107, 312. Bartholomew 1973:281, 1976:167; P. Lawrence 1978:115; Rupke 1983: chap. 12.

165. Gillispie 1959:147; K. Lyell 1881, 2:36; J. W. Clark and Hughes 1890, 1:427.

166. J. W. Clark and Hughes 1890, 2:86; Whewell 1832:118; 1837, 3:578–80.

ready faced Grant during committee meetings on the British Museum in 1836. And Buckland had helped Owen see off the threat at the Geological Society. None would have had any doubts about the value of Owen's science. With a fine attention to the details of succession and order, he had designed a system that would pack the same antitransformist punch as Lyell's, yet within a progressionist context. It met their needs precisely.

The background of economic gloom and Chartist agitation in the late 1830s only added to their fears. When the BAAS convened in Birmingham in 1839, the town had only been incorporated a year, and in the first council elections (December 1838) the Tories had been routed—"mangled and minced," as a radical editor put it.[167] The Birmingham Political Union held effective control, distributing jobs to its radical friends, to the disgust of Tories. Peel was shocked at the appointment of one of the authors of the Charter to the mayor's court. The Chartists had shifted their national convention to the town in the summer of 1839, before a local police force had been organized. Socialists too converged on the center, distributing half-a-million tracts, urging cooperation, female emancipation, and the abolition of marriage and property. Metropolitan police were drafted in, but in July, a month before Owen read his report, the Bull Ring Riots had broken out. The week of the meeting itself was a "feverish quiet," with peace ensured by "men in green and men in red, police staves and cavalry sabres."[168]

Given the social uncertainties of these years, Owen's work could only have been welcomed as a powerful defense of God and the divinely instituted social order. The managers prided themselves on having drawn "forth the man of genius and worth."[169] But his "worth," his scientific respectability, was guaranteed by his social credentials. Duty for Owen meant more than exposing the errors of a pernicious science; it involved active participation in the defense of the realm. Like Broderip, sentencing Chartists in court, and Buckland, who as the dean of Westminster in 1848 was prepared to cosh any Chartists who broke into the Abbey, Owen acknowledged the need to police the radical masses. In 1834 he enlisted in the Honourable Artillery Company, an ancient and self-financing volunteer regiment, composed mostly of merchants and urban gentry, who attended its Finsbury headquarters on a part-time basis. The corps served primarily as a backup for the civil powers during the working-class demonstrations. After the emergence of organized Chartism in London in 1840, the regiment was periodically called out by the

167. Fraser 1979:88–89; Holt 1938:228–29; Royle 1974:62.
168. Morrell and Thackray 1981:252.
169. Murchison and Sabine 1840:xl–xli.

Fig. 7.10. A satire on both the pretensions of the BAAS savants displaying their mechanical wares (here automaton constables) and the police authority itself (note the jeering people). This 1838 cartoon accompanied a report sending up the "Meeting of the Mudfog Association." By George Cruikshank, 1838. (From *Bentley's Miscellany*, 4:209).

Home Office, and Owen, despite the crush of college work, continued doing tours of watch duty.[170]

For their part, many radicals distrusted the clerisy and even the scientific "Parliament" itself. Coleridge's romanticism might have appealed to the Anglican patriots, but it did little for the London secularists and radical Dissenters. The managers' dislike of Benthamites and lecture hagglers resulted in many London teachers, particularly from the Godless College, boycotting the BAAS. Some who shunned the first meeting later came round (for example, Lindley). Others steadfastly refused to attend. Tellingly, Grant approved the annual meetings of German naturalists, but he believed that the BAAS could never "conduce to the advancement of science."[171] Wakley attended the Dublin meeting in 1835, but he misjudged

170. Jean Tsushima of the Honourable Artillery Company provided me with details of Owen's enlistment. See also R. S. Owen 1894, 1:167, 321; Raikes 1878–79, 2: chaps. 8 and 9. Goodway 1982 discusses the demonstrations. Broderip was correcting Owen's papers while sentencing Chartists: Broderip to Owen, 13 March 1848 (BMNH RO). E. Richards 1987:161–67 explores the sensitivity of Owen's High Church readers to the working-class threat.

171. Grant to Mantell, 16 July 1850 (ATL GM 83, folder 44). On the German naturalists: Grant 1833–34, 1:93.

the association's potential as a force for liberalism. By 1840, however, alerted by Owen's BAAS-backed "literary depredation" (his growing fat on "the toil-gotten facts of more humble, and . . . more useful workers in the field of science"), Wakley too was condemning the "voluminous twaddle of the *friends*" of the association. He now castigated its leaders as a horde of "ichneumon" wasps, sucking their industrious fellows dry in order to arrogate to themselves the title of "British '*savans*.'" The political parceling of funds called forth his venom: "How truly do we want an honest and upright board of science in this country, one in which admission should be regulated by labours alone, at whose portals nepotism and worldly interest might knock in vain."[172]

It is not surprising that Owen's anti-Lamarckian gambit paid off. He was preaching to the converted, an eminent congregation which shuddered at mob atheism and considered materialism the wellspring of discontent. Owen offered a new grindstone to hone an old political ax. For this he was lionized. The concluding anti-Lamarckian part of his report in 1841 was "generally acknowledged" to be "The Great Paper of the Year." The *Literary Gazette* rapturously endorsed Buckland's winding up encomium, when he called Owen "*a worthy successor of Cuvier* and an honour to his country and its science!" The paper made "The Fossil Reptiles of England" its lead story a week later, devoting almost a third of the issue to Owen's account. First and last the *Literary Gazette* emphasized the moral of Owen's report—that "older forms of organized beings" had not "graduated into later forms" and were not their "productive cause."[173] Owen's vision of nature devoid of "self-developing energies" was, as the *Literary Gazette* said, "so essential to science"—to the conservative science fostered by the managers. Nature sanctioned no upward delegation of power; it supported no radical ideal of inexorable democratic advancement. Rather it justified a "descensive" spiral of power and deference to traditional authority.

Such was the reception of the reptile report that the managers in 1841 voted £250 for its publication and another £200 to enable Owen to tackle the British fossil mammals.[174] The glowing press these reports received greatly strengthened the association's antiradical hand. At the same time Owen's elaborately reconstructed evidence against a materialist "produc-

172. "Mr. Nasmyth on the Development of the Teeth," *L* 1840–41, 2:841. Cf. "Meeting of the British Association in Dublin," *L* 1834–35, 2:699.

173. "The Fossil Reptiles of England," *Literary Gazette*, 14 August 1841, no. 1282:519; "Sketch of Society and British Association," ibid., 7 August 1841, no. 1281:510; also 21 August 1841, no. 1283:546. Rupke 1985:246. Broderip considered Owen's reptile paper his "opus magnum": Broderip to Buckland, 14 January 1842 (BL Add. MS 40,500, f. 247).

174. *Report BAAS 1841*, xxii.

tive cause" in nature was to provide a fundamental resource, enabling managers such as Sedgwick and Whewell three years later to check that new menace, the *Vestiges of Creation*.[175]

175. Discussed in E. Richards 1987; Desmond 1982; Brooke 1977b. Owen still considered the appearance of high-born animals in the fossil record before their simpler relatives the best argument against *Vestiges*, and he continued to ply Whewell with information on newly discovered fossils, like the "highly organized" Triassic Dicynodons from South Africa, to refute the book: Owen to Whewell, 22 February [1845] (TCL WW Add. MS a.210[88]).

8

Embryology, Archetypes, and Idealism:
New Directions in Comparative Anatomy

Journalists reported Owen's first Hunterian course in 1837 under the rubric of "College improvements." Many were adamant that the endowment of his chair was part of the council's strategy to preempt more drastic government action following the revelations of Warburton's committee. But while the radical press came hatchet in hand, it was to Owen's credit and the council's relief that the lectures were so well received. As a former critic admitted, they made "a striking contrast with those morbid ossific compositions which immediately preceded them."[1] Because of the college's civic position, the socially eminent were well represented at the lectures and orations. That tight-knit Oxford coterie—Buckland, Broderip, Enniskillen, and Egerton—was in constant attendance. Stranger Oxford faces were sometimes seen, with Bishop Wilberforce even escorting the Anglo-Catholic Henry Manning.[2] This Church-and-state presence in the front pews always galled the radicals. They noted the absurdity of "soliciting the attendance of such personages as the Duke of Wellington, Sir Robert Peel, and the Bishop of London" at the yearly addresses. "The circumstance of the gallant Duke being absolutely *asleep* during nearly the whole [of Brodie's 1837] oration, is a sufficient proof of *his* interest in the discoveries of John Hunter."[3] GPs cast an equally jaundiced eye on Owen's parting encomium on the council a few months later; a toadying act in "very bad taste," one called it.[4] And to the complaints that Owen's opening lectures on the skull in 1840 were elementary, Wakley piped up

1. "Working of the Improvements of the Royal College of Surgeons," *LMSJ* 1837, 11:174–76.

2. Pollock 1887, 1:273.

3. "The Hunterian Oration," *LMSJ* 1836–37, 10:774–75.

4. After the council had forced him to decline the Fullerian chair of physiology at the Royal Institution: "Royal College of Surgeons. Conclusions of Professor Owen's Lectures on Comparative Anatomy," *LMSJ* 1837, 11:378–79. I do not know whether the students attending Owen's courses came primarily from the hospitals, or also from the university and private schools. Few of any description evidently attended in the early years. This *LMSJ* source records that at the finish of the first course the theater was "somewhat more than half full." At the opening of the 1841 course, too, the "attendance was exceedingly thin; there were few

that since Owen was *au fait* with "our science," he must have been re-
freshing the nobles on the bones in their heads.[5] Sarcasm still ruled the
Lancet.

But consolidation in science was becoming a priority for moderates by
the late thirties. Now Owen's scientific industry harnessed to antiradical
ends was beginning to convince them, as much as the churchmen, that
investment in this sector could guarantee social stability and scientific
growth. Owen's appeal was not only to Tories, but increasingly now to
wealthy liberals. By 1840 they were already distancing themselves from
the *Lancet* demagogues and moving closer to the new men at the college.
The reviews, for example, now rated Owen England's premier philosophical anatomist. They considered that his *Odontography* finally dispelled
the myth of Britain's backwardness fostered by "native malcontents or foreign detractors," and they congratulated the college members for "having
contributed to develope his genius."[6] Interestingly, that new liberal con-
tender, the *Medical Times* (f. 1839),[7] looked not to the bureaucratic Cuvier as its Continental ideal but to the "venerable" Alexander von Hum-
boldt as proof that "conventional distinctions of rank" were now yielding
"to the loftier claims of inborn power and greatness"—that "the aristocracy of birth and fortune" was bowing to the "aristocracy of talent." And in
this new meritocracy it thought that Owen, if suitably financed "by the
munificent hand of his Sovereign and her government," might outshine
them all.[8] Owen conquered increasing areas of London's liberal-Peelite
middle ground in the 1840s, while slowly loosening his dependence on
Oxbridge patronage—to the extent that, as Evelleen Richards notes, by
the end of the decade he was worrying such older divines as Sedgwick
with his talk of nature's lawful progression.[9]

Owen's anti-Lamarckism was more than a reactionary or obstructive
doctrine; it insinuated itself into all aspects of his thought and shaped
much of his theoretical science. Here I examine its positive value, and
show how it guided his formulation of a new archetypal morphology and

---

members of the council, and only two or three visitors": *L* 1840–41, 2:64. Rupke (1985:240)
suggests more sanguinely that not uncommonly the theater was "filled to overflowing." Certainly in the later 1840s, even by the *Lancet*'s reckoning, attendances had picked up.

5. *L* 1840–41, 2:110–11.

6. "Owen on *Odontography*," *BFMR* 1840, 10:211.

7. The *Medical Times* (1839, 1:3) set out to strike at "the mammoth evils, which, in the
shape of monopoly and misgovernment, choke and almost destroy the profession." But after
the RCS charter of 1843 it settled down to become the standard middle-class liberal organ
(one that recognized Owen's potential: Rupke 1985:256–57).

8. "A History of British Fossil Mammalia," *MT* 1845, 12:46.

9. E. Richards 1987:151, 164, 166–67.

led to a series of anatomical "generalizations" that his patrons promoted as the most securely grounded in biology.[10] This ideal science itself then attracted the new anatomical Coleridgeans. Ironically many of these came from Knox's class, and included such students as John Goodsir, Edward Forbes, and Thomas Wharton Jones—all religious, all antitransmutationists, all enamored of the anatomical "Divine Idea," and all emasculating the master's savagely radical anatomy.[11] For them Owen's idealism provided a perfect countermeasure. In effect, his anatomy became a liberal-conservative paradigm in the mid-1840s; it provided a bulwark against the infidel relics of *idéologue* philosophy, against Lamarckian determinism, Unitarian pantheism, and the sectarian zoologies of the dying private schools. How Owen's antitransformist ideology helped structure this sophisticated rival morphology is the subject of this chapter.

Owen's Hunterian Lectures of 1837 and His Use of von Baer's Embryology

The Progress of knowledge will, of course, be impeded in proportion to the influence and popularity of those Teachers who for the sake of the small and transient reputation, gained by exciting the wonder of their hearers, or readers, advocate the baseless speculations of the transcendentalists, and indulge in exaggerated expressions of views to the development of which they are unable or unwilling to lend the co-operation of honest and unbiassed labours.

—Owen in his fourth lecture, 9 May 1837[12]

A characteristic of Owen's courses from the beginning was their denial of recapitulation and advocacy of Karl Ernst von Baer's rival embryology of divergence. In the 1970s Ospovat first broached the subject of Owen's use of von Baerian embryonic development "as the analogical basis for a new paleontological theory,"[13] one in which fossil lineages diverged from a common archetypal forerunner (rather than forming a linear series). More

10. Broderip and Owen 1853; Ospovat 1981: chap. 5.

11. Godlee 1921:102; W. Turner 1868, 1:24–27, 31–33, 119–21, 155–58; Mills 1984:380–85; Secord 1985c:192.

12. R. Owen, "Hunterian Lectures," Lectures 3 and 4, 6 and 9 May [1837], f. 97 (RCS, 42.d.4).

13. Ospovat 1976:2; R. Owen 1851. More generally on the changing ontogenetic concepts, see Gould 1977:52–63; Oppenheimer 1967:221–55, 295–307; Lenoir 1982; E. S. Russell 1916. Jacyna 1984a provides an indispensable study of the professional advantages of higher anatomy and embryology. Desmond 1985b and E. Richards 1987 for specific attempts to relate Owen's embryology to the social situation.

recently Richards has revealed just how tenaciously Owen defended "his property rights to the law of divergence"—how fiercely he upheld his priority claim against Carpenter and the French comparative anatomist Henri Milne Edwards. At the same time he virtually ignored Martin Barry, who had studied embryology in Berlin, Erlangen, and Heidelberg and in fact had done more than anyone to bring von Baer to the British.[14] There was nothing new in Owen's defense of what he considered his scientific real estate; his professional life was to be dogged by disputes and engulfed in an air of ill will because of the ham-fisted way he went about staking claims. However, on reading the manuscript of his course we can appreciate why he put such a high value on this particular piece of embryological "property"—why he was so receptive to von Baer's laws. In the lectures he persistently linked the issues of transcendental anatomy, embryology, and antitransmutation; his emphasis leaves no doubt that there was a strategic anti-Lamarckian payoff in the idea of embryonic divergence.

There is a rich irony in Owen's priority fights. Judging by the way he cannibalized Barry's articles and disgorged pieces whole in his 1837 lectures, he probably first learned of von Baer's embryology and its implications from this source. Writing in the *Edinburgh New Philosophical Journal* in 1836–37, Barry had deplored British "ignorance" of the great "*German* enterprise" in embryology. Our naturalists were still obsessed with the "twigs" of life, he said, and unaware of the embryological "branches" and "trunk that gave them forth." Even Geoffroy's emphasis on adult anatomy betrayed his poor appreciation of embryology. The new German "History of Development" was Barry's forte—especially von Baer's law, which stated that "a heterogeneous or special structure, shall arise out of one more homogeneous or general." Barry was so steeped in Baerian science that he even refused to use the terms *higher* and *lower* because they presupposed the old "'ascending' or 'descending' scale" of nature.[15] Von Baer had exploded that notion. Embryonic life was one of specialization away from a more general primordial germ; it was a process of differentiation and separation. As Barry pointed out, this allowed us to reconceptualize the unity of structure, moving it away from Geoffroy's older notion. This is undoubtedly what attracted Owen. Passages from Barry's papers—and not only Barry's, as we shall see—turned up verbatim in Owen's first lecture series. (This itself speaks volumes about his

14. E. Richards 1987:139–42. On the Carpenter priority dispute: Owen to Carpenter, 22 October 1851 (RCS RO 3: f. 366); Broderip and Owen 1853:55; Desmond 1982:92.

15. Barry 1837b:117–18, 139, 1837a:362; Jacyna 1984b:19. Also Ospovat 1976:10, 1981:130.

front-bench audience—old surgeons and civic functionaries. Few would actually have slept through his lectures, but most had little interest in Continental anatomy, and they were unlikely to detect the young professor's time-saving appropriations.) What did distinguish Owen's lectures was his characteristic use of Barry's works. He openly exploited them for antitransformist ends.

To understand the impact of this new morphology for Owen, we should recall the salient features of the existing science in London—the sort taught in the university and some private schools. It rested on three interrelated factors: a lineal progression of life from monad to man, a "unity of composition" holding throughout the series, and a recapitulation of the sequence during fetal development. A repetition in the human embryo of the adult forms of lower animals actually implied a single series hierarchically arranged; thus recapitulation was closely identified with lineal progression. Until the mid-1830s almost all the philosophical anatomists in London and Edinburgh were recapitulationists.[16] Some believed, following Geoffroy's disciple Etienne Serres, that the "formative force" in the fetus of the lower animals was less powerful. It allowed only a limited development of the organs, leaving the adults on their lowly rung.[17] In higher animals the force was stronger, causing embryonic growth to continue, advancing them farther up the scale. And monstrosities resulted when the force was pathologically weakened—resulting in the newborn or its organs being retarded on a level equivalent to some lower animal.

While following Serres, no London teacher (with one problematic exception) actually accepted a literal identity of the embryonic human organs with their adult counterparts among lower animals. Sharpey, for example, was careful to emphasize that these organs "*correspond*, not that they are identical."[18] The most favored word was *analogous*. And even then, higher anatomists made it clear that "this analogy is only seen between individual organs, not entire beings,"[19] and given the differential

16. This was true of Bennett, Grant, Quain, King, Anderson, and Grainger in London; and Fletcher, Sharpey, and (at this time) Knox in Edinburgh. For some early statements: Quain 1828:26–27; Grainger 1829:81; J. R. Bennett 1830:14–21; Grant 1833–34, 1:89; King 1834:7–8. Grant was also examining his students on recapitulation, questioning them, for instance, on the relationship of the circulatory system in embryonic mammals to its permanent form in the lower classes: *University of London. Questions on Comparative Anatomy. Saturday, January 8, 1831* (question 25), bound in *Grant on Zoological Subjects*, College Collection DG 76, UCL. Knox later developed rather idiosyncratic views (E. Richards 1988).

17. Serres 1830:48; Gould 1977:49; E. S. Russell 1916:81.

18. Sharpey 1840–41: 491; Anderson 1837:xvi; Quain 1828:27.

19. "Saint Hilaire's Treatise on Teratology," *BFMR* 1839, 8:5; "On Philosophical Anatomy," *MCR* 1837, 27:86; Quain 1828:27.

rate of development of the respective organs, as Bennett had argued, no embryo can ever "resemble in its *totality* an animal of a lower class."[20]

But the transformists' (rather ambiguous) language shows that they might have been prepared to go further and identify this lineal recapitulation with a literal ancestry. Geoffroy, in his major work on the environmental causes of fetal and fossil transmutation in 1833, wrote that an amphibian "is at first a fish under the name of tadpole," and he saw the process as a "transformation of one organic stage" into a superior one.[21] In the same tone and at the same time Grant talked of the "transient" organs of the human embryo as "repetitions of the permanent forms . . . in inferior classes."[22] What he meant by "repetitions" is highly problematic, as is Geoffroy's talk of embryonic transformations.[23] However, just as Geoffroy said that a frog is at first a fish, so Grant's language also implied a literal ancestral replay. Take his talks on Geoffroy's science to the fashionable audiences at the Royal Institution. At Faraday's request, Grant delivered ten Friday evening lectures between 1833 and 1841.[24] As open lectures preceding the soirée, they were supposed to be of an "easy and agreeable nature."[25] Typically, his were uncompromising and covered cherished themes: recapitulation, the progressive complication of organs, Geoffroyan vertebral osteology, and so on. As with all soirée talks, they were well attended.[26] Even a lecture on Geoffroyan esoterica, for instance that given on 6 June 1834 explaining "the Development of the Vertebral Column," had an audience of almost four hundred, though this was probably more a function of the soirée to come than its osteological hors d'oeuvre. It does however show the exposure Geoffroy's views were getting. Coming to the development of the nervous system (in February 1834), Grant certainly left the impression that he meant a literal recapitulation, for he too is reported to have "amused his audience . . . by informing them that at one period of foetal life, the brain of the future man is that of a tadpole, at another that of a fish, and subsequently that of a crocodile."[27]

20. J. R. Bennett 1830:12; J. Fletcher 1835–37, 1:15–16, 61–63.

21. Geoffroy 1833:82; E. S. Russell 1916:69.

22. Grant 1833–34, 1:89.

23. E. S. Russell 1916:69.

24. "Biographical Sketch," *L* 1850, 2:691; Grant to Faraday, 13 January 1837 (RI Faraday Folio II, f. 135).

25. Berman 1978:126.

26. Grant's Friday audiences ranged in size from 212 to 428 persons: data from RI Archives. For reports, *MG* 1832–33, 12:479; 1833–34, 13:927–28, 14:425–26; 1835–36, 17:831–32; 1836–37, 19:749–50; 1838–39, 23:840–41, 24:58–60. He also ran his stock extramural course on invertebrates here in 1834: Manager's Minutes 1832–53, 8: f. 303 (RI).

27. "Royal Institution," *MG* 1833–34, 13:927.

These glib statements by transmutationsists allow us to appreciate Owen's strategy in the 1837 Hunterian lectures. We will see how he used von Baer's law to curb this radical application of Geoffroy's "unity," destroying the foundation of the transformists' case. His ex cathedra pronouncements on the impossibility of transmutation in the lectures leave no doubt of his commitment to exposing the transformist associations of Geoffroyism. He was clearly worried by the extremists' use of transcendental anatomy. In trying to explain the conservativism of vertebrate structure,

some Anatomists, and especially those whose knowledge happened to be limited to a single great Division of the Animal Kingdom, were led to form Theories which are undoubtedly new, as regards their extravagance: assuming in these Speculations that Nature is restricted in the development of Animals to a supposed Unity of Composition;—a unity of Plan;—and a constancy of Connections; [*inserted:* & also a certain number and kind of component parts, which are all determined by an *a priori* theory] they have proposed most extraordinary Analogies.[28]

Three "extravagances" stood out in Owen's mind: the Geoffroyans' attempt to relate a lobster's shell rings to vertebrae, to establish an invertebrate-fish continuum, and to make a mammal's inner ear ossicles homologous with the opercular bones in a fish's gill covers. All of these examples reflected attempts by Geoffroy, Blainville, and their medical disciples on both sides of the Channel to construct a continuous series across Cuvier's supposedly discrete *embranchements*. Blainville had related what he called the "internal Osteozoa" (i.e., vertebrates) to the "external Osteozoa" (articulates) as early as 1816.[29] Like Geoffroy, Grant and his London pupils accepted a structural continuity between cephalopods and cartilaginous fishes. Within the Vertebrata itself, Blainville identified the fish's opercular plates with the reptile's lower jaw elements, while Geoffroy and Grant believed that these opercular bones were homologous to the mammalian inner-ear ossicles.[30] Owen had disputed the caphalopod-fish bridge as early as 1832 (in his pearly nautilus book). By the time he came to deliver his Barts lectures in 1835 he was also slating this operculum-ear analogy and invoking a rival adaptive explanation. He insisted that the bones of the gill cover were nothing more than dermal plates adapted "to the mode of Respiration peculiar to that Class."[31] In the forties, as his own

28. R. Owen, Hunterian Lectures 1 and 2, 2 and 4 May 1837, ff. 66–67 (BMNH MN 1828–41).

29. Appel 1980:301.

30. Ibid., 302–3; Geoffroy 1818–22, 1:15–30; Russell 1916:56; Grant 1833–34, 1:573–74; 1835–41:64–65. R. Owen (1846b:137–38, 1846c:232) cites Grant's Geoffroyan position on the opercular question.

31. R. Owen, Hunterian Lectures 1 and 2, 2 and 4 May 1837, f. 67 (BMNH MN 1828–41); R. Owen (1846b:137–38) recounts his 1835 views.

outlook became more rigorously homological, he was to abandon this line of reasoning.[32] Nonetheless, in the 1830s this was a typical ploy to discredit Geoffroy's wider analogies and make them inoperative in a transformist sense. The operculum-ear theory and similar strained relationships adopted by Geoffroy and his "followers" were, Owen said in 1837, "the result of an abuse of a sound and fruitful Principle [unity of organization], which has only suffered by an unwarrantable extent, and unjustifiable mode of its application."[33]

He finished his fourth lecture in 1837 with a review of the science since Hunter's time, which turned into an ill-concealed attack on these transcendental excesses. The Berlin physiologist Johannes Müller (on whose lectures Owen partly relied) had long "declared war" on *romantische Naturphilosophie* and mechanistic French physiology.[34] Owen too at this time shied away from the extremes of Green's *Naturphilosophie* and his uncritical support for German notions of the skeleton as "the repetition of a simple vertebra."[35] On the other hand, Owen was not prepared to follow Müller and denounce "the creative archetypes, the *eternal ideas* of Plato" as mere "fables."[36] He still accepted with Green that the elucidation of the Divine "Ideas, the archetypes and preexisting models"[37] on which the animal creation is based, was a major goal of science, and he still sought Coleridgean guidance on the meaning of life and law. He was however contemptuous of those German works in which "Metaphysical dogma" had run amuck, where "every part of a part must represent a whole," where "the head [is] a condensed representation of the intire [*sic*] body; . . . the nose is the thorax of the head; the jaws represent the extremities," and so on ad absurdum.[38] But his attack on the French was even more

32. By 1844 he was himself suggesting that the opercular bones were in fact diverging appendages of the tympano-mandibular arch, i.e., part of a cranial vertebra and therefore related to bones in other vertebrate skulls. W. B. Carpenter (1847:488) was much happier with this and actually considered it one of Owen's "most successful" homological determinations. Owen's admission of opercular homologies brought him much closer methodologically to the other morphologists. He still disagreed with them on details. (Not always by much: as Owen noted, Grant had floated the idea that the opercular plates might have been hemal arches, but of the occipital vertebra: Owen 1846b:138). Owen (1846c:231) now viewed this highly specific issue of the precise opercular affinities as "the chief battle-ground of homological controversy."

33. R. Owen, Hunterian Lectures 1 and 2, 2 and 4 May 1837, f. 67 (BMNH MN 1828–41).

34. Lenoir 1982:103.

35. Green 1840:57.

36. Müller 1837, 1:25.

37. Green 1840:xxv–vi.

38. Owen, Hunterian Lectures 3 and 4, 6 and 9 May [1837], ff. 94–95 (RCS 42.d.4).

pointed. After ridiculing those who postulated a single vertebral element "in every bone of the vertebrate body . . . in each ring of the Worm, and in every joint of the Lobster," he went on to criticize a priori determinations of cranial homologies, the seven-vertebra theory of the skull, the nine-element vertebral theory, and so on.[39]

He was not disputing the principle that the skull comprised modified vertebrae or that vertebrae were composed of standard components. Indeed, a succession of London teachers had proclaimed throughout the 1830s that the vertebral skull was "now entertained by all philosophic anatomists."[40] And in his own work on the vertebrate archetype Owen was to defend the notion of the skull composed of (four) cranial vertebrae. In the same way, concerning Geoffroy's theory of vertebral elements, reviewers had seen it as one of the best established doctrines, and Owen in 1839 was forced to agree.[41] No, what Owen was doing was singling out the dogmatic "Anatomist [who] from an *a priori* determination" was promoting a specific number. Recall that Grant was instilling Geoffroy's ideas on the seven-vertebra skull into his students at the university, while amusing more debonair audiences with them at the Royal Institution.[42] Since it was the Geoffroyans who opted for this configuration, Owen's attack on the seven-vertebra skull seems at first sight simple. But the situation was more complex. Owen implied that such "arbitrary" figures were suspect because they were not based on embryological researches—and he took this line because he wanted to endorse a particular sort of embryology for ideological reasons. Like Müller he was setting up embryology as "the final court of appeal"[43] in order to get a favorable verdict. He insisted that Geoffroy's "a priori" assumptions were "wholly unsupported by an examination into the primary formation of the cranial bones," which alone allow us "to determine how many [cranial vertebrae] are actually developed from the circumference of agglutinous *Chorda dorsalis;* the only true embryological condition of a Vertebrae."[44] Failure to appreciate the importance of embryology had even led Cuvier astray:

39. Ibid., f. 95; R. Owen 1839b:46.

40. W. B. Carpenter 1840–41, 2:57. All of London's higher anatomists accepted the theory of the vertebral skull. Turner in 1831 called it a "principle recognised by all modern authorities": *L* 1831–32, 1:188.

41. "Outlines of Comparative Anatomy," *MCR* 1835, 23:378; R. Owen 1839b:46.

42. "Development of the Vertebral Column," *MG* 1833–34, 14:425–26; Grant 1833–34, 1:539–41, 572–74, *passim;* 1835–37:57. R. Owen (1846c:232, 241n., 253) notes Grant's defense of Geoffroy's vertebral osteology.

43. E. S. Russell 1916:138.

44. Owen, Hunterian Lectures 3 and 4, 6 and 9 May [1837], f. 95 (RCS 42.d.4).

It was to obviate the retrograde tendencies of these Metaphysical or transcendental Theories of Animal Organization that Cuvier devoted his latest energies; and the abuse of the doctrine of Analogies perhaps led him by a natural reaction to under-rate its value.

The general laws of Animal Organization can never be developed from a consideration of the perfect or matured structure alone: and such *was* the general character of the knowledge from which Cuvier deduced his inferences.

The laws of Coexistence;—the adaptation of structure to function; and to a certain extent the elucidation of natural affinities may be legitimately founded upon the examination of fully developed species:—But to obtain an insight into the laws of development,—the signification or bedeutung of the parts of an animal body demands a patient examination of the successive stages of their development, in every group of Animals.[45]

This brings me to the crux, which is to demonstrate the benefit to Owen of von Baer's embryogeny—the antitransformist payoff in establishing that only the "truly philosophic inquiry" of the German embryologists could lay a permanent foundation for "a just and true theory of animal development and organic affinities."[46] Von Baer's science worked for Owen in two ways. First, it broke the recapitulatory crutch supporting serial transmutation. And second, it more subtly destroyed the possibility of homological (and therefore transmutational) relations between members of different *embranchements*. Take the latter case: in his 1843 course Owen used von Baer's science of fetal divergence to set strict limits to Geoffroy's "unity."[47] Owen ostensibly set out to test those limits. Knowing the precise resemblance between adult lower animals and "a higher organised animal . . . in its progress to maturity," he said, we will be able to judge how far unity of composition really holds. He proceeded to explain that, rather than recapitulate the entire inferior series, each embryo according to von Baer developed from a germ toward the characteristic organization of its *embranchement*—either vertebrate, mollusk, articulate, or radiate. In other words, it diverged toward its own singular archetypal structure. This meant that a mammalian embryo "does not represent all the inferior forms, nor acquire the organization of any of the forms which it transitorily represents." In a mammal's case, there is no repetition of the three invertebrate groups, so Lamarck's old-style series is certainly not recapitulated. As a result, Geoffroy's and Grant's search for taxonomic criteria that would allow them to establish a uniform succession of animal life was a quest for the transformist holy grail. Had, he said,

45. Ibid., ff. 96–97.

46. Ibid., f. 97.

47. Ospovat 1981:130–32. Also Jacyna 1984b:23–24; and on von Baer's science, Lenoir 1982:72–95; Gould 1977:52–63; E. S. Russell 1916: chap. 9.

the animal kingdom formed, as was once supposed, a single and continuous chain of being progressively ascending from the Monad to the Man, unity of organisation might then have been demonstrated to the extent in which the theory has been maintained by the disciples of the Geoffroyan school.[48]

The implication is this: because the structural unity between, say, squids and fishes, was restricted to the primary or germ stage of fetal development—after which there was a fundamental divergence—it was impossible for a squid to lengthen its gestation period and turn into a fish. By using embryology to split life into discrete types, each with its irreducible plan, Owen destroyed the transformists' continuum and dashed hopes for a wider unity.

To understand how von Baer was directly useful in breaking the recapitulatory support of transmutation, we have to return to Owen's 1837 lectures. His statements here reinforce the conclusion that he found von Baer essential for ideological reasons. The sequence of topics introduced in his fourth lecture was itself telling. He first spoke of the new embryology showing us the true path. In the next sentence he derided "those Teachers" who "advocate the baseless speculations of the transcendentalists." He clearly had in mind those who were extracting Lamarckian capital from the old embryology, for he complained that

the resemblance of the imperfect [i.e., fetal] condition of the organs of a higher species to the perfect [i.e., adult] conditions of corresponding organs in a lower organized species is misrepresented when it is stated that the Human Embryo *repeats in its development* the structure of any part of another animal; or that it *passes through the forms* of the lower classes; or when it is asserted that a Fish is an overgrown Tadpole.[49]

Owen was depicting the transformists as literal recapitulationists—as believing that a human baby really was the final transformation of a fetus that had been successively a fish, a reptile, and a lower mammal. He did this for good reason; by destroying recapitulation, he could now destroy its corollary, transmutation. So Owen was not attacking the London transcendentalist community—which was now quite large—only its Lamarckian extremists. In fact he was repeating what many morphologists had already said. They were themselves fed up with the accusations that "comparative anatomists are endeavouring to inculcate the absurd doc-

48. R. Owen 1843:367, 370. "On Formative and Structural Anatomy," *MCR* 1844, 40:18. Owen's von Baerian, antitransmutatory emphasis and restriction of unity between divisions to the germinal stage was reiterated in his published *Lectures* (1846b:10–11) and picked up in the reviews: "Owen's Lectures on Comparative Anatomy," *MCR* 1847, 6:154.

49. Owen, Hunterian Lectures 3 and 4, 6 and 9 May [1837], ff. 97–98 (RCS 42.d.4). Cf. Barry 1837a:347.

trine, that the human brain is really at one time exactly like that of a fish, of a reptile, or of a bird," rather than "in a somewhat similar condition."[50] In denying this identity, therefore, they too were adamant that there is no literal "metamorphosis" during gestation.[51] It is true that few of them had transformist tendencies. And even a transmutationist such as Grant often talked conventionally of the adult animal organs as simply "analogous" to the fetal human's[52]—even if the *Gazette* did once catch him telling his listeners that their brains had been those of crocodiles. Yet as Richards points out, a Lamarckian position probably demanded something closer to this identity. Logically, to create a new advanced species required an addition to the adult stage, a gestational prolongation up to the next level on the scale. The old adult form would then be packed back into the last period of ontogeny, becoming a literal ancestral rung that each embryo would thereafter be forced to climb to reach its new adult level.[53] Owen was evidently interpreting it like this. He treated Lamarckian recapitulation as a literal mutation during embryogeny, which repeated in microcosm the alleged metamorphoses of fossil ascent. These twin manifestations of transformism were linked in his rhetoric. As he put it:

The doctrine of Transmutation of forms during the Embryonal phases, is closely allied to that still more objectionable one, the transmutation of Species. Both propositions are crushed in an instant when disrobed of the figurative expressions in which they are often enveloped; and examined by the light of a severe logic.[54]

Of course he had given Lamarckism this meaning precisely in order to "crush" it. He now began this process using Müller's doctrine of the "organizing energy." Large parts of Owen's lecture entitled "Nature and Characters of Organized Beings" (11 May 1837) were lifted from Müller's *Elements of Physiology*. Müller had invoked a "rational creative force" to explain the building of organs and the "harmonizing" of their parts during embryonic growth. This force was already present in the germ and oversaw the building of the tissues, impressing a particular form for a special end.[55] Unlike a mental power, the organic force was not associated with any one organ, but was prior to all and resided in the whole; it was the cause rather than the result of organization. It thus acted "conformably to design, but without consciousness," "mental consciousness" being an "after product of development" and associated exclusively with the brain.

50. "Mr. Swan's Anatomy of the Nervous System," *MCR* 1837, 3:490.
51. King 1834:7–8.
52. Grant 1829:3.
53. E. Richards 1987:139.
54. Owen, Hunterian Lectures 3 and 4, 6 and 9 May [1837], ff. 98–99 (RCS 42.d.4).
55. Müller 1837:20–23.

Müller called it a "creative power modifying matter, blindly and unconsciously, according to the laws of adaptation."[56] The English, of course, modified it into a more cosmic "presiding hand of Intelligence," and some at least urged it against the pantheists with their self-organizing matter.[57] Owen kept closer to Müller's meaning. He too saw this "organizing energy" regulate structural growth to produce a well-adapted form. It did indeed operate "according to laws of Intelligence and Design; but we must not fall into the error of assigning to it the attributes of a conscious soul." It was an "unconscious power" and like other "imponderable agents" worked "according to determinate laws, which manifest in the highest degree the wisdom and design of the Law-giver."[58] While Müller at this point in his lectures simply dismissed transformism, Owen carefully drew out the implications. For him animal mutation, either in response to or in preparation for changing conditions, was precluded by the invariant action of this "unconscious power," which always functioned to build the same organic mechanism. "Each species of Organism is self-existent from the period of its Creation," he announced (quoting Müller without acknowledgment), and "maintains its specific character unchanged," finally disappearing "with the extermination of the reproductive Individuals;—for the Genus has no power to reproduce the Species, nor the Family the Genus."[59] What Owen was doing was invoking premises that outlawed all contemporary materialist theories: Geoffroy's fetal monstrosities, Grant's Lamarckism, Knox's generic reservoir, Tiedemann and Matthew's plastic power. Like Müller, he saw immutable laws make immutable species. And he explained this as a result of the "organizing energy" being limited. It could not be spontaneously stretched to change old organs or produce new species. That would be tantamount to investing the animal with self-evolving abilities.

The individuals of each species have a characteristic durability of Life;—the operation of the Organizing energy in them is limited. To suppose a power of prolonging the vital actions in an individual beyond the specific period, is to suppose that the organizing agent has the power to develope new organs, or to modify the old to such an extent as must cause a transmutation of the Species:—but this supposition cannot be maintained.[60]

With the energy exhausted at maturity, no power remained for a continued embryonic thrust. There was no means of producing a new species

56. Ibid., 25; Lenoir 1982:103–6; E. Benton 1974:29–31.
57. "On the Physiology of Man," *MCR* 1839, 30:453–54.
58. Owen, Hunterian Lecture 3, 11 May 1837, ff. 21, 30, 32 (BMNH MN 1828–41).
59. Ibid., ff. 32–33; cf. Müller 1837:25.
60. Ibid., ff. 36, 34.

"as an additional step once the normal course of development had been completed."[61] Nor was there any fossil evidence that this had ever occurred. Owen's study of ichthyosaurs, plesiosaurs, and dinosaurs in 1841 was designed to show that there was no uninterrupted ascent, the sort we might have expected had the "progressive development of animal organization ever extended beyond the acquisition of the mature characters of the individual."[62]

Owen was acutely aware that the transmutation of fossils and fetuses would obliterate individual existence. This presented a huge moral problem. Lamarckian claims that every man is once a fish and a reptile

imply that there exists in the Animal Sphere a scale of Structure differing *in degree* alone:—nay, they imply the possibility of an individual, at certain periods of its development, laying down its individuality, and assuming that of another Animal;—which would, in fact abolish its existence as a determinate concrete reality.[63]

Teleosaurs turning into crocodiles might not have been so immediately threatening. But a man as an "overgrown" ape—this did raise daunting problems. The same threat had been posed by the recapitulationists' study of human monsters, "arrested" in development and apparently possessing features of the lower animals. About this too Owen was scornful. He had shown in 1835 that even an idiot's skull, with a brain no larger than a chimpanzee's and mental abilities "scarcely more developed," was still human in all respects. And he believed that the "impassable generic distinctions between *Man* and the *Ape*" were already determined early in embryogeny.[64] In short, retardation left the idiot an undeveloped human, not an odd ape.

Owen's emphasis on developmental uniqueness was not catalyzed solely by his contact with the latest German embryology. Richards has shown how he was using Hunter's work to legitimate his views, particularly in the eyes of the surgeons (Owen was in the process of publishing some of Hunter's papers in 1837). He now used Hunter to undermine Geoffroy's environmental explanation of monstrosities, which had such brutalizing consequences for man. Hunter, working with pheasants, had recognized that deformities were not "a matter of mere chance," but neither were they related to changes in the environment, "for he observes that every species has a disposition to deviate from Nature is a manner

61. E. Richards 1987:149.
62. R. Owen 1841c:197.
63. Owen, Hunterian Lectures 3 and 4, 6 and 9 May [1837], ff. 97–98 (RCS 42.d.4); taken from Barry 1837a:347, 348.
64. R. Owen 1835a:372.

peculiar to itself." Hunter reasoned that the cause must therefore lie in the "original germ" and that "monsters are formed monsters from their very first formation."[65] This, again, enabled Owen to argue for some inner cause, a preordained change, initiated by the "organizing energy" already present in the embryonic germ. Deformities were programed at the outset. They were not caused by environmental factors (whatever the results of Geoffroy's experiments on egg incubation); nor were they the result of a retarded or accelerated development along some supposed common axis.[66] No ape could metamorphose into a man through a heroic push to the top of the scale. "Impassable" differences between man and ape were already potential in the germ, and its differentiation into a unique being began immediately.

This mated up with Barry's own expressions on individuality. While admitting that the germs "in all classes of animals, from Infusoria to Man," are "essentially the same in character," Barry meant no more than that they all had a "homogeneous general structure." There was no question of their identity; each from the first must have the ability to develop "into this or that individual"—its potential was quite unique. Therefore "not only is the human embryo at all periods of its existence a human embryo, but the human heart and brain, closely as they resemble corresponding organs in other Vertebrata at certain periods of development, are never any thing else than the heart and brain of Man."[67] So by 1837 three influences—Müller's "organizing energy," Barry's version of von Baer's laws, and Hunter's endogenous causation—provided Owen with a total alternative to the embryological equipment of the transformists. He rejected their single scale topped by man—the ultimate fossil and fetal "metamorphosis"—and denounced their environmental explanation of man as an "overgrown" ape.

On a social level, it is easy to understand the Lamarckian monster-makers' threat to the propertied class in a Chartist age. Not only would any compromise of human sovereignty undermine Christianity's cosmic status by submerging man in brute nature (a form of subversion the paupers pirating Holbach's atheistic books had long aimed at). But it opened a door to the antiproperty, antiindividual experiments of the burgeoning socialist movement (whose leaders on the Central Board, remember, were using Lamarckism to justify their communitarian policies). So Owen reassured his audience of political, Church, and medical leaders that "in-

65. Hunter 1840:25–26; Cross 1981:34–35. Owen's attempt to find in Hunter's writings antecedents of his own ideas helped greatly to legitimate them, for Hunter at this time was placed on a pedestal by Owen's superiors in the college (Jacyna 1983a).

66. E. Richards 1987:148–50.

67. Barry 1837a:347–48, 1837b:121.

dividualities . . . manifest themselves at very early periods of development [in von Baer's scheme], and cannot be laid aside."[68] Human uniqueness and accountability before God were not negotiable issues in biology,
and Owen's Anglican stand marked him off sharply from the atheist radicals with their Lamarckian, disestablishing, and democratic demands.
Whereas Grant refused to consider man as anything but the highest animal, Owen's Hunterian science underpinned a religion of submission to
the Divine Will.[69] It was designed to elucidate man's relationship to the
Almighty—to teach man

that if he were something less than a Deity he was something more than dust;—it
appears to have been essential that he should know that an Intelligence was superadded to matter, and had presided over its arrangements, but that the Omnipotence to whose Fiat both he and matter owed their existence, had in this world
endowed him alone with faculties to appreciate the Works around him; and would,
in another require from him a strict account of their Application.[70]

Of course, Owen was not the first to attribute a belief in a literal relationship between fetal and fossil development to the Lamarckians. Lyell
and Buckland had both worried over the problem. Lyell warned that Lamarck's followers considered man's fetal ascent a replay "as it were, of all
those transformations which the primitive species are supposed to have
undergone, during a long series of generations, between the present period and the remotest geological era."[71] Nor was Owen alone in rejecting
recapitulation for that reason. Von Baer himself probably deployed his
rival theory of divergence in the hope of scotching this serial trans-

68. Owen, Hunterian Lectures 3 and 4, 6 and 9 May [1837], f. 98 (RCS 42.d.4). Barry
1837b:141.
69. As Broderip recognized: Broderip to Buckland, 14 January 1842 (BL Add. MS 40,500,
f. 247).
70. Owen, Hunterian Lectures 1 and 2, 2 and 4 May 1837, f. 81 (BMNH MN 1828–41).
71. C. Lyell 1830–33, 2:62–64. Buckland returned repeatedly to this problem in the
early 1830s, as can be seen from his file marked "Species Change of Lamarck" (UMO WB).
He jotted:
    Case of Human Foetus. 1 like fish—2 like Bird—3 like Quadruped 4 like Man. it does
    not follow that modern Species as a Species has passed thro all these stages tho each
    individual has organs fitted to its several successive stages the foetal fish does not
    propagate non foetal Bird or Beast *wh* on Lamarcks theory it must do.
In another note he expressed the same sentiment:
    Absurd doctrines of Lamarck & Geoffroy. It is true these successive stages in the
    foetal state of each Individual Man like Shakespeare 7 ages of Human Life there are
    also stages of Progress before his Birth . . . but are we to infer from hence that in
    some former period the individual had reached its perfection & was born a Fish—
    that for a while such fishes lived & multiplied, & getting tired of their fishy State were
    transformed by the Effect of their own Appetency into Reptiles that the Reptiles tired

formism.[72] But Owen's position was tactically different from Lyell's. In fact Lyell in 1832 had been forced to accept the evidence for recapitulation (for lack of any alternative), so he was careful to deny that it had any relevance to fossil studies or indeed to Lamarckism. Because Owen could now reject linear recapitulation, it suddenly became useful again to link fossil and fetal development. By refuting recapitulation, he could now strip transmutation of its strongest embryological support, dashing radical hopes.

So Owen's abhorrence of transmutation in its brutalizing aspect—in its ability to obliterate human individuality and merge man with the beast— made von Baer's doctrine extremely attractive. In shifting embryology onto this new ground, Owen could set strict limits to Geoffroy's "unity," dispense with Lamarck's linear continuum, and stop animals illicitly crossing between Cuvier's *embranchements*. His antitransformist ideology made the new German embryology very profitable. And his resulting moral anatomy became a powerful weapon in the Anglican arsenal at a time of political crisis, when the clerisy were concerned to render scientific theory serviceable to the masters of civil order.

Peelite Patronage: The Financing of Acceptable Science

It would be indeed a national reproach & an irreparable loss to the World of Science if the possessor of such unrivalled talents & acquirements were obliged to descend to the Condition of a Bookseller's hack, when an addition of £300 a year would secure the devotion of his whole life, now in its prime, to the probable completion of a career of more original research, & more comprehensive views of the Totality of organized life than any human being has yet had talents combined with opportunities sufficient to accomplish, and I have peculiar satisfaction to add that his opinions on religious matters are sound & temperate; & that every new discovery he makes excites in him such feelings as a mind constructed like that of Paley is alone competent to enjoy.

—Buckland's carefully worded plea to Sir Robert Peel for a Civil List pension for Owen[73]

---

of Crawling at length by the mere wishing to fly were converted into Birds & hatched Broods of Young Birds.

Buckland's fears again show how useful Owen's new antirecapitulationist science was to be for the Oxford clerisy.

72. Oppenheimer 1967:230.
73. Buckland to Peel, 12 January 1842 (BL Add. MS 40,499, f. 250).

The clerisy portrayed Owen as the voice of moderation not only in his repudiation of radical science, but in religious matters too—in his rejection, at the other extreme, of the Puseyite "reptiles" or antiscience Anglo-Catholics.[74] Broderip agreed with Buckland that Owen was "so good, so modest, and with such a proper sense of true religion—none of [the] enthusiasm that is intoxicating the brains of many of your men at present."[75]

This image of moderation was cultivated by other groups among Owen's backers, particularly the "Germano-Coleridgeans" (as J. S. Mill called them) clustered round Whewell and the classical scholar Julius Hare at Trinity College, Cambridge. This group promoted their conserving reform and liberal Anglicanism as a midpath between a morally bankrupt Benthamism and "heartless" high Toryism.[76] Whewell was a typical member in his polymathic approach. Not merely a mineralogist, Kantian philosopher, Bridgewater author, and compiler of huge compendiums of inductive science, he also translated German and classical works and wrote hexameters. He learned his mineralogy in Germany, and his "Idealist reaction against Benthamite radicalism" hardened as he read Kant and Schelling in the original.[77] Coleridge had been deified at Trinity by the Apostles. This student society had been founded by those utilitarian renegades John Sterling and F. D. Maurice (the future Christian Socialist), both now converted to the idea of Coleridge's clerisy, both praising the new German divinity and historical criticism. In London, Sterling, Maurice, and their "band of Platonico-Wordsworthian-Coleridgeian-anti-Utilitarians" had gained control of the *Athenaeum* in 1828 for a short while and used it to fight Benthamism and "the dragon of materialism."[78] Sterling studied philosophy in Bonn in 1833. He became a curate a year later, resigned, and earned his living as an essayist. Maurice, who married Sterling's sister-in-law, was to become the Guy's Hospital chaplain in 1836 and professor of English literature at King's College, London, in 1840. These men were well aware of the atheistic and deistic leanings of the capital's radical activists. And they knew the political threat posed by Lamarckism, Maurice having actually been present at the debates between the Ben-

74. Owen to Baden Powell, 26 January 1850, discussed in Desmond 1982:46; Brooke 1979:40–41, 1977b:143; E. Richards 1987:161. Corsi 1988 deals with Powell's own anti-Puseyite stand and sympathy for Owen's science.

75. Broderip to Buckland, 14 January 1842 (BL Add. MS 40,500, f. 247).

76. Preyer 1981:44–45; Garland 1980:55–69; S. F. Cannon 1978:46–50. Mill 1962:108, 129.

77. Preyer 1981:43, 45; Becher 1986:61.

78. S. F. Cannon 1978:49; Preyer 1981:45.

thamites and Lamarckian socialists in the Co-operative Society in Red Lion Square in the mid-1820s.[79]

Among this Coleridgean intelligentsia, Owen was now lionized as the savior of anatomy. He had evicted the materialists from their morphological stronghold, and his ideal archetype and Platonic process were soundly Coleridgean, yet as challenging for the Apostles as the new German philology or historical criticism. Sterling delighted "in Owen, with all his enthusiasm for fossil reptiles." And in an age weary of radical declamations, his friend the essayist Thomas Carlyle—himself a slayer of materialist dragons—was "charmed" by Owen's "naturalness, and the simplicity he has preserved in a London atmosphere." Carlyle told Sterling in 1842 that he had learned more from Owen "than from almost any other man."[80] Whewell too had a long friendship with his townsman, backing Owen at the Geological Society and BAAS, and speaking to Peel on his behalf. Their early letters resounded with talk of the need for "an English Zoological Nomenclature," of Blainville's bias against Cuvier, of financing Owen's future work, and of the antitransformist meaning of new fossils. Owen plied Whewell with information for his *Philosophy of the Inductive Sciences* (1840), and after reading the proofs tactfully told him "how much we—i.e. the Cuvierian cultivators of Comparative Anatomy—are, and always must be, indebted to you for the clear statement of the scientific character of teleological reasoning"[81]—tactfully because he had already begun moving Whewell toward a more morphological explanation of structure.

Owen's social aspiration and training under Green showed in his early deference to Whewell and the Cambridge clergy. Through the 1830s Green continued to argue for a united professional clerisy in the "great seminaries of learning." Here turbulent innovation and "specious" utility were wisely rejected, and morality, "philanthropy, patriotism, and love of science" used to turn gentlemen to "the idea of abiding and fontal good." Through their moral bond, the barristers, clerics, and doctors of Coleridge's "learned class" were learning together to serve the nation—not one class or generation, "but the unity of the generations, the type of our inward humanity." Anglican clergymen, medical gentry, and landed mag-

---

79. Pankhurst 1954:97–99; Desmond 1987:105.

80. C. Fox 1972:113, 138. Also on Sterling: S. F. Cannon 1978:48–49; Garland 1980:64.

81. Based on Owen's letters to Whewell: TCL WW Add. MS. a 210; no. 55 (11 February 1839) on Blainville; 57 (7 March 1839) on nomenclature; 61 (26 March 1840) on "teleological reasoning"; 66 (4 September 1842) on the public dinner at Lancaster given by their townsmen to honor Owen and Whewell (also R. S. Owen 1894, 1:199); 88 (22 February 1845) on the value of the latest fossils in undermining transmutation.

istrates were thus "children of a common household" and professional custodians of English culture.[82]

Hence Owen's allegiance at this period to the Oxbridge teachers, Tory
peers, Broderip, and the bench. He turned to them in times of trouble or
when he needed to go over the council's head. And it was they who now
obtained Crown support for him, allowing him to complete his anatomical
renovation and check council criticisms that he was straying too far from
his official college business. Roy MacLeod long ago pointed out that the
financing of science still needed to be studied seriously, as did the role of
political paymasters in approving standards.[83] Matters of course have improved much, thanks mainly to Morrell and Thackray and their study of
the tactical allocation of BAAS funds. But specific cases still need to be
examined in depth, and Owen's shows how handsomely his cooperation
paid off in terms of career enhancement and cash injections. BAAS managers diverted considerable sums to Owen, and he was sensitive to the
spirit in which they were given. Hence his reports were strategic master-
pieces: scrupulously documented papers, often ending in attacks on Lamarckism and tacit support for the clerisy's socially stratified cosmos. It
was no coincidence that his social elevation followed on his BAAS and
Geological Society triumphs. He was now able to exploit his entrée into
privileged Oxbridge society provided by Whewell and Buckland. Both
had Peel's ear. With the fall of Melbourne's Whigs in 1841 and Peel's return to Downing Street, Whewell had himself been awarded the master-
ship of Trinity as a political gift. While pensions after the Civil List Act of
1837 were less obviously party gifts, Peel nonetheless relied on Whewell
and Buckland to advise him on the merits and respectability of scientific
claimants. These pensions were actually destined for persons "who have
either rendered a service to the public" or who "have rank and title and
no means of maintaining them to live at least in dignity."[84] While this referred to destitute gentlemen needing to keep up a respectable appearance, Owen saw no reason why it should not apply to a titular head of
science wanting an income commensurate with his station. This is how he
presented his case. He reported back to his wife from Cambridge two
days after Christmas 1841: "I have met, at Dr. Whewell's, the present
Chancellor of the Exchequer & Lord Brougham and have represented to
them my present anomalous position, holding a Cuvierian rank without
the means of doing it justice."[85] He acquainted them with the appalling

82. Green 1832:3, 32–36, 41–43, 1834:1–2, 5, 10, 1840:7.
83. MacLeod 1970:47.
84. Quoted ibid., 48.
85. R. Owen to C. Owen, 27 December [1841] (BL Add. MS 45,927, f. 38).

pay received by even the best "investigators of natural history" and put his case for £200 or £300 a year.

He capitalized on another aspect. The pensions also served—as Herschel put it to Peel—to relieve "men of a very high order of attainment, and who have distinguished themselves for original research, during those years *while their powers are still unimpaired and available for discovery.*"[86] It was to be an incentive to gentlemen of scientific rank to continue working free of worry. Hence Owen's phrasing in his follow-up letters to Buckland and Whewell in the new year. He intimated not only that his salary was incommensurate with that "position in society to which my scientific labours have raised me," but that he would be forced to shelve his great "national" works as a consequence. He had wanted to produce a multivolume treatise on comparative anatomy, the sort which Cuvier and Meckel promised but died before completing, with each book focusing on "one system of organs." But the costs were prohibitive. Publication of the pilot work, *Odontography,* had already forced him to economize, and there were still his son's school fees to consider (young Willie was destined for Westminster). He could take up Longman's lucrative offer and compile a dictionary of anatomy, but five years of "drudgery" would finish him for serious work. The alternative was Crown funding, and he now traded on his Oxbridge reputation to press his claim. He referred the chancellor Henry Goulburn (the Cambridge Tory M.P.) not to London physiologists (the leading investigators in their field), but to the Oxford and Cambridge professors of geology and medicine. Owen was of course supplying social references. Cambridge divines had convinced Owen that "science would be benefitted by the favourable consideration" of his claims, but they stood to benefit not a little from his exertions too. Owen ended on a patriotic note, claiming that his failure to produce a definitive treatise would be the country's loss, and "I am unwilling that England should lose the credit of producing that Work on Comparative Anatomy, which France & Germany have, as yet, failed in achieving."[87]

He was wedding his claim to national prestige, and Buckland (passing Owen's letter to Peel) amplified this in an accompanying note. He played on the premier's desire "to promote the scientific reputation of this Country," remarking on Owen's Continental standing:

At a meeting of the naturalists of Germany, at which I also attended, at Freyberg in 1839 Mr Owen was considered the first among many assembled representatives

86. Quoted in MacLeod 1970:50.

87. Owen to Buckland, 11 January 1842 (BL Add. MS 40,499, f. 252). He also mentioned his hoped-for success where Cuvier and Meckel had failed in a letter to Whewell, 25 February 1842 (TCL WW Add. MS a.210[64]).

of European science; and in our own Country his position is quite as high as that of Airey [sic], Faraday or Dalton . . . since the death of Cuvier even France herself has looked up to Owen as the only worthy successor of that great man.[88]

He emphasized Owen's age and qualificataions for executing his ambitious plans, and reminded Sir Robert that the BAAS thought enough of Owen to award him grants annually. But it was primarily Owen's position, potential, and religious respectability that Buckland played up. He exaggerated Owen's financial straits, and stressed the scandal if the English Cuvier were forced to become a pot-boiling hack for the sake of a few hundred pounds. Broderip too concentrated on these national and religious factors. Owen should not be left to eke out an existence when "he might be enlightening the world and raising the reputation of his country." He was so truly religious that Sir Robert, "when he thoroughly knows the case," will acknowledge that "duty and inclination" go hand in hand.[89]

Two weeks later Owen chivied Whewell with a letter recounting another threat to his "national" work. In 1842 the RCS Council appointed Owen the new conservator in succession to Clift. But with this came a dramatic increase in his duties and a demand that he concentrate more on college matters. Owen, though ambitious and making contacts in high places, was still of inferior status in the college. Given the rigid surgical class structure, he was seen to be rising above himself. The professor and conservator—there to do the bidding of the senior surgeons—was beginning to incite bitter feelings. The practical surgeons had intimated, Owen told Whewell in horror, that "I have done too little for the College, and too much for myself." He believed that the council, "incredible" as "it must appear," would "shackle" him into publishing exclusively and anonymously in the college *Transactions*.[90] Again, Owen alerted his friends among the Peelite gentry and teaching clergy, circulating a note of the council's demands to Whewell, Buckland, Egerton, and the Cambridge professor of anatomy William Clark. Dependent on college finances, he told Whewell, he would be forced to take on the conservator's menial tasks, leaving him no time to complete his great work. But if the government recognized his scientific freedom by conferring a pension, the old surgeons would "yield due deference." Owen was now tying his new anatomy and personal "independence" to Peelite financial security, putting pressure on his superiors through their own government patrons. Whe-

88. Buckland to Peel, 12 January 1842 (BL Add. MS 40,499, f. 250).
89. Broderip to Buckland, 14 January 1842 (BL Add. MS 40,500, f. 247).
90. Owen to Whewell, 22 January and 28 February 1842 (TCL WW Add. MS a.210[63 65]). Also Rupke 1985:252 on the RCS Museum Committee's criticism of Owen for spending too much time on fossil osteology.

well was asked to impress on the chancellor the importance of Owen's post-Cuvierian anatomy and gain him if not a pension, then at least an examiner's post in the college, which would boost his finances and put him into bargaining position with the council.[91]

When Peel came into office he found that the Civil List funds for that year had been exhausted by the outgoing government.[92] But he ensured that Owen was awarded a major share of the new grant in 1842—£200 of the £300 set aside for gentlemen of science.[93] Whewell, who had canvassed hard, told Owen that the "well deserved honour" would protect him from the "molestation" of less appreciative men (meaning the old surgeons).[94] But it could not protect him from the unappreciative radicals. In the House they raised Cain. Hume denounced such political retainers for their insidious effect, citing the case of the Poet Laureate Robert Southey, who had sunk into a reactionary old age on a Crown pension. Wakley, while not against pensions for savants, objected to Owen's. He protested that the college had ample funds to look after its own without raiding the public purse. But Peel defended it. He was well cued by Whewell and Buckland, announcing that the new Cuvier "would not have been able to continue" his anatomical work without Crown help and that this was a perfectly proper use of funds.[95] Wakley completely misjudged the mood. Given the perennial groaning in the press about "the niggardliness of Government in granting honours and pensions to *scientific* men," it was inevitable that his outburst would antagonize the moderates, who deplored his vendetta and applauded Peel for recognizing a philosophical anatomist of international stamp in their midst.[96] Indeed, the middle-brow *Medical Times* pooh-poohed £200 as the sort of "pittance" that an aristocrat would tip his French chef. Considering how much Parliament squandered annually "on the *Raptores* of the law, and the *Ferae* of our fleets," it was an insult to Owen and the real scientific "benefactors" of mankind.[97]

Owen was in little danger of becoming Longman's hack, but the additional income did enable him to pay off his draughtsmen, finish *Odontography* (published in three parts, in 1840–45), and bring out his *History of British Fossil Mammals* (twelve parts, 1844–46). The BAAS had already

91. Owen to Whewell, 25 February 1842 (TCL WW Add. MS a.210⁶⁴).

92. Peel to Buckland, 19 January 1842 (BL Add. MS 40,499, f. 254).

93. Peel to Owen, 1 November 1842 (BL Add. MS 40,518, f. 24); R. S. Owen 1894, 1:203–4. Owen to Peel, 1 November 1842 (BL Add. MS 40,518, f. 26).

94. Whewell to Owen, 9 November 1842 (BMNH RO); R. S. Owen 1894, 1:204–5.

95. *Hansard* 1844, 74:1478–80.

96. *MT* 1844, 10:188.

97. "A History of British Fossil Mammalia," *MT* 1845, 12:46.

contributed £250 toward the £1,000 costs of the latter. Owen now printed 350 copies and was "sanguine enough to expect no loss."[98] In an age when serious publishing cost an author money, his combined grants, pension, and pay (now about £700 a year in all)[99] allowed him to publicize his science effectively—and in a more lavish way than was possible for out-of-favor teachers tied to the university's laissez-faire pay system. The pension also guaranteed his freedom to work on "independent" projects, so he could develop his anti-Lamarckian paleontology while remaining an employee at the RCS. It also increased his political stature. By 1843 Owen had himself joined the scientific circle around the prime minister. He entertained Peel and Prince Albert at the Hunterian Museum, and in turn visited Peel's estate, Drayton Manor. So taken was Peel with his young guest that he wanted Owen's portrait by Pickersgill (see fig. 8.1) to hang alongside Cuvier's in his gallery.[100] The home secretary appointed Owen to the Health of Towns Commission, the premier sought his advice on publishing the zoologies of exploratory voyages, and within two years Peel had offered Owen a knighthood.[101]

The New Homological Anatomy

[A] master-mind has arisen among our ranks. . . . [Owen's "Report on the Archetype" is] unquestionably the most complete and philo-

98. R. S. Owen 1894, 1:207.

99. This breaks down as follows: Owen's £300 salary (which he had received since 1833) was being augmented at least by 1838 by a further "gratuity" of £100, to give him parity with Clift (MS, 13 December 1838, RCS Misc. 2). His pension of £200 and BAAS grant of probably £100 p.a. took the total to £700. Although the grants helped in the cost of publishing, this was still a risky business, as Owen complained (Spencer 1904, 1:393).

100. Owen to Buckland, 26 December [1844] (BL Add. MS 40,556, f. 294); R. S. Owen 1894, 1:244–47, and 211–12 on the prince slipping quietly into Owen's museum, as Buckland said, so as not to "excite jealousy among your inmates, whose company would not be desirable." Pickersgill arranged through Broderip for Owen to sit. On completion Broderip wrote to Buckland: "You know what a fine head our friend has with brains enough to fill two hats, and the painter has done justice to it." Peel's desire to own the portrait was "so gratifying to all dear Owen's friends": Broderip to Buckland, 27 December 1844 (BL Add. MS 40,556, f. 314). But if Peel did acquire the portrait, it was passed back to Owen later, for his family presented it to the National Portrait Gallery in 1893.

101. Though Owen is thought to have turned it down as a result of pressure from within the college (I. F. Lyle, Royal College of Surgeons, pers. comm.). He was, after all, an employee holding a teaching and curatorial post and supposedly at the council's beck and call. A knighthood might have made his superiors rather uncomfortable, particularly those still awaiting their own honors. For the radical spite at Owen's later "ludicrous appointment to the sewers and slaughter-houses": Grant to Mantell, 20 December 1849 (ATL GM 83, folder 44).

Fig. 8.1. A mezzotint of H. W. Pickersgill's portrait of Owen (1845). Sir Robert Peel planned to hang this picture next to Cuvier's (see fig. 2.4) in his private gallery. By W. Wakler, 1852. (Courtesy Wellcome Institute Library, London)

sophical work that has ever appeared in this or any other part of
Europe, in relation to organic science.

—*British and Foreign Medico-Chirurgical Review* swinging its weight
behind Owen[102]

It is to the recognition of this archetype that all the science of comparison must lead, and till it can be discovered, we can know nothing of the transcendent walk of nature and design, of her creations
of forms in variety.

—Geoffroyan Joseph Maclise in 1846[103]

Given the abhorrence of Lamarckism in polite society, we can see that
Owen was well placed to mediate between the medical moderates and
Oxbridge clerisy. He first attempted to move the High Churchmen in a
more morphological direction. As early as 1837, on receiving Whewell's
History of the Inductive Sciences, he began agitating for a more constructive Cambridge attitude toward the higher morphology, arguing for a
"general theory of animal organization" which took into account both
structural unity and adaptive specialization. Writing back to Whewell, he
even criticized Cuvier for his abandonment of any homological help and
his "desultory" descriptive approach. And Owen lauded Hunter for explaining the variation among organs by means of a transcendental plan
modified to meet specific "exigencies."[104] Owen presented his program as
an equal mix of transcendentalism and teleology, but he gave no idea of
the subordinate status he was eventually to give the latter.

Owen's subtlety in blending transcendentalism and teleology—plan
and adaptation—was unparalleled. Take one aspect singled out by reviewers. From compound bones he was able to isolate distinct "homological" and "teleological" components ("pedantic" terms, one journalist admitted, but "sufficiently expressive").[105] He showed, for example, that
while the occipital bone in the baby's skull ossifies from four centers, corresponding to the four separate bones in the echidna (making it "homolog-
ically compound" and a striking testament to the vertebrate unity of plan),
other types of compounding required a wholly different explanation. A
mammal's femur also ossifies from four primitive centers, but it is not four

102. "Owen and Maclise on the Archetype Skeleton," *BFMCR* 1848, 2:108.
103. Maclise 1846:300.
104. Owen to Whewell, 31 October 1837 (TCL WW Add. MS a.210⁵⁴); Rehbock 1983:79;
Rupke 1985:247. Owen in his letter also criticized Cuvier for selling Hunter short as a tactful
way of accusing Whewell of the same. In Broderip's words, Whewell was following "the
French in his blindness": Broderip to Owen, 20 October 1837 (BMNH RO).
105. "Owen's Lectures on Comparative Anatomy," *MCR* 1847, 6:153.

bones welded, for in reptiles it grows from only a single center. This is a case of fetal adaptation giving pliancy to the growing limb. With many young mammals needing to run soon after birth, it allowed a cushioning cartilate to develop between the ossification points. The bone was "teleologically" compounded, an adaptation to neonatal needs.[106]

Owen could still resort to a more Cuvierian functional "design" on occasions. But by the mid-1840s he was usually only doing so to scotch the Lamarckians' attempts to join the *embranchements* and establish a continuous succession of life subject to "self-creative forces."[107] In his 1844 lecture on the "teleology" of the fish skeleton, for instance, he again rounded on transmutationists for suggesting that the Devonian arthrodires (extinct armored fishes) were "transmuted Crustacea" and on Grant in particular for his "exaggerated" claim that the sturgeon's heavy scales were a sign of its inverterate origin. With no homologies possible between adult invertebrates and fishes, there could only be an adaptive similarity; he argued that the sturgeon's "scale armour" was ballast necessitated by its bottom-grubbing existence.[108] It was this "self-creating" alternative that galvanized Owen into taking a purely adaptive approach. He turned teleology against transmutationists who were still attempting to link the different *embranchements*. But while his chapter on teleology was extracted and reprinted, and even praised as an "eloquent discourse in Natural Theology,"[109] it could not disguise the general homological thrust of his *Lectures*. This thrust was overt in his BAAS "Report" on the vertebrate archetype in 1846. Owen's main concern by this time was to describe vertebrate unity in terms of a primal pattern. The report was a tour de force: in it Owen systematized the nomenclature of the homological bones throughout the vertebrate series—from man to fish—and searched out "what is truly constant and essential" in their character in order to build up a picture of the ideal archetype.[110]

Away from the personal propaganda—the puffing to obtain a pension and impress the Oxbridge Peelites—Owen was now heralded by the

106. R. Owen 1846b:36–40. Reviewers also reveled in his criticism of Cuvier's use of ossification centers as a means of determining the separate bones: Carpenter 1847:478–79; "Owen's Lectures on Comparative Anatomy," *MCR* 1847, 6:153, 158–59; "Owen and Maclise on the Archetype Skeleton," *BFMCR* 1848, 2:111–12.

107. R. Owen, Hunterian Lectures 1 and 2, ff. 66–67 (BMNH MN 1828–41); R. Owen 1846b:147; Desmond 1979:232–33.

108. R. Owen 1846b:146–49; cf. Grant 1833–34, 1:537.

109. "Owen's Lectures on Comparative Anatomy," *MCR* 1847, 6:166.

110. R. Owen 1846c:240. The "nonsense" of special names for human bones and need for a vertebrate-wide nomenclature reflecting the discoveries of homological anatomy were widely felt: Maclise 1846:300; "Owen and Maclise on the Archetype Skeleton," *BFMCR* 1848, 2:110.

medical pundits not as Cuvier's successor, but Geoffroy's. The reviews in the mid-1840s were still extolling Geoffroy's erratic genius, "as brilliant as it was bold," and his theories which "astonished and charmed." But more and more this was to provide a foil for Owen's "safe" science. They saw him founding a securer anatomy—one that incorporated the newer German embryology, eradicated the taint of transmutation, and abolished the absurdities, but one that still subordinated life's functional adjustments to the "essential" ground plan.[111] Not that this ideological safety left his work empirical or unadventurous. Reporters were already arguing that Owen's "great strength" lay in his logical grasp—giving the lie to later criticisms that he could only work from bone to bone.[112] And his major theoretical conclusion after reconfiguring the cranial vertebrae, that man's arms were "nothing else than 'diverging appendages' to his occipital bone," was well received in London—as could only have been the case in a city with a large "transcendental" community (see fig. 8.2).[113] It certainly delighted Carpenter, even if he expected the Philistines to snigger at the very idea of hands springing from the head.

At the same time, Owen's stand against Cuvier strengthened. Owen's insistence that similarity of function was powerless to explain the homologous bones of the wing, hand, and flipper—and his eventual comparison of final causes to barren vestal virgins—might have sounded passé to the bevy of local morphologists, who had always opposed Cuvier's extreme functionalism. But his detailed rebuttal of Cuvier's specific criticisms— particularly those concerning cranial homologies—was treated as definitive, and Cuvier's failure "as a fundamental guide" was trumpeted ever more loudly in the press.[114] As a result the seeds of the anti-Cuvierism that was to dominate the next generation were already germinating in the 1840s. Grainger in 1848 looked back to Cuvier's retarding influence.[115] And Owen's disciples charged the old surgeons who disliked higher anatomy to look again at Cuvier's *Leçons d'anatomie comparée;* without the

111. "Owen's Lectures on Comparative Anatomy," *MCR* 1847, 6:152–53, 155. Although Carpenter (1847:474–75) too praised Owen (1846b:21–23) for repudiating Geoffroy's "unphilosophical" comparison of the articulate's exoskeletal rings and fish vertebrae, he still believed that cuttlefish bone was "the true homologue" of the vertebrate endoskeleton.

112. "Owen's Lectures on Comparative Anatomy," *MCR* 1847, 6:166.

113. W. B. Carpenter 1847:489; R. Owen 1846c:271, 301, 338, 1849:69–70 on the hands as "parts of the head"; "Owen and Coote on the Homologies of the Vertebrate Skeleton," *BFMCR* 1849, 4:180–81; "Professor Owen on the Nature of Limbs," *MG* 1849, 44:291; "On the Nature of Limbs," *L* 1849, 1:617–18.

114. "Owen and Maclise on the Archetype Skeleton," *BFMCR* 1848, 2:108; "Owen and Coote on the Homologies of the Vertebrate Skeleton," *BFMCR* 1849, 4:181; R. Owen 1846c:241, 317, 1849:39–40 on vestal virgins.

115. Grainger 1848:14.

benefit of an archetypal guide the book was "well nigh useless" except as an "assemblage of anatomical facts."[116] Owen might have had severe disagreements with his Geoffroyan predecessors, but his triumphant anatomy had continued to "shake the throne of Cuvier" if not depose the Macedonian emperor.[117]

Owen's "Report on the Archetype" was aimed at the specialist and was necessarily technical. But his popular Royal Institution *Discourse on the Nature of Limbs* (1849)—where he showed that all vertebrate limbs were built to the same plan—still met complaints that it was "strong meat for the ladies."[118] And not only the ladies; students too were struggling with these morphological immensities. Owen's protégé Holmes Coote conceded that while they "show considerable interest in the subject . . . your work is rather beyond them." Coote, then working on the catalogs, proposed that he himself write a primer. "I should be only a labourer in your vineyard," he confessed, and it would be modeled *"after your archetype & with your* nomenclature for the use of the students of our school."[119] In the *Homologies of the Human Skeleton* (1849) he then condensed Owen's terminology, vertebral craniology, and identification of homologies in an effort to introduce them into "the general course of medical education."[120] But even an "Owen made easy" was still strong meat for some reviewers, who doubted whether Coote had actually managed to simplify matters much.[121]

With the "philosophic Naturalist" ordained by Coleridge as the high priest of the Logos, Owen was careful in the *Discourse* to interpret laws and causes as Divine "ministers"—like Green viewing them as "productive powers."[122] Owen's idealism makes the greatest sense against this Coleridgean background. Green in his 1840 oration had already explained

116. Coote 1849:8.

117. Maclise 1846:300; "Macedonian" metaphor: Limoges 1980:222.

118. Broderip to Owen, 2 February 1849 (BMNH RO). Owen himself admitted to R. Wagner (13 February 1849, RCS SC) that the "subject is better adapted for the character of mind and thought of a german Audience than for our matter-of-fact English."

119. Coote to Owen, 19 October 1848 (BMNH RO). Coote, a barrister's son and Westminster School educated, had been apprenticed to William Lawrence in the mid-1830s, but he became enamored of Owen's homological anatomy and went on to teach it as assistant surgeon at Barts.

120. Coote 1849:2, 8. Poor standards at the RCS remained an obstacle to this. As Coote complained to Owen (n.d., BMNH RO 8: f. 417): "The great difficulty, which I experience in the teaching of Homological Anatomy, is the necessity of keeping students down to the standard required by the Examining Board at the College of Surgeons."

121. "Professor Owen on the Nature of Limbs," *MG* 1849, 44:291–92; "Owen and Coote on the Homologies of the Vertebrate Skeleton," *BFMCR* 1849, 4:182.

122. R. Owen, "Notes and Annotations: Prof. Green's Lectures on Zoology & Comparative An.ʸ," f. 129 (RCS 275.b.21); R. Owen 1849:86.

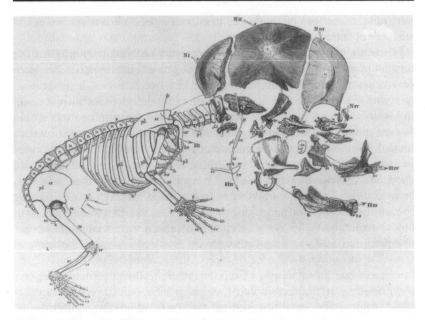

Fig. 8.2. A disarticulated baby's skull, showing Owen's breakdown of the four cranial vertebrae. N I–IV are the expanded neural spines of these four vertebrae, H I–IV the respective hemal arches. In Owen's view the "hands and arms are part of the head"—diverging appendages of the hemal arch of the first vertebra of the skull. (From Owen 1846, fig. 25)

the dual manifestation of Ideas. In one mode, they preceded nature, when "as thoughts of the Divine Intelligence" they remained Ideas, "the archetypes and preexisting models," showing that life as a power was "anterior in the order of thought to organization." But in another they realized the Creative design by acting on nature, manifesting "as a scheme of living forces," as "those energetic acts of Omnipotent wisdom" and Will which we call laws.[123] Owen's anatomical philosophy also encompassed these twin modes. The general vertebrate design pointed to a "predetermined pattern, answering to the 'idea' of the Archetypal World in the Platonic cosmogony."[124] And this vertebrate Archetype became incarnate through Creative Will; it was made flesh in a stream of increasingly specialized forms. There was always a strong emphasis on process in Coleridgean thought. Morphology for Owen was as much a study of its incarnation as of the Archetype. This "Greenian" heritage lends plausibility to

123. Green 1840:xxi, xxiv–vi, 19, 20, 24. This is derivative of Coleridge (Levere 1981:99–100).

124. R. Owen 1849:2; Owen to Maria [Owen], 7 November 1852 (RCS RO 3: f. 387).

the claim that Owen's statements on secondary causes were essentially theological and served a mediating function, allowing him to run with the Whewellian hares and hunt with the London hounds.[125] Certainly his "laws" were interpreted quite differently by the two groups. He made an idealist process attractive to the Germano-Coleridgeans, while in London he appeared to be climbing aboard the naturalistic bandwagon.

We clearly have to be careful in seeing Owen's eventual recourse to "natural laws" as somehow at the leading edge of naturalistic comparative anatomy. Even among Coleridgeans he was far from the vanguard. For a generation they had been talking of "law" in this way, as a causative agent "combining both power and intelligence."[126] True, Low Churchmen such as Sedgwick with no Coleridgean background completely misunderstood him—interpreting his "laws" too materialistically and censoring his apparent belief in a "natural" fossil development. The Anglo-Catholics went further and declared his expressions dangerous and demoralizing in the wake of the Chartist demonstrations and European revolutions of 1848.[127] But no Coleridgean eyebrows were raised, and Whewell, more and more the conservative Platonist, fell "gleefully" on Owen's archetypal offerings.[128] The Coleridgeans were quite aware of the meaning of Owen's science—and of the fact that it could be substituted mutatis mutandis for more heretical explanations. Edward Forbes actually tried (unsuccessfully) to convert Owen to the term *anamorphosis,* to be used "in opposition to *metamorphosis* [i.e., transmutation] as expressive of an *ideal change* & not of a *real or bodily change.*" This would allow the Platonists to steal a march on the materialists and describe the "changes *which take place throughout the members of a group* & which indicate progression towards a higher series of forms," but without implying genetic continuity.[129]

Given this Coleridgean context, Owen's Platonic "laws" actually cut the ground from under the medical self-rulers. The radical Dissenters and freethinkers viewed law not as a ruling Logos, a regal edict, but as an expression of the sovereign properties of matter. Owen, like Green and Colderidge, had only withering words for these "pantheists," for whom God "not only does, but *is* everything," for whom Will constitutes Nature.[130] He had still harsher words for the ultramaterialists, whose Lock-

125. Brooke 1979:41.

126. Green 1840:xxv.

127. E. Richards 1987:161–63.

128. Ruse 1979:123, 125; Yeo 1979:508–9; Ospovat 1981:142–43.

129. Forbes to Owen, 2 November 1846 (BMNH RO).

130. Coleridge 1913:44, 270; Green 1840:ix; Owen, Notebook 7 (January–May 1832), f. 64 (BMNH).

ean sensual limitations left them denying the possibility of a preexisting Archetype or its Creative Cause. He claimed that with Cuvierian adaptation failing as an explanation for structure, anatomists are left only with the Archetype indicating an anticipating intelligence, or some accidental harmony of atoms—and "from this Epicurean slough of despond every healthy mind naturally recoils."[131] Elsewhere he threw down a typical Coleridgean challenge: the historical progression of life is due either to the action of matter endowed with "vital properties" or to the "platonic *idea* or specific organizing principle."[132] On the one side we have a prefiguring Will and Divine delegation of power, and on the other a chance collusion or self-mustering of atomic forces (Coleridge's "convention" of sovereign atoms, his rabble democracy). Owen told his Royal Institution audience that the Archetype destroyed this atomistic atheism and illuminated the science of development. Fossil life "from the first embodiment of the Vertebrate idea under its old Ichthyic vestment, until it became arrayed in the glorious garb of the Human form" had advanced slowly, led by Divine "ministers," guided by the archetypal light.[133]

So it would be wrong to visualize Owen among the leading London naturalizers of nature, even if he were casting round for a physiological nontransmutatory explanation for successive development.[134] He has only been placed there by default, because of our ignorance of the welter of real anatomical materialists in the capital. The older Darwinian histories were quite remiss in this matter. They never ventured into medical lowlife, preferring to tackle only high-profile personnel. Yet many of the London schools provided a haven for French-inspired teachers, and for a generation it was they who had made the running—urging the primacy of materialistic morphological and paleontological laws. Owen was far from alone on the landscape. His works occupied only one social locale, albeit an expanding one. If there was a tension within his thought it was because he was looking two ways at once. He was engaged in dialogues with both the wealthy clerisy and the London materialists. His works were attempts to stake out a Coleridgean-Peelite middle position in anatomy and me-

131. R. Owen 1849:40, 85; Brooke 1977b:143; *BFMCR* 1849, 4:179–80.

132. R. Owen 1846c:339. His belief was actually more complicated. He postulated that a "polarizing force" is responsible for the repetition of parts on which unity is predicated and that a counteractive organizing principle works to specialize and adapt this segmental structure. The stronger the organizing principle the more the polarizing force is "subdued" and the further life specializes away from its archetypal antecedents. It is strongest in man, who has departed farthest from the archetype. The tensions within Owen's own thought on Platonism are nicely brought out by Ruse (1979:121–24). See also Rehbock 1983:80–84; and E. Richards 1987:151.

133. R. Owen 1849:86; Green 1840:xxv.

134. Desmond 1982: chap. 1; E. Richards 1987:129.

diate between the Cambridge Germanizers, corporation liberals, and medical moderates. He rejected the old corporation empiricism, provided a patriotic focal point in a preeminently French and German preserve, and offered a nonradical conception of nature's process. Positioning himself at the civic hub and organizing this new coalition, he assured his professional place, drew the patronage strings tighter to the Crown, and secured his freedom in the corporation fiefdom.

His strategy was a success. After the College of Surgeons received its new charter in 1843, the radicals ceased abusing him indiscriminately as a corporation lackey. Nor did his identification with the RCS surgeons experimenting with a limited democracy do him any harm in liberal London. Green and Brodie were already pointing the way forward, though not unopposed. Their "new school" of higher anatomy was disliked by the traditionalists, "old Surgeons many of them, who are unavoidably far behind the present state of Anatomical & Physiological Science"[135]—the practical men who wanted to rein in tight on the new conservator. Yet Green and Brodie saw the necessity of the right Romantic science for the "learned class" and continued to support Owen's work before the council.[136] They also wanted liberalization measures speeded up and the old guard ousted. True, their democratic concessions were strictly limited. In 1841, when Green finally signaled his acceptance of the "elective privilege," he still saw it restricted to the pure surgeons, who were to be constituted into a new body of fellows. Even then, only gentlemen of high *"moral character* and conduct" with an arts degree and six years (of largely hospital) experience were to be eligible.[137] Since professionalism for a Coleridgean implied a "deep sense of responsibility" and religious feeling, the medical portals had to remain guarded against militant materialists. In the end, the "elevating pursuits, scientific attainments, literary refinement, and moral excellence" of the gentry, with whom "it should be [the fellow's] ambition to associate," were to remain the criteria of acceptability.[138] So Owen's positions as a college liberal and client of the Cambridge Peelites were perfectly compatible, while the perfection of his improving patriotic morphology raised his stock among the medical moderates in town.

Sharpey, Carpenter, and Grainger recognized Owen's brilliance and applauded his search for a higher archetypal design. Grainger held up "the archetype vertebra as the significant symbol of one of the most sub-

---

135. Owen to Whewell, 28 February 1842 (TCL WW Add. MS a.210⁶⁵).

136. Ibid., and 25 February 1842 (TCL WW Add. MS a.210⁶⁴).

137. Green 1841:61–62. "Mr. Green and Medical Reform," *L* 1840–41, 1:803–5; "The Touchstone of Medical Reform," *MCR* 1841, 34:414–22.

138. Green 1841:15–19.

lime truths in nature, the all-pervading unity steadily gleaming forth, amidst the endless variety of adaptive forms."[139] Carpenter kept up a cordial if deferential correspondence. He purloined Bristol Institution specimens for Owen (including its large lungfish), requested testimonials (and a job as Owen's "aide"), and trained to town on the new Great Western line to read Owen his papers. Even if Carpenter was to settle into Huxley's Owen-hating and Owen-baiting camp in the 1850s, in the 1840s he was clearly trading on the Hunterian professor's scientific patronage. Not that Carpenter prostituted his philosophy. With all his *Review*'s enthusiasm for homological anatomy, it was careful to strip off the archetype's Platonic gloss. It still doubted Owen's claim that life's progression was inexplicable "by any known properties of matter," reminding him that "every advance in exact knowledge" has shown just "how large a share those forces [of common matter] have in the actions called vital." It implied that the time was approaching when the specialization of fossil life would find its material explanation.[140]

More radical anti-Platonists lampooned the Hunterian professor's heroic mystifications.[141] But most reviewers simply reinterpreted the archetype as a geometrical abstraction. Geoffroyans certainly subjected it to this kind of ideological translation. Joseph Maclise, a graduate of London University and the Paris schools, proclaimed that unity of organization was "admitted now-a-days as of little dispute as a mathematical axiom"— and that is what he made it (see figs. 8.3 and 8.4).[142] He attempted a geometrical breakdown of a typical vertebra—one from which all real vertebrae, indeed the whole skeletal frame, could be deduced. For him all bony elements were the products of a "metamorphosed original," and he

139. Grainger 1848:47–48. Carpenter to Owen, 23 August, 19 and 23 September, 13 November 1842 (BMNH RO). Desmond 1982:37–38ff. on Carpenter, and 55, 77 on Sharpey and the Archetype.

140. "Owen and Maclise on the Archetype Skeleton," *BFMCR* 1848, 2:108, 118.

141. The *Lancet*, although now admitting the technical excellence of Owen's work (e.g., 1847, 1:226; 1849, 1:95), still indulged in doggerel about his "myths and conundrums . . . leading the judgement astray" (1851, 1:314; Rupke 1985:252). Owen's classical allusions, his castigation of Democritans, Lucretians, and Epicureans (an understandable retreat into history, for as conservator in the Hunterian Museum he had to mix amicably with all sections of medical society), were matched by Grant (1861:3), who deplored the damage done by Plato in abandoning "legitimate" philosophy with its basis in the notion of inherently active matter, and in constructing a system of "supernatural agencies" to realize the "preconceived designs of his fancied agents." When Grant (p. vi) went on to praise Darwin for blowing aside the "pestilential vapours" of the mystical "species-mongers," Owen (1863:62, n. 1) took it personally.

142. Maclise 1847:iv. Maclise's father was a Scottish soldier-turned-shoemaker, his mother a Unitarian. He attended London University in 1834–35, then crossed the Channel to study at the Ecole Pratique and Hôpital de la Pitié.

argued that without first determining this archetypal pattern our "comparative research must be veiled in mystery." He doubted "whether absolute differences of form have any real existence, except as forms under metamorphosis," and meant this in a defiantly Geoffroyan sense. He dismissed Cuvier's classification and in his *Comparative Osteology* (1847) denied any

natural separation between presumed class and species. Conceit may establish its fancied line of classification as it chooses; but rigorous reasoning annihilates the tottering fabric. . . . Cuvier classifies species according to fancied diversities; but Geoffroy fuses all species into one line of extended analogies, and Nature herself responds to this latter interpretation. The doctrine of analogy transcends all bounds.[143]

But Owen was now sweeping all before him, and when Maclise disagreed over the number of skull vertebrae and the origin of arms, commentators considered Owen safer to follow.[144] In fact almost all the journals now fell in behind Owen on technical aspects. Carpenter considered him "second to none, either living or dead" in his knowledge and transcendental determinations.[145] Owen had already taken the standard in London for his brand of homological anatomy.

There was nothing sterile about his conservative Romantic anatomy. It attracted the Germano-Coleridgeans precisely because it was as challenging, imaginative, and unorthodox in its way as the new historical interpretations of the Bible. Green and Owen's imagery was the most sophisticated of the period—far in advance of the ossified fossil Lamarckism still being taught by Grant, with its undeviating upward drive.[146] Green's organic metaphors of the ramifying "ascent of animal life" were astonishingly vibrant. Life was

neither a scale, nor a ladder, nor a network; it is neither like the combination of a kaleidoscope, nor the pattern of a patchwork; it is no process by increase or superaddition:—but it is, as in all nature's acts, a growth, and the symmetry, proportion, and plan, arise out of an internal organizing principle [itself the "manifestation of a higher power acting in and by nature"]. This gradation and evolution of animated nature is not simple and uniform; nature is ever rich, fertile, and varied in act and product:—and we might perhaps venture to symbolize the system of

143. Maclise 1847:14; preceding quotes from Maclise 1846:299. Maclise here defined morphology as "the science of form passing through metamorphosis."
144. "Professor Owen on the Nature of Limbs," *MG* 1849, 44:290, 292; "Owen and Maclise on the Archetype Skeleton," *BFMCR* 1848, 2:119. Only the *Lancet's* Grantian reviewer really liked Maclise's book, unsurprisingly given its Geoffroyan bent: "Comparative Osteology," *L* 1847, 2:128–31.
145. W. B. Carpenter 1847:472.
146. Desmond 1984c:395–413.

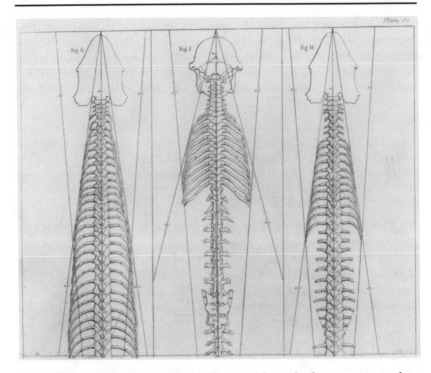

Fig. 8.3. Like other London University graduates, Maclise took a far more geometrical approach than Owen to the Archetype. Having determined the conformation of the archetypal vertebra, he discussed how the "production of species" resulted from a "metamorphosis" or change in size or shape of its various elements. (From Maclise 1847, pl. 54)

the animal creation as some monarch of the forest, whose roots, firmly planted in the vivifying soil, spread beyond our ken; whose trunk, proudly erected, points its summit to a region of purer light, and whose wide-spreading branches, twigs, sprays, and leaflets, infinitely diversified, manifest the energy of the life within. In the great march of nature nothing is left behind, and every former step contains the promise and prophecy of that which is to follow.[147]

Gone was the dread of a diverging, branching nature, with extinguished old lines and emergent new ones. Gone were the older taxonomists' doubts that this would make nature as bad as the Elizabethan architects who had taken "delight in the construction of 'galleries that lead to nothing.'"[148] To Green's organic metaphor Owen now added von Baerian foundations, a fetal divergence reinforcing the image of a ramifying, preordained nature. He then applied this embryological model to the fossil

147. Green 1840:109–10.
148. "M. Latreille's Familles Naturelles du Regne Animal," *Zool. J*. 1825, 2:428.

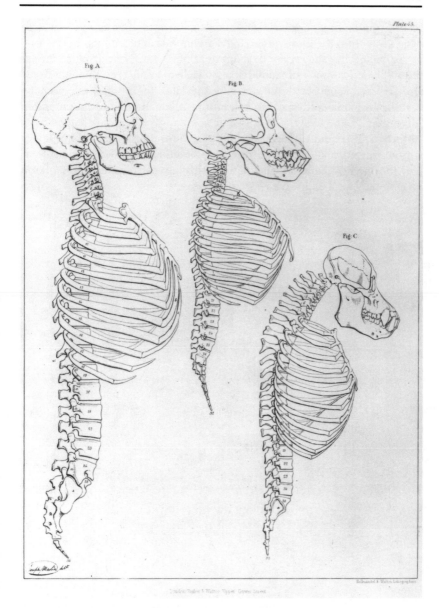

Fig. 8.4. For Maclise, like Geoffroy, there were no "absolute differences of form"; all species were fused "into one line of extended analogies." (From Maclise 1847, pl. 45)

record, generating an image of numerous lineages progressively special-
izing away from the more generalized archetypal antecedents. In this way
he explained the successive development of life, without invoking any of
the exploded notions of Lamarckian serialism. This tree metaphor was
already established in the reviews by the later 1840s, with a "common"
vertebrate trunk splitting into fish and reptilian stems and innumerable
"diverging" branches.[149] Coleridgean comparative anatomy might have
rested on antecedent Ideas, potentialities, and ministering laws, but this
antiradical morphology was never rigid or inflexible. With their tree of
life, the medical Coleridgeans had unwittingly established a metaphoric
image ready for a definitive Darwinian gloss.

149. W. B. Carpenter 1847:479. Ospovat 1981: chap. 5; Desmond 1982:92ff.; Bowler
1984b:122–26.

9

Grasping the Nettle:
Some Concluding Remarks

The calls for historians to grasp the nettle and recontextualize evolution
are growing.[1] One can see why. Its social basis has rarely been investigated, and yet a massive amount of energy is expended on every esoteric
snippet of Darwin's thought. We have brilliant studies of Darwin's day-to-
day jottings, but this close reading has caused a kind of myopia: the larger
picture has gone out of focus. To remedy this, we need studies of other
sorts of evolution in other sorts of contexts. We need to understand the
Vestiges' lawful appeal to the deists and Dissenters, Knox's racist theories,
the French approaches of the medical democrats, and not least the progressive anti-Malthusian sciences in the socialist underworld. If nothing
else, this will help us see Darwin's Malthusian appeal to the middle
classes in a fresh light. It is clearly time to pull away from Darwin and take
a broader view of science and society in his younger days.

The debates over anatomy and evolution that took place among the
medical community in the 1830s make a good starting point. At this time
many young doctors, Paris trained and *au fait* with French science, were
arguing for a law-bound, deterministic, even transforming nature. The
ruckus they caused has largely gone unnoticed by historians, whose sights
have been set on gentlemanly science. Yet studying this episode enables
us to get to the heart of British society, with all its problems, in a way that
must ultimately benefit the Darwin Industry. And if we are serious about
uncovering ideological concerns in science, this episode has enormous
potential. I have tried to show how the new anatomies were socially assessed, using a backdrop of professional injustice and radical correctives,
political expediency and Peelite compromise. This kind of approach is
particularly suited to the turbulent 1830s, a period when the institutions
of science were being reformed, and biological knowledge had a moral
and political signifiance.

The philosophical anatomy that swept through the medical schools at
this time was no homogeneous, neutral form of knowledge. Nor was it

1. Kohn 1985b:4; Lenoir 1987; Moore 1989b.

adopted indiscriminately by all London teachers. So we have to locate its adherents precisely and draw up a social topography of the science. We have to understand who imported it and why, how it was modified, and how each modification was rationalized in a social context. True, this sort of social approach pushes us harder and further into the literature; but to read a decade's worth of weekly medical journals brings science and society to life in a way that is impossible to imagine if we consult only contemporary textbooks. Nothing less can reveal why the Wakleyans, the private teachers, the corporation elite, and the Peelite gentry approached science with such different expectations. Or why the atheists used evolution to help smash Anglican supremacy; or how leading Whigs muted Geoffroy's doctrines; or how Coleridgean idealism was brought to the defense of Church and realm. Nor, without understanding how the rival sciences could be used to ratify different social orders, can we go very far in explaining the extraordinary rise and fall of some medical careers.

What I have done, in short, is assess the cultural implications of the competing comparative anatomies. This has enabled me to reinterpret the teachers' scientific strategems in social terms. Using this method, we do not need to sift history for "true" and "false" theories, or "good" and "bad" scientists. We can treat all the rival sciences equally by looking at their different embedding contexts. We can, in effect, do real justice to past scientists by exploring the dialectical relationship between their views on nature and their specific professional, religious, and class interests.[2]

A General Description of Events in 1826–46

[Geoffroy's theories] split the scientific world at once into two great parties; those who looked upon this great naturalist as an unerring guide, the Newton of transcendental anatomy, and those who considered him a mere visionary, ingenious indeed and full of talent, but still altogether a visionary.

—A reviewer in 1847 looking back over the previous generation[3]

Not long ago the kind of statement quoted in the epigraph would have been dismissed by historians. They simply doubted the first proposition—that Geoffroy had a group of British followers who accorded him the status of a Newton, despite the fact that even Owen complained publicly in the 1830s that Geoffroy's theories were carrying the day. Fritz

2. Berman 1974–75:32.
3. "Owen's Lectures on Comparative Anatomy," *MCR* 1847, 6:155.

Rehbock, for example, in the *Philosophical Naturalists* (1983), missed the large medical contingent of higher anatomists in London in the early 1830s and puzzled over Owen's claim.[4] It is by reconnecting higher anatomy to its medico-political base that we can appreciate just how many teachers actually supported Geoffroy, and why.

The reviewer sees one group hailing Geoffroy as the true lawgiver, another treating him as a "visionary"—in other words needing empirical correction and a new philosophical foundation. Actually the situation on the ground was more complicated; nonetheless, as a first approximation the introduction of the new morphology might be said to have fallen into two phases. The first, roughly from the mid-1820s to late 1830s, represented a little-recognized Geoffroyan phase. At this time the Edinburgh- and Paris-trained medical graduates were flocking to London, some to practice, others to staff the new university and private schools. From republican Paris they brought Geoffroy's, Serres's, and Lamarck's science, and they placed the emphasis on organic law, unity of plan, and above all on comparative studies. But London medicine was rigidly controlled by the corporations, and the "Scotch" and London Nonconformists found themselves locked out of the patronage network. All the marginal men suffered. A number of teachers went out of business as a result of the RCS's discriminatory legislation; a few became physical-force radicals, and some even went to court for their beliefs. The ensuing campaign for civil liberties meant that there was a ready-made "opposition" forum for Geoffroy's republican science in the private anatomy schools, London University, GPs' journals, and radical medical unions. The outsiders used its stern, lawful view of nature to vitiate the surgeons' Church-and-king natural theology. They portrayed the morphological laws as keys to a new medical science which vindicated their right to professional leadership.

The Paris-Edinburgh sympathies of these radicals go a long way to explain the widespread English belief that the growth of materialism was a Calvinist cancer.[5] In medicine the sciences of self-creation were cultivated almost exclusively by this "Scotch"-orientated group: Grant, Watson, Matthews, and Knox were all Edinburgh medical-educated, all enemies of priestcraft and corruption, all believers in the self-development of life. In London Grant's biology in particular was hitched to the environmentalist edifice constructed by Cobbett's and Bentham's medical dis-

4. Rehbock 1983:76.
5. Levere 1981:60; S. F. Cannon 1978:61. Badgered by Brougham into taking out shares for the new university, which was to crush "bigotry" (chap. 2), Burdett responded with the rather ungratifying observation that there was more "bigotry & intolerance" among the Scottish Presbyterians and English Methodists than the Anglican clergy: Burdett to Brougham, 12 August 1825 (UCL HB 20,031).

ciples. His Lamarckian anatomy, Southwood Smith's self-evolving creation, William Farr's environmental explanations of social ills, and Wakley's
cultural determinism all nestled together in the democratic press and
were all pounded in Tory prints. With the arrival of these men, then, we
witness the naturalization of creation, not as an upshot of a professionalizing campaign per se, but as part of the secularizing, Dissenting agitation
to redistribute medical privilege and democratize the institutions of science and state.

   This first phase began drawing to a close in the later 1830s with the
decline of Benthamism, the closure of the private schools, the dissolution
of the radical-liberal alliance, and the first corporation concessions. True,
Joseph Maclise in 1846 could still talk of all animals as "metamorphosed"
variations of one original and claim that interest in this subject had not
"worn off with the sunset of the great mind of Geoffroy St. Hilaire" (who
had died in 1844).[6] But by this time an idealistic science in the College of
Surgeons was well established and our second phase under way.

   I have studied Coleridgean morphology in this RCS context because
the political backdrop was so clear. Green and Owen's idealist anatomy
was developed in a corporation under attack—in a college whose council
was resisting the democrats' demands in an attempt to preserve its power
and privileges. The Coleridgeans had a profound contempt for the demagogues, the "Cobbetts and such like animals," and feared that democracy
would lead to mediocrity, as Lamarckism led to leveling.[7] Owen's morphology was antimaterialistic and reflected the new Peelite needs. It was
fashioned from German archetypal anatomy, von Baerian embryology,
and Coleridge's idealism and provided a sophisticated alternative to the
radicals' science of self-determination. It vindicated Green's noblesse
oblige view of nature—as a descensive spiral of power. Owen denounced
all talk of life's "self-developing energies" and ruthlessly targeted the Lamarckians' belief in mankind's ape ancestry and their attempts (using the
Stonesfield "opossum" and squid-fish homologies) to establish a continuum in nature. He openly admitted that the value of comparative anatomy, correctly understood, lay in its eradication of these false philosophies "of the origin of living species."[8]

   Understandably the rivalries between the teachers on either side of the
college portico intensified during the 1830s. It was an age when plural
livings were still essential to augment a professor's income. Thus the two
leading comparative anatomists in London, Owen and Grant, competed
directly for academic posts, institutional resources, and funding. I have

   6. Maclise 1846:299.
   7. Knights 1978:13, 61.
   8. R. Owen 1846b:3.

looked at the way their political differences manifested in their scientific and managerial stances in the learned societies. Owen's anti-Lamarckian idealism gave morphology a powerful new image, and its value to the captains of the Geological Society and British Association for the Advancement of Science was evident from their allocation of funds. It was sponsored by the younger liberal romantics at the College of Surgeons, the leading hospitals, and the Anglican King's College, gentlemen sickened by the radicals' bestial "Ouran Outang theology."

All in all, tackling comparative anatomy by separating the materialistic Geoffroyans from the idealistic Coleridgeans provides an extremely profitable approach to the subject. We can at last begin to explain how the political doctrines of the rival factions in the reform years "manifested in their views of nature."[9]

Connecting with the Social History of Science

> Every Englishman of the present day is . . . either a Benthamite or a Coleridgean.
>
> —J. S. Mill in 1840[10]

This analysis chimes well with current studies in the social history of science. As David Bloor notes, when Mill surveyed the scene, he saw it dominated by two influential giants, Bentham and Coleridge. They were "the teachers of the teachers." Yet their philosophies were designed to support such rival radical-Dissenting and Tory-Anglican superstructures that, as Mill said, "you might fancy them inhabitants of different worlds." Bloor believes that we ought therefore to find "these two currents of thought flowing through all the cultural products of the time," including science.[11] And we do. Of course Mill was concerned with high philosophy; on the ground things were messier, and the democrats' intellectual tools were actually honed from a variety of Cobbettian, Wakleyan, and Benthamite stones, while their eclectic arsenal often contained fragments from Paine and the Parisian *idéologues*. In fact the further we move from medical dissent toward the working-class radicals, the stronger the Paineite and cooperative components become, so that, at its extreme, gutterpress Lamarckism was tied tightly to socialist strategies and atheist demands.[12] However, in middle-class medicine the Coleridgean-

9. Jacyna 1983a:104.
10. Mill 1962:102–3.
11. Bloor 1983:605; Mill 1962:39, 101, 104.
12. Desmond 1987:91–104.

Benthamite divide is true enough for our purposes. The Benthamites were using the naturalistic sciences to legitimate a new social order, one based on democracy, Dissent, and trading wealth. To paraphrase Steven Shapin and Arnold Thackray: for a "degraded" medical class with no outlet to corporation power or the legislative process, the image of a democratic anatomy supporting a rational "Republic of Medicine" was appealing.[13]

That said, though, the relationship between the structure of anatomy and the needs of these social groups was always complex and changing. (Owen's eventual forging of a new liberal-conservative consensus makes this undeniable.) Judging from the shifting political affiliations, the social meanings of these sciences were local and contingent, and not logically inherent in the theories themselves.[14] The mechanistic anatomies were of transient use to the democrats in their efforts to oust the consultants, just as evolution was of transient use to the working-class atheists in a society dominated by Anglican priestly power. The radicals were attempting to take the moral high ground on a wholly naturalized landscape, one in which the elite's class-based privileges could no longer be justified.

The Historiography of Comparative Anatomy

The philosophical historian is still left wondering why British natural history took the turn it did between 1830 and 1860.

—James Moore, questioning how the rise of philosophical anatomy was related to the changes of "an industrial society undergoing political reform"[15]

One begins to see why philsophical anatomy had come to dominate large sectors of London medical science by the 1830s. And why it was only later that certain sanitized forms of it seeped into the respectable world of Rehbock's "philosophical naturalists." The medical manufactories were the original warehouses for this imported republican science. It was in these, as a consequence, that the laws of development were first taught. Knowing this, we can tackle a number of outstanding questions about "scientific naturalism" in the nineteenth century.

James Moore has already asked where this way of viewing the world originated, and whether it simply furnished "intellectuals with an arsenal

13. Shapin and Thackray 1974:10; Thackray 1974:678, 686–88; Berman 1978: chap. 4; Barnes and Shapin 1979:93–95; Inkster 1983:18.
14. For the agreement on this point among scholars from rival traditions: Bowler 1984a:257; and MacKenzie 1985:417.
15. Moore 1985c:451.

of 'ideological weapons'" or was somehow "related more organically to social and religious traditions."[16] Roger Cooter is right that an over-emphasis on the Darwin Industry has led to the false impression that a naturalistic worldview was ushered in by the *Origin of Species* in 1859. But phrenology was not the only "intellectual midwife" of a naturalistic style of thought;[17] nor was it the only sign that a deterministic lawful approach was favored by the reformers. The flowering of higher anatomy shows how widespread belief in a law-bound nature was, especially in the medical world. A naturalistic ideology flourished among the outsiders trying to get a foot in the corporation door. Whether they were radical Nonconformists attempting to wrest power from the Anglicans, with religious reasons for stripping nature of all spiritual connotation, or freethinking democrats naturalizing nature for leveling political purposes, the result was the same—a threat to the Paleyite elite's status as the interpreter of the natural order. And, nature having been stripped bare, it was left amenable to an evolutionary gloss. This context explains why a democratic, upwardly mobile evolution was largely restricted to the radical classroom and pauper press—at least until 1844, when Chambers finally gave it a theological gloss and middle-class cachet.

This radical Nonconformist critique also explains the rapid decline of the "design argument" in certain quarters. It used to be assumed that Darwin's *Origin* alone destroyed Paley's reasoning from animal design to a caring Designer. Thanks to the papers by John Brooke, we now know that natural theology was weakening and changing anyway before 1859. But the extent of its disappearance from the anatomy schools had not been appreciated. Nor had the timing of this event. The evidence suggests that Paley's approach to design was already fading from large areas of medicine in the 1820s and 1830s. Continentally inclined reductionists and secularists derided it as a crude sign of the clerical hold over British science, and many medical Nonconformists replaced it with a "higher design" based on unity of plan and the purposive action of a creative natural law.

We can go further in understanding how the new morphology was employed by the marginal men. Jacyna sees their rhetoric of natural law and anatomical order as an attempt to put comparative anatomy on an epistemological par with physics—it was a prestige-enhancing ploy. They could point to the productivity of their comparative approach: to the cell theory, reflex arc, laws of morphology and embryology. They presented themselves as the lay priests of a lawful anatomy, penetrating the "arcana of

16. Moore 1986a:62, 67.
17. Cooter 1984:87, 9–10.

life" using methodological tools exclusive to their sect.[18] Jacyna's emphasis is thus on professional benefits. The teachers were attempting to upgrade their science and to pick up converts by engaging in polemics with the Paleyites, whose emphasis on individual adaptations made them a perfect foil. This attempt by the "low status" teachers to raise their professional stock is the crucial insight in Jacyna's analysis. But it still raises the question of why they embraced Geoffroy's "filthy" views,[19] when—if they were intent on simply scrambling up the greasy pole—a traditional empiricism rubber-stamped by the RCS oligarchs, or an episcopally approved Paleyite teleology, might have seemed the easier way.

We get a clearer view of the answer when we pin the anatomical approaches more precisely to their institutional loci. Having disentangled the private, hospital, and university schools, we can see that Geoffroy's anatomy was taught primarily by lecturers who stood outside the corporate-Anglican-hospital power structure. Many of these teachers were simply barred from hospital posts by their lack of expensive apprenticeship, Oxbridge Anglican qualification for the fellowship of the RCP, or social status. They were business, religious, and often class rivals of the small group of hospital consultants running the corporations and ministering to the gentry. The de facto divide in the profession was by now between the consultants and the GPs; since the GPs were largely taught in the private and university schools, it was they who were the main recipients of Geoffroy's science. The GPs were politically powerless, and their agitation for recognition left the profession in turmoil in the 1820s and 1830s. The new knowledge was being sold on the strength that it would raise their lot. Philosophical anatomy, with its universal laws of organization and malformation, provided the up-and-coming GP with his scientific credentials, his excuse for demanding professional parity with the consultant. It was part of the movement to gain a wider franchise—to get GPs the vote at their own Royal College, a vote that would win their supporters and spokesmen power. So the scientific campaign was embedded in a much wider political program. As with radical-Dissenting interests in the country, the activists were intent on changing the fundamental distribution of power in medical society. They were bent on opening up the corporations, destroying the Paleyite and Creationist props of traditional authority, and curbing corruption in the parceling out of posts. Philosophical anatomy was being used to show the reformer as the real medical benefactor of industrial society.

No wonder the gentry, clergy, and corporation elite gravitated toward

18. Jacyna 1984b:41–46, 1984a:61–68.
19. Jacyna 1984a:64–65; Jacyna 1983b for an acute political analysis.

Owen's Coleridgean anatomy, which invalidated the democrats' self-creating sciences and reaffirmed the conservative principle of delegated Divine power. Historians of early Victorian Cambridge acknowledge that Sedgwick's and Whewell's hatred of Benthamism stemmed largely from the radicals' "cavalier" attitude toward "the safeguards of the established order."[20] But this fear has to be set squarely against the backdrop of radical demands for disestablishment, Church reform, the rights of Dissenters, and the abolition of tithe monies. Rarely has the Anglicans' antiradicalism been related to their anti-Lamarckian stance. I have tried to tease out this relationship. This is easier to do in a London context. Here, more than in the old university towns, the Oxbridge-educated physicians and the corporation elders were brought face to face with angry Quakers, Calvinists, and Methodists campaigning against the "fornicating" Church; here the London University's Francophiles were belittling the Bridgewater Treatises; the Wakleyans were goading the "'Church and State' bigots," and the GPs were demanding democratic rights. Here, in other words, the radical Lamarckian threat was overt and immediate, and called for a strong conservative response.

That response came from Owen and Green. And the context in which they elaborated their sciences enables us to understand the pure surgeons' use of their resources. Nicholas Rupke wonders why the RCS Council subsidized Owen's fossil and comparative studies rather than spending its money more legitimately on morbid anatomy and surgery.[21] It is a question at the center of the present study. There were important reasons for investment in this sector by a beleaguered college. Some were professional: Jacyna suggests that an "adulatory" approach to John Hunter in the yearly orations worked to the councillors' advantage. Hunter had been the doyen of eighteenth-century comparative anatomists; putting Hunterian science rather than craft at root of their business enabled the surgeons to assert their professional and intellectual parity with the physicians.[22] But there was another, flagrantly political, aspect to their spon-

20. Garland 1980:67; Preyer 1981:45.

21. Rupke 1985:252. The council was not sponsoring comparative anatomy because of "competition . . . from the London hospitals," as Rupke speculates (pp. 253–54). The councillors were hospital men; the overwhelming criticism before 1843 was exactly that council seats were the exclusive property of the hospital consultants. In reality, the surgeons were responding to radical attempts to break this monopoly.

22. Jacyna 1983a:87–98. Cf. Rupke (1985:254) who sees the baronets extolling a scientific anatomy because "it gave badly needed prestige" to the "surgeon-members." Like Jacyna I believe it had more to do with the elect's own interests. Most pure surgeons, after all, ignored the GPs' grievances and were desperate to keep a strong social demarcation between themselves and the license holders. For the consultants, a conservative scientific outlook was the mark of a leisured professional gentleman (although they were not unaware of its value

sorship. Even a cursory glance at the medical hacks, who abused the orators and accused the "rogues" of hijacking Hunter's museum, highlights
the tenseness of the period. Owen's own career was shaped by the council's reaction to the mass protests and to the threat from the new university. These events forced the pace of change—the college refurbishments
and Hunterian chair—as the council attempted to regain the initiative.
Coleridgean comparative anatomy in Owen's hands was part of the backlash against the corporation reformers pounding on the door; the compel-
ling anti-Lamarckian, antiradical conclusions of his fossil and zoological
researches served the council well. That Owen was championed by college romantics, the conservative gentry, and the Oxbridge dons shows
how crucial this kind of science was to a clerisy fighting to hold its ground.

The Changing Social Relations of the London Teachers after 1843

Coleridge died in 1834 believing that Church and sound science were
doomed before the arrayed forces of Dissenters and freethinkers. It was
not to be. With the passing of the "reform mania"[23] in the 1840s, Owen's
archetypal anatomy came into the ascendant. Indeed, following the rechartering of the College of Surgeons in 1843, a new—third—phase
opened which saw a scientific regrouping on the center ground. The liberal takeover of London science in the later 1840s is important to docu-
ment because some of the older institutional and intellectual "network"
models have been criticized for their rigidity. As M. J. S. Hodge says, we
need to take account of the "wider, shifting interests and ideologies to
which those institutions owe their foundation and transformation" during
periods of social change.[24] This emphasis on scientific progress in a transforming society is extremely important. More recent studies of London's
scientific institutions have certainly been sensitive to their changing proprietorial ideologies and scientific production.[25] Here too, although I have
broken down medical teaching into discrete institutional sectors, I have
noted the switching allegiances: for example, as the private schools became radicalized in the early 1830s, and again as the College of Surgeons

---

in indoctrinating the materialistic lower orders). Given the social and professional divide
between the consultants and GPs, we should be cautious in suggesting that one group had
the other's interests at heart.

23. "Medical Reform," *MG* 1841–42, 29:118; Halévy 1950:133, 152, on Coleridge.

24. Hodge 1985:239.

25. Berman 1978; MacLeod 1983; Desmond 1985a; Secord 1986b; Allen 1986.

was reformed in the 1840s. It was after the latter that the moderate reformers effectively joined with Owen to create a new liberal consensus in morphological science.

In the early 1840s plans were already afoot for the reform of both Royal Colleges. As always, the *Gazette* best caught the mood of the conservative reformers. With Peel back in power in 1841, it saw the time as ripe for a series of sensible reforms, not the one-faculty nonsense suited to an "England revolutionized," with its "annual elections, annual canvassings, annual ballotings, and universal suffrage" demanded by the "democratic brawlers." The corporations had to "steal a march" on Cobbett's men and initiate their own "sound, moderate, practical, well-digested medical reform."[26] The *Gazette* argued that to meet the more "liberal" needs of society, the RCP fellowship restrictions must be eased and the RCS be given a new electoral college.

The physicians did actually make reform plans in 1840–41. Catalyzed by the spate of radical bills before Parliament supporting state control of medicine, the physicians proposed to give the Dissenters a larger say and the licentiates the power to nominate candidates for the fellowship. The proposals actually came to nothing, but from this time leading Dissenters were elected to the fellowship.[27] Far and away the most important concession, however, was the new charter for the College of Surgeons in 1843. A college liberal such as Green, who had wanted the newly created fellowship as a pure-surgeon's prerogative, now found himself outflanked; the first three hundred fellows embraced a wider selection of appointees than he had anticipated. Included, of course, were members of Astley Cooper's coterie and the more respectable teachers, comparative anatomists, and provincials.[28] But some private school proprietors (Carpue, Lane, and Grainger) were appointed. Even the wily Wardrop was nominated in an attempt to mollify the malcontent. The council was enlarged to twenty-four members, although as a concession—presumably to get the measure through—the existing councillors were allowed to keep their seats for life. The council was no longer self-electing, nor were the new councillors to be life tenured; the fellows were to vote three members on and off yearly. "Old Corruption" was finally being checked.

26. "Medical Reform," *MG* 1841–42, 29:117–20.

27. G. Clark 1964–72, 2:702–12. Had the reforms gone through, the Scots and indeed all holders of British medical degrees would have been admitted as licentiates without further examination.

28. *L* 1841–42, 2:553. Comparative anatomists included Owen, Gideon Mantell, Newport, Mayo, McWhinnie, Langston Parker, and Solly. The charter and list are published in *L* 1843–44, 1:94–98, 411–18; and in *Plarr's Lives* 1930, 1:ix–xxiii.

The emerging liberal coalition reflected these developments. Owen was himself losing sympathy with the intransigent old surgeons. Like Gladstone, once the hope of the unbending Tories, he was beginning to embrace the new Liberalism. Nor was one side only ceding ground. The moderates, their professional aspirations being met as the hospitals and Royal Colleges opened their doors, were themselves moving to meet the college romantics.

Events at the university were to strengthen this coalition. Attempts to gain the Gower Street school a degree-awarding charter had always been stopped by a formidable phalanx of opponents: Oxford and Cambridge, the Royal Colleges, hospitals, and private teachers.[29] So the Whigs engineered a unique compromise. In 1836 a charter was granted to a new examining and degree-conferring board called the University of London (the renamed University College and King's College were to be the teaching institutions). The new board was inevitably dubbed the "Whig Government University." So it must have appeared with a senate nominated by the government and with the secretary of state responsible for the by-laws. The Tories were suspicious, portraying the senators as "mere tools in the hands of their masters in Downing-street" and science as subject "to the minister of the day."[30] They need not have been; the senate was packed with alert liberal conservatives. Anglicanism was represented by an influential group of Cambridge men: a quarter of the senators were Trinity College educated, and some were Coleridgeans.[31] Unlike the Godless College, with its Edinburgh-Paris traditions, this looked more like a tentacle of Trinity on the Thames, and Cambridge Whiggery did not have much appeal to Cobbett's men. There were even two liberal bishops and four fellows of the College of Physicians. Some of these nominations the *Lancet* considered frankly "disgraceful."[32] But it was a successful formula for the parliamentary Whigs, working to create a larger liberal consensus. It was now the senate that had the privileges—the power to confer degrees. And already by 1840 the leading London hospitals had joined the federal university as constituent schools.

This appeared very "un-English" to some, but then, as a German visitor recognized, it marked the beginning of the "modern bureaucratically

29. Bellot 1929: chap. 7. Even some of the university's own teachers opposed it. The more radical ones, like Grant (1833b:17), decried all privileges as "absurd vanities" and wanted no monopoly for the university, which they considered would only lead to further protectionism and sectarianism in the profession.

30. "The Radical University," *MG* 1836–37, 19:464; also 422–23, 504; "Charter of the University of London," *L* 1836–37, 1:491–94; Harte 1986:73–74, 79–88.

31. Harte 1986:86, 91–92.

32. "New Metropolitan University," *L* 1836–37, 1:465.

centralized state-system" in education.[33] The great liberal imperial university of midcentury was in the making. The university now brought the formerly rival teachers under the same administrative umbrella. They were constrained by the same examination requirements. As a result the business antagonisms of the 1820s and 1830s began fading as the educational standardization took effect. By the early 1840s the main benefactors of the changes, men such as Sharpey, Grainger, and Carpenter (later a registrar of the university), were moving closer to political favorites such as Owen and Roget (a senator and examiner) to forge a new center-ground coalition, which Owen came to dominate through his sheer industry and output.

The surgeons' concessions hastened the dissolution of the radical-liberal alliance, bringing the crisis to a head for the ultraradicals. They had had very little institutional power of their own. Their London College of Medicine had come to nothing; although it was still issuing diplomas in 1835, it had already "died of inanition."[34] Many radical schools had declined or collapsed during the thirties: Brookes's, Carpue's, and Dermott's all closed. With them went their house organ, the *London Medical and Surgical Journal*, to be replaced effectively by the liberal *Medical Times* in 1839.

The impact of the concessions can best be gauged on the British Medical Association, the most successful of the pressure groups. This had always been an ad hoc coalition of GP representatives, disaffected professionals, and doctrinaire radicals. By 1840 cracks in the consensus were becoming evident, and the differences between men such as Hall, Grant, and Grainger were accentuating as they responded to the corporation initiatives. Grant was consciously a martyr to the cause, refusing to submit to the RCP for a license to practice and thereby sacrificing potential earnings. When Lord John Russell received a British Medical Association deputation at the Home Office, he was astonished at Grant's predicament: here was "one of the most distinguished naturalists of the present day," licensed in Edinburgh, yet unable to "prescribe in London for a single patient."[35] The bittterness showed in Grant's basting of the College of Physicians and his admission in 1841 that for fourteen years he had been prevented by law from writing a prescription even "to save a brother's life."[36] As a result in the 1840s he was having to scrimp and scratch

33. Bellot 1929:248; *MG* 1836–37, 19:501; Harte 1986:96; Thornton 1974:65.

34. "The London College of Medicine," *Brit. Med. J.* 1926, 1:8; E. Epps 1875:270. Of the schools, the one in Aldersgate Street survived until 1848, and Sydenham College continued until 1849.

35. *L* 1837–38, 2:412–13; P. B. Granville 1874, 2:274–75; Grant 1841:81.

36. Grant 1841:10.

around, often surviving on less than £100 a year. His living conditions contrasted with the more compromising Hall's (who might have "valued fame far more than money," but who was nevertheless earning over £2,200 per annum in the 1830s from his practice among the rich).[37] The contrast illustrates the enormous financial consequences of a doctrinaire commitment. Hall was essentially seeking the abolition of privilege to clear a career path for innovators like himself; Grant was sacrificing his prospects on the altar of political change. Such diverging motives made it inevitable that the first corporation concessions would split the union.

Hall was offered a fellowship in the RCP in 1841, and he accepted despite being warned by the Lancet that he was approaching a "precipice."[38] He remained Grant's friend and even continued to abuse the College Council. But such collaboration rankled enough for Grant to lament that the "dearest companions" of youth could later become "the most inveterate of foes" because of these "invidious distinctions."[39] Compromise, however, was the order of the day. The outsiders were now penetrating the reorganized hospitals. Grainger, whose school had only been founded because of the closed-door policy of the Guy's governors, finally shut his Webb Street classrooms in 1842 to teach general anatomy at St. Thomas's, where he was joined by Carpenter and other Gower Street graduates. But the crunch for the British Medical Association really came the following year when Grainger's men accepted their fellowships in the College of Surgeons.[40] The college had successfully cut off the BMA's moderate head. It was impossible for a union that had become preoccupied with corporation bashing to continue in its old form, and it amalgamated with the new National Association of General Practitioners (f. 1844).

The doctrinaires for their part viewed the RCS rechartering with "astonishment, indignation, and disgust," believing that it was designed to forestall more sweeping reforms.[41] A mass meeting was called at the Crown and Anchor, and activists set about organizing resistance. But its effectiveness was limited by infighting and the loss of liberal support. The extremists had boxed themselves into a corner, and many were suffering.

37. C. Hall 1861:65, 69, 120, 122, 210.

38. "Reform in the College of Physicians," L 1840–41, 2:233–34; G. Clark 1864–72, 2:721.

39. L 1837–38, 1:66.

40. Of the BMA's managers, the London University clinical surgeon Robert Liston had already taken a council seat in 1840; Grainger and Pilcher followed in 1846 and 1849. In 1843 Grainger, Pilcher, and Charles Henry Harrison were made fellows. Waddington (1984:73, 79) discusses the BMA's amalgamation with the National Association of General Practitioners.

41. "New Charter of the College of Surgeons," L 1843–44, 1:23; also 125, 618–22.

Dermott's case was tragic if not typical. After his school closed in the late thirties, he had continued to teach at his house in Charlotte Street. Here he set up his Medical Protection Assembly in response to the college reforms, only to see it allegedly hijacked by Wakley in an "act of the basest treachery."[42] In 1845 his premises were condemned by the commissioner for metropolitan improvements, and he died of Bright's disease two years later.

The compromisers on the other hand were faring well. By 1842 Hall was the Gulstonian lecturer at the RCP, and he signaled his establishment success in 1848 by sending his son to Cambridge. Grainger capped his career that year as Hunterian orator at the College of Surgeons, acknowledging in his speech the completion of that "vast revolution" in science which had led in twenty years to the establishment of higher anatomy and higher design.[43] Grainger's praise for Owen shows how strongly the new alliance had been sealed. Owen's mediation between the moderates, romantics, and liberal conservatives proved highly successful. He carried teachers such as Carpenter, Grainger, and Sharpey with him on matters of the vertebrate archetype and higher design (though not his idealism), while maintaining the patronage strings to dons such as Whewell and Buckland. After 1843 the growing scientific prestige of the College of Surgeons was largely due to Owen's superb technical anatomy. His own social rise in the 1840s was meteoric: on a Crown pension at age thirty-eight, an intimate of Peel's and Prince Albert's at forty, and a palace favorite presented with Sheen Lodge by Queen Victoria at forty-eight.

The Decline of a Radical

[The doctor's disaffection] has soured him to the extent, we are sorry to say, of prejudicing his judgment.

—*Medical Times* on Grant's increasingly despondent writings[44]

While Owen enjoyed considerable success, the radicals who refused to compromise went into precipitous decline. They quickly lost ground to the new men, becoming further marginalized by the liberal-Peelite pact and the college concessions. Materialist transmutation, Geoffroy's serieswide unity, and recapitulation were now abandoned by the liberals as the intellectual accoutrements of social extremism. The failure of Lamarckism was the failure of the ultraradical movement. It became one more *ism* on

42. "Mr. Dermott," *MT* 1847, 16:619.
43. Grainger 1848:12–14, 48; C. Hall 1861:155–59, 168; Manual 1980:145.
44. "Reviews," *MT* 1841, 5:79.

the ultras' self-help list, stacked ignominiously between phrenology and Mesmerism, alongside socialism and d'Holbach's ravings.[45]

Many radicals suffered personally in this period, and of the mighty who fell few crashed so resoundingly as Knox and Grant. In both cases the fall told dramatically (if differently) on their sciences. Evelleen Richards relates Knox's increasingly pessimistic worldview directly to his professional decline. After the Burke and Hare scandal, Knox never dissected a human again, and in the mid-1830s, when Edinburgh University made its own anatomy course compulsory, his class finally collapsed. He continued to torment the university teachers and town councillors, only to have Kirk and council close ranks ever more firmly against him. Finally in 1842, following the death of his wife, the man who had once conducted Edinburgh's largest anatomy class left the city for an itinerant life of "hack journalism." He pandered to the public taste for sensationalism, trying to keep "one step ahead of pauperism" and to provide for his children (farmed out to a nephew).[46] Richards emphasizes the degree to which Knox's views now diverged from those of the London Wakleyans. They remained true to the *idéologues'* social and biological environmentalism, and insisted on state control of medicine and an expansion of the civic improvement schemes. He saw progress as a utopian dream; he deplored this state instrusion, resented the Sanitary Commission's Benthamite power, and demanded deregulation and laissez-faire—even crying off legislation against prostitution, which, he believed, infringed the rights of women. His "political nihilism" became acute in the forties. He sank into a dark fatalistic view of the human condition. The efforts of the social reformers, he gloomily predicted, would always be checked by the "iron-clad laws of human nature"—by race hatreds, religious bigotry, and colonial exploitation.

Professionally ruined (the Edinburgh College of Surgeons took away his teaching diploma in 1847), he began elaborating his cynical, deterministic, racial theories (interpreting civilization—and even the 1848 revolutions—as a series of racial struggles). He expunged the last vestiges of progressivism and environmentalism from his ethnology. He dropped all talk of arrested developments; he spurned the animal series and human perfectibility, and with these went any ranking of races that could justify oppression and slavery. Knox's "peculiar" radicalism left him supporting an anti-imperialist charter. He saw the mental characters of the different

45. J. F. C. Harrison 1987 on the way many demagogues channeled their energies into the deviant sciences after the political failure of the forties.

46. E. Richards 1988. "Dr. Knox, of Edinburgh," *MT* 1844, 10:246. Richardson (1987: chap. 6 and pp. 95–96) reassesses Knox's culpability in the Burke and Hare scandal.

races as invariant, not subject to environmental shaping or transformation. Yet mankind for Knox *remained* the product of a "self-creating" nature. He never lost his faith in life as "a property inherent in matter," or in the blood relationship of all organisms. But now he invoked a nontransmutatory mechanism, seeing the new races arise from the "generic embryo" rather than by a transformation of the adult characters of their parents. In some ways his denial of environmental control and Lamarckian transmutation allowed him to keep pace with Owen and Goodsir's idealist science (even if Owen was busily stripping the genus of all species-creating potential). But unlike these "low transcendentalists," as Knox sniffily called them, he continued to use naturalistic doctrines to buttress his anti-Church polemics. There is no doubt in Richards' mind that Knox's "pessimistic vision of human history," in which progress was utopian, reform useless, and racial conflict inevitable, directly affected his science.[47] His anatomy became antienvironmentalist; his ethnology eschewed all racial ranking; and he "rejected a reforming and improving Lamarckism" for a racially fixed alternative. But although his "moral anatomy" undermined all that the old *idéologue* optimists stood for (and it was later exploited by right-wing racists for this reason), he never relinquished a naturalistic outlook or dropped his opposition to the "Bilgewater" Creationists.

Contrast this with Grant's decline at the same time. He was bleeding to death from self-inflicted wounds, particularly from the massive social hemorrhaging caused by his steadfast refusal to kowtow to the College of Physicians or cooperate with even the reformed Royal Society. Grant remained a medical democrat committed to state control and complete suffrage. He retained the backing and (with the wolf at the door) financial help from the Wakleyans. When the wolf bit, Grant's science took a fatalistic turn quite unlike Knox's.

Grant's disillusionment was already evident in 1841. Reviewers considered his sarcastic BMA speech that year "a clever work by a disappointed man."[48] The *Outlines of Comparative Anatomy,* being printed in parts, terminated abruptly in 1841, left in midair with no introduction or conclusion. He ceased publishing from then on, something later Darwinians were at a loss to explain. He had always been an intensely private man (to the extent of routinely refusing dinner invitations), and it was common

47. E. Richards 1988; Biddis 1976:249; Knox 1852:206, n. 3. Knox rejected recapitulation in embryology at the same time that he discarded its equivalent social expression, that is, the belief that higher civilizations had passed through the stages of barbarism represented by today's "primitive" societies. On the latter see Burrow 1966:12ff.

48. "Reviews," *MT* 1841, 5:79.

knowledge that his "solitary life . . . was not without its privations."[49] But his affectations were becoming worse: he appeared eccentric and old-fashioned to a rising generation. His dress dated with his lectures: his students laughed at his frayed black swallowtail coat and white choker and complained that they were being dosed with an equally outdated Restoration anatomy. By the fifties his lectures—still couched in terms of animate matter, spontaneous generation, transformist ascent, and nebular development—were themselves fossil relics. His students now found him a "dry, melancholic, disappointed, humorous man," if still "full of burning zeal" for tidal life. Although he was dean of the medical faculty at University College in 1847, his finances teetered, and his poor living conditions began to tell. The Euston street in which he lived sank into "a slum of the worst description" (and was cleared of everybody else by the authorities). But he refused to move, telling well-wishers that the world was "chiefly composed of knaves and harlots," and he "would as lief live among the one as the other."[50]

Wakley rated him "the most self-sacrificing, and the most unrewarded man in the profession."[51] Self-sacrifice was certainly part of it, but Grant's problems were also to do with the university's joint-stock arrangements. His finances had always been precarious; from the very first he had suffered "anxiety" because of his low fees and had talked of his subsidizing an "unprotected" (i.e., noncompulsory) subject.[52] Because a professor in Gower Street was expected to live off his fees, Grant's earnings depended on the number of courses he gave and the students they attracted. So he stretched his lecture load to capacity to make ends meet. He delivered three courses comprising a total of two hundred lectures a year from 1830, and he was still delivering this number in 1850.[53] But he could only make £115 a year on average (and that before the proprietor's share was deducted to pay the dividend), leaving him the poor relation of such colleagues as Turner and Sharpey. This sum fell far short of the £300 or so required to maintain a typical middle-class lifestyle. Not that Grant's was typical; he had no wife, family, or retinue of servants to support, and he lived a Spartan existence. It was all made worse by his unworldiness

49. *L* 1874, 2:322; Poore 1901; Barlow 1958:49. On his traveling with packed sandwiches and refusing dinner invitations: Clarke 1874b:563–64; Grant to P. B. Ayres, 11 May 1852 (WI).

50. Beddoe 1910:33; Poore 1901; Schafer 1901; Clarke 1874c:277–78.

51. *L* 1846, 1:418.

52. Grant to Horner, 16 November 1830, 12 March 1831 (UCL CC P128, 2397).

53. Grant to C. C. Atkinson, 3 December 1842 (UCL CC); "Biographical Sketch," *L* 1850, 5:690.

where money was concerned. He had never really known his own worth, and although he had traveled the country lecturing in the various literary and philosophical societies his charges were considered wholly "inadequate as even a business remuneration."[54]

The problem of his research drying up also has an institutional explanation in part. Brewster saw the insidious consequences of this self-financing system for teachers of optional subjects. He put it starkly: either a savant teaches serious science and forfeits high class fees, or he becomes a "commercial speculator" and abandons his researches.[55] The loss of research seemed less of a drawback to the Benthamites, who viewed the school as a teaching factory, where wealth was generated by churning out the greatest number of trained students. Authorship was an unnecessary luxury, incidental to the main purpose. To this extent the university's role was like that of the Society for the Diffusion of Useful Knowledge which, as Lord John Russell explained, was "for the distribution and not for the discovery of Knowledge."[56]

But Grant refused to become Brewster's circus stuntman, and he always demanded standards before profit. Again, this was not only financially self-defeating, but it showed how far his perceptions of the school's role differed from those of the stockholders. Even when the university looked as though it was foundering in the early days, he insisted that cheapening degrees to raise fortunes was ill conceived and "calculated to sink our titles and dignities and our vaunting Establishment into contempt with all sensible and reflecting men." Vulgarization was tantamount to perpetuating "that monastic ignorance which has so long degraded the Universities of England."[57] With the collapse of the more appreciative private schools, a large part of his additional income dried up, throwing him

54. Clarke 1874b:563–64. Grant's fee of five guineas per lecture at the provincial institutes had to cover his train fare, food, and printed prospectus. Even when his hosts offered to reimburse his fares, he refused; nor would he allow them to wine and dine him afterward, preferring a bag of sandwiches as he set up his diagrams.

On pay and middle-class needs: J. F. C. Harrison 1979:131; Kitson Clark 1962:119; and Hays 1983:103 on the widespread undercapitalization of London's scientific institutions. Grant's situation was not uncommon. Many holders of "unprotected" chairs in London, those who were not gentlemen of means, found themselves in straitened circumstances. James Rennie's chair of natural history at King's College actually collapsed in 1834, and in 1839 he was trying to obtain money from the council to enable him to emigrate: Council Minutes, Vol. C, f. 135 (King's College Archives). For a fuller study of the psychological and pecuniary collapse (and suicide) of an orthodox blue-collar jobbing scientist, see Secord 1985b. Grant's uncompromising attitude simply exacerbated an already difficult situation.

55. Brewster 1830:326.

56. Grobel 1932, 1:20.

57. Grant to Horner, 5 November 1830 (UCL CC P130).

back on his scanty course fees.[58] When these then plummeted disastrously in 1850, the college was finally forced to step in and give him a stipend of £100 per annum—far less than those it had awarded to younger professors of compulsory subjects. Wakley of course used Grant's rescue from the edge of "penury" to slap down the "pretended patrons of learning"—the college councils and managers of science—in this "great and wealthy empire."[59] As always he saw the situation in cold political terms: as an exploitation of scientific labor and a guarding of corporation interests. But Grant's own intransigence clearly had a lot to do with it.

Grant's uncompromising attitude showed outside the university. As a dyed-in-the-wool academic, he was incapable of acceding to the demands of, say, the popularist Royal or London institutions for crowd-pulling dilettantism; nor was he equipped to meet those other contemporary requirements, piety and utility.[60] His anatomy was too abstruse and his paleontology too dry. To an extent, his leveling evolutionary views and anti-Church materialism made him unwelcome in the learned societies. But in the long run it was his political intransigence that counted most against him; he had closed off too many avenues to establishment patronage. His opposition to monopoly and corruption cost him dear in resources, institutional power, and professional privilege. He shut himself off from BAAS funding and out of the Zoological Society museum; he was unable to practice and unsuitable for a Civil List pension.[61] Nonetheless Wakley's doleful rhetoric reawakened the idea of a public presentation for this "greatest of living Comparative Anatomists" (the hype at least never let up). A campaign was launched, organized by Grant's pupils, coordinated through the *Lancet*, and in 1853, among a gathering of old radicals, savants, and students, the sixty-year-old Grant was presented with an inscribed microscope and a life annuity of £50.[62]

58. And it was not only that the private schools were collapsing. After 1837 the council actively discouraged his teaching in the nearby Sydenham College (where he had taught yearly), fearing that the competition would harm the university: Grant to C. C. Atkinson, 22 September 1837 (UCL CC 4166).

59. "The Life and Labours of Professor Grant," *L* 1850, 2:711. His yearly earnings from fees were: 1850, £39; 1851, £90; 1852, £33; 1853, £39; 1854, £41; 1855, £66 (Professors' Fees Book, College Collection, UCL). The college had awarded Sharpey an annual income of about £600: *MT* 1846, 14:25.

60. Brewster 1830:326; Hays 1974:146–47, 151–54, 160. D. W. Taylor 1971:258–59 lists Sharpey's tiny published output.

61. In 1854 the college was still trying to obtain a government pension for Grant, reminding Sir James Graham of Owen's and Newport's awards (Newport had just died): J. Wood to Graham, 12 May 1854 (BL Add. MS 43,191, f. 212, also f. 210), but to no avail (f. 217). Of course in radical eyes Newport had not been "the most worthy scion" of the university to receive a pension in the first place: *L* 1850, 2:711.

62. Speeches were made by Hall, Bowerbank, Webster, Sharpey, and Edwin Lankester:

Fig. 9.1. Thomas Wakley late in life. (From *Illustrated London News*, 1862, 40:610; collection of the author)

To the end he continued shutting doors on himself. With Hall and Wakley, for instance, he carried on needling the Council of the Royal Society. The old sores here had never been allowed to heal. In 1846 the obstetrician Robert Lee charged the society once more with illegality, this time

"Testimonial to Dr. Grant," *L* 1853, 1:140–42. On the run-up: *L* 1851, 1:33, 60, 99; *MT* 1851, 23:408–9; 1852, 25:99, 321–22; 1853, 27:145–46.

over its 1845 award to the young physiologist Thomas Beck, for a paper which, against the rules, had not been published in the *Philosophical Transactions*, and which in part merely disputed Lee's work. Wakley again stepped in to lash Roget, demanding his removal ("nineteen years of place, and salary, and favouritism [are] enough").[63] And the reformers called for a censure motion on the "Physiological Committee, which has caused the Society to stink in the nostrils of all decent people." The *Lancet* urged the appointment of Grant and Hall to the council, ostensibly as an act of justice, but these men of "inflexible honesty," mandated in this way, would of course be responsible to the radical lobby. As always, while the case for appointing radicals was on the face of it based on the need to curb "scientific Jesuitry," the move was ultimately designed to shift power from the Roget-Children salon to the medical marketplace, far further than the moderates were prepared to allow.

This "extraparliamentary opposition" to the scientific "Lords" helped catalyze the Fellows into action. A series of sweeping reforms were ushered in: the Physiology Committee was dissolved, Roget retired (none too gracefully) in 1848, the president stepped down, and even a triennial presidency was mooted (if not carried into law). But though, in Roy MacLeod's scenario, the society was switching its allegiance from Church and Crown to commerce and utility, it was never a radical victory. The incoming council included new bourgeois blood—the physicist William Grove and Edward Forbes—but these young men looked forward a decade to T. H. Huxley and John Tyndall rather than back to the old agitators. Ultimately, constitutional reforms here, as at the Zoological Society and College of Surgeons, only served to isolate the old doctrinaires further. Radicalism was a spent force; the Chartist disintegration of 1848 was a sign of the times. Not that Grant was capable of accepting an appointment now—the wounds cut too deep. The years of struggle had left their mark. A conciliatory hand was held out, but to no avail. Invited by the new secretary Thomas Bell to join the purged Committee of Zoology and Animal Physiology in 1849, he refused, deploring its "secret and invidious functions" and seeing no honor in serving on such a body.[64]

Grant's unwillingness to sit on the Physiology Committee, take out a

63. "The Royal Society," *L* 1846, 1:635–36; also 391–92, 418–19; Emblen 1970:249–54, 330, n. 29; MacLeod 1971:89–90, 1983:72; M. B. Hall 1984:83–88. Owen's role, as an expert witness for the Royal Society Council, needs to be more clearly defined. In 1847, for instance, he refereed and rejected Robert Lee's paper on the cardiac ganglia and wrote a controversial "Report": Roget to Owen, 24 January 1848 (BL Add. MS 39,954, f. 91). R. Owen 1848b:82–84.

64. Grant to Bell, 5 March 1849, in *L* 1850, 1:88. Hall, on the other hand, did join the RS Council in 1850. MacLeod 1983:57, 72. On the changes see also Crosland 1983:179–83.

license, or remain in the Zoological Society highlights the disastrous personal repercussions of his campaign for management reform. The radicals promoting Lamarckian mobility, devolution of power, and destruction of monopoly were always at their best as "extraparliamentary" activists. In practice Grant's stand, like Wakley's, was so idealistic as to be at times self-defeating. By contrast, the reformation of scientific management that occurred between 1835 and 1848 was a victory for liberal compromise and, like Parliament's, stopped well short of fierce democracy.

The old radical ostentatiously turned his back on British science. After Geoffroy's death in 1844, Grant began visiting Germany and Holland yearly, touring museums and learning Dutch.[65] He took a keen interest in philology, attending the conventions of Dutch and Belgian philologists, and early in the fifties he was elected a corresponding member of a number of Continental societies. His lectures reflected this Continental bias. Even applying to the British Museum for a Swiney lectureship on geology in 1852 he considered that he was placing himself "in the midst of the Philistines."[66] And in his resulting course on transformist paleontology, French and German authors were his predominant source. Nowhere was there the sort of evocative patriotic language used by a geological imperialist like Roderick Murchison. The clerical geologists at home were passed over in silence and Owen's paleontology was ignored.

Grant's disillusionment spilled over into his cosmological speculations. Unlike Knox, now despairing of progress and doom-mongering about future racial wars, Grant remained faithful to his radical roots and Wakleyan patrons. His paleozoology course represented the scientific swan-song of the old anti-Malthusians. It centered on the naturalistic "march of development"—an upward sweeping, self-generated ascent, fed from below, with whole faunas progressing on a wave front. They moved together, in harmony, not through the stronger members killing the weaker off (as the Malthusians would have it). Grant's zoology ruled out Providence and priestcraft; it provided a climatic cause for that "mystery of mysteries," the emergence of new species, and explained the changing form and physiology from fish to man in terms of the cooling conditions of the planet.[67] Yet his misanthropy removed any real optimism. He relegated Man to an insignificant paleontological position and projected his imminent ice-death on a refrigerating globe—the first of a series of extinctions

65. Grant to Mantell, 16 July 1850, 29 September 1851 (ATL GM 83, folder 44). "Biographical Sketch," L 1850, 2:691; Sharpey 1874:ix. He was made a corresponding member of the (French) Société de Biologie in 1851 and of the Société Royale des Sciences de Liege in 1852: uncataloged box marked "Robert Grant: Diplomas etc," UCL.

66. Grant to P. B. Ayres, 11 May 1852 (WI). Secord 1982 on Murchison.

67. Grant to Babbage, 30 April 1856 (BL Add. MS 37,196, f. 489).

Fig. 9.2. Robert Grant as an old man. (Courtesy British Museum [Natural History])

that were to extend in reverse order back to the monad. The millennium was to arrive with no hope of the Divine Society; the barren earth was to go on cycling indefinitely in the stillness of space.[68] Unlike contemporary

68. Grant, Palaeozoology lectures, ff. 23, 254–55 (BL Add. MS 31,197); Desmond 1984c. Clerics like Whewell offset any fear of a planetary freeze-up by invoking the coming Divine

cosmic optimists (typified by Herbert Spencer and Robert Chambers), Grant ended on a disconsolate note. He added a new taxonomic category, the "Metazoic" period, not for the reception of man, but for the lifeless period following "the decease of the last abortive nucleus of an aquatic cell."

His materialism now plumbed the cynical depths despised by the younger teachers. Like some German materialist, he examined his students on the "forces" that had evolved and effaced "the temporary organic film of our planet,"[69] and was unabashed about bringing his austere chemical reductionism into the classroom. (Not that the wave of émigrés' sons at University College after the revolutions of 1848—young Alexander Herzen among them—would have noticed how un-British it all was.) Grant continued to deride Owen's Platonic fancies and joke about Providence in class. But his atheistic science remained locked in a defunct world of medical manufactories and radical unions. It was all too much for the managers of the Royal Institution, who refused to have anything more to do with him.[70] Grant's vision became as bleak as Knox's. For both men, massive disillusionment and a deconsecrated Calvinism shaped their science of fatalism and despair. Grant's pessimistic projection was simply more cosmic in scope; for him, planetary death was to be accompanied by a paleontological dirge.

For the up-and-coming professionals his pessimism was incongruous in the boom years of Pax Britannica and sterling imperialism. Grant's foreign sympathies isolated him further. He remained committed to Parisian science, even though the Muséum itself was now in decline. By contrast the English celebrating at the Crystal Palace in 1851 saw their technology and arts leading the world. In the 1850s there was a new national pride in English achievement; the economy was on the upturn, social conditions were improving, and Huxley's young bloods were taking over, ushering in an age of civic pride, professional science, and social "equipoise."[71]

Grant had become an émigré in his own land. He was safely ignored by Huxley and the young Turks of zoology. The first "English Cuvier"— the man who had promised so much—had become a mere "shadow of a reputation" to a rising generation.[72] It was a common radical tragedy.

Society, but Grant would have scoffed at the idea (Desmond 1984c:406). For the understanding of solar heat in the 1850s: A. J. L. F. James 1982:163, 173–74.

69. Zoology Examination Papers, 1857–58, pp. 1, 6; 1858–59, p. 5, in *Grant on Zoological Subjects*, College Collection DG 76, UCL.

70. J. Barlow to Owen, n.d. (BMNH RO 2: f. 220); Managers' Minutes, 11: f. 146 (RI). Also Grant to Barlow, 20 May 1856 (RI General Archives, box 14, file 142).

71. Burn 1965; Desmond 1982; on Paris: Limoges 1980:213–21.

72. Forbes to Huxley, 16 November 1852 (IC THH).

Afterword
Putting Darwin in the Picture

It might seem surprising in a book on comparative anatomy and evolution in the 1830s to find no mention of Charles Darwin. But my primary concern has been public science in London's medical institutions, and Darwin is tangential in this context. Still, some words should be said, because my study could have implications for three (partially connected) problem areas of the Darwin Industry. These are Darwin's cooling relations with Grant, his own potential audience in 1837–42, at the time he developed and sketched out his evolutionary theory, and his failure to publish for twenty years.

Given the new historiographical emphasis on Grant in the Darwin Industry, it is particularly important to reexamine their political and social differences. It has long been known that Grant was the first transmutationist Darwin met (indeed one of the few he ever met before the 1850s) and that, even as an old man, Darwin could still remember the day when Grant first burst forth in praise of Lamarck.[1] Only recently however has the real nature of Grant's importance as Darwin's teacher been tackled. Phillip Sloan, in a pioneering study of Darwin's second year at Edinburgh (1826–27), has shown how he was influenced by, and subsequently modified, Grant's approach to the problem of the generation of zoophytes. Indeed M. J. S. Hodge now talks of Darwin as a "lifelong generation theorist" and of two "inheritances"—Grantian and Lyellian—as "by far the dominant determinants" of his biological theorizing.[2] Clearly, this provides new scope for a social enquiry. Hodge himself argues that reconstructions of Darwin's science will henceforth have to negotiate the partial Grantian-generational debt, and I intend to take account of it here.

Like his grandfather, father, and brother, Charles went to Edinburgh to study medicine (abortively, as it happened). In 1826 the modern Athens with its rich Firth of Forth fauna, colonial connections, university chair of natural history, active societies, and Parisian ties was a main cen-

1. Barlow 1958:49.
2. Hodge 1985:207, 238, 1983:25–26, 76–77; Sloan 1985:72. Also K. S. Thomson and Rachootin 1982:27. Kohn (1980:80–81) argues for the centrality of reproduction in Darwin's earliest transmutatory theories.

ter for the study of marine invertebrates in Britain. Darwin made good use of his university time. He dined with professors and other friends of his father's, immersed himself in the thriving student zoological community, collected along the shores with older students like Grant and Coldstream, and sat on the Plinian Society Council only a week after joining the society in 1826.[3] Perhaps we should not give too much credence to stories of his naïveté; true, he was young, only turning seventeen in 1826. And in letters home he might have spoken of himself as a "boy" needing scolding for reading two novels at once, and he might have written to his sister Caroline for advice on the best Gospel to read.[4] But in student society he probably presented quite a different persona. He owned a cabinet, was *au fait* with insect classification, and competent enough to comment on the Plinian papers, including one on the principles of classification.[5] As Sloan says, Darwin's autobiographical recollection of this period as one of theoretical sterility needs substantial revision.[6]

Darwin and Grant both remembered their Edinburgh friendship and days together foraging along the estuaries.[7] They had met at least by November 1826, when Darwin enrolled in Jameson's class and joined the Plinian Society.[8] Grant's influence during these months may be treated on two levels—technical and theoretical. On the technical side, Grant was adept at microscopic and sectioning techniques; his Wernerian Society talks, for example on the ova of the marine polyp *Flustra*, were illustrated by drawings made at × 240 magnification.[9] He guided Darwin's researches, dissected alongside him, gave him offprints,[10] and initiated him into the writings of Lamarck, J. V. F. Lamouroux, and other Continental authorities on the polyps. He took Darwin to Wernerian meetings; Grant was on its council, and he read papers on the octopus, sea pen, and *Flus-*

3. Ashworth 1935:102; Plinian Society Minutes, 1: f. 35 (EUL Dc.2.53); F. Burkhardt and Smith 1985–86, 1:28–29.

4. F. Burkhardt and Smith 1985–86, 1:25, 28–29, 39.

5. He spoke after Browne's paper on the cuckoo and Ainsworth's on classification (de Beer 1964:26–27).

6. Sloan 1985:73.

7. Grant 1861:v–vi; and "Biographical Sketch," *L* 1850, 2:693; Barlow 1958:49–51.

8. Ashworth 1935:101; Plinian Society Minutes, 1: f. 35 (EUL Dc.2.53).

9. Wernerian Society Minutes, 1: f. 272 (EUL Dc.2.55).

10. Darwin owned eight of Grant's *EJS, ENPJ,* and *EPJ* offprints (now in CUL), two of them inscribed, in addition to a copy of Schweigger's "Observations on the Anatomy of the Corallina Opuntia" (1826), undoubtedly obtained through Grant. Grant knew Schweigger, translated and communicated the paper to Jameson's *Journal* with covering notes, and probably distributed the offprints. He bound one into his own collection (UCL Zoology Store, R 920 GRA).

tra there. These papers became Darwin's guide. By March 1827 Darwin was already focusing on problems raised by Grant's work, recording his own observations on the ova of these zoophytes, helping "to generalise" Grant's "law" of their free-swimming existence.[11] Grant also encouraged Darwin to exhibit his *Flustra* specimens at Plinian meetings and routinely incorporated Darwin's discoveries—for example of the skate-leech *Pontobdella*'s eggs—announcing them in his own talks and papers.[12]

But Sloan has shown that there was a more significant scientific side to the debt. As Grant in London was shortly to use organs in their simplest state (in the embryo or in animals low in the scale) to fathom the evolution of their complex condition in the higher forms, so Sloan depicts him in Edinburgh using these primitive zoophytes as "simplifying paradigmatic models" from which the universal laws of form and generation could be deduced.[13] Sloan suggests that Darwin also adopted Grant's "analytic" attitude, accepting that a study of the lowliest invertebrates could reveal the laws of the highest importance concerning generation. This would partly explain why he devoted so much energy on the *Beagle* voyage to observing the corallines and zoophytes. (Darwin circumnavigated the globe on the surveying ship H.M.S. *Beagle* in 1831–36, traveling as a self-financed naturalist and gentleman companion to the captain, Robert FitzRoy.) Although in 1827 Darwin rejected Grant's transformism, universal monadism, and unitary point of origin for plants and animals, late in the voyage (1836) he did move closer to Grant's position on the last issue: he came to accept that the coralline algae (plants) reproduced asexually in a similar way to the encrusting Flustrae and certain coral animals. Sloan believes that this common reproductive model was a prerequisite for Darwin's accepting transmutation itself, which he did shortly after. Such an analysis, of course, gives an added interest to studies of Darwin's Notebook speculations in 1837. Not only in generational matters did they retain "many marks of this 'Grantian' heritage,"[14] but early in the B Notebook Darwin picked over (often soon to reject) other aspects of the radical Grantian model: monads, spontaneous generation, the Italian Giovan

11. C. R. Darwin Notebook March 1827, f. 6 (CUL); printed as "On the Ova of Flustra," in Barrett 1977, 2:288. On Darwin's reading of Lamarck: F. N. Egerton 1976.

12. Wernerian Society Minutes, 1: f. 272; Plinian Society Minutes, 1: f. 57 (EUL). It was common practice for students (e.g., Coldstream) and experienced naturalists (Jameson, Brewster, Fleming) to present Grant with specimens for dissection. This itself was tacit acknowledgment of his expertise on the subject. So it is not surprising that Darwin gave Grant his *Pontobdella* eggs; Grant (1827b) duly published a notice in the *Edinburgh Journal of Science*, congratulating his "zealous young friend" for first recognizing them.

13. Sloan 1985:77.

14. Ibid., 86, 89–91, 111.

Brocchi's theory of species senescence and extinction (which Grant had discussed and dropped), and a cooling earth.[15]

However, despite a certain continuity at the generational level, Darwin's transmutationism was by now a much larger self-sustaining exercise, embracing Lyellian, biogeographical, behavioral, and eventually Malthusian aspects, all of which caused it to diverge quickly and profoundly from Grant's model. Nonetheless, Sloan's emphasis on their common "generational" starting point highlights even more the problem of their eventual estrangement. After returning from his voyage, Darwin moved to London, living first in Great Marlborough Street (1837–39) and then Upper Gower Street (1839–42). We know Grant and Darwin reestablished contact, because Grant in 1836 offered to examine Darwin's corals from the voyage.[16] But even if he saw Darwin's corallines, he did not publish a description. Darwin was looking for speed and efficiency in the zoologists to whom he farmed out specimens. In 1836 Grant had a full teaching load and was in the midst of publishing his *Outlines*, so his pace at least would not have seemed to Darwin to be slackening. On the other hand, with Darwin's own theoretical work centering on these zoophytes, one can understand why he might not have wanted Grant to take them over. Darwin was now himself a competent (and competing?) zoophytologist and possibly reserved the corals for his own projected work on reef formation. Anyway, that seems to have been the end of their relationship. Darwin made no mention of Grant in the Transmutation Notebooks. He scribbled "Nothing" on each of the first four numbers of Grant's *Outlines* (the last one remains uncut: he did not even bother to read it). Why this lack of dialogue with the one man who had such an abundant enthusiasm for transmutation—the man who had originally guided him, shared his obsession for *Flustra* and the larger laws of life, and was teaching in the university's South Cloisters, not a stone's throw from Darwin's Gower Street house?

The lack of hard evidence prevents any pat answer. However, we can reconstruct the personal, professional, and social differences that provided the setting if not substance of their deteriorating relations. There is some suggestion that their friendship was already strained in Edinburgh precisely because Darwin was working in Grant's "domain." The elderly Darwin evidently remembered being upbraided by Grant one day for trenching on his "subject" (the generation of *Flustra*); the memory was

15. E.g., in B15–23: Barrett et al. 1987:174–76; Kohn 1980:74–76, 156, n. 14; Herbert 1980:63; Corsi 1978:238.

16. F. Burkhardt and Smith 1985–86, 1:512. Herbert 1974:241–45. For his own work: C. Darwin 1842. Darwin's copy of Grant's *Outlines* is now in Cambridge University Library.

bitter enough for Darwin to talk of his "contempt for all such little feelings."[17] This rather stark recollection late in life does not sit well with
Sloan's analysis or the Plinian record, which shows that Grant urged
young Darwin forward at meetings with his "discoveries." Nonetheless
some altercation apparently left a lasting impression. If it involved some
breach of patron-client etiquette or Darwin's failure to acknowledge
Grant's publishing seniority it could be squared with the Plinian record.
The issue does, though, raise the whole question of professional caution.
Like many gentlemen naturalists, Grant and Darwin had a strong sense
of scientific property. Grant's was only too evident in his ferocious descent
on those "parasites" and "plagiarists" Newport and Roget six years later—
attacks that had political overtones, with Grant's testimony helping to
shape the radical stance toward these betrayers of the third estate.
Equally, Darwin's Notebook persona is highly proprietorial: his mechanism for species change is prominently stamped "my theory." He was reluctant to tip his hand for fear that public transmutationists—meaning
after 1844 the *Vestiges'* author—might capitalize on his theory.[18] So he was
probably loath to let Grant hear of it, particularly if it had evolved out of
Grant's own "program" and there had already been a fracas over Grant's
proprietorial rights.

The issues dominating politics and medicine discussed in this book can
also shed light on the problem of Darwin's audience and his failure to
publish. Even in Edinburgh Darwin knew that much of the Lamarckian,
materialist, phrenological ground was held by radicals. He heard the debates; his note on *Flustra* was read at the same Plinian meeting as
W. A. F. Browne's "censured" paper on mind and matter, which generated a heated discussion (involving Grant). But it was in London that
Grant really became known as a radical activist: "whenever a good, honourable, generous, and liberal cause was in agitation," Marshall Hall once
said, there you "would find the name of Professor Grant."[19] The trouble
was these good, honorable causes were often not Darwin's. When he reestablished contact in 1836, Grant was denouncing corporation monopoly

17. Late in life Darwin told his daughter Henrietta that on announcing his sighting of
cilia on *Flustra's* ova, he was rebuked by Grant for trespassing on his terrain (Jespersen
1948–49:164–65; de Beer 1964:27). What really happened is difficult to tell. The evidence
is unfortunately scrappy. Darwin only recalled the incident half a century after the event;
some years later Henrietta jotted down an account for Francis Darwin, who was editing the
Life (1887). The scrap of paper was brought to light by O. J. R. Howarth earlier this century
but has now been lost again (Sydney Smith, pers. comm.).

18. F. Darwin 1887, 2:122.

19. *L* 1840–41, 1:117.

on a BMA platform. He had become the darling of the Wakleyans, in league with Church haters, disestablishers, and democrats. He stood condemned by medical Tories for his classroom politics, for his attacks on the Royal Colleges and the "monastic ignorance" of the English universities. Only a year before he had caused a furor at the Zoological Society with his demands for retrenchment and accountability, and the reverberations were still being felt. Darwin would not have wanted his evolutionary views associated with this fierce radicalism; indeed his mature Malthusian theory supported a far less destructive social program.

The Darwin family's antiradical Broughamite and Unitarian Whiggism (Darwin's mother and wife Emma were part of a wealthy Unitarian family, daughters of Josiah Wedgwood's patriarchal pottery dynasty) is becoming clearer with the publication of the *Correspondence*. Darwin was pushed further in an antiradical direction by his switch to Cambridge and a potential clerical career (1828–31). He seems to have taken the move from Edinburgh, where his grandfather's *Zoonomia* was praised by Grant,[20] to Cambridge, where Paley's *Natural Theology* treated it as a principal target, completely in his stride, which surely says something about his growing social "ambivalence."[21] By 1837 he was carrying a diverse amount of political baggage around with him. And this interplay between the family's Whiggism, his Cambridge patronage, and the widespread fear of the urban radicals provides a possible clue to his publishing delay.

The idea of ordination in the Church of England was not so odd in Darwin's case. Nor was it strange that his freethinking father packed him off to Cambridge. The Darwin-Wedgwood family typified the wealthier aspects of the provincial Whig squirearchy, not least in its practical attitude toward the social and recreational benefits of a Church career. The curacy was a safety net for second sons, preventing them from turning into wastrels on the family fortune. Charles seemed suited as a sporting-naturalist gent. It was, after all, "a vocation with modest demands that left room for much else besides—a little shooting, a little drinking, a little doubt, and, if one liked, a good deal of natural history."[22] But once again Darwin changed direction. While his Cambridge friends dutifully rusticated in their country vicarages—leading men to heaven, one confessed, without really knowing the way[23]—Darwin put off the day of his ordination and joined Castlereagh's naval nephew Robert FitzRoy in 1831 on the *Beagle*'s more certain journey. Brother Erasmus, taken with the dissipations and political life of London, worried at Charles's returning to a "hor-

20. Grant 1814:8, 1861:v.
21. Hodge 1985:211.
22. Moore 1985a:442.
23. F. Burkhardt and Smith 1985–86, 1:159; Moore 1985a:441–42.

rid little parsonage in the desert."[24] Of course Darwin never did take holy orders. But James Moore, in his study of Darwin's "vocation" as the squire-naturalist of Downe, argues the advantage of a pastoral move: by emulating the respectable life of "the clerical naturalists of Cambridge" (Down House had indeed been a parson's residence), he could keep his heretically transformed Creation in the contemplative sphere.[25] And yet, by moving farther from the potential audience for his new theory—those shallow, fashionable Londoners, as Sedgwick called them—Darwin seemed only to be piling on the incongruity.

A crucial need for quiet respectablity dominated Darwin's life. He remained an indomitable Whig, supporting Brougham and the Reform Bill. (On board ship he received a stream of gossip on its seesawing fate interspersed with comments on the "ruin" the "Borough mongers & Bishops" would visit on the country if it failed.) Yet like Uncle Josiah Wedgwood (Whig M.P. for Shropshire), the whole family loathed the "fierce & licentious" radicals.[26] The news from home was full of foreboding and fears that the radicals were "gaining strength." In the clubs and learned societies these worries were just as pronounced. Darwin on his return now steered an antiradical and storm-free course suited to a gentleman drawing on father's bank (Charles and Emma married in 1839 on a combined allowance of £1,300 a year). He hated loudmouth radicalism. He was shocked by the "snarling" as Grant's bulldog (Wakley) drew the Zoological "junto's" blood. Indeed it is clear that Grant's whole lifestyle in London, his union activities, medical leveling, and guinea-grabbing teaching occupation, were totally alien to a leisured gentleman on a family allowance. Darwin rarely ventured to Bruton Street, preferring the decorum of the Geological Society, where he became the secretary during Whewell's presidency in 1838–39. Tellingly, he was one of the "elite" Geological Club diners at the Crown and Anchor in December 1838, summoned by Buckland to greet Lord Brougham and watch Grant's radical paleontology being given the coup de grace. Of course, Darwin had no more sympathy for Parisian serialism than Owen and Whewell ("in my theory," he had scribbled the month before, "there is no absolute tendency to progression").[27] Nor was his Malthusian evolution anything like Grant's modified Lamarckism. He

24. F. Burkhardt and Smith 1985–86, 1:259.

25. Moore 1985a:459–63. Darwin engaged for years in what Kohn (1985:245) calls a fruitful "internal dialogue." Desmond (1982:30) describes Sedgwick's dim view of *Vestiges*' London reception.

26. F. Burkhardt and Smith 1985–86, 1:299, 212, 302, 372–73; on the allowance, Freeman 1982:11. Also Schweber 1980:257–59. Uncle Jos was "greatly revered" by Darwin as "the very type of an upright man" (Barlow 1958:56).

27. N47: Barrett et al. 1987:576.

was already explaining the newer romantic anatomy naturalistically. He had adopted a branching concept of nature and realized that it was "absurd to talk of one animal being higher than another." He could explain rudimentary organs as the remnants of once-functional structures, and man's vertebral skull as a sign of his ancestry from a mollusk with vertebrae "& no head-!!"[28]

Walter Cannon once suggested that Darwin opted for a "complete system of utilitarian materialism"—one that could embrace the origin of structure, habits, reason, beauty, and morality—precisely to meet the challenge posed by his romantic compatriots.[29] All-inclusiveness was now essential because the Coleridgeans had denied that the sentiments of beauty and morality were emergent faculties, born of the baser instincts, or that they were susceptible to any utilitarian explanation. If Cannon is right, then there was an even greater irony in Darwin's joining this polite company convened to bury Grant's Lamarckian fossil "opossum." Here Darwin was, with his social equals, most of them Oxbridge conservatives, some of them Anglican divines. Yet he was hatching an evolutionary replacement for their entire worldview. He too was (privately) calling this little "opossum" "the father of all Mammalia" and in the process speculating on the Creationist "fabric" falling, even as Sedgwick talked of society crumbling as a consequence.[30] In short his science equally undermined the Anglican social ideals of his powerful patrons. It could well have been the stuff of inner conflict.

Where did Darwin's doctrines place him politically, given contemporary alignments? It is well known that he made Malthus's pessimistic view of human nature and doctrine of progress through "painful struggle" the backbone of his biological theory.[31] Where Malthus had seen population increases lead to competition, endemic starvation, and wars (undermining the Jacobins' rosy view of inevitable progress), Darwin saw a cutthroat competition benefit each species through the survival of improved stock. At the same time he naturalized Paley; adaptation became the yardstick of improvement, and Sedgwick's caring Father collapsed into a piece of self-regulated selection. Like his "revered" Uncle Jos, turning his work force into machines and devising better machines to replace them, Dar-

28. B74, 84, 99, E.89: Barrett et al. 1987:189, 192, 195, 420. Ospovat 1981:28–30.

29. W. F. Cannon 1976:378–81.

30. B87–88, C76: Barrett et al. 1987:192, 263. Darwin was actually using Owen's work to show that reptiles "formerly might have approached nearer to the Mammalian type" and have given rise to the monotremes and marsupials (F. Burkhardt and Smith 1985–86, 2:106, also 415).

31. Young 1985:188, chap. 2; Ospovat 1981:62ff.; Bowler 1984b:94–99; Shapin and Barnes 1979:128–33.

win was recasting nature as a self-improving workshop. All this puts him
on the ideological perimeter of the Broughamite-manufacturing camp—
far from the center, to be sure, for he was naturalizing Malthus and Paley
in a way that the old-school Whigs (whose SDUK magazines were staple
fare for Darwin's sisters)[32] could never have condoned. Nor would moderate Broughamites of the Bell-Roget camp have endorsed Darwin's reductionism; indeed, it would have had a much stronger appeal to the Unitarian Left opposing Bell's group on the SDUK Committee. Silvan
Schweber makes a convincing case for Darwin's familiarity with the political economists holding an individualist credo—those for whom social
progress depended on competition and the struggle for personal gain.
Uncle Jos toed a Broughamite line in the House. He was the brother-in-
law of Bentham's friend Sir James MacKintosh, whom Darwin "much es-
teemed," having first met him at the Wedgwood manor in 1827. Before
the *Beagle* voyage Darwin again dined with MacKintosh and referred to
him extensively in the transmutation notebooks. Through such Whig intellectuals Darwin gained a broad understanding of the social advantage
deriving from individual struggle. Also, as Schweber shows, the economists' division of labor provided a political model for his later theory of
divergence, which explained the competitive division of varieties and
their exploitation of separate niches in nature's marketplace.[33]

The above is a deliberately simplistic reading. Even as a short sum-
mary it is too crude and takes no account of the dialectical richness of
Darwin's thought. David Kohn and M. J. S. Hodge are right: a full study
is now needed of "the institutional and ideological content of Darwin's
entire notebook zoonomical program,"[34] taking in his metaphors of
struggle and competition, and exploring beyond his cultural free-market

32. F. Burkhardt and Smith 1985–86, 1:299. McKendrick (1961:34) on Josiah Wedgwood,
Sr., as a factory organizer, creating a disciplined work force and division of labor—in short,
making such "*machines* of the Men as cannot err."

33. Schweber 1980:256–59, 277, 1985:44–47, 52–56, 64–65; F. Burkhardt and Smith
1985–86, 1:97, 245; Barlow 1958:56; and Manier 1978:141, 163. Darwin made the competitive social views of the utilitarians central to his science of human progress. "There should
be open competition for all men," he pleaded in the *Descent of Man* (quoted in Greene
1977:2–3, 24), "and the most able should not be prevented by laws or customs from succeeding best." As we have seen over and over again in the text, he was far from alone in this view.
In the 1830s such calls were rampant among the reformers attacking aristocratic Anglican
privilege, and social evolutionists such as Matthew and Southwood Smith saw nature sanction their utilitarian rallying cry in just the same way. The language was absolutely de rigueur
among the free traders trying to open up society when Darwin was devising his theory.
Moore (1986b) gives the most incisive portrayal to date of Darwin as a Social Darwinian,
importing his politics into nature and extracting "scientific" conclusions comforting to a liberal Englishman of his class.

34. Hodge and Kohn 1985:205.

bias to issues such as colonialism and antislavery. The latter is certainly relevant. Darwin's abhorrence of slavery reinforced his Broughamite sympathies. (Brougham, the member for Yorkshire's Dissenting industrialists, had fought the 1830 election on an abolitionist platform, and he was still stirring passions with his speeches on West Indian slavery in 1838.)[35] A hatred of slavery colored Darwin's ethical appreciation of animals, as it did for many others at this time. "Animals—whom we have made our slaves we do not like to consider our equals," he jotted. "Do not slave holders wish to make the black man other kind?"[36] In the same way, Quakers juxtaposed ethical, antislavery, and egalitarian sentiments. John Epps could talk in the same breath of an abominable slavery and the equality of animal life. For many radical Nonconformists animals had a mind and suffered pain, and Darwin now used this kinship in pain to draw evolutionary conclusions, jotting breathlessly:

if we choose to let conjecture run wild then animals our fellow brethren in pain disease death & suffering & famine; our slaves in the most laborious work, our companion in our amusements. they may partake, from our origin in one common ancestor we may be all netted together.[37]

Gruber imagined Darwin worrying over the outcry against putting the "rudiments of human mentality in animals."[38] But the situation was not so simple. Robert Richards has shown how natural theologians were already divided over the issue of consciousness in animals. True, many a Cambridge don restricted reason to man, whereas the sensationalists (from old Jacobins like Godwin and Erasmus Darwin, through the Regency republicans, to the 1830s radicals and transmutationists) denied this mental chasm. (They interpreted instincts as intelligent responses to the environment that had become habitual.) But others who influenced Darwin— John Fleming, Brougham, and the old Bridgewater author William Kirby—also accepted a creative ascent of reason from animals to man. Richards' point is that Darwin's delay cannot be attributed to his fear of a basting for giving a mind to the brute.[39] A second point emerges from Richards' work: Brougham's *Dissertations* convinced Darwin that some

35. Stewart 1986:242–45, 286, 334; New 1961:283–304; F. Burkhardt and Smith 1985–86, 1:337–38 and *passim*. The Darwins and Wedgwoods had long been active in the antislavery movement. Darwin even exploded with rage in 1845 on hearing Lyell report uncritically the views of the Carolina slave owners he had met on his American trip (Colp 1986:24–25).

36. B231: Barrett et al. 1987:228; Gruber and Barrett 1974:41, 65–68.

37. B232: Barrett et al. 1987:228–29; C. U. M. Smith 1978:246–64.

38. Gruber and Barrett 1974:202, chap. 11.

39. R. J. Richards 1981:199–218, 227. Schweber 1980:231–32 deals with the influence of Brougham's quantitative approach to biology and its derivation "from Utilitarian political economy."

instincts (like that of a wasp sealing a grub into its egg cell as future food for the larva) could not possibly have developed from an intelligent habit. In other words Brougham moved Darwin away from older sensationalist notions. Darwin was forced to use natural selection to explain instinctive acts—to search for evidence that instincts (like organs) vary slightly, giving selection something to work on. So again, Darwin appears in an ultra-Broughamite tradition, rather than a radical *idéologue* one.

But Darwin's delay is still a problem. The point is not that Brougham saw "human reason prefigured in the mind of a worm."[40] Any justification Darwin could have extracted from his lordship's work would have been grossly oveshadowed by his own belief that man's mind had evolved from the worm's through a self-adjusting Malthusian mechanism. This was the crux, making it difficult to believe that Darwin could have presented his 1842 *Sketch,* even as a piece of unfinished Broughamite business, without causing massive controversy. By making mind and morality subject to "self-developing" forces, he threatened those ideals so important to Owen, Lyell, and the clerisy—man's dignity and individuality. But this aside, when we do identify the London groups that were standing on soapboxes arguing the case for a conscious, suffering creation, we see that Darwin might still have had cause for concern. Animal awareness, pain, and immortality were not neutral ethical or religious issues. They attracted tenacious Methodist and Dissenting support, and many Christian radicals went along with atheists like Elliotson, imputing consciousness and pain even to the primitive polyps. As Epps said (with suitable emphasis), Elliotson in his *Human Physiology* maintained "what every true philosopher must maintain, *that animals enjoy* MIND."[41] The Dissenting medical schools were buzzing with such notions. And we have seen how they were intertwined with anti-Anglican politics and aired in the radical press, where they engendered furious responses from the clergy. Despite Darwin's Broughamite respectability and reformed Malthusian views, he ran the obvious risk of being lumped with these "fierce & licentious" radicals. So I agree with Gruber that there would have been an establishment outcry. But it was not only Darwin's science that would have been impugned. These contingent political aspects suggest that he himself would have been accused of social abandon.

We are beginning to see where Darwin's medley of ideas would have located him in the clerisy's eyes. His naturalism typified that of arriviste professional and industrialist groups busy curbing monarchic caprice in

40. R. J. Richards 1981:227.

41. J. Epps 1828:118. Durant 1985a and J. Browne 1985 for studies of Darwin's "denial of dualism" and his belief in the continuity of mind and expression in man and animals.

society and nature. For the oligarchs of new wealth a legislative harness on the absolute monarchy guaranteed John Bull's freedoms and obviated the need for social upheaval. Darwin himself read the books of civil and natural history together; as he put it, a lawful zoological and social development "baffles idea of revolution."[42] Regular laws kept the querulous canaille in their place; they also prevented a capricious deity from interfering in earthly matters. Consider Darwin's complaint that law ruled the heavens, but Creation was contained on the earth: "We can allow satellites, planets, suns, universe, nay whole systems of universe to be governed by laws, but the smallest insect, we wish to be created at once by special act."[43] This kind of catastrophic interruption was increasingly deplored in a reformed society. "No, no," the *Medico-Chirurgical Review* snapped at the sight of Kirby's manikin nature dancing to the tunes of Creative whim: man and the "elements" were bound by the same hierarchy of "general laws." Like the new Dissenting professionals disavowing Anglican interpretations of nature, Darwin was also invoking natural law. Like them, too, he absorbed Comte's positivism ("Zoology itself is now purely theological"), and he followed Harriet Martineau's arguments for moral relativism.[44] That he almost got Martineau for a sister-in-law only shows how close the family (or at least brother Erasmus) was now to the heart of popular Broughamism.

Having considered the reformed Malthusian aspects of Darwin's science we now move on to the question of his potential audience in the late 1830s and 1840s. Sandra Herbert and Martin Rudwick have both had difficulty locating one for Darwin's covert science. But they were looking among the gentlemanly naturalists and elite of the learned societies.[45] Yet we should not have expected supporters in the star chambers, among the Oxbridge divines of the Geological Society, the "closet taxonomists" of the Linnean Society, or Kirby's "*patriot* Zoologists."[46] It will pay to look beyond these groups to the back benches; an audience implies listeners, buyers of books, and sponsors of science, not only its gentlemen producers. Aspects of Darwin's science could have appealed to many of the moderate reformers discussed in the present study—to the medical Unitarians, Dissenting industrialists, and urban Benthamites trying to root out

42. E6e; Barrett et al. 1987:398; Moore 1986a:53.

43. N36, also B101: Barrett et al. 1987:573, 195; Ospovat 1981:30–33.

44. N12: Barrett et al. 1987:566–67; F. Burkhardt and Smith 1985–86, 1:345–46, 518–19, 524. Erasmus was quite smitten by Martineau; in the 1830s she was "a great Lion in London," patronized by Brougham, who persuaded her to write on the Poor Laws. Charles met her in 1836, but wondered about the prospect of so fiercely independent a sister-in-law.

45. Herbert 1977:189; Rudwick 1982:203.

46. Desmond 1985a:164–68.

the old monopolies and encourage laissez-faire. Had he published his *Sketch* in 1842 it would have sat comfortably beside Southwood Smith's *Divine Government,* with its leitmotiv of individual striving, competition, and social improvement. So we can say something more than that Darwinism had a "uniquely British character."[47] Its appeal would have been to specific class fractions. Darwin had no truck with Coleridgean idealism. He once wrote, after listening to his freethinking brother: "Plato says in Phaedo that our *'necessary ideas'* arise from the preexistence of the soul, are not derivable from experience.—read monkeys for preexistence."[48] Now, no Coleridgean patriot could have substituted an ape ancestry for the sublime Platonic process, but Darwin was not writing on the side of Church privilege and Tory corporation authority. Just as certainly, though, committed socialists and antiworkhouse radicals who decried the "revolting" Malthusian Poor Laws as anti–working class could not have written Darwin's *Sketch* either. To have done so would have been to condemn the cooperative movement and condone the workhouses and callousness of the new capitalists. The artisan atheists were fashioning their own rival social product—a Jacobin Lamarckism, environmentally controlled, and driven from below, which resulted in an upward-pushing inevitable progress toward a fully cooperative society.[49] This, too, was totally unlike anything Darwin envisaged. He was articulating something quite different—a Malthusian science for the rising industrial-professional middle classes.

We find potentially supportive Broughamites and utilitarians turning

47. Schweber 1980:198; Moore 1985a:438.

48. M128: Barrett et al. 1987:551.

49. Desmond 1987:79. Two decades later (in 1858) the phrenologist and naturalist-explorer Alfred Russel Wallace did postulate a selectionist theory, although the extent to which it differed from Darwin's is still debated. For example, Kottler (1985) discusses Wallace's preferences for enviromental selection rather than individual competition, and also for an un-Darwinian "group selection." Bowler (1976a) and J. Browne (1983:181, 193) also look at Wallace's and Darwin's differences over the analogy of artifical and natural selection. Wallace, though a poor lawyer's son, had breezed in and out of Mechanics' Institutes and socialist Halls of Science in the later thirties, come under Combe's spell, and been converted to the idea of evolutionary development by the *Vestiges* in 1845. The apparent paradox of a socialist striking up a "Darwinian" position in 1858 is dealt with in two crucial studies. Durant (1979) explores Wallace's short period of false consciousness, when he was buffeted by countervailing Spencerian winds, and traces his return in the 1860s to a more ideologically consonant spiritualist-socialist evolutionary posture. R. Smith (1972) tackles the phrenological, spiritualistic, and reformist aspects of Wallace's scientific thought, and exposes his fundamental disagreement with Darwin over the meaning of causality and the function of selection, which in Wallace's case was to realize "the ideal of perfect man." On the other side, Darwin's growing opposition to social environmentalist views (especially those of female emancipationists) is examined in E. Richards 1983:67ff., 1989.

up increasingly in the London societies as the thirties and forties progressed. There were the making of support groups even in the Geological society in the 1830s, judging by the scatter of men like Warburton, the Unitarian mining entrepreneur John Taylor, and the London professors. They could have used a naturalistic theory based on competition and the rewarding of talent. Couple these industrial patrons and academics with the young Geological Survey professionals (who sat as a party in the society's rooms from the mid-1840s and later supported the *Origin*), and we can see that there was a potential professional audience.[50] The back benches accommodated a number of different groups, including in the Zoological Society medical radicals and empire builders, merchants and M.P.s, some of whom welcomed a more naturalistic science. As we saw earlier, many of the Zoological Society's management reformers and radical anatomists left in 1835, taking potential support for an 1842 *Sketch* with them (although a Benthamite opposition did remain active throughout the decade). A coalition of improving lawyers, physicians, and engineers had greater success a decade later in taking control of the Botanical Society. By the mid-1840s the society's genteel floral concerns had given way to more pressing Benthamite ones, with discussions switching to the potato blight, adulteration of sugar, and use of peat charcoal to purify London's sewage.[51] So these societies had their utilitarian cadres, groups of improving professionals who were possible converts to Darwin's brand of science.

Gruber suggests that Darwin's mental anguish might have been partially alleviated had he allied himself with Mill's utilitarians at this point.[52] He might indeed have found a more sympathetic audience. But it was one that was envious of traditional Anglican power, and there lies the trouble. The Benthamites who might have championed an 1842 *Sketch* were those who were antagonistic to the Cambridge-educated gentry. Darwin at this time risked his evolutionary naturalism being incorporated into their strategies to weaken Church and corporation authority. The Benthamites' attacks on the Church and the Establishment were deplored in the Tory press, which raised the specter of a Church in ruins and all moral safeguards shattered. Such a hypothetical move into Mill's camp could seriously have affected Darwin's respectable standing. Not that such a move appeared remotely likely, judging by Moore's evocation of Darwin as the "squarson"-naturalist of Down.

Yet any fears Darwin might have had on this point would have been

50. Secord 1986b:224, 260–61.
51. Allen 1986:41–42.
52. Gruber and Barrett 1974:71.

dwarfed by the prospect of ultraradical appropriation, especially during the period of Chartist violence (1839–42). There was a real risk of this. Even if a Malthusian *Origin* was more acceptable to the middle-class utilitarians, nothing could have stopped the anti-Malthusians—who included many extremists—from cannibalizing it for their own ends. There is reason to suppose that Darwin's mass audience would have included a large proportion of these groups, lured by his scientific naturalism.[53] The anti-Whig wing was crowded with social Lamarckian environmentalists: statisticians compiling their ledgers of death, radical anatomists, and anti–Poor Law activists in the press, private schools, and university. Even more worrying, outside medicine, the atheists writing in the illegal *Oracle of Reason* (1841–43) and the socialists congregating in Red Lion Square were thrusting materialism and transmutation into a confrontationist class context. As the illegal street prints show, the ultraradicals were adept at selective cannibalizing. Little escaped their democratic depredations. Lawrence, Southwood Smith, Lyell—all were grist for their materialist mills.[54] So even though the street evolutionists hated the Malthusian weak-to-the-wall thesis, many would still have reveled in the sight of the Anglicans' interfering Deity bound up by law. When the *Origin* was published a similar sort of selective endorsement did actually occur. Grant for one delighted in Darwin's destruction of the "species mongers," yet he continued to refine his rival radical science, with its "live" matter, spontaneous generation, and continuous origination of discrete evolutionary trees.[55] He remained oblivious to the real meaning of Darwin's Malthusian mechanism.

Darwin continually broached the question of materialism in his notes. He read the phrenological works and accepted mind as a "function" of the brain. Indeed his M Notebook was an exploration of the hereditarian and environmental cause of thought and behavior—even of reverence ("love of the deity effect of organization, oh you Materialist!").[56] He must therefore have realized how ripe his theory was for exploitation by the extremists. He had gone far in echoing the sorts of materialist metaphors wielded by the deists, atheists, and radical Dissenters in the 1830s. Few would have been surprised by his bon mot: "Why is thought. being a secretion

53. Cooter 1979, 1984 for extended discussions of naturalistic science and the working classes.

54. Desmond 1987:99–101.

55. Grant 1861:v–9. Few of the Edinburgh savants active in the 1820s that we have discussed ever really escaped the science of the period. Looking back from the 1860s, Knox, like Symonds, saw little that was new or worthwhile after 1830, and unlike Grant he treated the *Origin* quite dismissively (Symonds 1871:8; E. Richards 1988).

56. C166, N5: Barrett et al. 1987:291, 564.

of brain, more wonderful than gravity a property of matter? It is our arrogance . . . our admiration of ourselves."[57] Flamboyant atheists like Elliotson, outspoken materialists like Dermott, and the artisan agitators had all resorted to this kind of crude reductionism. Indeed, it was the sort of slogan he had heard bandied around by the Plinian phrenologists in 1827. All in all, it is difficult to believe that he was blind to the danger of falling into the extremists' hands—extremists who were using materialism to underwrite their anticlerical, antistate programs.

One begins to appreciate why in 1838 Darwin began devising ways of camouflaging his materialism. Bad dreams were not the only sign that he feared a heated reaction. He began dwelling on the prospect of persecution and collected quotes on the "opposition of divines to progress of knowledge."[58] He had no illusions about the reception he could expect from the clerisy: his grandfather's own books had been assailed by the *Anti-Jacobin*, the Lawrence episode was notorious, Elliotson's cynical "Spinozaism" was still subject to furious attacks in the Tory press, and Darwin in his teens had witnessed the "censored" Plinian debates.[59]

Gruber suggests that Darwin feared being persecuted as an atheist. At one level this is perfectly probable, even though we know from the study of the London anatomists that mental materialism did not imply atheism. But Darwin was too worldly-wise not to realize the wider social implications. Being labeled a materialist or atheist carried with it a much more damaging class indictment (as Lawrence found out). By "netting" man and ape together in a materialist evolutionary sweep Darwin invited being identified with Dissenting or atheistic lowlife, with activists campaigning against the "fornicating" Church, with teachers in court for their politics, with men who despised the "political archbishops" and their corporation "toads." Ultimately Darwin was frightened for his respectability. These fears of a fierce reaction were justified. Sedgwick—with whom Charles had ridden on a geological tour in 1831—envisaged the social fabric being torn apart by the laboring unions armed with godless developmental theories. For a gentleman called to dine with the Oxbridge elite, priming itself to defend man's "dignity" against the radical anti-Creationists, publishing a theory of evolution would have been tantamount to treachery. Like Lawrence's crime in 1819, Darwin's in 1842 would have been treated as a betrayal of the clerisy. His "crisis" might have been part of a "culture-

57. C166: Barrett et al. 1987:291. I agree with Moore (1985a:452) that "neither professionally not politically . . . was it prudent for Charles to disclose his thoughts."

58. N19e, M57, C123: Barrett et al. 1987:568, 532, 276; Gruber and Barrett 1974: chap. 2.

59. Gruber and Barrett 1974:39, 43, 47, 204–5; McNeil 1987: chap. 3; Schweber 1977:310–14; Desmond 1984b:199–200.

wide phenomenon rooted in the basic tendencies of an industrializing society,"[60] but it was still the stuff of nightmares. He might have been formulating the future hegemonic science of industrial capitalism. But for the present it threatened the existing elite, the Church and corporation men, who were resisting calls for continued reform, suspicious, like Coleridge, of an excessive, greedy capitalism destroying the old harmony.

Lawrence had been "frightened" into withdrawing his book and keeping his Sabbath, leaving the radicals to profit from their pirate operations. Darwin was evidently not about to follow suit. In 1844 he left his wife £400 with instructions to publish on the event of his death.

60. Moore 1986a:66.

Appendix A
Comparative Anatomy Teachers
in London in the 1830s

Before 1837, comparative anatomy teaching at the Royal College of Surgeons was sporadic, even though a number of Hunterian professors of comparative anatomy had been appointed since 1810, including **Everard Home** (1810, 1813, 1822), **Astley Cooper** (1814–15), **William Lawrence** (1816–19), **Benjamin Brodie** (1820–21), **Joseph Henry Green** (1824–28), and **Herbert Mayo** (1829–30). In addition, the Hunterian Orations sometimes centered on comparative anatomy, the most memorable being **Anthony Carlisle**'s in 1826, with its disastrous appraisal of oyster "design." Comparative anatomy teaching in the hospitals was very infrequent, but **John Flint South** lectured yearly on the subject at St. Thomas's after 1825.

In 1826, on the founding of the London University, a chair was established in comparative anatomy. This was taken by **Robert Grant,** who started teaching in autumn 1828 (1828–74). In the early 1830s Grant's was still the only full course on offer in town. In 1832–33 **Charles Bell** delivered short design-oriented lectures on the Hunterian preparations at the RCS. In 1834 St. Bartholomew's instituted a course, taught first by **Richard Owen** (1834–35), then by his protégé **Arthur Farre** (1835–40), followed in 1840 by **Andrew McWhinnie.** In 1835 Grant's pupil **William John Little** started a course at the London Hospital in the Whitechapel Road (1835–41 or later), as did **Thomas Bell** the same year at Guy's (1835–37), where he was succeeded in 1837 by **Thomas Wilkinson King.** Owen's pupil **Thomas Rymer Jones** delivered a trial course at King's College in spring 1836 and was appointed to the new chair of comparative anatomy that August, while Owen himself took the first permanent Hunterian chair at the RCS the same year and started teaching in 1837. **Thomas Wharton Jones** was appointed lecturer on comparative anatomy at Charing Cross Hospital in 1837, and in 1840 **Samuel Solly** took over the teaching at St. Thomas's.

The private anatomy schools also catered to this increasing interest in comparative anatomy. Grant taught the subject at Aldersgate Street (1835–38), Great Windmill Street (1836–37), and the new Sydenham College in Gower Street (1837–38). **Henry William Rush** delivered a sixteen-lecture course at the Westminster School of Medicine (1835–36), the radical **Thomas King** taught it at Blenheim Street (1836–37), and the fiery **George Dermott** delivered addresses on the science at his Gerrard Street school (for instance, in 1836).

In 1833 the Royal Institution founded the Fullerian professorship of physiology. This was a triennial appointment, and its early incumbents were all anatomical physiologists or comparative anatomists. The first four professors were **Peter**

Mark Roget (1833), Grant (1837), Rymer Jones (1840), and **William Benjamin Carpenter** (1844).

From this list it can be seen that the science flowered in the mid-1830s, but the only full-time academic comparative anatomists were Grant, Owen, and Rymer Jones. The rest derived their living from teaching anatomy, medicine, or surgery, or from a practice.

(This appendix was compiled from the course listings published at the beginning of each academic year in the *Lancet,* with additional information from printed lectures and biographies. The details of the Hunterian professorships were supplied by the Royal College of Surgeons; those of Rymer Jones's courses come from the Council Minutes, Kings College archive.)

Appendix B
Biographical List of
British Medical Men

This list covers the main surgeons, physicians, medical teachers, publishers, and graduates discussed in the text. (Those mentioned only once or in passing are excluded.) It is deliberately restricted to medically trained men. I have refrained from integrating them with the gentlemen naturalists usually targeted in histories of British biology in order to keep the spotlight on the science generated in the various medical milieux.

Consultants honored with a baronetcy before 1835 are titled Sir. An asterisk (*) indicates a philosophical anatomist (or sympathizer). A double asterisk (**) indicates a Coleridgean idealist or supporter of Richard Owen's archetypal anatomy.

John Abernethy (1764–1831). Influential teacher of anatomy, and the assistant (1787) and full surgeon (1815–27) at St. Bartholomew's Hospital. He was close to the elderly John Hunter and, as lecturer at the RCS in 1814–17, he expounded Hunter's vitalistic views. President of the RCS in 1826. Unoriginal in research and often blunt to the point of rudeness, he was nonetheless a riveting lecturer, whose students included Lawrence, Brodie, Owen, Farre, and many others.

John Anderson.* A one-time clinical clerk at Guy's Hospital. He summarized the latest Continental transcendental anatomy in a series of papers to the Physical Society at the Hospital in 1835–36. They were published as *Sketch of the Comparative Anatomy of the Nervous System* (1837).

John Barclay (1758–1826). A minister of the church who took an Edinburgh M.D. in 1796. He founded a successful Surgeons' Square school, was a committed antimechanist, and emphasized comparative and veterinary anatomy. Grant lectured for him in 1824–25, and Knox bought out the school the following year.

Martin Barry (1802–55).* Embryologist, educated in Edinburgh, Paris, and Germany (M.D. 1833). Influential in introducing von Baer's nonrecapitulatory embryology to the British. He obtained his fellowship of the Royal Society for work on the fertilization of the ovum. In 1845 he became house surgeon to the Royal Maternity Hospital, Edinburgh.

Sir Charles Bell (1774–1842). An Episcopalian minister's son. Bell was an Edinburgh-educated Whig who worked closely with Brougham and Horner in the SDUK and London University. He bought the Great Windmill Street School (1812) and was professor of anatomy and surgery at the RCS (1824). His RCS lectures on comparative anatomy (1832–33) were traditionally Paleyite; he wrote a Bridgewater Treatise and with Brougham edited Paley's *Natural Theology*. He dined with Cuvier in London in 1830, but he loathed Geoffroy's science and gave up his chair at London University (1828–30) because of the school's disorganization and Geoffroyan emphasis. He was knighted in 1831 and died the professor of surgery at Edinburgh (1836–42).

Thomas Bell (1792–1880). Dental surgeon at Guy's Hospital (1817–61), where he also lectured on comparative anatomy. A good administrator in the Royal Society (secretary, 1848–53). While professor of zoology at King's College, London (1836), he wrote on British mammals and described the reptiles from Darwin's *Beagle* voyage.

Edward Turner Bennett (1797–1836). A surgeon trained by Joshua Brookes; active in the Zoological Society administration (secretary, 1831–36).

George Bennett (1804–93). Plymouth-born naturalist-traveler. He studied medicine (member RCS, 1828) and assisted Owen in the RCS Museum before sailing for the Far East. He returned to England in 1831 exhibiting a native girl, and showed a cannibal's skull at the Phrenological Society. He shipped Owen platypus specimens from Australia (1833) and wrote *Wanderings in New South Wales* (1834). Returning to live in Australia, he managed the Sydney Museum, had a large practice, married three times, and described many new species.

James Richard Bennett (d. 1831).* Dublin B.A. (1817). He then studied medicine in the city before founding an English school in the Hôpital de la Pitié in Paris (1822–25). On its closure, he came to London to teach in Carpue's Dean Street school. He was demonstrator (1828) and professor (1830) of anatomy at London University. He helped found the medical school in the French style, taught philosophical anatomy, and tormented the incumbent Pattison over his old-fashioned approach. Bennett was the students' favorite; on his premature death in the spring of 1831 (described as a "public calamity" by the *Lancet*), they collected sixty guineas for a bust.

George Birkbeck (1776–1841). Edinburgh M.D., 1799; professor at the Andersonian University, Glasgow. Financial backer and first president of the London Mechanics' Institute (1824), where he took the "dangerous" step of dissecting cadavers before a working-class audience. Physician to the Aldersgate Street General Dispensary and councillor at London University.

Sir Benjamin Collins Brodie (1783–1862). Charterhouse and Oxford educated, and Everard Home's apprentice at St. George's Hospital, where Brodie himself became surgeon (1822). He was professor of comparative anatomy at the RCS in 1820–21. Brodie was a Foxite Whig and a family friend of Lord Holland; his noblesse oblige was typical of the RCS's reformers. He was made sergeant-surgeon to the king in 1832 and a baronet in 1834. He was a councillor of the RCS (1829–62) and president in 1844.

Joshua Brookes (1761–1833). Studied under William Hunter and at the Hôtel Dieu in Paris. A successful teacher and comparative anatomist, he built a huge museum in Blenheim Street and pioneered techniques for preserving specimens. His protégés included Dermott, Youatt, T. Bell, and E. T. Bennett. But he was perpetually grimy and disdained by the sergeant-surgeons. After the RCS refused to accept summer certificates his school declined, and he retired in 1826; his collection was auctioned off (1828–30), and he died apparently penniless in 1833.

William Alexander Francis Browne (1805–85). He left Edinburgh University in 1826 a materialist, phrenologist, and one of Combe's coterie. He traveled in France (1826–30) and was a mad doctor at Montrose and (in 1839) Crichton Asylum, where his humane treatment gained him widespread praise. In 1858 he became the first commissioner of the Scottish Lunacy Board.

Sir Anthony Carlisle (1768–1840). A student of John Hunter's and a surgeon at Westminster Hospital (1793–1840). He was a courtier knighted in 1820 by the Prince Regent. As an RCS councillor (1815–40), he opposed any democratization of the college. He was the Hunterian orator (1820, 1826) and president of the RCS in 1828 and 1837.

William Benjamin Carpenter (1813–85).* Bristol-educated Unitarian. He studied at London University (1834–35) and Edinburgh (1835–39), and taught physiology (1840–44) at Bristol Medical School before taking the Fullerian chair at the Royal Institution (1844). He wrote for the *British and Foreign Medical Review* and edited the *British and Foreign Medico-Chirurgical Review*. His *Priciples of General and Comparative Physiology* went through four editions between 1839 and 1854. He was a powerful London University advocate; he became professor of forensic medicine, university registrar (1856), and a member of the senate. His interests ranged from foraminifera and *Eozoön* through the unconscious brain to the temperance movement. He gave qualified support to Darwin, stood on the rim of Huxley's circle, and was president of the BAAS in 1872.

Joseph Constantine Carpue (1764–1846). Jesuit educated and cultured, Carpue was a surgeon at the National Vaccine Institution before opening his own anatomy school in Dean Street, Soho, in 1800 (where he taught using chalk diagrams). The school suffered in the late 1820s as a result of the RCS bylaws, and Carpue testified against the college before Warburton's 1834 committee.

William Clift (1775–1849). A ploughboy and carpenter who became John Hunter's amanuensis and eventually the conservator at the RCS Hunterian Museum (1799–1842). A "bright little bald-headed man," "kindly-hearted" and helpful.

Owen boarded with the Clifts in the early 1830s and married their daughter Caroline Amelia in 1835.

John Coldstream (1806–63). The evangelical son of a Leith merchant. He continued his interest in the local zoophytes in Edinburgh's Plinian and Wernerian societies. After taking his M.D. (1827) he suffered a religious breakdown. Being reborn, he went on to devote his life to the Edinburgh Medical Missionary Society. He became an elder of the Free St. Andrew's Church.

John Conolly (1794–1866). A soldier turned physician (Edinburgh M.D., 1821). He resigned his chair of medicine at London University (1828–30) during the riots and supported Bell's faction inside the SDUK. He wrote extensively on insanity and, as physician to Hanwell Asylum (1839–44), dispensed with all mechanical restraints.

Sir Astley Paston Cooper (1768–1841). Son of a Norfolk curate. He was apprenticed to his uncle at Guy's and Henry Cline at St. Thomas's. As surgeon and lecturer at Guy's and St. Thomas's, he developed a huge patronage network, placing his protégés in key posts (even starting a medical school for his nephew Bransby Cooper at Guy's). He earned large sums in private practice. At the RCS he was the professor of comparative anatomy (1814–15), a councillor (1815–41), and president in 1827 and 1836. After his death the college bought many of his anatomical specimens.

Holmes Coote (1817–72).** Educated at Westminster School before being apprenticed to William Lawrence. He worked under Owen on the RCS Hunterian catalogs in the 1840s and popularized Owen's work on the archetype in 1849. At St. Bartholomew's he rose from assistant surgeon (1954) to surgeon.

David Davis (1777–1841). Glasgow-educated professor of midwifery at London University. Wakley's crony and a radical supporter of the London College of Medicine. An eminent obstetrician, he was present at the birth of the future Queen Victoria.

George Darby Dermott (1802–47). Rough-cut son of a Methodist preacher. He was Brookes's favorite pupil and taught anatomy and surgery in a succession of his own schools in Soho, the most important being that in Gerrard Street (1829–37). Among the best of the private teachers, he trained students quickly and cheaply. He was also the most uncompromising. He despised the "half-educated and under-bred" surgeon-baronets and the feeling was reciprocated. He died of Bright's disease at the age of forty-five.

Nathaniel Eisdell.* Student demagogue; failed by Pattison in his LU course, yet awarded Grant's first gold medal (1830). A supporter of Bennett and Grant's new approach and leader of the student agitation to unseat Pattison in 1830. A British Medical Association councillor in 1839.

John Elliotson (1791–1868). Educated at Edinburgh, Cambridge, and Guy's (M.D. 1821), Elliotson had the same sort of reductionist attitudes as his contemporaries Southwood Smith and Lawrence. But his bravura pushed him over the top. His sartorial innovations (he wore trousers from 1826) and technical gadgetry (use of a stethoscope) were less shocking than his phrenology and mesmerism. In 1831 he took the chair of medicine at London University, but lost it in 1838 as a result of his mesmeric experiments on hospital patients.

John Epps (1805–69). Born into a Calvinist middle-class family, but repudiated his father's vindictive God and became a Quaker. Epps was a member of Combe's phrenological circle at Edinburgh (M.D. 1827), then a lecturer in the Aldersgate Street, Windmill Street, and Gerrard Street schools. He defined *radical* as "fair play and an open field for all," and his *Diary* reads like a dictionary of the reform movements of his day: Catholic emancipation, antislavery, women's rights, medical democracy, disestablishment, the Charter, the Anti–Corn Law League.

William Farr (1807–83). Son of a farm laborer and taught by a Dissenting minister. Visiting Paris, he observed the treatment of the wounded in 1830. He studied at London University in 1830–32 and wrote for the *Lancet*. He was a social Lamarckian, and at the General Register Office (1839–79) used statistics to show the environmental causes of disease in order to increase government action. He was president of the Statistical Society (1871) and commissioner for the census of 1871.

Arthur Farre (1811–87). Educated at Cambridge (M.D. 1841) and St. Bartholomew's, where in 1835 he succeeded Owen as lecturer on comparative anatomy. He was professor of obstetric medicine at King's College, London (1842–62), and practiced among the highest classes, becoming physician extraordinary to Queen Victoria in 1875.

John Fletcher (1792–1836).* A London merchant's son who found the counting house intolerable and switched to medicine (Edinburgh M.D. 1816). After practicing in London, Oxford and Edinburgh, he joined the Argyle Square Medical School (1828–36). His lectures, published as *Rudiments of Physiology* (1835–37), were a tour de force of philosophical anatomy. He died prematurely in 1836.

Edward Forbes (1815–54).** At Edinburgh (1831–36) he was influenced by Knox. He studied deep-sea organisms on an expedition to the Mediterranean (1841–42). After his father's bankruptcy he taught botany at King's College, London (1842), and worked for the Geological Survey (1844). Forbes was exuberant, and totally idiosyncratic in his Platonic views of the diversity of past life. He died shortly after taking Jameson's chair of natural history at Edinburgh in 1854.

John Goodsir (1814–67).** Educated at Edinburgh and encouraged by Knox. He was influenced by the German morphology of Carus and Owen, and by Coleridge's philosophy. He was religious, conservative, lackluster, and a close friend of Forbes's. Conservator to the Edinburgh RCS (1841) and the University Museum (1843), he finally took the Edinburgh chair of anatomy in 1846. Later he strongly opposed Darwinism.

Richard Dugard Grainger (1801–65).* Tall, stooping grammar-school-educated son of a Birmingham surgeon. He ran the Webb Street school for twenty years before joining St. Thomas's to teach anatomy (1842–60). Gained a fellowship of the Royal Society for his study in 1837 of the spinal cord, which supported Hall's reflex work. He was active in the Christian Medical Association and, like Carpenter, saw nature's unity of plan as proof of a unity of design. A medical and social reformer, he acted as an inspector for the Children's Employment Commission (1841) and for the Board of Health (1849). On the RCS Council 1846–50.

Robert Edmond Grant (1793–1874).* After his Edinburgh M.D. (1814) he spent many years in Paris, Italy, and Germany on a legacy (his father, a writer to the signet, died in 1808). He specialized in sponges (coining the name Porifera), lectured for Barclay (1824), and held the London University chair of comparative anatomy for life (1827–74). He was elected fellow of the Royal Society (1836), Fullerian professor to the Royal Institution (1837), dean of the University College London Medical Faculty (1847), and Swiney lecturer on geology to the British Museum (1853). He imported Lamarck's, Geoffroy's, and Blainville's science, but after Geoffroy's death (1844) he increasingly visited the German and Dutch museums and came to talk more like a German materialist. He was a radical supporter of Wakley and the British Medical Association, and even though an anti-Malthusian progressive transmutationist he did praise Darwin in 1860. A retiring bachelor, Grant was always formally clad in tail coat, choker, and low tied shoes. His financial plight eased after 1850, when University College gave him a stipend, his students bought him an annuity, and his last surviving brother bequeathed him some property. Although stone deaf, he was still delivering five lectures a week at age eighty and teasing students about Providence. His bound volumes of letters (known to have contained many from Geoffroy, Cuvier, Blainville, etc.) have not survived.

Augustus Bozzi Granville (1783–1872). M.D. Pavia (where he was jailed as a student insurgent). He heard Cuvier and Geoffroy in Paris in 1816. In London he was physician at the Westminster Dispensary (1818) and president of the Westminster Medical Society (1829). Like Farr, he studied the statistics on health and death among the working classes and was a strong reformer. He kept the Royal Institution out of debt (secretary 1832–52), attacked the Royal Society placemen (1830), and delivered the first British Medical Association oration (1838). Granville was urbane, known to foreign royalty, and well traveled (visiting Russia twice); in London he worked tirelessly for Italian independence.

Joseph Henry Green (1791–1863).** A merchant's son who was apprenticed to his uncle Henry Cline and became surgeon to St. Thomas's (1820–52). Green was educated partly in Hanover and returned to study philosophy in Berlin (1817). As Coleridge's disciple he delivered *Naturphilosophie* lectures at the RCS (1824–27). In the college (council 1835, president 1849 and 1858) he took a paternalistic attitude to reform. He relinquished his chair of surgery at King's College, London (1830–36), to devote himself to collecting Coleridge's fragments. His turgid Hunterian Oration, *Vital Dynamics* (1840), applied Coleridgean idealism to comparative anatomy. He was president of the General Medical Council in 1860.

George James Guthrie (1785–1856). An army surgeon (1801–15) who served in the Peninsular campaign and became surgeon at the Westminster Hospital (1827–43). He had a military bearing, wrote on gunshot wounds, and, on the RCS Council (1824–56), opposed all democratic reforms (including the 1843 charter). He was president of the college in 1833, 1841, and 1854.

Sir Henry Halford (1766–1844). Oxford-educated physician who inherited a large estate and was knighted as the king's favorite physician in 1809. Long-standing president of the RCP (1820–44) who denounced innovations in both physic and politics.

Marshall Hall (1790–1857). From a radical Methodist, cotton-manufacturing family. Trained in Edinburgh (M.D. 1812), Paris, and Germany. Hall was urbane and had a lucrative practice; he taught at Aldersgate Street, Webb Street, Sydenham College, and (1842–46) St. Thomas's. When his theory of the reflex arc (announced to the Zoological Society in 1832) was attacked, he was volubly championed by his BMA colleagues (Grainger, Pilcher, Grant, Wakley). A professed "Ultra-Whig," he harangued the oligarchs, nonetheless joining them when invited: fellow of the Royal College of Physicians 1841, Gulstonian lecturer at RCP 1842, Royal Society Council 1850. He toured America in 1853.

Sir Everard Home (1756–1832). Educated at Westminster School, then studied under John Hunter (who married his sister), succeeding him as surgeon to St. George's Hospital in 1798. Entrusted with the Hunterian manuscripts, Home systematically plagiarized and burned them. He was the first president of the RCS (1822).

Joseph Hume (1777–1855). Scottish-educated assistant surgeon to the East India Company (1799), who spoke Hindi and amassed a fortune abroad. Tory M.P. (1812), but later a leading radical in the House (1818–55). He helped repeal the Combination Acts (1824) and spoke for suffrage, emancipation, and retrenchment. He chaired Wakley's great public meetings for medical reform.

John Hunter (1728–93). Starting as a cabinetmaker, Hunter rose to become the respected surgeon extraordinary to George III and the doyen of anatomists. He built up an enormous museum at his school in Leicester Square and wrote on the teeth, circulation, digestion, inflammation, and comparative anatomy. His preparations and manuscripts passed to the nation under the trusteeship of the RCS after his death.

Robert Jameson (1774–1854). An Edinburgh-trained assistant surgeon in Leith before studying mineralogy in Freiberg under Abraham Gottlob Werner. Jameson was Regius Professor of Natural History at Edinburgh (1804–54), where he amassed a huge collection of museum specimens (many sent by former students in the colonies). He founded the *Edinburgh Philosophical Journal* in 1819.

James Johnson (1778–1845). Ulster Protestant and the youngest son of a farmer. He took to the seas as surgeon's mate in 1798 and visited Canada, Egypt, India and China, working his way up to flag surgeon and publishing *Oriental Voyager* and *On Tropical Climates*. He got his "Scotch degree" in 1813 and became a licentiate of the London RCP in 1820. In 1825 he retired with dyspepsia and wrote numerous books on health while editing the *Medico-Chirurgical Review.*

Thomas Rymer Jones (1810–80).** Educated at Guy's Hospital and in Paris. A close family friend of the Owens' and professor of comparative anatomy at King's College, London (1836–74). He was Fullerian professor at the Royal Institution in 1841.

Thomas Wharton Jones (1808–91).** Knox's assistant c. 1827, and lecturer in comparative and human anatomy at Charing Cross Hospital (1837–51). He was nominated a fellow of the RCS in 1844, and Fullerian professor at the Royal Institution in 1851. He specialized in ophthalmic medicine. He was religiously orthodox and interpreted unity of plan in terms of unity of design; he thoroughly opposed Darwin's doctrine of the descent of man from the apes.

Thomas King (1802–39).* M.D. Paris (1828) and house surgeon to the Hôtel Dieu. Returned to Britain as lecturer on general and comparative anatomy at the Aldersgate Street and Blenheim Street schools. Thoroughly imbued with French philosophical anatomy. A militant radical, summonsed by the RCS Council for "rioting" in the theater in 1831.

Robert Knox (1793–1862.)* Knox's father had been a member of the Jacobin "Friends of the People" (a movement crushed in 1792). He was also a Freemason (the object of Tory vituperation), and young Knox, after graduating (Edinburgh M.D. 1814) and serving with the expeditionary forces in South Africa, enrolled in the Paris Lodge of the Freemasons in 1822. From France he brought back Geoffroy's "new philosophy," teaching it in Barclay's school after 1825. This school became the best attended in the city, although his classes dwindled after the Burke and Hare scandal. Knox skewered the city elders and made many enemies through his outspoken atheism. He left for London in 1842 but found it impossible to obtain a post. Eventually he became anatomist to the London Cancer

Hospital in 1856. By this time his growing antienvironmentalist, racially deterministic views made him the darling of a reactionary anti–"rights of man" clique in the Ethnological Society. They engineered an honorary fellowship for him in 1860 and appointed him honorary curator just before his death. Knox saw nothing new in Darwinism.

Samuel Lane (1803–92). Owner of the Grosvenor Place School of Anatomy, founded (1830) near St. George's Hospital after the governors refused to allow a medical school on hospital grounds.

William Lawrence (1783–1867). An apprentice of Abernethy's, Lawrence was surgeon to the hospitals of Bridewell and Bethlem (1815) and St. Bartholomew's (1824). His *Lectures* (1819) were declared blasphemous and went through eight pirate editions. He helped Wakley found the *Lancet* and spoke at the mass meetings for medical reform (1826), although he quieted down considerably after being elected to the RCS Council in 1828 and specialized in eye surgery. He became president of the RCS (1846), sergeant-surgeon to Queen Victoria (1857), and was knighted in 1867.

Robert Lee (1793–1877). An Edinburgh- (M.D. 1814) and Paris-educated obstetrician who lectured at the Webb Street School and St. George's Hospital (1835–66). Lee published important papers on the uterus (fellow of the Royal Society 1830) but was snubbed by the Royal Society Council. He became fellow of the RCP in 1841 and the Royal Society's Croonian lecturer in 1862.

John Lindley (1799–1865). Professor of botany at London University (1829–60) and an officer at the Horticultural Society (1822–62). Author of numerous books on botany and classification.

Joseph Maclise (c. 1815–c. 1880).* The son of a Scottish shoemaker and brother of the painter Daniel Maclise. He studied at London University (1834–35) and in Paris. A partisan morphologist, even in the late 1840s, who saw Geoffroy "shake the throne of Cuvier" and sweep aside his "tottering fabric."

Patrick Matthew (1790–1874). Edinburgh-educated arboriculturalist, owner of an estate near Perth, and author of *Naval Timber and Arboriculture* (1831), with its evolutionary appendix. He published many radical social tracts attacking the nobility. He was a free trader and the Perth delegate to the Chartists' General Convention in London in 1839 (where the physical-force radicals labeled him a middle-class traitor and he resigned). To establish unbridled competition he wanted both hereditary power and poor-law charity scrapped. Like Darwin he interpreted nature in terms of a Malthusian-like struggle (his *Emigration Fields* in 1839 pointed the poor who lost out toward the colonies), as well as racial domination by the best adapted. This led to certain similarities in their evolutionary expressions, which were recognized in 1860 after Darwin published.

Alexander Nasmyth (1789–1848). Surgeon-dentist to Queen Victoria. Member of RCS 1831; fellow of RCS 1844. He announced the cellular structure of the tooth pulp (and fought a priority dispute with Owen), and sectioned fossil teeth.

George Newport (1803–54). A former wheelwright's apprentice helped through his London University courses (1832) by the professors waiving their fees. A brilliant dissector, he worked on the motor nerve tracts of the articulates (taking the Royal Society's Royal Medal in 1836), on butterflies, and the fertilization of the ovum. He became president of the Entomological Society (1844) and was awarded a Civil List pension in 1847.

Richard Owen (1804–92).** Studied briefly at Edinburgh (1824), then at St. Bartholomew's. Clift's assistant in the RCS (1827), Hunterian professor (1836), and conservator (1842). Gained a fellowship of the Royal Society in 1834 for his work on monotremes and marsupials. Patronized by the royal family, who presented him with a cottage in Richmond Park. Also a favorite of Peel's (who put him on the Civil List), although Owen later became a Gladstonian Liberal. He was Fullerian professor at the Royal Institution in 1858, and president of the BAAS (1858). As superintendent of the natural history departments of the British Museum (from 1856), he saw them transferred to the new purpose-built museum in South Kensington (1881). Owen published voluminously on paleontology after midcentury: on the dinosaurs, *Archaeopteryx*, moas, early mammals, and the mammallike reptiles. But he was personally quirky and disliked by the young bloods. Opposing natural selection with his brand of quasi-naturalistic "continuous creation," he was rounded on by Darwin's Bulldogs. He retired from the British Museum in 1883 and was knighted the following year.

Granville Sharp Pattison (1791–1851). Irascible Glasgow-educated anatomy lecturer. Indicted for body snatching when twenty-three, accused of malpractice at twenty-six, and hounded from city to city after his affair with Andrew Ure's wife. A lecturer in Philadelphia and Baltimore (where he ended up dueling) before being recruited to London University in 1827. Here he fell foul of the Francophiles. After his sacking, he returned in 1831 to Philadelphia, finally settling for a chair at New York University.

George Pilcher (1801–55).* Lecturer on anatomy and aural surgery at his brother-in-law Richard Grainger's Webb Street school. He was a Benthamite whose lectures were based on Blainville and Continental sources. When the school closed he taught surgery in St. George's Hospital (1843), and he joined the RCS Council (1849–55).

Jones Quain (1796–1865).* Cork born, educated at Trinity College, Dublin (M.B. 1820). He studied in Paris and lectured in Aldersgate Street (1825–31) before accepting the chair of general anatomy at London University (1831–35). His *Elements of Anatomy* (1828) was a standard text.

Henry Riley (1797–1848).* He graduated M.D. in Paris in the mid-1820s. His Geoffroyan lectures (1831–33) at the Bristol Institution were the first heard in the city. He was physician at Bristol's St. Peter's Hospital (1832) and Infirmary (1834–47), and he taught at Bristol Medical School until 1846. He supported the founding of the local Mechanics' Institute and horticultural and zoological societies.

Peter Mark Roget (1779–1869).* Son of a Genevan pastor in Soho and nephew of Sir Samuel Romilly. Educated at Edinburgh (M.D. 1798) and under Abernethy, and well placed in Whig society as physician to the marquis of Lansdowne. A long-standing secretary to the Royal Society (1827–49). He was the first Fullerian professor at the Royal Institution (1834–37). His *Animal and Vegetable Physiology* (1834) embraced the new Geoffroyan thinking. He was influential in the SDUK and (although not Oxbridge educated) specially appointed a fellow of the RCP in 1831. He was a senator of the University of London (1836) and examiner in comparative anatomy (1839–42). He started compiling his *Thesaurus* after retiring in 1840.

Michael Ryan (1794–1840). Educated in Ireland, studied medicine in Dublin and Edinburgh (M.D.), and became interested in obstetrics. Practiced in Kilkenny before moving to London, where he taught at Dermott's Gerrard Street school and edited the *London Medical and Surgical Journal*. He was a member of the London College of Medicine, Westminster Medical Society, and BMA.

William Sharpey (1802–80).* Studied under Barclay and Knox in Edinburgh, Brookes in London, as well as in Germany and at the Hôtel Dieu in Paris. He held the physiology chair at London University (1836–74). Influential as a physiologist (even if he published little) and as a secretary of the Royal Society (1853–71).

Thomas Southwood Smith (1788–1861). Unitarian money kept him at Edinburgh (M.D. 1816). Here he wrote *Divine Government,* which had a strong social Lamarckian emphasis. He worked in the East End hospitals in the 1820s, became Bentham's physician, and helped found the *Westminster Review*. He publicly dissected Bentham in Webb Street (1832) during a gothic electrical storm. Active on the commissions investigating child labor, sanitary and factory conditions, and in schemes to provide housing for the poor. Head of the General Board of Health in Whitehall (1848–54).

John Flint South (1797–1882). The son of "a sound Tory" druggist; an Anglican (although tolerant of religious differences). South was the lecturer on anatomy at St. Thomas's (1825–41). Here he also delivered comparative anatomy lectures, supporting Cuvier's science and nomenclature. He left vivid cameo portraits of the medical men of his day (Feltoe 1884).

John Addington Symonds (1807–81).* Left Edinburgh (M.D. 1828) enamored of philosophical anatomy. He became physician to the Bristol General Hospital

(1831) and lectured at Bristol Medical School until 1836. He was refined, man-nered, and "by connections and conviction a Liberal." He later accepted Darwin-ism (which he found compatible with a creating Deity) and the higher biblical criticism.

Alex Thomson. * London University student demagogue expelled in 1830 during the riots to unseat Pattison. He supported Bennett's French approach and lashed the "prudish dames of the Council." His father, Anthony Todd Thomson, the professor of materia medica, was one of those who recommended Pattison's dis-missal.

Benjamin Travers (1783–1865). One of Astley Cooper's apprentices. A specialist in eye surgery; surgeon at St. Thomas's from 1815; RCS Council member (1830–58) and president (1847).

Charles Augustus Tulk (1786–1849).** Swedenborgian mystic, idealist, and friend of Coleridge who audited courses at London University (1828–32). His son Alfred translated the romantic *Naturphilosophen* Lorenz Oken into English.

Edward Turner (1798–1837).* Edinburgh- (M.D. 1819) and Göttingen-educated chemist. Professor of chemistry at London University (1828–37) who worked on the atomic weights of the elements.

Thomas Wakley (1795–1862). Devon born; a pub fighter and apothecary's ap-prentice as a teenager. He heard Green lecture at St. Thomas's, but learned most from the Webb Street school. Befriended by Cobbett and Hume, he founded the *Lancet* (1823) and was continually sued by the hospital consultants (for libel, pi-racy, and reporting their bungled operations). He took the Finsbury seat (1835–52) and spoke in the House against the poor laws, police bills, newspaper tax, Lord's Day observance, and for Chartism, the Tolpuddle Martyrs, Irish indepen-dence, and medical reform. He argued for medical coronerships and was himself elected coroner for West Middlesex (1839), where he raised hackles by holding inquests into those who died in custody.

James Wardrop (1782–1869). An ophthalmic surgeon who taught with Lawrence at Aldersgte Street. His faked "Intercepted Letters" gave the young *Lancet* its cutting edge. Wardrop was the king's surgeon in 1828, but he declined a knight-hood, moved out of royal circles, and continued to harangue the old surgeons. Like Lawrence he was rehabilitated, becoming a fellow of the RCS in 1843.

Hewett Cottrell Watson (1804–81). Reacted against his father—an ultra-Tory Stockport J.P.—and at Edinburgh (1828–31) turned radical freethinker and phre-nologist. In 1833 he settled in Thames Ditton, south of London. He studied the environmental causes of plant distribution and accepted the mutability of species, welcoming Chambers's *Vestiges* and Darwin's *Origin*.

George Webster (d. 1875). Educated in Edinburgh (member RCS 1815) and Aberdeen (M.D. 1829). He was an army surgeon before joining Hall in practice in Dulwich (1815). He was concerned with the suffering of the young poor and founded the BMA in 1836.

William Youatt (1776–1847). Intended for the Nonconformist ministry but studied under Brookes and became a veterinary surgeon. Founder of the *Veterinarian* (1828), author of SDUK books on domestic breeds, and veterinary surgeon to the Zoological Society and SPCA. His lectures at London University (1830–35) helped raise the scientific status of a degraded profession.

Abbreviations

Archives

ATL   Alexander Turnbull Library, National Library of New Zealand
        GM   Gideon Mantell Family Papers
BL   British Library, London
BMNH British Museum (Natural History), London
        RO   Richard Owen correspondence
        MN   Richard Owen, Manuscript Notes and Synopses of Lectures
             (Owen Collection 38).
CUL   Cambridge University Library
EUL   Edinburgh University Library
GS   Geological Society of London
IC   Imperial College, London
        THH T. H. Huxley correspondence
LS   Linnean Society of London
        WS   William Swainson correspondence
RCS   Royal College of Surgeons of England
        RO   Richard Owen correspondence
        SC   Stone Collection
RI   Royal Institution, London
RS   Royal Society
        JH   John Herschel correspondence
TCL   Trinity College Library, Cambridge
        WW William Whewell correspondence
UCL   University College London
        HB   Henry Brougham correspondence
        CC   College correspondence
        SDUK Society for the Diffusion of Useful Knowledge correspon-
             dence
UMO   Arkell Library, University Museum, Oxford
        WB   William Buckland papers
WI   Wellcome Institute for the History of Medicine, London
ZS   Zoological Society of London
        MC   Minutes of Council
        MM Minutes of Meetings

Scientific and Medical Institutions

BAAS   British Association for the Advancement of Science
BMA   British Medical Association
LCM   London College of Medicine
LU   London University
PMSA   Provincial Medical and Surgical Association
RCP   Royal College of Physicians
RCS   Royal College of Surgeons
SDUK   Society for the Diffusion of Useful Knowledge
SPCA   Society for the Prevention of Cruelty to Animals

Medical Journals

BFMCR   *British and Foreign Medico-Chirurgical Review*
BFMR   *British and Foreign Medical Review*
L   *Lancet*
LMSJ   *London Medical and Surgical Journal*
MCR   *Medico-Chirurgical Review*
MG   *London Medical Gazette*
MT   *Medical Times*
There were two identically titled *LMSJ*s after 1834 as a result of the editor (Ryan) and publisher (Renshaw) of the original *LMSJ* rowing and going their own ways. Ryan's *LMSJ* ran in 1834–37, Renshaw's in 1834–35. Unless otherwise stated, I refer to Ryan's.

Scientific Journals

EJS   *Edinburgh Journal of Science*
EPJ   *Edinburgh Philosophical Journal*
ENPJ   *Edinburgh New Philosophical Journal*
Report BAAS   *Report of the British Association for the Advancement of Science*

Parliamentary Papers

Report SCME   Report from the Select Committee on Medical Education . . . Part 1. Royal College of Physicians. Part 2. Royal College of Surgeons, London (Parliamentary Papers, 13 August 1834).
Report SCBM   Report from the Select Committee on British Museum (Parliamentary Papers, 14 July 1836).

Bibliography

Anonymous reviews, editorials, and press articles are cited fully in the notes, rather than being included in the bibliography.

Abernethy, J. 1825. *Physiological Lectures Addressed to the College of Surgeons*. London: Longman.

Allen, D. E. 1978. *The Naturalist in Britain: A Social History*. Harmondsworth: Penguin.

———. 1986. *The Botanists: A History of the Botanical Society of the British Isles through 150 Years*. Winchester: St. Paul's Bibliographies.

Anderson, J. 1835–36. A Paper on the Application of Comparative Anatomy to Human Embryology, *MG* 17:72–79, 103–9.

———. 1837. *Sketch of the Comparative Anatomy of the Nervous System: with Remarks on Its Development in the Human Embryo*. London: Sherwood, Gilbert, and Piper.

———. 1839–47. Nervous System, Comparative Anatomy of. In Todd 1836–59, 3:601–26.

Appel, T. 1980. Henri de Blainville and the Animal Series: A Nineteenth-Century Chain of Being. *J. Hist. Biol.* 13:291–319.

———. 1987. *The Cuvier-Geoffroy Debate: French Biology in the Decades before Darwin*. New York and Oxford: Oxford University Press.

Appleyard, R. T., and T. Manford. 1980. *The Beginning: European Discovery and Early Settlement of Swan River Western Australia*. Perth: University of Western Australia Press.

Ashworth, J. H. 1935. Charles Darwin as a Student in Edinburgh, 1825–1827. *Proc. Roy. Soc. Edin.* 55:97–113.

Ayers, P. B. 1841–42. Memorial to the Senate of the University of London. *L* 2:315–19.

Balfour, J. H. 1865. *Biography of the Late John Coldstream*. London: Nisbet.

Barclay, J. 1822. *An Inquiry into the Opinions, Ancient and Modern, concerning Life and Organization*. Edinburgh: Bell and Bradfute.

———. 1827. *Introductory Lecture to a Course of Anatomy*. Edinburgh: Maclachlan and Stewart.

Barlow, N., ed. 1958. *The Autobiography of Charles Darwin*. New York: Norton.

Barnes, B. 1977. *Interests and the Growth of Knowledge*. London: Routledge and Kegan Paul.

———. 1982. *T. S. Kuhn and Social Science*. London: Macmillan.

Barnes, B., and D. Bloor. 1982. Relativism, Rationalism and the Sociology of Knowledge. In Hollis and Lukes 1982:21–47.

Barnes, B., and S. Shapin, eds. 1979. *Natural Order: Historical Studies of Scientific Culture*. Beverly Hills and London: Sage.

Barrett, P. H., ed. 1977. *The Collected Papers of Charles Darwin*. 2 vols. Chicago: University of Chicago Press.

Barrett, P. H., P. J. Gautrey, S. Herbert, D. Kohn, and S. Smith, eds. 1987. *Charles Darwin's Notebooks, 1836–1844*. Cambridge: British Museum (Natural History)/Cambridge University Press.

Barry, M. 1837a. Further Observations on the Unity of Structure in the Animal Kingdom, and on Congenital Anomalies, including "Hermaphrodites"; with Some Remarks on Embryology, as facilitating Animal Nomenclature, Classification, and the Study of Comparative Anatomy. *ENPJ* 22:345–64.

———. 1837b. On the Unity of Structure in the Animal Kingdom. *ENPJ* 22:116–41.

Bartholomew, M. 1973. Lyell and Evolution: An Account of Lyell's Response to the Prospect of an Evolutionary Ancestry for Man. *Brit. J. Hist. Sci.* 6:261–303.

———. 1976. The Non-Progress of Non-Progression: Two Responses to Lyell's Doctrine. *Brit. J. Hist. Sci.* 9:166–74.

———. 1979. The Singularity of Lyell. *Hist. Sci.* 17:276–93.

Bastin, J. 1970. The First Prospectus of the Zoological Society of London: New Light on the Society's Origins. *J. Soc. Biblphy Nat. Hist.* 5:369–88.

———. 1973. A Further Note on the Origins of the Zoological Society of London. *J. Soc. Biblphy Nat. Hist.* 6:236–41.

Becher, H. W. 1984. The Social Origins and Post-Graduate Careers of a Cambridge Intellectual Elite, 1830–1860. *Vict. Stud.* 28:97–127.

———. 1986. Voluntary Science in Nineteenth Century Cambridge University to the 1850s. *Brit. J. Hist. Sci.* 19:57–87.

Beddoe, J. 1910. *Memories of Eighty Years*. Bristol: Arrowsmith.

Bell, C. 1833. *The Hand: Its Mechanism and Vital Endowments as Evincing Design*. London: Pickering.

———. 1833–34. Lectures on the Hunterian Preparations. *L* 1:279–85, 313–19, 486–92, 912–19, 962–69; 2:216–21, 265–71, 346–52, 414–16, 745–51, 794–806, 824–29, 875–87.

———. 1838 [1827–29]. *Animal Mechanics, or, Proofs of Design in the Animal Frame*. In *Library of Useful Knowledge. Natural Philosophy, IV*. Reprint. London: Baldwin and Cradock.

Bell, G. J., ed. 1870. *Letters of Sir Charles Bell*. London: Murray.

Bellot, H. H. 1929. *University College London 1826–1926*. London: University of London Press.

Bennett, E. T. 1831. Evidences in Proof of Certain Statements Contained in the "Gardens and Menagerie of the Zoological Society Delineated." *Mag. Nat. Hist.* 4:199–206.

Bennett, G. 1835. Notes on the Natural History and Habitat of the *Ornithorhynchus paradoxus*, Blum. *Trans. Zool. Soc.* 1:229–58.

Bennett, J. R. 1830. *Lecture Introductory to the Course of General Anatomy: Delivered in the University of London, on Wednesday, October, 6, 1830.* London: Taylor.

Benton, E. 1974. Vitalism in Nineteenth-Century Scientific Thought: A Typology and Reassessment. *Stud. Hist. Phil. Sci.* 5:17–48.

Benton, M. 1982. Progressionism in the 1850s: Lyell, Owen, Mantell and the Elgin Fossil Reptile *Leptopleuron (Telepeton). Arch. Nat. Hist.* 11:123–36.

Berman, M. 1974–75. "Hegemony" and the Amateur Tradition in British Science. *J. Soc. Hist.* 8:30–50.

———. 1978. *Social Change and Scientific Organization: The Royal Institution, 1799–1844.* London: Heinemann.

Biddis, M. D. 1976. The Politics of Anatomy: Dr. Robert Knox and Victorian Racism. *Proc. Roy. Soc. Med.* 69:245–50.

Blainville, H. M. D. de. 1838a. Doutes sur le prétendu Didelphe fossile de Stonefield [*sic*]. *Comptes Rendus de l'Académie des Sciences* 7:402–18.

———. 1838b. Nouveaux doutes sur le prétendu Didelphe de Stonesfield. *Comptes Rendus de l'Académie des Sciences* 7:727–36.

———. 1839–40. Comparative Osteography. Ed. R. Knox. *L* 1:137–45, 185–92, 217–22, 297–307.

Blake, C. C. 1870–71. The Life of Dr. Knox. *J. Anthropol.* 1:332–38.

Bloor, D. 1981. Hamilton and Peacock on the Essence of Algebra. In Mehrtens, Bos, and Schneider 1981:202–31.

———. 1982. Durkheim and Mauss Revisited: Classification and the Sociology of Knowledge. *Stud. Hist. Phil. Sci.* 13:267–97.

———. 1983. Coleridge's Moral Copula. *Soc. Stud. Sci.* 13:605–19.

Blundell, J. 1825–26. Introductory Physiological Lecture. *L* 9:113–25.

Bompas, G. C. 1888. *Life of Frank Buckland.* London: Smith, Elder.

Bory de Saint-Vincent, J. B. 1827. Orang. In *Dictionnaire classique d'histoire naturelle,* 12:261–85. Paris.

Bourdier, F. 1969. Geoffroy Saint-Hilaire versus Cuvier: The Campaign for Paleontological Evolution (1825–1838). In Schneer 1969:36–61.

Bowerbank, J. S. 1864–82. *A Monograph of the British Spongiadae.* 4 vols. London: Ray Society.

Bowler, P. J. 1975. The Changing Meaning of "Evolution." *J. Hist. Ideas* 36:95–114.

———. 1976a. Alfred Russel Wallace's Concepts of Variation. *J. Hist. Med.* 31:17–29.

———. 1976b. *Fossils and Progress: Paleontology and the Idea of Progressive Evolution in the Nineteenth Century.* New York: Science History Publications.

———. 1984a. E. W. MacBride's Lamarckian Eugenics and Its Implications for the Social Construction of Scientific Knowledge. *Ann. Sci.* 41:245–60.

———. 1984b. *Evolution: The History of an Idea.* Berkeley: University of California Press.

———. 1988. *The Non-Darwinian Revolution: Reinterpretation of a Historical Myth.* Baltimore: Johns Hopkins University Press.

Bradfield, B. T. 1968. Sir Richard Vyvyan and the Country Gentlemen, 1830–1834. *English Hist. Rev.* 83:729–43.

Bradley, I. 1976. *The Call to Seriousness: The Evangelical Impact on the Victorians.* London: Cape.

Brewster, D. 1830. Decline of Science in England. *Quart. Rev.* 43:305–42.

———. 1833–34. Whewell's *Astronomy, and General Physics. Edin. Rev.* 58:422–57.

———. 1834. Dr. Roget's Bridgewater Treatise. *Edin. Rev.* 60:142–79.

———. 1845. Vestiges of the Natural History of Creation. *North Brit. Rev.* 3:470–515.

Broderip, W. J. 1827. Observations on the Jaw of a Fossil Mammiferous Animal, Found in the Stonesfield Slate. *Zool. J.* 3:408–18.

———. 1836. The Zoological Gardens-Regent's Park. *Quart. Rev.* 56:309–32.

Broderip, W. J., and R. Owen, 1851–52. Progress of Comparative Anatomy. *Quart. Rev.* 90:363–413.

———. 1853. Generalizations of Comparative Anatomy. *Quart. Rev.* 93:46–83.

Brodie, B. 1840. Medical Reform. *Quart. Rev.* 67:53–79.

Brook, C. 1945. *Battling Surgeon.* Glasgow: Strickland Press.

Brooke, J. H. 1977a. Natural Theology and the Plurality of Worlds: Observations on the Brewster-Whewell Debate. *Ann. Sci.* 34:221–86.

———. 1977b. Richard Owen, William Whewell, and the *Vestiges. Brit. J. Hist. Sci.* 10:132–45.

———. 1979. The Natural Theology of the Geologists: Some Theological Strata. In Jordanova and Porter 1979:39–64.

———. 1985. The Relations between Darwin's Science and His Religion. In Durant 1985b: 40–75.

Brookes, J. 1828. *A Catalogue of the Anatomical and Zoological Museum of Joshua Brookes.* London: R. Taylor.

———. 1830. *Museum Brookesianum: A Descriptive and Historical Catalogue of the Remainder of the Anatomical and Zoological Museum of J. Brookes.* London: R. Taylor.

Brougham, H. 1827. Natural Theology. *Edin. Rev.* 46:515–26.

———. 1828. *A Discourse of the Objects, Advantages, and Pleasures of Science.* London: Baldwin and Cradock.

———. 1835. *A Discourse of Natural Theology.* London: C. Knight.

———. 1839. *Dissertation on Subjects of Science Connected with Natural Theology: Being the Concluding Volumes of the New Edition of Paley's Work.* 2 vols. London: C. Knight.

Brougham, H., and C. Bell. 1836. *Paley's Natural Theology, with Illustrative Notes . . . to Which are Added Supplementary Dissertations, by Sir Charles Bell.* 2 vols. London: C. Knight.

Brown, F. K. 1961. *Fathers of the Victorians: The Age of Wilberforce.* Cambridge: Cambridge University Press.

Browne, J. 1983. *The Secular Ark: Studies in the History of Biogeography.* New Haven: Yale University Press.

————. 1985. Darwin and the Expression of the Emotions. In Kohn 1985b: 307–26.

Browne, W. A. F. 1836. Observations on Religious Fanaticism. *Phrenol. J.* 9: 288–302, 532–45, 577–603.

————. 1837. *What Asylums Were, Are, and Ought to Be*. Edinburgh: Black.

Buckland, W. 1824. Notice on the Megalosaurus or Great Fossil Lizard of Stonesfield. *Trans. Geol. Soc.* 1:390–96.

————. 1837. *Geology and Mineralogy Considered with Reference to Natural Theology*. 2d ed. 2 vols. London: Pickering.

Budd, S. 1977. *Varieties of Unbelief: Atheists and Agnostics in English Society, 1850–1960*. London: Heinemann.

Buffetaut, E. 1987. *A Short History of Vertebrate Palaeontology*. London: Croom Helm.

Burkhardt, F. H., and S. Smith, eds. 1985–86. *The Correspondence of Charles Darwin*. Vol. 1, 1985; vol. 2, 1986. Cambridge: Cambridge University Press.

Burkhardt, R. W. 1977. *The Spirit of System: Lamarck and Evolutionary Biology*. Cambridge: Harvard University Press.

Burn, W. L. 1965. *The Age of Equipoise: A Study of the Mid-Victorian Generation*. New York: Norton.

Burns, J. H. 1962. *Jeremy Bentham and University College*. London: Athlone Press.

Burrow, J. W. 1966. *Evolution and Society: A Study in Victorian Social Theory*. London: Cambridge University Press.

Bushnan, J. S. 1837. *The Philosophy of Instinct and Reason*. Edinburgh: A. and C. Black.

Bynum, W. F., and R. Porter, eds. 1985. *William Hunter and the Eighteenth-Century Medical World*. Cambridge: Cambridge University Press.

————, eds. 1987. *Medical Fringe and Medical Orthodoxy, 1750–1850*. London: Croom Helm.

Cannon, S. F. 1978. *Science in Culture: The Early Victorian Period*. New York: Dawson.

Cannon, W. F. 1976. The Whewell-Darwin Controversy. *J. Geol. Soc.* 132:377–84.

Cantor, G. N. 1975. The Edinburgh Phrenology Debate, 1803–1828. *Ann. Sci.* 32:195–218.

Cardwell, D. S. L. 1972. *The Organization of Science in England*. London: Heinemann.

Carlile, R. 1821. An Address to Men of Science. In Simon 1972: 91–137.

Carlisle, A. 1826. *The Hunterian Oration, Delivered before the Royal College of Surgeons in London, on Tuesday, February 14th, 1826*. London: Booth.

Carnarvon, Earl of. 1837. Presidential Address. In *Eleventh Annual Report*, SPCA, 15–22.

Carpenter, L. 1822. *A Discourse on Divine Influence and Conversion*. Bristol: R. Hunter.

Carpenter, W. B. 1835–36. On the Structure and Functions of the Organs of Respiration. *West of England Journal of Science and Literature* 1:217–28, 279–87.

———. 1837. On Unity of Function in Organized Beings. *ENPJ* 23:92–114.

———. 1838a. Macilwain's *Medicine and Surgery*. *BFMR* 6:98–113.

———. 1838b. On the Differences of the Laws Regulating Vital and Physical Phenomena. *ENPJ* 24:327–53.

———. 1838c. Physiology an Inductive Science. *BFMR* 5:317–42.

———. 1838d. Powell's *Natural and Divine Truth*. *BFMR* 5:548–49.

———. 1839. *Prize Thesis: Inaugural Dissertation on the Physiological Inferences to be Deduced from the Structure of the Nervous System in the Invertebrated Classes of Animals*. Edinburgh: Carfrae.

———. 1839–47. Life. In Todd 1836–59, 3:141–60.

———. 1840a. Dubois and Jones on Medical Study: Principles of Medical Education. *BFMR* 10:175–203.

———. 1840b. Dubois on Medical Study: Preliminary Education. *BFMR* 9:411–31.

———. 1840c. *Remarks on Some Passages in the Review of "Principles of General and Comparative Physiology," in the Edinburgh Medical & Surgical Journal*. Bristol: Philip and Evans.

———. 1840–41. Lectures on the Functions of the Nervous System. *MG* 27:777–84, 858–67, 938–45; 28:57–63, 460–68, 518–27, 601–6, 633–37, 709–17, 778–82, 810–14, 890–97, 934–39.

———. 1841a. Alison, Bushnan, Swainson on Instinct. *BFMR* 11:90–103.

———. 1841b. *Principles of General and Comparative Physiology*. 2d ed. London: Churchill.

———. 1843. *Animal Physiology*. London: Orr.

———. 1845a. Natural History of Creation. *BFMR* 19:155–81.

———. 1845b. *Zoology: Being a Systematic Account of the General Structure Habits, Instincts, and Uses of the Principal Families of the Animal Kingdom*. 2 vols. London: Orr.

———. 1847. Professor Owen on the Comparative Anatomy and Physiology of the Vertebrate Animals. *BFMR* 23:472–92.

———. 1851. *On the Use and Abuse of Alcoholic Liquors, in Health and Disease*. 2d ed. London: C. Gilpin.

———. 1888. *Nature and Man: Essays Scientific and Philosophical*. London: Kegan Paul, Trench.

Carus, C. G. 1827. *An Introduction to the Comparative Anatomy of Animals*. Trans. R. T. Gore. 2 vols. London: Longman.

———. 1846. *The King of Saxony's Journey through England and Scotland in the Year 1844*. Trans. S. C. Davison. London: Chapman and Hall.

Cathcart, C. W. 1882. Some of the Older Schools of Anatomy Connected with the Royal College of Surgeons, Edinburgh. *Edin. Med. J.* 27:769–81.

Chambers, R. 1844. *Vestiges of the Natural History of Creation*. 2d ed. London: Churchill.

———. 1884. *Vestiges of the Natural History of Creation*. 12th ed. London and Edinburgh: W. and R. Chambers.

Chapman, R. G., and C. T. Duval, eds. 1982. *Charles Darwin, 1809–1882: A Centennial Commemorative*. Wellington, New Zealand: Nova Pacifica.

Children, J. G. 1823. *Lamarck's Genera of Shells*. London.

Chippendale, J. 1838–39. Experiments on Animals. *L* 1:357–58.

Chitnis, A. C. 1970. The University of Edinburgh's Natural History Museum and the Huttonian-Wernerian Debate. *Ann. Sci.* 26:85–94.

———. 1973. Medical Education in Edinburgh, 1790–1826, and Some Victorian Social Consequences. *Med. Hist.* 17:173–85.

Clark, G. K. 1962. *The Making of Victorian England*. London: Methuen.

Clark, G. N. 1964–72. *A History of the Royal College of Physicians of London*. 3 vols. Oxford: Clarendon Press.

Clark, G. T. 1835–36. An Introduction to Zoology. *West of England Journal of Science and Literature* 1:19–44, 101–11.

Clark, J. 1869. *A Memoir of John Conolly*. London: Murray.

Clark, J. W., and T. M. Hughes, eds. 1890. *The Life and Letters of the Reverend Adam Sedgwick*, 2 vols. Cambridge: Cambridge University Press.

Clarke, J. F. 1874a. *Autobiographical Recollections of the Medical Profession*. London: Churchill.

———. 1874b. The Late Professor Grant. *MT* 2:563–64.

———. 1874c. Robert Edmond Grant, *MT* 2:277–78.

Cobban, A. 1981. *A History of Modern France, Volume 2: 1799–1871*. Harmondsworth: Penguin.

Cohen, S., and A. Scull, eds. 1983. *Social Control and the State: Historical and Contemporary Essays*. Oxford: Martin Robertson.

Coldstream, J. 1836. Acalepha. In Todd 1836–59, 1:35–46.

Coldstream, J. P. 1877. *Sketch of the Life of John Coldstream, M.D., F.R.C.P.E.: The Founder of the Edinburgh Medical Missionary Society*. Edinburgh: Maclaren and Macniven.

Cole, R. J. 1952. Sir Anthony Carlisle, F.R.S. (1768–1840). *Ann. Sci.* 8:255–70.

Coleridge, S. T. 1913. *Aids to Reflection*. London: Bell.

———. 1917. *The Table Talk and Omniana of Samuel Taylor Coleridge*. Oxford: Oxford University Press.

———. 1949. *The Philosophical Lectures*. London: Pilot Press.

———. 1972. *On the Constitution of the Church and State According to the Idea of Each*. London: Dent.

Colp, R. 1986. "Confessing a Murder": Darwin's First Revelations about Transmutation. *Isis* 77:9–32.

Combe, A. 1838. Remarks on the Fallacy of Professor Tiedemann's Comparison of the Negro Brain and Intellect with Those of the European. *BFMR* 5:585–89.

Combe, G., ed. 1850. *The Life and Correspondence of Andrew Combe, M.D.* Edinburgh: Maclachan and Stewart.

Conolly, J. 1828. Nervous System. *Edin. Rev.* 47:441–81.

———. 1831. Dr. Conolly's Address to His Pupils. *MG* 8:61–62.

Conolly, J., et al. 1830. *Statements Respecting the University of London, Prepared, at the Desire of the Council, by Nine of the Professors*. London.

Conry, Y. 1980. L'idée d'une "Marche de la Nature" dans la biologie prédarwinienne. *Rev. d'Hist. des Sciences* 33:97–149.

Conybeare, W. D. 1835–36. Essay Introductory to Geology. *West of England Journal of Science and Literature* 1:1–19, 89–100.

Cooper, A. 1840. *On the Anatomy of the Breast*. London: Longman.

Coote, H. 1849. *The Homologies of the Human Skeleton*. London: Highley.

Cooter, R. J. 1976. Phrenology and British Alienists, c. 1825–1845. *Med. Hist.* 20:1–21, 135–51.

———. 1979. The Power of the Body: The Early Nineteenth Century. In Barnes and Shapin 1979:73–92.

———. 1984. *The Cultural Meaning of Popular Science: Phrenology and the Organization of Consent in Nineteenth-Century Britain*. Cambridge: Cambridge University Press.

Cope, Z. 1959. *The Royal College of Surgeons of England: A History*. London: Anthony Blond.

———. 1965. Extracts from the Diary of Thomas Laycock, Chiefly Written While He Was a Medical Student, 1833–35. *Med. Hist.* 9:169–76.

———. 1966. The Private Medical Schools in London (1746–1914). In Poynter 1966:89–109.

Corrigan, P., ed. 1980. *Capitalism, State Formation and Marxist Theory: Historical Investigations*. London: Quartet.

Corsi, P. 1978. The Importance of French Transformist Ideas for the Second Volume of Lyell's Principles of Geology. *Brit. J. Hist. Sci.* 11:221–44.

———. 1988. *Science and Religion: Baden Powell and the Anglican Debate, 1800–1860*. Cambridge: Cambridge University Press.

Country Gentleman, A. 1826. *The Consequences of a Scientific Education to the Working Classes of This Country Pointed Out; and the Theories of Mr. Brougham on That Subject Confuted; in a Letter to the Marquess of Lansdown* [*sic*]. London: T. Cadell.

Cowan, C. F. 1969. Notes on Griffith's *Animal Kingdom of Cuvier* (1824–1835). *J. Soc. Biblphy Nat. Hist.* 5:137–40.

Cowen, D. L. 1969. Liberty, Laissez-Faire and Licensure in Nineteenth Century Britain. *Bull. Hist. Med.* 43:30–40.

Cowherd, R. G. 1956. *Politics of English Dissent*. New York: New York University Press.

Cranefield, P. F. 1974. *The Way In and the Way Out: Francois Magendie, Charles Bell and the Roots of the Spinal Nerves*. New York: Futura.

Crosland, M. 1983. Explicit Qualifications as a Criterion for Membership of the Royal Society: A Historical Review. *Notes and Records of the Royal Society* 37:167–87.

Cross, S. J. 1981. John Hunter, the Animal Oeconomy, and Late Enlightenment Physiological Discourse. *Stud. Hist. Biol.* 5:1–110.

Crouch, E. A. 1827. *An Illustrated Introduction to Lamarck's Conchology Contained in His Histoire Naturelle des Animaux sans Vertebres*. London: Longman.

Cullen, M. J. 1975. *The Statistical Movement in Early Victorian Britain*. Hassocks, Sussex: Harvester.

Curwen, E. C. 1940. *The Journal of Gideon Mantell, Surgeon and Geologist*. Oxford: Oxford University Press.

Cuvier, G. 1813. *Essay on the Theory of the Earth*. Ed. R. Jameson. Edinburgh: Blackwood.

———. 1825a. Nature. In *Dictionnaire des sciences naturelles*, 34:261–68. Paris.

———. 1825b. *Recherches sur les Ossemens fossiles*. 3rd ed. 5 vols. Paris.

———. 1835. Eloge de M. de Lamarck. *Mémoires de l'Académie Royale des Sciences* 13:1–31.

Darwin, C. 1842. *The Structure and Distribution of Coral Reefs: Being the First Part of the Geology of the Voyage of the "Beagle."* London: Smith, Elder.

Darwin, F., ed. 1887. *The Life and Letters of Charles Darwin*. 3 vols. London: Murray.

Dean, D. R. 1981. "Through Science to Despair": Geology and the Victorians. *Ann. N.Y. Acad. Sci.* 360:111–36.

de Beer, G. 1964. *Charles Darwin: Evolution by Natural Selection*. New York: Doubleday.

Dempster, W. J. 1983. *Patrick Matthew and Natural Selection*. Edinburgh: Paul Harris.

Dermott, G. D. 1828–29. On the Organic Materiality of the Mind: The Immateriality of the Soul, and the Nonidentity of the Two. *L* 1:39–43, 2:230–34.

———. 1830. *A Discussion on the Organic Materiality of the Mind, the Immateriality of the Soul, and the Non-Identity of the Two*. London: Callow and Wilson.

———. 1833. *A Lecture Introductory to a Course of Lectures on Anatomy, Physiology, and Surgery, Delivered at the School of Medicine and Surgery, Gerrard Street, Soho*. London: J. Fellowes.

———. 1834–35. Preservation of the Medical Committee Papers from the Fire at Westminster. *L* 1:160–62.

———. 1835a. Defence of the Concours. *LMSJ* 6:662–63.

———. 1835b. Mr. Dermott on Hospitals and Dispensaries. *LMSJ* 6:20–21.

———. 1835c. Mr. Dermott's Introductory Lecture to His Course of Anatomy, Physiology, and Surgery, Delivered at the Gerrard-street School, Westminster Dispensary; Session 1834–35. *LMSJ* 6:362–67.

———. 1835–36. The Medical Profession as It Is, and as It Ought to Be. *LMSJ* 8:62–64, 92–94, 159–60, 190–92, 315–17.

Desmond, A. 1979. Designing the Dinosaur: Richard Owen's Response to Robert Edmond Grant. *Isis* 70:224–34.

———. 1982. *Archetypes and Ancestors: Palaeontology in Victorian London, 1850–1875*. London: Blond and Briggs. Chicago: University of Chicago Press, 1984.

———. 1984a. Interpreting the Origin of Mammals: New Approaches to the History of Palaeontology. *Zoo. J. Linn. Soc.* 82:7–16.

———. 1984b. Robert E. Grant: The Social Predicament of a Pre-Darwinian Transmutationist. *J. Hist. Biol.* 17:189–223.

———. 1984c. Robert E. Grant's Later Views on Organic Development: The Swiney Lectures on "Palaeozoology," 1853–1857. *Arch. Nat. Hist.* 11:395–413.

———. 1985a. The Making of Institutional Zoology in London, 1822–1836. *Hist. Sci.* 23:153–85, 223–50.

———. 1985b. Richard Owen's Reaction to Transmutation in the 1830's. *Brit. J. Hist. Sci.* 18:25–50.

———. 1987. Artisan Resistance and Evolution in Britain, 1819–1848. *Osiris* 3:77–110.

———. 1989. Lamackism and Democracy: Corporations, Corruption, and Comparative Anatomy in the 1830s. In Moore 1989b.

Di Gregorio, M. A. 1982. In Search of the Natural System: Problems of Zoological Classification in Victorian Britain. *Hist. Phil. Life Sciences* 4:225–54.

Dobson, J. 1954. *William Clift*. London: Heinemann.

Donajgrodzki, A. P., ed. 1977. *Social Control in Nineteenth Century Britain*. London: Croom Helm.

D'Oyly, G. 1819. Abernethy, Lawrence, &c. on the Theories of Life. *Quart. Rev.* 22:1–34.

Drummond, J. L. 1841. Thoughts on the Equivocal Generation of Entozoa. *Ann. Mag. Nat. Hist.* 6:101–8.

Dubois, C. 1824. *An Epitome of Lamarck's Testacea: Being a Free Translation of That Part of His Work, De l'Histoire Naturelle des Animaux sans Vertebres*. London: Longman.

Durant, J. R. 1979. Scientific Naturalism and Social Reform in the Thought of Alfred Russel Wallace. *Brit. J. Hist. Sci.* 12:31–58.

———. 1985a. The Ascent of Nature in Darwin's *Descent of Man*. In Kohn 1985b: 283–306.

———. 1985b. *Darwinism and Divinity: Essays on Evolution and Religious Belief*. Oxford: Blackwell.

Durey, M. J. 1975. Bodysnatchers and Benthamites: The Implications of the Dead Body Bill for the London Schools of Anatomy, 1820–42. *London Journal* 2:200–225.

———. 1983. Medical Elites, the General Practitioner and Patient Power in Britain during the Cholera Epidemic of 1831–2. In Inkster and Morrell 1983:257–78.

Edmonds, J. M. 1975. The First Geological Lecture Course at the University of London, 1831. *Ann. Sci.* 32:257–75.

Edsall, N. C. 1971. *The Anti–Poor Law Movement, 1834–44*. Manchester: University of Manchester Press.

Egerton, F. 1842. Address. In *Report BAAS 1842*, xxxi–xxxvi.

Egerton, F. N. 1976. Darwin's Early Reading of Lamarck. *Isis* 67:452–56.

———. 1979. Hewett C. Watson, Great Britain's First Phytogeographer. *Huntia* 3:87–102.

Eiseley, L. 1961. *Darwin's Century: Evolution and the Men Who Discovered It*. New York: Anchor Books.

Elkana, Y., ed. 1974. *The Interaction between Science and Philosophy*. Atlantic Highlands, N.J.: Humanities Press.

Elliotson, J. 1831–32. Dr. Elliotson's Defence of Phrenology. *L* 1:287–94, 357–64.

———. 1835. *Human Physiology*. London: Longman.

Emblen, D. L. 1970. *Peter Mark Roget: The Word and the Man*. London: Longman.

Epps, E., ed. 1875. *Dairy of the Late John Epps*. London: Kent.

Epps, J. 1828. Elements of Physiology. *MCR* 9:97–120.

———. 1829. Essay on the Gradual Development of the Nervous System, from the Zoophyte to Man; Read before the London Phrenological Society, December 1st, 1828. *LMSJ* 2:35–46, 150–57, 231–38, 503–7.

———. 1832. *The Life of John Walker*. 2d ed. London: Whittaker.

———. 1834. *The Church of England's Apostacy*. London: J. Dinnis.

———. 1836. *Internal Evidences of Christianity Deduced from Phrenology*. 2d ed. London: E. Palmer.

Estlin, J. P. 1813. *Discourses on Universal Restitution, Delivered to the Society of Protestant Dissenters in Lewin's Mead, Bristol*. London: Longman.

Evidence, Oral and Documentary, Taken and Received by the Commissioners Appointed by His Majesty George IV July 23 1826 . . . Visiting the Universities of Scotland. 1837. 4 vols. Parliamentary Papers, 35.

Eyler, J. M. 1979. *Victorian Social Medicine: The Ideas and Methods of William Farr*. Baltimore: Johns Hopkins University Press.

Fairholme, E. G., and W. Pain. 1924. *A Century of Work for Animals: The History of the RSPCA, 1824–1924*. London: Murray.

Farber, P. L. 1976. The Type-Concept in Zoology during the First Half of the Nineteenth Century. *J. Hist. Biol.* 9:93–119.

Farley, J. 1972. The Spontaneous Generation Controversy (1700–1860): The Origin of Parasitic Worms. *J. Hist. Biol.* 5:95–125.

Farr, W. 1839–40. Medical Reform. *L* 1:105–11.

Farre, A. 1837. Observations on the Minute Structure of Some of the Higher Forms of Polypi, with Views of a More Natural Arrangement of the Class. *Phil. Trans. Roy. Soc.*, 387–426.

Feltoe, C. L., ed. 1884. *Memorials of John Flint South*. London: Murray.

Figlio, K. M. 1976. The Metaphor of Organization: An Historiographical Perspective on the Bio-Medical Sciences of the Early Nineteenth Century. *Hist. Sci.* 16:17–53.

Finlayson, G. B. A. M. 1969. *England in the Eighteen Thirties*. London: Arnold.

Fish, R., and I. Montagu. 1976. The Zoological Society and the British Overseas. In Zuckerman 1976: 17–48.

Fleming, J. 1822. *The Philosophy of Zoology: or a General View of the Structure, Functions, and Classification of Animals*. 2 vols. Edinburgh: Constable.

———. 1826. The Geological Deluge, as Interpreted by Baron Cuvier and Professor Buckland, Inconsistent with the Testimony of Moses and the Phenomena of Nature. *EPJ* 14:204–39.

———. 1828a. *A History of British Animals*. Edinburgh: Bell and Bradfute.

———. 1828b. On the Value of the Evidence from the Animal Kingdom, tending to Prove that the Arctic Regions Formerly Enjoyed a Milder Climate Than at Present. *ENPJ* 6:227–86.

———. 1829. Systems and Methods in Natural History. *Quart. Rev.* 41:302–27.

———. 1859. *The Lithology of Edinburgh*. Edinburgh: Kennedy.

Fletcher, H. R. 1969. *The Story of the Royal Horticultural Society, 1804–1968*. London: Oxford University Press.

Fletcher, J. 1834–36. Substance of a Course of Lectures on Physiology. *LMSJ* vols. 6 (1834–35) through 8 (1835–36); 31 lectures.

———. 1835–37. *Rudiments of Physiology*. 3 pts. Edinburgh: Carfrae.

———. 1836. *Discourse on the Importance of the Study of Physiology as a Branch of Popular Education*. Edinburgh: Black.

Flourens, M. J. P. 1834–35. Lectures on Human Embryology Illustrated by Comparative Anatomy. *L* 2:273–75, 305–9, 369–73, 401–5, 433–37, 561–66.

Flower, W. H. 1894. Richard Owen. *Proc. Roy. Soc. London* 55:i–xiv.

Forbes, E. 1839. On Two British Species of Cydippe. *Ann. Nat. Hist.* 3:145–50.

Fox, C. 1972. *The Journals of Caroline Fox, 1835–71*. Ed. Wendy Monk. London: Elek.

Fox, R., and G. Weisz, eds. 1980. *The Organization of Science and Technology in France, 1808–1914*. Cambridge: Cambridge University Press.

Fraser, D. 1976. *Urban Politics in Victorian England*. Leicester: Leicester University Press.

———. 1979. *Power and Authority in the Victorian City*. Oxford: Blackwell.

———, ed. 1982. *Municipal Reform and the Industrial City*. Leicester: Leicester University Press.

Freeman, R. B. 1982. The Darwin Family. *Biol. J. Linn. Soc.* 17:9–21.

French, R. D. 1975. *Antivivisection and Medical Science in Victorian Society*. Princeton: Princeton University Press.

Fullarton, J. 1831. Reform in Parliament. *Quart. Rev.* 45:252–339.

Gallistel, C. R. 1981. Bell, Magendie, and the Proposal to Restrict the Use of Animals in Neurobehavioral Research. *Amer. Psych.* 36:357–60.

Garland, M. M. 1980. *Cambridge before Darwin: The Ideal of a Liberal Education, 1800–1860*. Cambridge: Cambridge University Press.

Gash, N. 1978. After Waterloo: British Society and the Legacy of the Napoleonic Wars. *Trans. Roy. Hist. Soc.* 28:145–57.

Geison, G. L. 1978. *Michael Foster and the Cambridge School of Physiology: The Scientific Enterprise in Late Victorian Society*. Princeton: Princeton University Press.

Gellner, E. 1964. *Thought and Change*. London: Weidenfeld and Nicholson.

Genty, M. 1935. A French Physiologist in England in the Year 1822. *Brit. Med. J.* 2:1271.

Geoffroy St. Hilaire, E. 1812. Tableau des quadrumanes. *Annales du Muséum d'Histoire Naturelle* 19:85–122.

———. 1818–22. *Philosophie anatomique*. 2 vols. Paris.

———. 1825. Recherches sur l'Organisation des Gavials. *Mémoires du Muséum d'Histoire Naturelle* 12:97–155.

———. 1826. Sur un appareil glanduleux récemment découvert en Allemagne dans l'Ornithorhynque. *Annales des Sciences Naturelles* 9:457–60.

———. 1828. Rapport fait à l'Academie Royale des Sciences sur un Mémoire de M. Roulin. *Mémoires du Muséum d'Histoire Naturelle* 17:201–29.

———. 1829. Considérations sur des oeufs d'ornithorinque, formant de nouveaux documens pour le question de la classification des Monotremes. *Annales des Sciences Naturelles* 18:157–64.

———. 1830. *Principes de philosophie zoologique.* Paris: Pichon.

———. 1833. Divers mémoires sur de grands sauriens. *Mémoires de l'Académie Royale des Sciences de l'Institut de France* 12:1–138.

———. 1836a. Considérations sur les singes les plus voisins de l'homme. *Comptes Rendus de l'Académie des Sciences* 2:92–95.

———. 1836b. Etudes sur l'orang-outang de la ménagerie. *Comptes Rendus de l'Académie des Sciences* 3:1–9.

———. 1837. Du sivatherium de l'Himalaya. *Comptes Rendus de l'Académie des Sciences* 4:77–82.

———. 1838. De quelques contemporains des crocodiliens fossiles des ages antédiluviens. *Comptes Rendus de l'Académie des Sciences* 7:629–33.

Gerstner, P. A. 1970. Vertebrate Paleontology, an Early Nineteenth-Century Transatlantic Science. *J. Hist. Biol.* 3:137–48.

Gillispie, C. C. 1959. *Genesis and Geology: A Study in the Relations of Scientific Thought, Natural Theology, and Social Opinion in Great Britain, 1790–1850.* New York: Harper.

Godlee, R. J. 1921. Thomas Wharton Jones. *Brit. J. Ophthalmol.* 97–117, 145–56.

Gold, V. 1973. Samuel Taylor Coleridge and the Appointment of J. F. Daniell, F.R.S. as Professor of Chemistry at King's College London. *Notes and Records of the Royal Society* 28:25–29.

Goodfield-Toulmin, J. 1969. Some Aspects of English Physiology, 1780–1840. *J. Hist. Biol.* 2:283–320.

Gooding, D., T. Pinch, and S. Schaffer, eds. 1988. *The Uses of Experiment.* Cambridge: Cambridge University Press.

Goodway, D. 1982. *London Chartism, 1838–1848.* Cambridge: Cambridge University Press.

Gordon, E. O. 1894. *The Life and Correspondence of William Buckland.* London: Murray.

Gordon-Taylor, G., and E. W. Walls. 1958. *Sir Charles Bell: His Life and Times.* London: Livingstone.

Gould, S. J. 1977. *Ontogeny and Phylogeny.* Cambridge: Harvard University Press.

Grainger, R. D. 1829. *Elements of General Anatomy.* London: Highley.

———. 1837. *Observations on the Structure and Functions of the Spinal Cord.* London: Highley.

———. 1842–43a. The Functions of Animal Life Only Chemico-Physical Processes. *L* 1:171–72.

———. 1842–43b. Illustrations of the Medical Uses of Comparative Anatomy. *L* 1:93.

———. 1842–43c. Inquiries in Physiology. *L* 1:172–73.

———. 1842–43d. Mr. Grainger's Address. *L* 1:228–32.

———. 1848. *Observations on the Cultivation of Organic Science*. London: Highley.

Grant, R. E. 1814. *Dissertatio Physiologica Inauguralis, de Circuitu Sanguinis in Foetu*. Edinburgh: Ballantyne.

———. 1826a. Notice of a New Zoophyte (Cliona celata, Gr.) from the Firth of Forth. *ENPJ* 1:78–81.

———. 1826b. Observations on the Nature and Importance of Geology, *ENPJ* 1:293–302.

———. 1826c. On the Structure and Nature of the Spongilla friabilis. *EPJ* 14:270–84.

———. 1826d. Remarks on the Structure of Some Calcareous Sponges. *ENPJ* 1:166–71.

———. 1827a. Notice of Two New Species of British Sponges. *ENPJ* 2:203–4.

———. 1827b. Notice Regarding the Ova of the Pontobdella muricata, Lam. *EJS* 7:121–25.

———. [?] 1827c. Of the Changes Which Life Has Experienced on the Globe. *ENPJ* 3:298–301.

———. 1827d. On the Structure and Characters of the Octopus ventricosus, Gr. (Sepia octopodia, Pent.), a Rare Species of Octopus from the Firth of Forth. *ENPJ* 2:309–17.

———. 1828. Observations on the Generation of the *Lobularia digitata*, Lam. (*Alcyonium lobatum*, Pall.). *EJS* 8:104–10.

———. 1829. *Essay on the Study of the Animal Kingdom*. 2d ed. London: Taylor.

———. 1830a. Baron Cuvier. *Foreign Review and Continental Miscellany* 5:342–80.

———. 1830b. On the Egg of the Ornithorhynchus. *ENPJ* 8:149–51.

———. 1830c. Zoophytology. *Edinburgh Encyclopaedia* 18:838–46.

———. 1833a. On the Anatomy of the *Loligopsis guttata*, Grant, and *Sepiola vulgaris*, Leach. *Proc. Zool. Soc.* 1:90–91.

———. 1833b. *On the Study of Medicine: Being an Introductory Address Delivered at the Opening of the Medical School of the University of London, October 1st, 1833*. London: Taylor.

———. 1833c. On the Zoological Characters of the Genus *Loligopsis*, Lam., and Account of a New Species from the Indian Ocean. *Proc. Zool. Soc.* 1:26–7.

———. 1833d. *Outline of a Course of Lectures on the Structure and Classification of Animals, to Be Delivered to the Members of the Zoological Society of London, in Their Museum, to Commence on Tuesday the 15th of January, 1833, and to Continue on the Succeeding Tuesdays and Thursdays, at Half-past Seven o'clock p.m.* London: Mills, Jowett, and Mills.

———. 1833–34. Lectures on Comparative Anatomy and Animal Physiology. *L* vols 1 (1833–34) through 2 (1833–34); 60 lectures.

———. 1834. On a Fossil Tooth Found in a Red Sandstone above the Coal Formation in Berwickshire. *ENPJ* 16:38–43.

————. 1835a. On the Anatomy of the *Sepiola vulgaris*, Leach. *Trans. Zool. Soc.* 1:77–86.

————. 1835b. On the Nervous System of Beroë Pileus, Lam., and on the Structure of Its Cilia. *Trans. Zool. Soc.* 1:9–12.

————. 1835–41. *Outlines of Comparative Anatomy.* 4 pts. London: Bailliere.

————. 1836. Animal Kingdom. In Todd 1836–59, 1:108–17.

————. 1837–38a. Further Observations on Dr. Hall's Statement regarding the Motor Nerves of Articulata. *L* 1:897–900.

————. 1837–38b. Reply to Mr. Newport's Insinuations respecting the Writings of Dr. Marshall Hall and Dr. Grant. *L* 1:746–48.

————. 1838. Antediluvian Remains at Stourton Quarry. *Liverpool Mercury*, 24 April; also *Mag. Nat. Hist.* 1839, 3:43–48.

————. 1839. *General View of the Characters and the Distribution of Extinct Animals.* London: Bailliere.

————. 1841. *On the Present State of the Medical Profession in England.* London: Renshaw.

————. 1842. On the Structure and History of the Mastodontoid Animals of North America. *Proc. Geol. Soc.* 3:770–71.

————. 1844. On the Structure and History of the Polygastric Animalcules. *Transactions of the British and Foreign Institute* 1:353–58.

————. 1846. Dr. Roget's Bridgewater Treatise. *L* 1:445–46.

————. 1861. *Tabular View of the Primary Divisions of the Animal Kingdom.* London: Walton and Maberly.

Granville, A. B. 1830. *Science without a Head; or the Royal Society Dissected.* London: T. Ridgway.

Granville, P. B. 1874. *Autobiography of A. B. Granville.* 2 vols. London: Henry S. King.

Green, J. H. 1831. *Distinction without Separation: In a Letter to the President of the College of Surgeons on the Present State of the Profession.* London: Hurst, Chance.

————. 1832. *An Address Delivered in King's College, London, at the Commencement of the Medical Session, October 1, 1832.* London: Fellowes.

————. 1834. *Suggestions Respecting the Intended Plan of Medical Reform Respectfully Offered to the Legislature and the Profession.* London: Highley.

————. 1840. *Vital Dynamics: The Hunterian Oration before the Royal College of Surgeons in London 14th February 1840.* London: Pickering.

————. 1841. *The Touchstone of Medical Reform in Three Letters Addressed to Sir Robert Harry Inglis Bart., M.P.* London: Highley.

————. 1865. *Spiritual Philosophy: Founded on the Teaching of the Late Samuel Taylor Coleridge.* 2 vols. London: Macmillan.

Greene, J. C. 1961. *The Death of Adam: Evolution and Its Impact on Western Thought.* New York: Mentor.

————. 1977. Darwin as a Social Evolutionist. *J. Hist. Biol.* 10:1–27.

Greenough, G. B. 1819. *A Critical Examination of the First Principles of Geology; in a Series of Essays.* London: Strahan and Spottiswoode.

Gregory, F. 1977. *Scientific Materialism in Nineteenth Century Germany*. Dordrecht: Reidel.

Griggs, E. L., ed. 1956–71. *Collected Letters of Samuel Taylor Coleridge*. 6 vols. Oxford: Clarendon Press.

Grobel, M. C. 1932. The Society for the Diffusion of Useful Knowledge, 1826–1846. 4 vols. Ph.D. diss., University of London.

Gruber, H. E., and P. H. Barrett. 1974. *Darwin on Man: A Psychological Study of Scientific Creativity, Together with Darwin's Early and Unpublished Notebooks*. New York: Dutton.

Gunther, A. E. 1978. John George Children, F.R.S. (1777–1852) of the British Museum. Mineralogist and Reluctant Keeper of Zoology. *Bull. Brit. Mus. (Nat. Hist.) Hist. Ser.* 6:75–108.

———. 1980. *The Founders of Science at the British Museum, 1753–1900*. Halesworth, Suffolk: Halesworth Press.

Haines, B. 1978. The Inter-Relations between Social, Biological, and Medical Thought, 1750–1850: Saint-Simon and Comte. *Brit. J. Hist. Sci.* 11:19–35.

Halévy, E. 1950. *The Triumph of Reform, 1830–1841*. London: Ernest Benn.

———. 1972. *The Growth of Philosophic Radicalism*. London: Faber.

Hall, C., ed. 1861. *Memoirs of Marshall Hall*. London: Bentley.

Hall, M. 1837–38. Letter from Dr. Marshall Hall. *L* 1:748–49.

———. 1840–41. A Letter on Medical Reform. *L* 1:886–89.

Hall, M. B. 1984. *All Scientists Now: The Royal Society in the Nineteenth Century*. Cambridge: Cambridge University Press.

Hall, V. M. D. 1979. The Contribution of the Physiologist, William Benjamin Carpenter (1813–1885), to the Development of the Principle of the Correlation of Forces and the Conservation of Energy. *Med. Hist.* 23:129–55.

Hamburger, J. 1963. *James Mill and the Art of Revolution*. New Haven: Yale University Press.

Harlan, A. H. 1914. *History and Genealogy of the Harlan Family*. Baltimore: Lord Baltimore Press.

Harlan, R. 1841. A Letter from Dr. Harlan, Addressed to the President, on the Discovery of the Remains of the *Basilosaurus* or *Zeuglodon*. *Trans. Geol. Soc.* 6:67–68.

Harris, R. W. 1969. *Romanticism and the Social Order, 1780–1830*. London: Blandford Press.

Harrison, B. 1967. Religion and Recreation in Nineteenth-Century England. *Past and Present* 36:98–125.

———. 1973. Animals and the State in Nineteenth-Century England. *English Hist. Rev.* 88:786–820.

Harrison, J. F. C. 1979. *Early Victorian Britain, 1832–51*. London: Fontana.

———. 1987. Early English Radicals and the Medical Fringe. In Bynum and Porter 1987: 198–215.

Hart, J. 1965. Nineteenth-Century Social Reform: A Tory Interpretation of History. *Past and Present* 31:39–61.

Harte, N. 1986. *The University of London, 1836–1986*. London and Atlantic Highlands, N.J.: Athlone Press.

Hawkins, T. 1834. *Memoirs of Ichthyosauri and Plesiosauri, Extinct Monsters of the Ancient Earth*. London: Relfe and Fletcher.

Hays, J. N. 1964. Science and Brougham's Society. *Ann. Sci*. 20:227–41.

———. 1974. Science in the City: The London Institution, 1819–1840. *Brit. J. Hist. Sci*. 7:146–62.

———. 1981. The Rise and Fall of Dionysius Lardner. *Ann. Sci*. 38:527–42.

———. 1983. The London Lecturing Empire, 1800–50. In Inkster and Morrell 1983:91–119.

Hepton, D. 1984. *Methodism and Politics in British Society, 1750–1850*. London: Hutchinson.

Herbert, S. 1974. The Place of Man in the Development of Darwin's Theory of Transmutation. Part 1. To July 1837. *J. Hist. Biol*. 7:217–58.

———. 1977. The Place of Man in the Development of Darwin's Theory of Transmutation, Part 2. *J. Hist. Biol*. 10:155–227.

———. 1980. *The Red Notebook of Charles Darwin*. London: British Museum (Natural History).

Hetzel, P. J. 1842. *Scènes de la vie privée et publique des animaux vignettes par Grandville*. Paris: J. Hetzel et Paulin.

Hodge, M. J. S. 1971. Lamarck's Science of Living Bodies. *Brit. J. Hist. Sci*. 5:323–52.

———. 1972. The Universal Gestation of Nature: Chambers' *Vestiges* and *Explanations*. *J. Hist. Biol*. 5:127–51.

———. 1983. Darwin and the Laws of the Animate Part of the Terrestrial System (1835–37): On the Lyellian Origins of His Zoonomical Explanatory Program. *Stud. Hist. Biol*. 6:1–106.

———. 1985. Darwin as a Lifelong Generation Theorist. In Kohn 1985b: 207–43.

Hodge, M. J. S., and D. Kohn, 1985. The Immediate Origins of Natural Selection. In Kohn 1985b: 185–206.

Hollis, M., and S. Lukes, eds. 1982. *Rationality and Relativism*. Oxford: Blackwell.

Hollis, P. 1970. *The Pauper Press: A Study in Working-Class Radicalism of the 1830s*. Oxford. Oxford University Press.

Holloway, S. W. F. 1964. Medical Education in England, 1830–1858: A Sociological Analysis. *History* 49:299–324.

Holt, R. V. 1938. *The Unitarian Contribution to Social Progress in England*. London: George Allen and Unwin.

Houghton, W. E. 1979. The Westminster Review, 1824–1900. In W. E. Houghton, ed., *The Wellesley Index to Victorian Periodicals, 1824–1900*, 3:529–58. Toronto: University of Toronto Press.

Hunter, J. 1840. *Observations on Certain Parts of the Animal Oeconomy. With Notes by Richard Owen*. Philadelphia: Haswell.

Inkster, I. 1979. London Science and the Seditious Meetings Act of 1817. *Brit. J. Hist. Sci*. 12:192–96.

———. 1983. Introduction: Aspects of the History of Science and Science Culture in Britain, 1780–1850 and Beyond. In Inkster and Morrell 1983: 11–54.

Inkster, I., and J. Morrell, eds. 1983. *Metropolis and Province: Science in British Culture, 1780–1850*. London: Hutchinson.

Jacyna, L. S. 1981. The Physiology of Mind, the Unity of Nature, and the Moral Order in Late Victorian Thought. *Brit. J. Hist. Sci.* 14:109–32.

———. 1983a. Images of John Hunter in the Nineteenth Century. *Hist. Sci.* 21:85–108.

———. 1983b. Immanence or Transcendence: Theories of Life and Organization in Britain, 1790–1835. *Isis* 74:311–29.

———. 1983c. John Goodsir and the Making of Cellular Reality. *J. Hist. Biol.* 16:75–99.

———. 1984a. Principles of General Physiology: The Comparative Dimension to British Neuroscience in the 1830s and 1840s. *Stud. Hist. Biol.* 7:47–92.

———. 1984b. The Romantic Programme and the Reception of Cell Theory in Britain. *J. Hist. Biol.* 17:13–48.

———. 1987. Medical Science and Moral Science: The Cultural Relations of Physiology in Restoration France. *Hist. Sci.* 25:111–46.

James, A. J. L. F. 1982. Thermodynamics and Sources of Solar Heat, 1846–1862. *Brit. J. Hist. Sci.* 15:155–81.

James, K. W. 1986. *"Damned Nonsense!"—The Geological Career of the Third Earl of Enniskillen*. Ulster: Ulster Museum.

James, W. 1855. *Memoir of John Bishop Estlin*. London: C. Green.

Jespersen, P. H. 1948–49. Charles Darwin and Dr. Grant. *Lychnos*, 159–67.

Johnson, R. 1977. Educating the Educators: "Experts" and the State, 1833–39. In Donajgrodzki 1977:77–107.

Jones, T. Rymer. 1841. *A General Outline of the Animal Kingdom, and Manual of Comparative Anatomy*. London: van Voorst.

Jordanova, L. J. 1984. *Lamarck*. Oxford: Oxford University Press.

Jordanova, L. J., and R. S. Porter, eds. 1979. *Images of the Earth: Essays in the History of the Environmental Sciences*. Chalfont St. Giles, Bucks.: British Society for the History of Science Monograph.

Keith, P. 1831. Of the Conditions of Life. *Phil. Mag.* 10:32–40.

Kelly, T. 1957. *George Birkbeck: Pioneer of Adult Education*. Liverpool: Liverpool University Press.

King, T. 1834. *The Substance of a Lecture, Designed as an Introduction to the Study of Anatomy Considered as the Science of Organization; and Delivered at the Re-Opening of the School, Founded by the Late Joshua Brookes, Esq. in Blenheim Street, October 1st, 1833*. London: Longman.

Kirby, W. 1835. *On the Power Wisdom and Goodness of God as Manifested in the Creation of Animals and in Their History Habits and Instincts*. 2 vols. London: Pickering.

Knight, C., ed. 1841–44. *London*. 6 vols. London: C. Knight.

Knight, D. 1987. Background and Foreground: Getting Things in Context. *Brit. J. Hist. Sci.* 20:3–12.

Knights, B. 1978. *The Idea of the Clerisy in the Nineteenth Century*. Cambridge: Cambridge University Press.

Knoepflmacher, U. C., and G. B. Tennyson, 1977. *Nature and the Victorian Imagination*. Berkeley: University of California Press.

Knox, R. 1824. On the Osseous, Muscular, and Nervous Systems of the *Ornithorynchus paradoxus*. *Memoirs of the Wernerian Natural History Society* 5:161–74.

———. 1831. Observations on the Structure of the Stomach of the Puruvian Lama; to Which Are Prefixed Remarks on the Analogical Reasoning of Anatomists, in the Determination A Priori of Unknown Species and Unknown Structures. *Trans. Roy. Soc. Edin*. 11:479–96.

———. 1839–40. Inquiry into the Present State of Our Knowledge Respecting the Orang-Outang and Chimpanzee. *L* 2:289–96.

———. 1843. Contributions to Anatomy and Physiology. *MG* 2:463–67, 499–502, 529-32, 537–40, 554–56, 586–89, 860–62.

———. 1850. *The Races of Men: A Fragment*. London: Renshaw.

———. 1852. *Great Artists and Great Anatomists: A Biographical and Philosophical Study*. London: van Voorst.

———. 1855. Contributions to the Philosophy of Zoology. *L* 1:625–27; 2:24–26, 45–46, 67–68, 162–64, 186–88, 216–18.

Kohlstedt, S. G. 1983. Australian Museums of Natural History: Public Priorities and Scientific Initiatives in the 19th Century. *Historical Records of Australian Science* 5:1–29.

Kohn, D. 1980. Theories to Work by: Rejected Theories, Reproduction, and Darwin's Path to Natural Selection. *Stud. Hist. Biol*. 4:67–170.

———. 1985a. Darwin's Principle of Divergence as Internal Dialogue. In Kohn 1985b: 245–47.

———, ed. 1985b. *The Darwinian Heritage*. Princeton: Princeton University Press.

Kottler, M. J. 1985. Charles Darwin and Alfred Russel Wallace: Two Decades of Debate over Natural Selection. In Kohn 1985b: 367–432.

Lamarck, J. -B. -P. -A. 1809. *Philosophie zoologique*. 2 vols. Paris: Dentu.

Latreille, P.-A. 1825. *Familles naturelles du règne animal*. Paris.

Laudan, R. 1987. *From Mineralogy to Geology: The Foundations of a Science, 1650–1830*. Chicago: University of Chicago Press.

Lawrence, P. 1977. Heaven and Earth: The Relation of the Nebular Hypothesis to Geology. In Yourgrau and Breck 1977: 253–81.

———. 1978. Charles Lyell versus the Theory of Central Heat: A Reappraisal of Lyell's Place in the History of Geology. *J. Hist. Biol*. 11:101–28.

Lawrence, W. 1816. *An Introduction to Comparative Anatomy and Physiology: Being the Two Introductory Lectures Delivered at the Royal College of Surgeons, on the 21st and 25th of March, 1816*. London: J. Callow.

———. 1834. *The Hunterian Oration, Delivered at the Royal College of Surgeons, on the 14th of February, 1834*. London: Churchill.

———. 1844. *Lectures on Comparative Anatomy, Physiology, Zoology, and the Natural History of Man*. London: Taylor.

Lefanuc, W. R. 1937. British Periodicals of Medicine: A Chronological List. *Bull. Inst. Hist. Med*. 5:735–61.

————. 1964. Sir Benjamin Brodie F.R.S. (1783–1862). *Notes and Records of the Royal Society* 19:42–52.

Lenoir, T. 1982. *The Strategy of Life: Teleology and Mechanics in Nineteenth Century German Biology*. Dordrecht: Reidel.

————. 1987. Essay Review: The Darwin Industry. *J. Hist. Biol.* 20:115–30.

Lessertisseur, J., and F. K. Jouffroy. 1979. L'idée de série chez Blainville. *Rev. d'Hist. Sci.* 32:25–42.

Levere, T. H. 1981. *Poetry Realized in Nature: Samuel Taylor Coleridge and Early Nineteenth Century Science*. Cambridge: Cambridge University Press.

Lewes, C. L. 1898. *Dr. Southwood Smith: A Retrospect*. Edinburgh: W. Blackwood.

Leys, R. 1980. Background to the Reflex Controversy: William Alison and the Doctrine of Sympathy before Hall. *Stud. Hist. Biol.* 4:1–66.

Limoges, C. 1980. The Development of the Muséum d'Histoire Naturelle of Paris, c. 1800–1914. In Fox and Weisz 1980: 211–40.

Lindley, J. 1834–35. Address Delivered at the Commencement of the Medical Session, 1834–35, on Wednesday, Oct. 1st, 1834. *L* 1:86–95.

Little, E. M. 1932. *History of the British Medical Association*. London: British Medical Association.

London Medical Directory. 1845. London: Mitchell.

Lonsdale, H. 1870. *A Sketch of the Life and Writings of Robert Knox*. London: Macmillan.

Lord, P. B. 1834. *Popular Physiology: Being a Familiar Explanation of the Most Interesting Facts Connected with the Structure and Function of Animals, and Particularly of Man*. London: John W. Parker.

Loudon, I. S. L. 1981. The Origins and Growth of the Dispensary Movement in England. *Bull. Hist. Med.* 55:321–42.

Lovett, W. 1920. *Life and Struggles of William Lovett in His Pursuit of Bread, Knowledge, and Freedom*. 2 vols. London: Bell.

Lubenow, W. C. 1971. *The Politics of Government Growth: Early Attitudes toward State Intervention, 1833–1848*. Newton Abbot, Devon: David and Charles.

Lyell, C. 1827. State of the Universities. *Quart. Rev.* 36:216–68.

————. 1830–33. *Principles of Geology*. 3 vols. London: Murray.

Lyell, K., ed. 1881. *Life Letters and Journals of Sir Charles Lyell, Bart.* 2 vols. London: Murray.

McCalman, I. 1984. Unrespectable Radicalism: Infidels and Pornography in Early Nineteenth Century London. *Past and Present* 104:74–110.

MacCulloch, J. 1823. French Geology of Scotland. *Edin. Rev.* 38:413–37.

McEvoy, J. G., and J. E. McGuire. 1975. God and Nature: Priestley's Way of Rational Dissent. *Hist. Stud. Phys. Sci.* 6:325–404.

McFarland, T. 1969. *Coleridge and the Pantheist Tradition*. Oxford: Clarendon Press.

McKendrick, N. 1961. Josiah Wedgwood and Factory Discipline. *Hist. J.* 4:30–55.

MacKenzie, D. A. 1981. *Statistics in Britain, 1865–1930*. Edinburgh: Edinburgh University Press.

———. 1985. The Political "Implications" of Scientific Theories: A Comment on Bowler. *Ann. Sci.* 42:417–20.

McLachlan, H. 1934. *The Unitarian Movement in the Religious Life of England. 1. Its Contribution to Thought and Learning, 1700–1900*. London: George Allen and Unwin.

MacLeod, R. 1970. Science and the Civil List, 1824–1914. *Technology and Society* 6:47–55.

———. 1971. Of Medals and Men: A Reward System in Victorian Science, 1826–1914. *Notes and Records of the Royal Society* 26:81–105.

———. 1973. Statesmen Undisguised. *Amer. Hist. Rev.* 78:1386–1405.

———. 1981. Introduction: On the Advancement of Science. In MacLeod and Collins 1981:17–42.

———. 1982. On Visiting the "Moving Metropolis": Reflections on the Architecture of Imperial Science. *Historical Records of Australian Science* 5:1–16.

———. 1983. Whigs and Savants: Reflections on the Reform Movement in the Royal Society, 1830–48. In Inkster and Morrell 1983:55–90.

MacLeod, R., and P. Collins, eds. 1981. *The Parliament of Science: The British Association for the Advancement of Science, 1831–1981*. Northwood, Middlesex: Science Reviews.

Maclise, J. 1846. On the Nomenclature of Anatomy (Addressed to Professors Owen and Grant). *L* 1:298–301.

———. 1847. *Comparative Osteology: Being Morphological Studies to Demonstrate the Archetype Skeleton of Vertebrated Animals*. London: Taylor and Walton.

McMenemey, W. H. 1966. Education and the Medical Reform Movement. In Poynter 1966:135–54.

McNeil, M. 1987. *Under the Banner of Science: Erasmus Darwin and His Age*. Manchester: Manchester University Press.

Manier, E. 1978. *Young Darwin and His Cultural Circle*. Dordrecht: Reidel.

Manuel, D. E. 1980. Marshall Hall F.R.S. (1790–1857): A Conspectus of His Life and Work. *Notes and Records of the Royal Society* 35:135–66.

———. 1987. Marshall Hall (1790–1857): Vivisection and the Development of Experimental Physiology. In Rupke 1987: 78–104.

Marshall, P. H. 1984. *William Godwin*. New Haven and London: Yale University Press.

Maulitz, R. C. 1981. Channel Crossing: The Lure of French Pathology for English Medical Students, 1816–36. *Bull. Hist. Med.* 55:475–96.

Mazumdar, P. M. H. 1983. Anatomical Physiology and the Reform of Medical Education: London, 1825–1835. *Bull. Hist. Med.* 57:230–46.

Medvei, V. C., and J. L. Thornton, eds. 1974. *The Royal Hospital of Saint Bartholomew, 1123–1973*. London: St. Bartholomew's Hospital.

Mehrtens, H., H. Bos, and I. Schneider, eds. 1981. *Social History of Nineteenth Century Mathematics*. Boston: Birkhauser.

Mendelsohn, E. 1974. Revolution and Reduction: The Sociology of Methodological and Philosophical Concerns in Nineteenth Century Biology. In Elkana 1974: 407–26.

Mill, J. S. 1962. *Mill on Bentham and Coleridge*. Ed. F. R. Leavis. London: Chatto and Windus.

Miller, D. P. 1983. Between Hostile Camps: Sir Humphry Davy's Presidency of the Royal Society of London, 1820–1827. *Brit. J. Hist. Sci.* 16:1–47.

Milligan, Dr. 1838. Climate and Diseases of Swan River. *MCR* 29:281–82.

Mills, E. I. 1984. A View of Edward Forbes, Naturalist. *Arch. Nat. Hist.* 11:365–93.

Mitchell, P. C. 1929. *Centenary History of the Zoological Society of London*. London: Zoological Society of London.

Moore, J. R. 1982. 1859 and All That: Remaking the Story of Evolution and Religion. In Chapman and Duval 1982:167–94.

———. 1985a. Darwin of Down: The Evolutionist as Squarson-Naturalist. In Kohn 1985b: 435–81.

———. 1985b. Evangelicals and Evolution: Henry Drummond, Herbert Spencer, and the Naturalisation of the Spiritual World. *Scot. J. Theol.* 38:383–417.

———. 1985c. Review of *The Philosophical Naturalists* by P. F. Rehbock. *Ann. Sci.* 42:449–51.

———. 1986a. Crisis without Revolution: The Ideological Watershed in Victorian England. *Revue de Synthèse* 4:53–78.

———. 1986b. Socializing Darwin. In L. Levidow, ed., *Science as Politics*. London: Free Association Books.

———. 1989a. Of Love and Death: Why Darwin "Gave Up Christianity." In Moore 1989b.

———, ed. 1989b. *History, Humanity, and Evolution: Essays in Honor of John C. Greene*. Cambridge: Cambridge University Press.

Morgan, S. E. de. 1882. *Memoir of Augustus de Morgan*. London: Longmans, Green.

Morrell, J. B. 1971a. Individualism and the Structure of British Science in 1830. *Hist. Stud. Phys. Sci.* 3:183–204.

———. 1971b. Professors Robison and Playfair, and the *Theophobia Gallica*: Natural Philosophy, Religion, and Politics in Edinburgh, 1789–1815. *Notes and Records of the Royal Society* 26:43–63.

———. 1972. Science and Scottish University Reform: Edinburgh in 1826. *Brit. J. Hist. Sci.* 6:39–56.

———. 1975. The Leslie Affair: Career, Kirk, and Politics in Edinburgh in 1805. *Scot. Hist. Rev.* 54:63–82.

———. 1976. London Institutions and Lyell's Career, 1820–41. *Brit. J. Hist. Sci.* 9:132–46.

Morrell, J., and A. Thackray. 1981. *Gentlemen of Science: Early Years of the British Association for the Advancement of Science*. Oxford: Clarendon Press.

Morris, R. J., ed. 1986. *Class, Power and Social Structure in British Nineteenth-Century Towns*. Leicester: Leicester University Press.

Moss, A. W. 1961. *Valiant Crusade: The History of the R.S.P.C.A.* London: Cassell.

Moyal, A. M. 1975. Sir Richard Owen and His Influence on Australian Zoological and Palaeontological Science. *Records of the Australian Academy of Sciences* 3:41–56.

Mudie, R. 1825. *The Modern Athens: A Dissection and Demonstration of Men and Things in the Scotch Capital.* London: Knight and Lacey.

Müller, J. 1837. *Elements of Physiology.* Trans. W. Baly. London: Taylor and Walton.

Murchison, R. I., and E. Sabine. 1840. Address. In *Report BAAS 1840,* xxxv–xlviii.

Murphy, H. R. 1955. The Ethical Revolt against Christian Orthodoxy in Early Victorian England. *Amer. Hist. Rev.* 60:800–817.

Napier, M. 1879. *Selection of the Correspondence of the Late Macvey Napier.* London: Macmillan.

Nasmyth, A. 1839. *Researches on the Development, Structure, and Diseases of the Teeth.* London: Churchill.

―――. 1841. *Three Memoirs on the Development and Structure of the Teeth and Epithelium, Read at the Ninth Meeting of the BAAS, Held at Birmingham, in August, 1839.* London: Churchill.

Negus, V. 1966. *History of the Trustees of the Hunterian Collection.* Edinburgh and London: E. and S. Livingstone.

Neuburger, M. 1953. C. G. Carus on the State of Medicine in Britain in 1844. In Underwood 1953, 2:263–73.

Neve, M. 1983. Science in a Commercial City: Bristol, 1820–60. In Inkster and Morrell 1983:179–204.

New, C. W. 1961. *The Life of Henry Brougham to 1830.* Oxford: Clarendon Press.

Newbould, I. D. C. 1980. Whiggery and the Dilemma of Reform: Liberals, Radicals and the Melbourne Administration, 1835–9. *Bull. Inst. Hist. Res.* 53:229–41.

Newman, C. 1957. *The Evolution of Medical Education in the Nineteenth Century.* London: Oxford University Press.

Newport, G. 1837–38a. Mr. Newport's Reply to Prof. Grant and Dr. Marshall Hall. *L* 1:812–17.

―――. 1837–38b. On the Anatomy of Certain Structures in Myriapoda and Arachnida, Which Have Been Thought to Have Belonged to the Nervous System. *MG* 21:970–73.

Norton, B. 1983. Review of *Statistics in Britain* by D. MacKenzie. *Brit. J. Hist. Sci.* 16:304–6.

Ogilby, W. 1839. Observations on the Structure and Relations of the Presumed Marsupial Remains from the Stonesfield Oolite. *Proc. Geol. Soc.* 3:213.

Oldroyd, D. R. 1986. Grid/Group Analysis for Historians of Science. *Hist. Sci.* 14:145–71.

Oldroyd, D. R., and I. Langham, eds. 1983. *The Wider Domain of Evolutionary Thought.* Dordrecht: Reidel.

Oppenheimer, J. M. 1946. *New Aspects of John and William Hunter*. London: Heinemann.

———. 1967. *Essays in the History of Embryology and Biology*. Cambridge: MIT Press.

Ospovat, D. 1976. The Influence of Karl Ernst von Baer's Embryology, 1828–1859: A Reappraisal in Light of Richard Owen's and William B. Carpenter's "Palaeontological Application of von Baer's Law." *J. Hist. Biol*. 9:1–28.

———. 1977. Lyell's Theory of Climate. *J. Hist. Biol*. 10:317–39.

———. 1978. Perfect Adaptation and Teleological Explanation: Approaches to the Problem of the History of Life in the Mid-Nineteenth Century. *Stud. Hist. Biol*. 2:33–56.

———. 1981. *The Development of Darwin's Theory: Natural History, Natural Theology, and Natural Selection, 1838–1859*. Cambridge: Cambridge University Press.

Outram, D. 1978. The Language of Natural Power: The Funeral *Eloges* of Georges Cuvier. *Hist. Sci*. 16:153–78.

———. 1980. Politics and Vocation: French Science, 1793–1830. *Brit. J. Hist. Sci*. 13:27–43.

———. 1984. *Georges Cuvier: Vocation, Science and Authority in Post-Revolutionary France*. Manchester: Manchester University Press.

Owen, R. 1830. On the Anatomy of the Orang Utan (Simia Satyrus, L.). *Proc. Comm. Sci. Corres. Zool. Soc*. 1:4–5, 9–10, 28–29, 67–72.

———. 1832a. [The Mammary Gland of *Echidna Hystrix*.] *Proc. Comm. Sci. Corres. Zool. Soc*. 2:179–81.

———. 1832b. *Memoir on the Pearly Nautilus*. London: Taylor.

———. 1832c. On the Mammary Glands of the *Ornithorhynchus paradoxus*. *Phil. Trans. Roy. Soc*., 517–38.

———. 1834. On the Ova of the *Ornithorynchus paradoxus*. *Phil. Trans. Roy. Soc*., 555–66.

———. 1835a. On the Osteology of the Chimpanzee and Orang Utan. *Trans. Zool. Soc*. 1:343–79.

———. 1835b. On the Young of the *Ornithorhynchus paradoxus*, Blum. *Trans. Zool. Soc*. 1:221–8.

———. 1836. Acrita. In Todd 1836, 1:47–49.

———. 1839a. Note sur les différences entre le *Simia morio*, d'Owen, et le *Simia Wurmbii* dans la période d'adolescence, decrit par M. Dumortier. *Comptes Rendus de l'Académie des Sciences* 8:231–36.

———. 1839b. Report on British Fossil Reptiles. In *Report BAAS 1839*, 43–126.

———. 1840–45. *Odontography; or, a Treatise on the Comparative Anatomy of the Teeth; Their Physiological Relations, Mode of Development, and Microscopic Structure, in the Vertebrate Animals*. 2 vols. London: Bailliere.

———. 1841a. Observations on the *Basilosaurus* of Dr. Harlan (*Zeuglodon cetoides*, Owen). *Trans. Geol. Soc*. 6:69–79.

———. 1841b. Observations on the Fossils Representing the *Thylacotherium Prevostii*, Valenciennes, with Reference to the Doubts of Its Mammalian and

Marsupial Nature Recently Promulgated; and on the *Phascolotherium Bucklandi. Trans. Geol. Soc.* 6:47–65.

———. 1841c. Report on British Fossil Reptiles. Pt. 2. In *Report BAAS 1841*, 60–204.

———. 1842. Report on the British Fossil Mammalia. Pt 1. In *Report BAAS 1842*, 54–74.

———. 1843. *Lectures on the Comparative Anatomy and Physiology of the Invertebrate Animals*. London: Longman.

———. 1846a. *A History of British Fossil Mammals and Birds*. London: van Voorst.

———. 1846b. *Lectures on the Comparative Anatomy and Physiology of Vertebrate Animals. Pt. I: Fishes*. London: Longman.

———. 1846c. Report on the Archetype and Homologies of the Vertebrate Skeleton. In *Report BAAS 1846*, 169–340.

———. 1848a. Broderip's Zoological Recreations. *Quart. Rev.* 82:119–42.

———. 1848b. Report of the Physiological Committee of the Royal Society on Dr. Robert Lee's Paper, Entitled "On the Ganglia and Nerves of the Heart." *MG* 6:82–4.

———. 1849. *On the Nature of Limbs*. London: van Voorst.

———. 1851. Lyell: On Life and Successive Development. *Quart. Rev.* 89:412–51.

———. 1863. *Monograph on the Aye-aye*. London: Taylor and Francis.

———. 1871. *Monograph of the Fossil Mammalia of the Mesozoic Formation*. London: Palaeontographical Society.

Owen, Rev. R. S. 1894. *The Life of Richard Owen*. 2 vols. London: Murray.

Page, L. E. 1969. Diluvialism and Its Critics in Great Britain in the Early Nineteenth Century. In Schneer 1969: 257–71.

Paine, T. 1819. *The Age of Reason*. London: Carlile.

Paley, E., ed. 1830. *The Works of William Paley*. 6 vols. London: Rivington.

Pankhurst, R. K. P. 1954. *William Thompson (1775–1833): Britain's Pioneer Socialist, Feminist, and Co-Operator*. London: Watts.

Parker, S. L. 1830–31. Lectures on Comparative Anatomy, as Illustrative of General and Human Physiology. *MG* 7:1–6, 65–70, 97–104, 129–33, 161–66, 225–33, 289–94, 353–61, 449–56.

Parsons, F. G. 1932–36. *The History of St. Thomas's Hospital*. 3 vols. London: Methuen.

Pattison, F. L. M. 1987. *Granville Sharp Pattison: Anatomist and Antagonist, 1791–1851*. Edinburgh: Conongate.

Pattison, G. S. 1831. *Professor Pattison's Statement of the Facts of His Connexion with the University of London*. London: Longman.

Peel, J. D. Y. 1971. *Herbert Spencer: The Evolution of a Sociologist*. London: Heinemann.

Pereira, J. 1835–36. Lectures on Materia Medica, or Pharmacology, and General Therapeutics . . . Lecture VIII. Characteristics of the Two Great Kingdoms of Nature. *MG* 17:241–51.

Peterson, M. J. 1978. *The Medical Profession in Mid-Victorian London*. Berkeley: University of California Press.

————. 1984. Gentlemen and Medical Men: The Problem of Professional Recruitment. *Bull. Hist. Med*. 58:457–73.

Philips, D. 1983. "A Just Measure of Crime, Authority, Hunters and Blue Locusts": The "Revisionist" Social History of Crime and the Law in Britain, 1780–1850. In Cohen and Scull 1983:50–74.

Pilcher. G. 1840–41. Course of Lectures on the Anatomy, Physiology, and Diseases of the Ear . . . Delivered at the Webb-Street School of Medicine, Southwark. *L* 1:521–25, 665–68, 841–44; 2:145–49, 241–44, 353–57, 465–70, 769–75, 814–19.

Plarr's Lives of the Fellows of the Royal College of Surgeons of England. 1930. 2 vols. London: Royal College of Surgeons.

Playfair, J. 1812. Geographie mineralogique des environs de Paris. *Edin. Rev*. 20:369–86.

————. 1813–14. Cuvier on the Theory of the Earth. *Edin. Rev*. 22:454–75.

Pollock, W. F. 1887. *Personal Remembrances of Sir Frederick Pollock*. 2 vols. London: Macmillan.

Poore, G. V. 1901. Robert Edmond Grant. *University College Gazette* 2 (34):190–91.

Porter, R. 1973. The Industrial Revolution and the Rise of the Science of Geology. In Teich and Young 1973: 320–43.

————. 1978. Gentlemen and Geology: The Emergence of a Scientific Career, 1660–1920. *Hist. J*. 21:809–36.

————. 1980. *The Making of Geology: Earth Sciences in Britain, 1660–1815*. Cambridge: Cambridge University Press.

————. 1985. William Hunter: A Surgeon and a Gentleman. In Bynum and Porter 1985: 7–34.

Powell, B. 1838. *The Connexion of Natural and Divine Truth; or, The Study of the Inductive Philosophy Considered as Subservient to Theology*. London: Parker.

Power, D'A. 1895. The Rise and Fall of the Private Medical Schools in London. *Brit. Med. J*. 1:1388–91, 1451–53.

Poynter, F. N. L. 1962. Thomas Southwood Smith, the Man (1788–1861). *Proc. Roy. Soc. Med*. 55:381–92.

————, ed. 1966. *The Evolution of Medical Education in Britain*. London: Pitman Medical.

Prévost, C. 1825. Observations sur les schistes oolithiques de Stonesfield en Angleterre dans lesquelles ont trouvés ossemens fossiles de mammiferes. *Ann. Sci. Nat*. 5:389–417.

Preyer, R. O. 1981. The Romantic Tide Reaches Trinity. *Ann. N.Y. Acad. Sci*. 360:39–68.

Prichard, A. 1894. The Early History of the Bristol Medical School. In *Bristol Medical School*. Bristol: Arrowsmith.

Quain, J. 1828. *Elements of Descriptive and Practical Anatomy*. London: Simpkin and Marshall.

————. 1831. *Lecture Introductory to the Course of Anatomy and Physiology, Delivered at the Opening of Session, 1831–32*. London: Taylor.

Rachootin, S. P. 1985. Owen and Darwin Reading a Fossil: *Macrauchenia* in a Boney Light. In Kohn 1985b: 155–83.

Rae, I. 1964. *Knox the Anatomist*. Edinburgh: Oliver and Boyd.

Raffles, S. 1830. *Memoir of the Life and Public Service of Sir Thomas Stamford Raffles*. London: Murray.

Raikes, G. A. 1878–79. *The History of the Honourable Artillery Company*. 2 vols. London: R. Bentley.

Raumer, F. von. 1836. *England in 1835*. 3 vols. London: Murray.

Raymond, J., and Pickstone, J. V. 1986. The Natural Sciences and the Learning of the English Unitarians. In B. Smith 1986: 129–64.

Rehbock, P. F. 1983. *The Philosophical Naturalists: Themes in Early Nineteenth-Century British Biology*. Madison: University of Wisconsin Press.

————. 1985. John Fleming (1785–1857) and the Economy of Nature. In Wheeler and Price 1985:129–40.

Report from the Select Committee on British Museum. 1836. Parliamentary Papers, 14 July 1836.

Report from the Select Committee on Medical Education . . . Part 1. Royal College of Physicians. Part 2. Royal College of Surgeons, London. 1834. Parliamentary Papers, 13 August 1834.

Richards, E. 1983. Darwin and the Descent of Woman. In Oldroyd and Langham 1983:57–111.

————. 1987. A Question of Property Rights: Richard Owen's Evolutionism Reassessed. *Brit. J. Hist. Sci.* 20:129–71.

————. 1988. The "Moral Anatomy" of Robert Knox: A Case Study of the Interplay between Biological and Social Thought in the Context of Victorian Scientific Naturalism. *J. Hist. Biol.* (forthcoming).

————. 1989. Huxley and Woman's Place in Science: The "Woman Question" and the Control of Victorian Anthropology. In Moore 1989b.

Richards, P. 1980. State Formation and Class Struggle, 1830–48. In Corrigan 1980:49–78.

Richards, R. J. 1979. Influence of Sensationalist Tradition on Early Theories of the Evolution of Behavior. *J. Hist. Ideas* 40:85–105.

————. 1981. Instinct and Intelligence in British Natural Theology: Some Contributions to Darwin's Theory of the Evolution of Behavior. *J. Hist. Biol.* 14:193–230.

————. 1982. The Emergence of Evolutionary Biology in the Early Nineteenth Century. *Brit. J. Hist. Sci.* 15:241–80.

Richardson, R. 1987. *Death, Dissection and the Destitute*. London and New York: Routledge and Kegan Paul.

Rieppel, O. 1984. Atomism, Transformism and the Fossil Record. *Zool. J. Linn. Soc.* 82:17–32.

Ritvo, H. 1987. *The Animal Estate: The English and Other Creatures in the Victorian Age*. Cambridge, Mass., and London: Harvard University Press.

Robb-Smith, A. H. T. 1966. Medical Education at Oxford and Cambridge prior to 1850. In Poynter 1966: 19–52.

Roberts, R. S. 1966. Medical Education and the Medical Corporations. In Poynter 1966:69–88.

Robertson, J. 1835–36. Dr. Elliotson on Life and Mind. *MG* 17:203–10, 251–57.

Roget, P. M. 1826. *An Introductory Lecture on Human and Comparative Physiology*. London: Longman.

———. 1834. *Animal and Vegetable Physiology Considered with Reference to Natural Theology*. 2 vols. London: Pickering.

———. 1838. *Treatises on Physiology and Phrenology*. Edinburgh: Black.

———. 1846. The Proceedings of the Royal Society. *L* 1:420.

Roulin, M. -D. 1829. Inquiries Respecting Certain Changes Observed to Have Taken Place in Domestic Animals, Transported from the Old to the New Continent. *ENPJ* 7:326–38.

Royal College of Surgeons. 1835. *Directions for Collecting and Preserving Animals; Addressed by the Board of Curators of the Museum of the Royal College of Surgeons in London to Professional, Scientific, and Other Individuals with an Invitation for Contributions to the Museum*. London: Taylor.

Royle, E. 1974. *Victorian Infidels: The Origins of the British Secularist Movement, 1791–1866*. Manchester: Manchester University Press.

Rubinstein, W. D. 1983. The End of "Old Corruption" in Britain, 1780–1860. *Past and Present* 101:55–86.

Rudwick, M. J. S. 1972. *The Meaning of Fossils: Episodes in the History of Palaeontology*. London: Macdonald; New York: American Elsevier.

———. 1975. Charles Lyell, F.R.S. (1797–1875) and His London Lectures on Geology, 1832–33. *Notes and Records of the Royal Society* 29:231–63.

———. 1982. Charles Darwin in London: The Integration of Public and Private Science. *Isis* 73:186–206.

———. 1985. *The Great Devonian Controversy: The Shaping of Scientific Knowledge among Gentlemanly Specialists*. Chicago: University of Chicago Press.

Rupke, N. 1983. *The Great Chain of History: William Buckland and the English School of Geology (1814–1849)*. Oxford: Clarendon Press.

———. 1985. Richard Owen's Hunterian Lectures on Comparative Anatomy and Physiology, 1837–55. *Med. Hist.* 29:237–58.

———, ed. 1987. *Vivisection in Historical Perspective*. London: Croom Helm.

Ruse, M. 1979. *The Darwinian Revolution: Science Red in Tooth and Claw*. Chicago: University of Chicago Press.

Rush, H. W. 1835. *A Lecture, Introductory to the Science of Comparative Anatomy*. London.

Russell, C. A. 1983. *Science and Social Change, 1700–1900*. London: Macmillan.

Russell, E. S. 1916. *Form and Function: A Contribution to the History of Animal Morphology*. London: Murray.

Ryan, M. 1829. Physiology, Phrenology, Materialism, Immateriality of the Mind. *LMSJ* 3:44–51.

———. 1836–37. Introductory Lecture to a Course on the Principles and Practice of Medicine, Delivered in the Hunterian School of Medicine, Great Windmill-Street. - Session 1836–37. *LMSJ* 10:406–17.

Sandwith, T. 1827. A Comparative View of the Relations between the Development of the Nervous System and the Functions of Animals. *Phrenol. J.* 4:479–94.

Saunders, L. J. 1950. *Scottish Democracy, 1815–1840: The Social and Intellectual Background.* Edinburgh: Oliver and Boyd.

Schafer, E. A. 1901. William Sharpey. *University College Gazette* 2 (36):215.

Schaffer, S. 1984. Priestley's Questions: An Historiographic Survey. *Hist. Sci.* 22:151–83.

———. 1989. The Nebular Hypothesis and the Science of Progress. In Moore 1989b.

Scherren, H. 1905. *The Zoological Society of London: A Sketch of Its Foundation and Development, and the Story of Its Farm, Museum, Gardens, Menagerie and Library.* London: Cassell.

Schneer, C. J., ed. 1969. *Toward a History of Geology.* Cambridge: MIT Press.

Schumacher, I. 1973. Die Metamorphose-Theorie Friedrich Tiedemanns. In *Festschrift für Claus Nissen,* 584–92. Weisbaden.

Schweber, S. S. 1977. The Origin of the *Origin* Revisited. *J. Hist. Biol.* 10:229–316.

———. 1980. Darwin and the Political Economists: Divergence of Character. *J. Hist. Biol.* 13:195–289.

———. 1985. The Wider British Context in Darwin's Theorizing. In Kohn 1985b: 35–69.

Schweigger, A. 1826. Observations on the Anatomy of the Corallina Opuntia, and Some Other Species of Corallines. *ENPJ* 1:220–24.

Secord, J. A. 1982. King of Siluria: Roderick Murchison and the Imperial Theme in Nineteenth-Century British Geology. *Vict. Stud.* 25:413–42.

———. 1985a. Darwin and the Breeders: A Social History. In Kohn 1985b: 519–42.

———. 1985b. John W. Salter: The Rise and Fall of a Victorian Palaeontological Career. In Wheeler and Price 1985: 61–75.

———. 1985c. Natural History in Depth. *Soc. Stud. Sci.* 15:181–200.

———. 1986a. *Controversy in Victorian Geology: The Cambrian-Silurian Dispute.* Princeton: Princeton University Press.

———. 1986b. The Geological Survey of Great Britain as a Research School, 1839–1855. *Hist. Sci.* 24:223–75.

———. 1988. Extraordinary Experiment: Electricity and the Creation of Life in Victorian England. In Gooding, Pinch, and Schaffer 1988:337–83.

———. 1989. Behind the Veil: Robert Chambers and *Vestiges.* In Moore 1989b.

Sedgwick, A. 1834a. Address to the Geological Society. *Proc. Geol. Soc.* 1:187–212.

———. 1834b. Address to the Geological Society. *Proc. Geol. Soc.* 1:281–316.

———. 1845. Vestiges of the Natural History of Creation. *Edin. Rev.* 82:1–85.

Seed, J. 1982. Unitarianism, Political Economy and the Antinomies of Liberal Culture in Manchester, 1830–50. *Soc. Hist.* 7:1–25.

————. 1986. Theologies of Power: Unitarianism and the Social Relations of Religious Discourse, 1800–50. In Morris 1986: 108–56.

Serres, E. R. A. 1824–26. *Anatomie comparée du Cerveau*. 2 vols. Paris: Gabon.

————. 1830. Anatomie transcendante: Quatrième mémoire, Loi de symétrie et de conjugaison du système Sanguin. *Ann. Sci. Nat*. 21:5–49.

Shapin, S. 1975. Phrenological Knowledge and the Social Structure of Early Nineteenth-Century Edinburgh. *Ann. Sci*. 32:219–43.

————. 1979a. Homo Phrenologicus: Anthropological Perspectives on an Historical Problem. In Barnes and Shapin 1979:41–67.

————. 1979b. The Politics of Observation: Cerebral Anatomy and Social Interests in the Edinburgh Phrenology Disputes. In Wallis 1979:139–78.

————. 1982. History of Science and Its Sociological Reconstructions. *Hist. Sci*. 20:157–211.

————. 1983. "Nibbling at the Teats of Science": Edinburgh and the Diffusion of Science in the 1830s. In Inkster and Morrell 1983:151–78.

Shapin, S., and B. Barnes. 1977. Science, Nature and Control: Interpreting Mechanics' Institutes. *Soc. Stud. Sci*. 7:31–74.

————. 1979. Darwin and Social Darwinism: Purity and History. In Barnes and Shapin 1979: 125–42.

Shapin, S., and A. Thackray. 1974. Prosopography as a Research Tool in History of Science: The British Scientific Community, 1700–1900. *Hist. Sci*. 12:1–28.

Sharpey, W. 1840–41. Anatomy and Physiology. *L* 1:73–78, 142–47, 281–85, 425–28, 489–93.

————. 1874. Dr. Robert Edmond Grant. *Proc. Roy. Soc. London* 23:vi–x.

Sheets-Pyenson, S. 1981. War and Peace in Natural History Publishing: *The Naturalist's Library*, 1833–43. *Isis* 72:50–72.

Simon, B., ed. 1972. *The Radical Tradition in Education in Britain*. London: Lawrence and Wishart.

Singer, C., and S. W. F. Holloway. 1960. Early Medical Education in England in Relation to the Pre-History of London University. *Med. Hist*. 4:1–17.

Sloan, P. R. 1985. Darwin's Invertebrate Program, 1826–1836: Preconditions for Transformism. In Kohn 1985b: 71–120.

Smith, B., ed. 1986. *Liberty, Truth, Religion: Essays Celebrating Two Hundred Years of Manchester College*. Oxford: Manchester College.

Smith, C. U. M. 1978. Charles Darwin, the Origin of Consciousness, and Panpsychism. *J. Hist. Biol*. 11:245–67.

Smith, R. 1972. Alfred Russel Wallace: Philosophy of Nature and Man. *Brit. J. Hist. Sci*. 6:177–99.

————. 1973. The Background of Physiological Psychology in Natural Philosophy. *Hist. Sci*. 11:75–123.

————. 1977. The Human Significance of Biology: Carpenter, Darwin, and the *vera causa*. In Knoepflmacher and Tennyson 1977: 216–30.

Smith, T. S. 1827a. Human and Comparative Physiology. *West. Rev*. 7:416–44.

————. 1827b. Life and Organization. *West. Rev*. 7:208–26.

————. 1829–30. *Animal Physiology*. London: Baldwin and Cradock.

————. 1830. Phenomena of the Human Mind. *West. Rev.* 13:265–92.

————. 1832. *A Lecture Delivered over the Remains of Jeremy Bentham, Esq., in the Webb-Street School of Anatomy & Medicine, on the 9th of June, 1832.* London: E. Wilson.

————. 1866. *The Divine Government.* 5th ed. London: Trübner.

Smith, T. S., and J. Bentham. 1829. Anatomy. *West. Rev.* 10:116–48.

Solly, S. 1836. *The Human Brain, Its Configuration, Structure, Development, and Physiology; Illustrated by References to the Nervous System in the Lower Orders of Animals.* London: Longman.

Southwell, C. 1842. Is There a God? *Oracle of Reason,* 19 March.

Spencer, H. 1904. *An Autobiography.* 2 vols. London: Williams and Norgate.

Sprigge, S. S. 1899. *The Life and Times of Thomas Wakley.* London: Longman.

Stanley, Lord, et al. 1835. *Statement by the President and Certain Members of the Council of the Zoological Society, in Reply to Observations and Charges Made by Colonel Sykes and Others, at the General Meeting of the Society, on the 29th of April Last, and at the Monthly Meeting on the 2nd of the Same Month.* London: William Nichol.

Statement by the Council of the University of London. 1827. London: Longman.

Stevenson, L. G. 1956. Religious Elements in the Background of the British Anti-Vivisection Movement. *Yale J. Biol. Med.* 29:125–57.

Stewart, R. 1986. *Henry Brougham, 1778–1868: His Public Career.* London: Bodley Head.

Struthers, J. 1867. *Historical Sketch of the Edinburgh Medical School.* Edinburgh: Maclachlan and Stewart.

Sulloway, F. J. 1982. Darwin's Conversion: The *Beagle* Voyage and Its Aftermath. *J. Hist. Biol.* 15:325–96.

Swainson, W. 1834. *A Preliminary Discourse on the Study of Natural History.* London: Longman.

Swan, J. 1835. *Illustrations of the Comparative Anatomy of the Nervous System.* London: Longman.

Symonds, J. A. 1871. *Miscellanies by John Addington Symonds M.D.* London: Macmillan.

Taylor, D. W. 1971. The Life and Teaching of William Sharpey (1802–1880), "Father of Modern Physiology in Britain." *Med. Hist.* 15:126–53, 241–59.

Taylor, M. A., and H. S. Torrens. 1986. Saleswoman to a New Science: Mary Anning and the Fossil Fish Squaloraja from the Lias of Lyme Regis. *Dorset Natural History and Archaeological Society* 108:135–48.

Teich, M., and R. M. Young, eds. 1973. *Changing Perspectives in the History of Science.* London: Heinemann.

Temkin, O. 1977. Basic Science, Medicine, and the Romantic Era. In O. Temkin, ed., *The Double Face of Janus and Other Essays in the History of Medicine,* 345–72. Baltimore and London: Johns Hopkins University Press.

Thackray, A. 1974. Natural Knowledge in Cultural Context: The Manchester Model. *Amer. Hist. Rev.* 79:672–709.

Thomis, M. I., and P. Holt. 1977. *Threats of Revolution in Britain, 1789–1848*. London: Macmillan.

Thompson, E. P. 1980. *The Making of the English Working Class*. 3rd ed. London: Gollancz.

Thompson, W. 1824. *An Inquiry into the Principles of the Distribution of Wealth*. London: Longman.

———. 1826. Physical Argument for the *Equal* Cultivation of All the Useful Faculties or Capabilities, of Men and Women. *Co-Operative Magazine and Monthly Herald* 1:250–58.

Thomson, A. 1830. A Sop for Cerberus! *LMSJ* 5:437–56.

Thomson, K. S., and S. P. Rachootin. 1982. Turning Points in Darwin's Life. *Biol. J. Linn. Soc.* 17:23–37.

Thomson, S. C. 1942. The Great Windmill Street School. *Bull. Hist. Med.* 12:377–91.

———. 1943. The Surgeon-Anatomists of Great Windmill Street School. *Bull. Soc. Med. Hist. Chicago* 5:55–75.

Thornton, J. L. 1974. The Medical College from Its Origins to the End of the Nineteenth Century. In Medvei and Thornton 1974:43–77.

Tiedemann, F. 1834. *A Systematic Treatise on Comparative Physiology*. Trans. J. M. Gully and J. H. Lane. London: Churchill.

———. 1836. On the Brain of the Negro, Compared with That of the European and the Orang-Outang. *Phil. Trans. Roy. Soc.*, 497–527.

Todd, R. B., ed. 1836–59. *The Cyclopaedia of Anatomy and Physiology*. 5 vols. London: Sherwood, Gilbert and Piper.

Traill, T. S. 1823. British Museum. *Edin. Rev.* 28:379–98.

———. 1837. Address. In *Report BAAS 1837*, xxv–xlii.

Turner, J. 1980. *Reckoning with the Beast: Animals, Pain, and Humanity in the Victorian Mind*. Baltimore: Johns Hopkins University Press.

Turner, W., ed. 1868. *The Anatomical Memoirs of John Goodsir*. Edinburgh: Adam and Charles Black.

Underwood, E. A., ed. 1953. *Science Medicine and History*. 2 vols. London: Oxford University Press.

Valenciennes, A. 1838. Observations sur les machoires fossiles des couches oolithiques de Stonesfield. *Comptes Rendus de la Académie des Sciences* 7:572–80.

Vidler, A. R. 1974. *The Church in an Age of Revolution*. Harmondsworth: Penguin.

Vigors, N. A. 1830. An Address Delivered at the Sixth and Last Meeting of the Zoological Club of the Linnean Society of London, on the 29th of November, 1829. *Mag. Nat. Hist.* 3:201–26.

Vincent, D. 1981. *Bread, Knowledge and Freedom: A Study of Nineteenth-Century Working Class Autobiography*. London: Europa.

Waddington, I. 1984. *The Medical Profession in the Industrial Revolution*. Dublin: Gill and Macmillan.

Wakley, T. 1834–35. An Address from Mr. Wakley to the Electors of the Great Metropolitan District of Finsbury. *L* 1:533–36.

Wallis, R., ed. 1979. *On the Margins of Science: The Social Construction of Rejected Knowledge*. Sociology Review Monograph No. 27. Keele, Staffordshire.

Waterhouse, G. R. 1841. Description of a New Genus of Mammiferous Animal from Australia, Belonging Probably to the Order Marsupialia. *Trans. Zool. Soc.* 2:149–54.

Webster, C., ed. 1981. *Biology, Medicine and Society, 1840–1940*. Cambridge: Cambridge University Press.

Weindling, P. J. 1979. Geological Controversy and Its Historiography: The Prehistory of the Geological Society of London. In Jordanova and Porter 1979:249–71.

———. 1981. Theories of the Cell State in Imperial Germany. In Webster 1981:99–155.

Weissenborn, W. 1838. On Spontaneous Generation. *Mag. Nat. Hist.* 2:369–81.

Wells, K. D. 1971. Sir William Lawrence (1783–1867): A Study of Pre-Darwinian Ideas on Heredity and Variation. *J. Hist. Biol.* 4:319–61.

———. 1973. The Historical Context of Natural Selection: The Case of Patrick Matthew. *J. Hist. Biol.* 6:225–58.

Wheeler, A., and J. Price, eds. 1985. *From Linnaeus to Darwin: Commentaries on the History of Biology and Geology*. London: Society for the History of Natural History.

Whewell, W. 1832. Lyell's Geology. *Quart. Rev.* 47:103–32.

———. 1837. *History of the Inductive Sciences*. 3 vols. London: Parker.

Whitley, G. P. 1975. *More Early History of Australian Zoology*. Sydney: Royal Zoological Society of New South Wales.

Wiener, J. H. 1969. *The War of the Unstamped: The Movement to Repeal the British Newspaper Tax, 1830–1836*. Ithaca, N.Y.: Cornell University Press.

———. 1983. *Radicalism and Freethought in Nineteenth-Century Britain: The Life of Richard Carlile*. Westport, Conn.: Greenwood Press.

Wilde, C. B. 1982. Matter and Spirit as Natural Symbols in Eighteenth-Century British Natural Philosophy. *Brit. J. Hist. Sci.* 15:99–131.

Williams, G. A. 1974. The Infidel Working Class, 1790–1830. *Bull. Soc. Stud. Lab. Hist.* 29:8–9.

Wilson, L. 1972. *Charles Lyell. The Years to 1841: The Revolution in Geology*. New Haven: Yale University Press.

Winsor, M. P. 1976. *Starfish, Jellyfish, and the Order of Life: Issues in Nineteenth Century Science*. New Haven and London: Yale University Press.

Woodward, H. 1893. Sir Richard Owen. *Geol. Mag.* 10:45–54.

Wright, A. 1970. *The Chartist Movement in Scotland*. Manchester: Manchester University Press.

Yeo, R. 1979. William Whewell, Natural Theology and the Philosophy of Science in Mid-Nineteenth-Century Britain. *Ann. Sci.* 36:493–516.

———. 1984. Science and Intellectual Authority in Mid-Nineteenth Century Britain: Robert Chambers and *Vestiges of the Natural History of Creation*. *Vict. Stud.* 28:5–31.

————. 1986. The Principle of Plenitude and Natural Theology in Nineteenth-Century Britain. *Brit. J. Hist. Sci.* 19:263–82.

Youatt, W. 1831–32. Introductory Lecture on Veterinary Medicine and Surgery. *L* 1:78–82.

————. 1835–36a. Account of the Habits and Illness of the Late Chimpanzee. *L* 2:202–6.

————. 1835–36b. Life and Death of the Chimpanzee. *MG* 18:214–16.

Young, R. M. 1985. *Darwin's Metaphor: Nature's Place in Victorian Culture.* Cambridge: Cambridge University Press.

Yourgrau, W., and A. D. Breck., eds. 1977. *Cosmology, History, and Theology.* New York: Plenum Press.

Zuckerman, S., ed. 1976. *The Zoological Society of London, 1826–1976 and Beyond.* London: Academic Press.

Index